INTERNATIONAL CENTRE FOR MECHANICAL SCIENCES

COURSES AND LECTURES - No. 302

# NONSMOOTH MECHANICS AND APPLICATIONS

EDITED BY

J.J. MOREAU
UNIVERSITY OF MONTPELLIER

P.D. PANAGIOTOPOULOS
ARISTOTLE UNIVERSITY
AND
R.W.T.H.

SPRINGER-VERLAG WIEN GMBH

Le spese di stampa di questo volume sono in parte coperte da contributi
del Consiglio Nazionale delle Ricerche.

This volume contains 77 illustrations.



**Originally published by Springer Verlag Wien-New York in 1988**

**ISBN 978-3-211-82066-7** **ISBN 978-3-7091-2624-0 (eBook)**
**DOI 10.1007/978-3-7091-2624-0**

# PREFACE

*This book is devoted to problems of Mechanics involving nonsmooth relations.*

*In place of classical derivatives current concepts of Nonsmooth Analysis are used, such as subdifferentials, generalized gradients, fans or quasidifferentials. The corresponding problems may take the form of variational or hemivariational inequalities or of differential and integral inclusions. Other mechanical topics investigated in the present volume lead to the determination of saddle points for non-differentiable functions.*

*In Dynamics, velocity is not supposed to be a smooth function of time, but only to have locally bounded variation, so that acceleration is a vector-valued measure on the concerned interval of time. Evolution is thus governed by measure differential equations or by measure differential inclusions. The formalism used makes easy to discretize for numerical treatment.*

*Generally, the practicef Nonsmooth Analysis requires the handling of multivalued mappings (i.e. set-valued mappings, also called multifunctions). In Mechanics, this concept has for long been implicit in such classical domains as Plasticity of Dry Friction. Its use in a formalized way provides a new insight into these subjects and concurrently proves to be operationally efficient. The same is true for a number of mechanical or thermodynamical topics.*

*Chapter I, by J.J. Moreau, concerns the Dynamics of systems with a finite number of degrees of freedom, involving unilateral contact whereupon possible friction is assumed to obey Coulomb's law. New formulations of the corresponding evolution problems are developed which lead to effective numerical algorithms. These numerical techniques in particular prove effective in the handling of the catastrophic events which may be manifested in the presence of dry friction.*

*Chapter II, by P.D. Panagiotopoulos, deals with mechanical problems admitting variational formulations in terms of Hemivariational Inequalities. The notions of substationarity and quasidifferentiability are introduced and some new nonclassical variational principles in inequality form are presented.*

*Chapter III, by M. Frémond, develops the use of nonsmooth potentials in reversible or irreversible Thermodynamics. This provides new approaches to subjects such as Phase Changes, Heterogeneous Media, Soil Freeze, Adhesions of Solids or Shape-memory Alloys.*

*Chapter IV, by J. Haslinger, is devoted to the Numerical Analysis of Unilateral Problems of Deformable Bodies and Shape Optimization. The approach is based on the formulation of the contact boundary conditions, with friction taken into account, through saddle point properties or variational and quasi variational inequalities.*

*Chapter V, by P.M. Suquet, emphasizes first that discontinuous solutions should normally be expected in the Yield Design (Limit Analysis) of Plastic Structures and in the quasi-static evolution of Elastoplastic Bodies. The treatment of a bond between two deformable solids as the limit, in a mathematical sense, of a thin layer with nonlinear behaviour is presented. The chapter introduces the use of the concept of epi-convergence in homogenization, i.e. the derivation of macroscopic constitutive laws for microscopically heterogeneous bodies.*

*Finally, in Chapter VI, J.J. Telega, applies duality methods to the nonlinear boundary value problems one meets when studying the equilibrium or the quasistatic evolution of elastic and inelastic bodies submitted to unilateral contacts. More elaborate models of friction than Coulomb's law are presented and also the homogenization of microfissured elastic solids and plates, with unilaterality is considered.*

*In short, the book demonstrates how several currently developed mathematical tools of Nonsmooth Analysis effectively work in Mechanics and Engineering Science, with definite advantages from the standpoint of fundamental understanding as well as practical computation. It gathers in one volume both some innovative ideas and a good deal of unpublished material.*

*J.J. Moreau, P.D. Panagiotopoulos*

## CONTENTS

Page

Preface

Unilateral Contact and Dry Friction in Finite Freedom Dynamics
*by J.J. Moreau* .......... 1

Nonconvex Superpotentials and Hemivariational Inequalities. Quasi Differentiability in Mechanics
*by P. D. Panagiotopoulos* .......... 82

Contact with Adhesion
*by M. Fremond* .......... 177

Approximation of Contact Problems. Shape Optimization in Contact Problems
*by J. Haslinger* .......... 223

Discontinuities and Plasticity
*by P.M. Suquet* .......... 279

Topics on Unilateral Contact Problems of Elasticity and Inelasticity
*by J.J. Telega* .......... 341

# UNILATERAL CONTACT AND DRY FRICTION IN FINITE FREEDOM DYNAMICS

J.J. Moreau
Université des Sciences et Techniques du Languedoc
Montpellier, France

ABSTRACT

An approach to the dynamics of mechanical systems with a finite number of degrees of freedom, involving unilateral constraints, is developed. In the n-dimensional linear spaces of forces and velocities, some classical concepts of Convex Analysis are used, but no convexity assumption is made concerning the constraint inequalities. The velocity is not supposed to be a differentiable function of time, but only to have locally bounded variation, so the role of the acceleration is held by a n-dimensional measure on the considered time interval. Dynamics is then governed by measure differential inclusions, which treat possible velocity jumps on the same footing as smooth motions. Possible collisions are described as soft, thus dissipative. Friction is taken into account under a recently proposed expression of Coulomb's law. These formulations have the advantage of generating numerical algorithms of time-discretization, able to handle, in particular, the nonsmooth effects arising from unilaterality and from dry friction.

# 1. INTRODUCTION

Usual mechanisms consist of parts which, at the first level of investigation, are treated as perfectly rigid bodies. The mechanism operation rests on the fact that some of these parts may come into contact or get loose from each other, but none of them can overlap. Similarly, the parts may touch the external bodies which support the mechanism, but can never encroach upon the region of space they occupy.

In terms of the parameters $q^1,\dots, q^n$, making an element of $\mathbb{R}^n$ denoted in the sequel by q, which are used to locate every position of the considered system, the above impenetrability properties may as a rule be expressed by a set of inequalities, say $f_\alpha(q)\leq 0$, $\alpha\in\{1,\dots,\nu\}$. Each of these inequalities corresponds to what is traditionally called a *unilateral constraint*. Naturally, the description of a constraint in Mechanics does not reduce to the geometric restriction it imposes to the system possible positions. Predicting the system behaviour always requires some additional information about the *forces of constraint* or *reactions* needed by the system dynamics, for the geometric conditions to be satisfied at every instant.

Constraints whose geometric effect is expressed by equalities are, in contrast, said *bilateral*. They are commonly realized by the conjunction of several unilateral constraints and, in practice, this may leave a residual *looseness* whose investigation has primary importance in some applications.

However omnipresent unilateral constraints are in machines, the place they receive in the books on Classical Mechanics is very modest. Here is the traditional approach of the situation.

Starting with a position of the system in which some of the contacts are effective (i.e. the corresponding inequalities hold as equalities) and with velocities compatible with the persistence of these contacts, the subsequent motion is calculated under the tentative assumption that all the said contacts do persist. The calculation is identical to what is more familiarly done in the case of bilateral constraints, with friction possibly taken into account. At every instant of the calculated motion, the respective contact forces are evaluated. As long as the direction of each of these vectors is found compatible with the unilaterality of the corresponding contact, the calculated solution is accepted (rigorously, this does not dispense from investigating also the possiblity of contact breaking, since the uniqueness of solution to an initial value problem of Dynamics is not granted in general).

But if the above calculation yields, after an instant $\tau$, a non feasible value for some of the contact forces, the tentative assumption has to be rejected from this instant onward, and other types of motions, in which some of the contacts initially present get loose, are to be tested in the same way. The number of combinations to be tried may be high, if many unilateral contacts are involved. In practical situations, such instants as $\tau$ make a finite set, but this cannot be asserted in general.

It was not before the first quarter of this century that E. Delassus (cf.[1] for the frictionless case) observed that, contrary to what had been formerly believed, the contacts which get loose at time $\tau$ are not necessarily those for which the calculation, performed at $t>\tau$, yields contact forces of non feasible direction. Delassus' papers seem today rather intricate; a clearer account of his arguments may be found in [2] . More recently [3][4], the same question has been revisited, by the means of elementary Convex Analysis and Quadratic Programming.

The present lectures develop a novel approach to the dynamics of systems involving unilateral constraints. Here are the dominent features.

1° The function $t \to q(t) \in \mathbb{R}^n$ describing the investigated motion on a time interval I, with initial instant $t_0$, is not a priori assumed differentiable everywhere. Instead, one supposes that q equals the time integral of a *velocity function* $t \to u(t) \in \mathbb{R}^n$ with *locally bounded variation* on the interval; notation: $u \in lbv(I, \mathbb{R}^n)$. Classically, such a function u may have discontinuities but, for every t in the interior of I, the existence of the *right-limit* $u^+(t)$ and of the *left-limit* $u^-(t)$ is secured (see Sec.2 below for a convention concerning the case $t = t_0$). These limits equal the respective one-side derivatives of the function $q : I \to \mathbb{R}^n$ at point t.

2° In view of these discontinuities, the existence of the acceleration $q'' = u'$ cannot be expected everywhere. But, with every $u \in lbv(I, \mathbb{R}^n)$, one classically associates an $\mathbb{R}^n$-valued measure [5][6] on the interval I, called in the sequel the *differential measure* [7] of u and denoted by du.

The function u is *locally absolutely continuous* if and only if the vector measure du possesses a *density function*, say $u'_t \in \mathcal{L}^1_{loc}(I, dt; \mathbb{R}^n)$, relative to the Lebesgue measure on the interval I. We denote the latter measure by dt ; this is in fact the differential measure of the real function $t \to t$, which evidently belongs to $lbv(I, \mathbb{R})$. A function u of this sort may constitute a solution to a differential equation, in the classical sense of Caratheodory.

Here is another special case: suppose that, for some $\tau \in \text{int } I$, one has $u^-(\tau) \neq u^+(\tau)$. Then, the $\mathbb{R}^n$-valued measure du possesses at point $\tau$ an *atom* with value $u^+(\tau) - u^-(\tau)$. This value is an element of $\mathbb{R}^n$ that we shall call the *jump* of u at instant $\tau$.

In general, a function $u \in lbv(I, \mathbb{R}^n)$ may be a solution to some *measure differential equation*, a notion about which the reader could find some information in [8].

3° Velocity functions with locally bounded variation make the setting in which we develop the *Nonsmooth Dynamics* of mechanical systems with a finite number of degrees of freedom. This is governed by an extension of Lagrange equations that we introduce in Sec.7. It includes as a special case the traditional equations of the Dynamics of Percussions. Concerning the connection of this general formulation with the classical principles of Dynamics, some details may be found in [9].

4° The set of inequalities $f_\alpha \leqslant 0$ (with $f_\alpha \in C^1$ and $\nabla f_\alpha \neq 0$) defines in $\mathbb{R}^n$ the *feasible region*, denoted by $\Phi$ and assumed in the sequel *independent of time*. If a motion $t \rightarrow q(t)$ is described in the above terms and if $q(t) \in \Phi$ for every $t$, one elementarily finds (see Sec.2) that $u^+(t)$ belongs to a certain polyhedral conic convex subset of $\mathbb{R}^n$, denoted by $V(q(t))$. This is the *tangent cone* to the region $\Phi$ at point $q(t)$, equal in particular to the whole of $\mathbb{R}^n$ when $q(t)$ is interior to $\Phi$. Actually, a cone denoted by $V(q)$, and its *polar cone* $N(q)$ in the sense of the standard scalar product of $\mathbb{R}^n$, will in the sequel be defined even for $q \notin \Phi$. When $q \in \Phi$, the cone $N(q)$ is nothing but the (outward) *normal cone* to $\Phi$ at this point (reduced to $\{0\}$ if $q$ is interior to $\Phi$).

5° The mechanical formulation of unilateral constraints has to encompass the geometric condition $\forall t \in I : q(t) \in \Phi$, together with some infomation about the associated *forces of constraint*. In the framework of traditional (smooth) Analytical Dynamics, this system of forces is represented, for every $t$, by its covariant components, say $r_1,\ldots, r_n$, relative to the generalized coordinates in use. This makes an element of $\mathbb{R}^n$ that we shall denote by $r$

The simplest case is that of *frictionless contact*. This classically means that the force of constraint at every possible point of contact is normal to the concerned bodies, with direction agreeing with unilaterality. One elementarily finds (see Sec.5 below) that, if all the considered unilateral constraints are of this sort, the whole information about them lets itself be summarized into the writing

$$\forall t \in I: \quad q(t) \in \Phi \quad \text{and} \quad -r(t) \in N(q(t)). \tag{1.1}$$

(About the concept of a frictionless contact in the case of a less regular feasible region than above, see [10].)

Starting from (1.1), a decisive observation is made in Sec.5, namely that, for smooth motions, it implies the *stronger* assertion

$$-r(t) \in \partial\psi_{V(q(t))}(u^+(t)). \tag{1.2}$$

According to the usual notations of Convex Analysis, the right-hand member equals the normal cone at point $u^+(t)$ to the convex subset $V(q(t))$ of $\mathbb{R}^n$.

In addition, it is established that, if the initial data satisfy $q(t_0) \in \Phi$, then (1.2), assumed to hold for (dt-almost) every $t$, *secures that* $q(t)$ *will remain in* $\Phi$.

The advantage of (1.2) over (1.1) lies in the following. First, as we shall develop in Sec.5, this writing directly suggests some algorithms of time-discretization for computing the solutions to initial value problems. Secondly, by entering the velocity into the contact law, it paves the way to the consideration of friction. Furthermore, it is easily generalized to Nonsmooth Dynamics.

6° The function $t \to r(t) \in \mathbb{R}^n$ which, in the traditional Lagrange equations, represents the forces of constraints has, in Nonsmooth Dynamics, to be replaced by an $\mathbb{R}^n$-valued measure on the time-interval I, called the *contact impulsion* and denoted by $dR$. For smooth motions, this measure admits the above function as its density relative to Lebesgue measure. A priori, there exists an infinity of representations of a vector measure, such as $dR$, in the form $dR = R'_\mu d\mu$, where $d\mu$ is a nonnegative real measure and $R'_\mu$ a vector-valued locally $d\mu$-integrable density function. We shall admit, as the law of frictionless contact in Nonsmooth Dynamics, the following generalization of (1.2), to be satisfied for every $t$ in I,

$$-R'_\mu(t) \in \partial\psi_{V(q(t))}(u^+(t)). \qquad (1.3)$$

Because the right-hand side is a cone, one shows that *this condition is indifferent to the choice of the base measure* $d\mu$. Furthermore, the existence of a function $R'_\mu$ verifying (1.3) implies that $u^+(t)$ belongs to $V(q(t))$. Through Prop.2.4 below, this ensures $q(t) \in \Phi$ for every $t$, provided the initial data satisfy $q(t_0) \in \Phi$, .

Assertion (1.3) about the contact impulsion makes the definition of the class of unilateral constraints that we call *frictionless and soft*. When transported into the equality of $\mathbb{R}^n$-valued measures, which governs Nonsmooth Dynamics, it generates a *measure differential inclusion*. The existence of solutions to the resulting initial value problems has so far been established only in some special cases [11][12][13] and is currently under investigation. The velocity jumps possibly occuring in such solutions are of the sort the author has previously called "standard inelastic shocks"[14]

[15]. These are *dissipative*, so the corresponding evolution problems are essentially different from those one meets when the possible bounces are assumed "elastic" [16][17][18][19]. A synthetic view may be gained from the *energy balance* drawn in Sec.10 below. The replacement of $u^+$ in (1.2) by some weighted mean of $u^+$ and $u^-$ results in the introduction of a "dissipation index" $\delta$, with zero value in an elastic bounce, while the softness case corresponds to $\delta = 1$.

7° *Dry friction* at a point of contact will be described by some extension of Coulomb's law to possibly anisotropic surfaces. The traditional formulation of this law rests on the decomposition of a contact force into its normal and tangential components; the formulation then consists of two separate statements respectively pertaining to zero and nonzero sliding velocity. In some of the author's early papers [20][21][22], it has been observed that, as soon as the normal component is treated as known, these pair of statements lets itself be synthetically expressed as a law of resistance deriving from a "pseudopotential". This in turn may be transcribed into a variational inequality [23], reflecting a "principle of maximal dissipation".

By a law of resistance, we mean a relation (in the present case, nonsmooth and not expressible through a single-valued function) between the contact force and the *sliding velocity*. Recall that significant mathematical and numerical papers have, in recent years, been devoted to problems which instead involve a "pseudo-friction" law. These problems are developed in the framework of small deviations and the sliding velocity vector is replaced by the tangential relative *displacement* of the contacting bodies. The status of such a pseudo-friction, compared to proper friction, is similar to that of Hencky plasticity with respect to proper plastic flow rules.

The present lectures rest on a newer formulation of the possibly anisotropic Coulomb law, avoiding the decomposition of the contact force [24][25]. Similarly to what has been shown for the frictionless case, these formulations suggest numerical algorithms of time discretization. Furthermore, the resulting relation being conic with regard to the contact force, it admits an extension to Nonsmooth Dynamics, independent, as before, of the choice of a base measure $d\mu$.

8° Here again, the possible nonsmooth motions are found to be governed by some measure differential inclusions. These differential inclusions are applied in Sec. 15 to the dynamics of velocity jumps.

Singularities in the dynamics of systems involving Coulomb friction used to be a matter of controversy during the first quarter of this century. It was observed that some initial value problems could admit several solutions or no solution and also that the behaviour of the investigated system depended on its constants on a discontinuous way. At the time, these findings were considered by such authors as P.Painlevé as contradicting the very bases of Physics. In modern views, nothing looks paradoxical in that, so there only remains of all these discussions the assertion, first made by L.Lecornu [26], that, in the presence of dry friction, velocity jumps are not necessarily the consequences of collisions.

It is shown by an example that the numerical techniques we propose can handle these *frictional catastrophes* without difficulty.

9° These lectures are restricted, for brevity, to time-independent constraints. However, by changing the reference frame and introducing adequate fictitious forces, one is able to apply the proposed methods to the motion of a small object lying on a vibrating table or, in the course of an earthquake, on the ground surface. An example is displayed, exhibiting some unexpected features.

10° For better agreement with the behaviour of real systems, one is commonly led to apply the traditional law of Coulomb with different values of the friction coefficient, depending on whether the sliding velocity vanishes or not. This distinction made between the "static" an "dynamic" friction coefficients seems, at first glance, to destroy the unity brought into the formulations by the use of Convex Analysis. Actually, it is shown in Sec.17 that, far from beeing a mere empirical alteration of these formulations, such a distinction is inherently involved in the consequent developments. The numerical techniques proposed in these lectures are able to handle it without causing any computing problem. In fact, whether the sliding velocity exactly vanishes or not at the end of a time-step is explicitely determined by the algorithms; so the friction coefficient for the next step may be adjusted accordingly.

Let us close this introduction by aknowledging that Coulomb's law can provide only a rather crude approximation of the reality of dry friction (a recent review of the subject may be found in [27]). Also, the collisions affecting parts in real machines cannot be expected to fall exactly under one of the categories respectively described as "soft" or "elastic". And it is unlikely that any definite value of the "dissipation index" could be identified on a clear basis. But a fact of life is that, in most engineering situations, the higher order information needed for more accurate description is not available. So one has to be content with some moderately precise calculation, accounting at least for the main features of phenomena. In three years of experiments, the approach we propose has proved to be very workable. Because of their theoretical consistency and numerical stability, the described algorithms seem to be "robust" enough for accepting in the future various empirical alterations, aimed at improving their power of prediction.

No allusion is made in these lectures to the contact between *deformable bodies*, currently a very active domain of research. The reader will find references to this subject in other parts of this volume. In what concerns computation, since the spatial discretization of a continuous medium, using for instance a finite element scheme, generates a finite-dimensional space of positions, the design of numerical procedures may take an inspiration from the methods presented here (see e.g. [28]). But some fundamental differences between continuum dynamics and finite freedom dynamics have to be kept in mind. Because, in continuous media, every contact particle has zero mass, the concept of a soft contact, as opposed to an elastic contact, becomes unsignificant (it only stays as an option in constructing numerical algorithms). Possible dissipation reenters the scene through the constitutive laws which govern the behaviour of the concerned bodies. In elastic bodies, shock waves are expected to originate from boundary impacts. It is only when the time taken by these waves to travel the whole system is short, with respect to some other typical time values, that the treatment of deformable systems may be strictly conducted in the lines of finite freedom dynamics. Actually, most papers on continuous systems so far are restricted to *quasistatic* evolution problems, i.e. the terms involving inertia are neglected.

# 2.DIFFERENTIAL PROPERTIES

Let a mechanical system have a finite number n of degrees of freedom; every possible position of it may be located through the value it imparts to $q=(q^1,...,q^n)$, an element of some open subset $\Omega$ of $\mathbb{R}^n$. This holds at least locally; in other words, $q^1,...,q^n$ are local coordinates in the manifold of the system possible positions.

One defines a motion by making q depend on time. If the derivatives $q'^i$ of the n functions $t \to q^i$ exist at an instant $\tau$, we shall refer to the element $q'=(q'^1,...,q'^n)$ of $\mathbb{R}^n$ as the velocity of the system at this instant.

Motions will be studied on some time interval I, containing its origin $t_0$ but nonnecessarily closed nor bounded from the right. We shall not suppose the function $q:I \to \mathbb{R}^n$ derivable everywhere. Instead, we assume the existence of a *velocity function* $u:I \to \mathbb{R}^n$ such that

$$\forall t \in I: \qquad q(t)=q(t_0)+\int_{t_0}^{t} u(\tau)\,d\tau. \tag{2.1}$$

This makes sense as soon as u is locally Lebesgue-integrable on I. More specially, we shall suppose that the function u has *locally bounded variation* on I, i.e. it has bounded variation on every compact subinterval of I; notation: $u \in lbv(I,\mathbb{R}^n)$. This secures that, at every $\tau$ in the interior of I, the *right-limit* $u^+(\tau)$ and the *left-limit* $u^-(\tau)$ exist.

By convention, for the initial instant $t_0$, the left-limit $u^-(t_0)$ is understood as equal to $u(t_0)$. This is more than a notational trick; such a writing actually reflects the significance we generally mean to give to the *initial condition* $u(t_0)=u_0$ of an evolution problem. It is intimated that investigation begins at $t_0$, but that the mechanical system was already in existence before. By $u_0$ is imparted some abridged information about the system history, precisely the left-limit $u^-(t_0)$.

Symmetrically, if I possesses a right end, say $t_r$, and contains it, the writing $u^+(t_r) = u(t_r)$ will prove convenient.

From (2.1) it results that the function q possesses at every $\tau > t_0$ a *left-derivative* $q'^-(\tau)$, equal to $u^-(\tau)$ and, at every $\tau$ different from the possible right end of I, a *right-derivative* $q'^+(\tau)$, equal to $u^+(\tau)$.

In addition to the constraints which have permitted the q parametrization, we assume that the system is submitted to some *unilateral constraints* whose geometric effect is expressed by a finite set of inequalities

$$f_\alpha(q) \leq 0, \qquad \alpha \in \{1,2,\ldots,\nu\}. \tag{2.2}$$

The functions $f_\alpha : \Omega \to \mathbb{R}$ are supposed $C^1$ with respective gradients $\nabla f_\alpha = (\partial f_\alpha / \partial q^1, \ldots, \partial f_\alpha / \partial q^n)$ different from zero, at least in a neighbourhood of the corresponding hypersurface $f_\alpha = 0$.

*Inequalities* (2.2) *define the* feasible region $\Phi$ *of* $\Omega$ *; for brevity, we assume that the functions* $f_\alpha$ *do not depend on time,* thereby leaving aside the possibility of moving constraints.

Through the chain rule, the existence of one-side derivatives for the functions $t \to q^i(t)$ implies the same for $t \to f_\alpha(q(t))$. Consequently, if a motion verifies $f_\alpha(q(t)) \leq 0$ for every t, then at any instant $\tau$ such that $f_\alpha(q(\tau)) = 0$, one readily finds $u^+(\tau).\nabla f_\alpha(q(\tau)) \leq 0$ and $u^-(\tau).\nabla f_\alpha(q(\tau)) \geq 0$ (the dot refers to the usual scalar product of $\mathbb{R}^n$).

Generally, let us put:

NOTATION 2.1 *For every* $q\in\Omega$ *define*

$$J(q) := \{\alpha\in\{1,\dots,\nu\} : f_\alpha(q)\geq 0\} \tag{2.3}$$

*and*

$$V(q) := \{v\in\mathbb{R}^n : \forall\alpha\in J(q),\ v.\nabla f_\alpha(q)\leq 0\} \tag{2.4}$$

*(observe that* $V(q)$ *equals the whole of* $\mathbb{R}^n$ *if* $J(q)=\emptyset$ *).*

Using as above the one-side derivatives, one obtains:

PROPOSITION 2.2 *If a motion* $t\to q(t)$ *agrees with the set of constraint inequalities* (2.3), *i.e.* $q(t)\in\Phi$ *for every* $t$, *then*

$$\forall t\in \operatorname{int} I: \quad u^+(t)\in V(q(t)) \quad \textit{and} \quad u^-(t)\in -V(q(t)).$$

REMARK 2.3 In existential studies as well as in numerical algorithms, the definition (2.4) of $V(q)$ will commonly be invoked with $q\notin\Phi$. Then, the following is useful:

PROPOSITION 2.4 *Let the function* $t\to q(t)$ *be associated with some* $u\in\mathcal{L}^1_{loc}(I,\mathbb{R}^n)$ *through* (2.1). *Suppose that* $q(t_0)\in\Phi$ *and that, for Lebesgue-almost every* $t\in I$, *one has* $u(t)\in V(q(t))$. *Then* $q(t)\in\Phi$ *for every* $t$.

PROOF. Let us suppose the existence of some $\tau\in I$, with $q(\tau)\notin\Phi$ and look for contradiction. There exists $\alpha\in\{1,\dots,\nu\}$ such that $f_\alpha(q(\tau))>0$. The set $\{t\in I: t\leq\tau$ and $f_\alpha(q(\tau))\leq 0\}$ is nonempty (it contains $t_0$); let $\sigma$ denote its l.u.b.. Due to the continuity of $f_\alpha$, one has $f_\alpha(q(\sigma))=0$. Since $f_\alpha$ is $C^1$, the function $t\to f_\alpha(q(t))$ is absolutely continuous on $[\sigma,\tau]$ ; after expressing its derivative by the chain rule, one may write

$$f_\alpha(q(\tau)) = \int_\sigma^\tau u(t).\nabla f_\alpha(q(t))\,dt.$$

In view of the definition (2.4) of V, the integrand should be $\leq 0$ for Lebesgue-almost every t, hence $f_\alpha(q(\tau))\leq 0$, which is a contradiction. ■

If u has locally bounded variation, it belongs to $\mathcal{L}^1_{loc}(I,\mathbb{R}^n)$ and the set of its discontinuity points is countable, hence Lebesgue-negligible. Thus, in using the above Proposition, one may replace u in the assumption $u(t)\in V(q(t))$ by $u^+$ or $u^-$ or any weighted mean of them.

REMARK 2.5 The subset $V(q)$ of $\mathbb{R}^n$ is a *closed convex cone*. In case $q\in\Phi$, this coincides with what is usually called the *tangent cone* to the region $\Phi$ at point q (equal, in particular, to the whole of $\mathbb{R}^n$ if $q\in \text{int}\,\Phi$). On the contrary, if $q\notin\Phi$, one commonly agrees to say that the tangent cone to $\Phi$ at this point is empty; *so is not* $V(q)$.

Some caution is needed when interpreting the concept of a tangent cone. Let $q_0\in\Phi$ and $v\in\mathbb{R}^n$; in view of Prop.2.2, for the existence of a mapping $q:I\to\Phi$ such that $q(t_0)=q_0$ and $q'^+(t_0)=v$, it is necessary that $v\in V(q_0)$; a counter-example may be found in [15], showing that this is not sufficient. However, if in addition one assumes $\text{int}\,V(q_0)\neq\emptyset$, then existence is secured [29]. Through classical Convex Analysis, the latter assumption is equivalent to the *polar cone* of $V(q_0)$ having a compact basis; this is the convex cone generated in $\mathbb{R}^n$ by the elements $\nabla f_\alpha(q_0)$, with $\alpha\in J(q_0)$, so the assumption amounts to assert the existence of a hyperplane in $\mathbb{R}^n$, not containing the origin, which intersects all the half-lines generated by these elements. We shall meet this cone again in Sec.5.

REMARK 2.6 A deeper insight into the situation could be gained by considering the *differential manifold* $\mathcal{P}$ of the system positions, without preference to any peculiar system of local coordinates. A motion is the conceived as a mapping $p:I\rightarrow\mathcal{P}$. The (possibly one-side) velocity of the system at some instant $\tau$ equals, by definition, the (possibly one-side) derivative of this mapping, an element, say $p'(\tau)$, of the *tangent space* $\mathcal{P}'_{p(\tau)}$ to $\mathcal{P}$ at point $p(\tau)$. The real numbers $q'^{i}(\tau)$ considered in the foregoing equal the components of $p'(\tau)$, relative to the base induced in $\mathcal{P}'_{p(\tau)}$ by the local coordinate system in use. Inequalities (2.2) are imparted a coordinate-free meaning provided one understands the functions $f_\alpha$ as $C^1$ mappings of $\mathcal{P}$ to $\mathbb{R}$, without reference to any choice of local coordinates. Then, by the gradient $\nabla f_\alpha(p)$ is meant an element of $\mathcal{P}'^{*}_{p}$, the *cotangent space* to $\mathcal{P}$ at point p. Also in this linear space, the local coordinate system induces a base; the partial derivatives precedingly invoked equal the components of $\nabla f_\alpha(p)$ relative to this base. In Definition (2.4), the Euclidean scalar product of $\mathbb{R}^n$ should then be replaced by the bilinear form $\langle .,. \rangle$ which puts the linear spaces $\mathcal{P}'_p$ and $\mathcal{P}'^{*}_{p}$ in duality. Thereby, for every position p, a pair of mutually polar convex cones is defined in these dual linear spaces, without reference to any choice of local coordinates.

The concept of a convex cone in $\mathcal{P}'_p$ or in $\mathcal{P}'^{*}_{p}$ is meanigful, in view of the linear structure of these tangent spaces. Concerning, on the contrary, the feasible region, the differentiable manifold $\mathcal{P}$ cannot in general support any convexity assumption. If such an assumption is made, it only refers to some peculiar coordinate system. However, let us mention the following special case.

For the treatment of dynamical problems concerning a scleronomic system, the expression $\frac{1}{2}A_{ij}(q)q'^i q'^j$ of the kinetic energy has to be introduced. This is a positive definite quadratic form in $q'$ and, classically, by putting $ds^2 = A_{ij}(q)\, dq^i dq^j$, one equips the differential manifold $\mathcal{P}$ with a Riemannian metric independent of the coordinates in use. So this metric is intrinsically associated with the dynamical structure of the mechanical system. Now, it may happen that some local coordinates exist, such that the coefficients $A_{ij}(q)$ are constant in q; so is the case, for instance, if the system consists of a single rigid body performing only motions parallel to a fixed plane. Under such circumstances, the curvature of $\mathcal{P}$ is zero; in other words, this manifold is locally Euclidean. Then, at least in sufficiently small regions, the concept of the convexity of a subset of $\mathcal{P}$ becomes mechanically meaningful. The mathematical paper [16] was precisely based on the convexity of the feasible region.

Anyway, the writing in (2.1) makes sense only as far as the functions $t \to q$ and $t \to u$ take their values in a fixed linear space, namely $\mathbb{R}^n$ for the present. On the contrary, in the differential geometric setting, the velocity at time t would be an element of the tangent space $\mathcal{P}'_{p(t)}$, which depends on t through the unknown mapping $p: I \to \mathcal{P}$.

# 3.KINEMATICS

In all the sequel, each of the inequalities $f_\alpha \leq 0$ will be understood as characterizing the system positions agreeing with the mutual impenetrability of a certain pair of rigid bodies. For instance, let us drop the subscript $\alpha$ and assume that condition $f \leq 0$ expresses that some rigid part $\mathcal{B}_1$ of the system does not overlap a given external obstacle $\mathcal{B}_0$, fixed relative to the reference frame in use. The impenetrability of two rigid bodies $\mathcal{B}_1$ and $\mathcal{B}_2$, which both are constituents of the system, would finally result in the same formalism (see [30], where the case of an external obstacle with prescribed motion is also considered).

Equality $f(q)=0$ means that, in the position $q$ of the system, the part $\mathcal{B}_1$ touches $\mathcal{B}_0$ . We shall always assume that contact takes place *through a single particle* of $\mathcal{B}_1$ , which in general depends on $q$ , say $\mathcal{M}_1(q)$. The respective boundaries of contacting bodies will be supposed to permit the definition of a *common tangent plane* at $\mathcal{M}_1(q)$ to these boundaries. This does not preclude edges or vertices; one of the bodies may even reduce to a single particle, provided the boundary of the other is a smooth surface.

*Let* $\mathcal{N}_q$ *denote the unit vector, normal to this tangent plane and directed toward* $\mathcal{B}_1$ .

As usual, the primitive constraints of the system, i.e. the constraints which have permitted the parametrization through $(q^1,...,q^n)$, are assumed smooth enough for the following to hold. Let a motion be described by giving $q$ as a function of $t$. For every $t$ such that the (possibly one-side) derivatives $q'^1,...,q'^n$ exist, every particle, say $\mathcal{M}$ , of the system possesses a (possibly one-side) velocity vector, relative to the reference frame in use. Calculating this vector yields an expression $\mathcal{V}(\mathcal{M}, q, q')$ , affine with regard to $q'$.

For brevity, we shall restrict the sequel to the *scleronomic* case, i.e. the primitive constraints do not depend on time; then the above expression is *linear* in $q'$. Let us apply this to the contact particle $\mathcal{M}_1(q)$ and put the notation

$$G_q q' := \mathcal{V}(\mathcal{M}_1(q), q, q'). \tag{3.1}$$

For every q such that $f(q)=0$, this introduces the mapping $q' \to \mathcal{G}_q q'$, linear of $\mathbb{R}^n$ to the Euclidean linear space $\mathcal{E}_3$ of the vectors of physical space.

In the case of a pair of bodies which both are parts of the system, a linear mapping similar to $\mathcal{G}_q$ would express from q' the *relative velocity*, at a possible contact point, of one of this part with respect to the other.

The writing in (3.1) is not restricted to motions agreeing with the impenetrability constraint. We now are to take this constraint into account. Let a value of q correspond to contact, i.e. $f(q)=0$, and let $v \in \mathbb{R}^n$. Let a motion start from this position q at some time $\tau$, with right-velocity $q'^+$ equal to v at this instant. Evaluating f at all subsequent positions, one obtains a function of time whose right-derivative at $\tau$ may be expressed through the chain rule, namely $v.\nabla f(q)$. Assume $v.\nabla f(q)<0$; then instant $\tau$ is followed by a nonzero time interval over which $f<0$, i.e. $\mathcal{B}_1$ and $\mathcal{B}_0$ break contact. This implies that, at $\tau$, the right-velocity $\mathcal{G}_q v$ of the contact particle $\mathcal{M}_1(q)$ of $\mathcal{B}_1$ verifie $\mathcal{N}_q.\mathcal{G}_q v \geqslant 0$; otherwise the motion of this particle would require of $\mathcal{B}_1$ to overlap $\mathcal{B}_0$, so making $f \geqslant 0$.

This shows that, for $v \in \mathbb{R}^n$ and for any fixed q satisfying $f(q)=0$, one has the implication

$$v.\nabla f(q)<0 \quad \Rightarrow \quad \mathcal{N}_q.\mathcal{G}_q v \geqslant 0. \tag{3.2}$$

Let us introduce now the linear mapping $\mathcal{G}_q^* : \mathcal{E}_3 \to \mathbb{R}^n$, the *transpose* of $\mathcal{G}_q$ in the sense of the Euclidean autodualities of $\mathcal{E}_3$ and $\mathbb{R}^n$; then $\mathcal{N}_q.\mathcal{G}_q v = v.\,\mathcal{G}_q^* \mathcal{N}_q$. Recall that we have assumed $\nabla f \neq 0$; through a unilateral version of the Lagrange multiplier theorem, implication (3.2) yields:

PROPOSITION 3.1 *For every* q *verifying* $f(q)=0$, *there exists* $\lambda \geqslant 0$ *such that*

$$\mathcal{G}_q^* \mathcal{N}_q = -\lambda \nabla f(q). \tag{3.3}$$

REMARK 3.2 We shall later need that the element $\mathcal{G}_q^* \mathcal{N}_q$ of $\mathbb{R}^n$ differ from zero, i.e. $\lambda>0$. It is a general fact that the kernel of $\mathcal{G}_q^*$ equals the subspace of $\mathcal{E}_3$ orthogonal to the range $\mathcal{G}_q(\mathbb{R}^n)$ of $\mathcal{G}_q$. In particular, when $\mathcal{G}_q(\mathbb{R}^n)$ is the whole of $\mathcal{E}_3$, the kernel of $\mathcal{G}_q^*$ reduces to $\{0\}$ and this secures $\mathcal{G}_q^* \mathcal{N}_q \neq 0$. But, in some usual applications, $\mathcal{G}_q(\mathbb{R}^n)$ will be a strict subspace of $\mathcal{E}_3$; for instance, if the primitive constraints allow $\mathcal{B}_1$ to perform only motions parallel to some fixed plane, then $\dim \mathcal{G}_q(\mathbb{R}^n) = 2$. What precedes shows that $\mathcal{G}_q^* \mathcal{N}_q = 0$ if and only if $\mathcal{G}_q(\mathbb{R}^n)$ is contained in the two-dimensional subspace of $\mathcal{E}_3$ consisting of the vectors parallel to the common tangent plane to contacting bodies.

REMARK 3.3 So far, $\mathcal{G}_q$ has been defined only for such q that $f(q)=0$. In computation, as well as in existential studies, it will prove useful to extend the definitions of $\mathcal{G}_q$ and $\mathcal{N}_q$, in a smooth arbitrary way, to the whole range $\Omega$ of the local coordinates in use, or at least to some neighbourhood of the hypersurface $f=0$. This extension may additionally be required to preserve the property (3.3).

REMARK 3.4 Let a motion comply with condition $f \leq 0$ at every time. Consider an instant of contact, i.e. at which $f(q)=0$, and suppose that the *two-sided* derivative $q'$ exists at this instant. Since the latter equals the common value of $q'^{+}$ and $q'^{-}$, the observations made in Sec.2 imply $q'.\nabla f(q)=0$. In view of Proposition 3.1, this yields $\mathcal{N}_q . \mathcal{G}_q q' = 0$, expressing that the velocity of the particle $\mathcal{M}_1(q)$ of $\mathcal{B}_1$ through which this body touches the fixed obstacle $\mathcal{B}_0$ is a vector parallel to the common tangent plane.

Similar reasoning would apply to a pair of moving parts of the system: if contact holds at some instant and if, at this instant, the time-derivative $q'$ exists, in the ordinary two-side sense, then the relative velocity of one of these bodies with respect to the other, at the contact point, is a vector parallel to the common tangent plane. The same fact is classically established in elementary Kinematics, under the stronger assumption of permanent contact. One refers to this relative velocity as the *sliding velocity* of the first body upon the second.

# 4. LAGRANGE EQUATIONS

Let a motion of the system be described under the form (2.1). It will be said *smooth* if the velocity function u is *locally absolutely continuous*, i.e. absolutely continuous on every compact subinterval of I. This implies the existence of the derivative $u'(t)=q''(t)$ for Lebesgue-almost every t. We shall refer to the element $q''$ of $\mathbb{R}^n$ as the *acceleration* of the system.

Such a motion agrees with Dynamics if and only if the function $t \rightarrow q(t)$ verifies the Lagrange differential equation

$$\frac{d}{dt}(\partial \mathcal{E}_c/\partial q'^i) - \partial \mathcal{E}_c/\partial q^i = c_i . \tag{4.1}$$

Here $\mathcal{E}_c(q,q')$ denotes the expression of the kinetic energy; since we assume the system scleronomic, this is

$$\mathcal{E}_c(q,q') = \frac{1}{2} A_{ij}(q)\, q'^i q'^j , \tag{4.2}$$

a positive definite quadratic form in $q'$.

By $c_i$ are denoted the covariant components of the totality of the forces acting on the system. These possibly comprise some given forces, whose covariant components $F_i$ are known functions of time, position and velocity, and also comprise the a priori unknown *reactions* or *constraint forces*, involved in the constraints that the system experiences. As usual, when Lagrange equations are applied, we shall assume that the primitive constraints, i.e. those which have permitted the q parametrization, are perfect, in the sense that the corresponding reactions have zero covariant components. But we shall have to take into account the reactions of the superimposed unilateral constraints.

The left-hand side of (4.1) may be developed into

$$A_{ij}q''^j + (A_{ij,k} - \tfrac{1}{2} A_{jk,i})\, q'^j q'^k, \tag{4.3}$$

where $A_{ij,k}$ denotes the partial derivative of $A_{ij}$ with regard to $q_k$. Therefore, (4.1) takes on the form $A_{ij}q''^j = K_i + r_i$ ; here the term $r_i$ refers to the totality of the reactions of superimposed constraints, while $K_i$ is a known function of t, q, q', equal to $F_i$ minus all the terms in (4.3) which involve $q'^j q'^k$. Since the matrix A is invertible, we may finally write this down as

$$q'' = A^{-1}K + A^{-1}r. \tag{4.4}$$

It has been assumed that each inequality $f_\alpha(q)=0$ expresses the contact between some pair of rigid bodies. Let us denote by $r_1^\alpha,\ldots,r_n^\alpha$ the covariant components of the corresponding reaction, making an element of $\mathbb{R}^n$ denoted by $r^\alpha$. Formally, this term will be introduced also when the said contact is not in effect, so we shall state

$$f_\alpha(q) < 0 \quad \Rightarrow \quad r^\alpha = 0. \tag{4.5}$$

To fix the ideas, suppose, as in Sec.3, that the contact expressed by $f_\alpha(q) = 0$ takes place between some part, say $\mathcal{B}_\alpha$, of the system and an unmoving external obstacle. Suppose that the contact action on $\mathcal{B}_\alpha$ results in a single force $\mathcal{R}^\alpha$, applied to a particle of this body denoted by $\mathcal{M}_\alpha(q)$. By definition, the covariant components of this force make the element $r^\alpha$ of $\mathbb{R}^n$ such that

$$\forall v \in \mathbb{R}^n \quad \mathcal{R}^\alpha . \mathcal{V}(\mathcal{M}_\alpha(q), q, v) = r^\alpha . v \,. \tag{4.6}$$

On the left-hand side, the dot refers to the Euclidean scalar product of $\mathcal{E}_3$, on the right-hand side to the standard scalar product of $\mathbb{R}^n$. As before, due to scleronomy, $\mathcal{V}(\mathcal{M}_\alpha(q), q, v)$ is a linear expression in v, say $G_q^\alpha v$. Then, by introducing the transpose mapping $G_q^{\alpha *} : \mathcal{E}_3 \rightarrow \mathbb{R}^n$, one equivalently writes down (4.6) as

$$r^\alpha = G_q^{\alpha *} \mathcal{R}^\alpha . \tag{4.7}$$

# 5. SMOOTH FRICTIONLESS MOTIONS

In this Section, we shall assume that the possible unilateral contacts are *frictionless*. Under the preceding notations, this means that, for every $\alpha$ such that $f_\alpha(q) = 0$,

$$\exists \rho \in \mathbb{R} : \mathcal{R}^\alpha = \rho \, \mathcal{N}_q^\alpha , \tag{5.1}$$

where $\mathcal{N}_q^\alpha$ denotes the common normal unit at $\mathcal{M}_\alpha(q)$ to the contacting bodies, *directed toward* $\mathcal{B}_\alpha$.

We shall assume in addition that the contact is *unilateral without adhesive effect*, i.e. $\rho \geq 0$.

Through Prop.3.1, conditions (4.7) and (5.1) imply

$$\exists \mu_\alpha \geq 0 : r^\alpha = - \mu_\alpha \nabla f_\alpha(q). \tag{5.2}$$

*Note that the repetition of a Greek index will never be understood as implying summation.*

Provided that $G_q^{\alpha *} \mathcal{N}_q^\alpha \neq 0$ (see Remark 3.2), it may more precisely be checked that (5.2) is *equivalent* to the existence of $\mathcal{R}^\alpha$ agreeing with the above assumptions. And, in view of the convention made in (4.5), this equivalence remains valid for every $q \in \Phi$ if one stipulates that $\mu_\alpha = 0$ when $f_\alpha(q) < 0$. Furthermore, in formulating our evolution problem under the geometric condition $q \in \Phi$ , it is immaterial to state that (5.2) also holds for non feasible q.

Therefore, a value of the total reaction term $r = \sum r^\alpha$ is compatible with the stated laws of contact if and only if it satisfies

$$r \in - N(q), \tag{5.3}$$

where N(q) denotes *the convex cone generated in* $\mathbb{R}^n$ *by the elements* $\nabla f_\alpha(q)$, $\alpha \in J(q)$ (see Notation 2.1). According to an usual convention, if $J(q) = \emptyset$ this cone consists of the zero of $\mathbb{R}^n$. In all cases V(q), as defined in (2.4), and N(q) make a pair of *mutually polar* or *conjugate* cones. When q belongs to the feasible region, N(q) is nothing else than the (outward) *normal cone* to $\Phi$ at point q, but what we are denoting here by N(q) also makes sense and is nonempty for $q \notin \Phi$.

Eliminating r between (5.3) and the Lagrange equations, as they are displayed in Sec.4, one obtains that a smooth motion of the system agrees with all the mechanical conditions stated, if and only if the *differential inclusion*

$$-A(q)\, q'' + K(t,q,q') \in N(q) \tag{5.4}$$

is satisfied Lebesgue-a.e. in I, together with the geometric condition of the unilateral constraints,

$$\forall t \in I : \quad q(t) \in \Phi. \tag{5.5}$$

The Proposition below marks a turning point, regarding all our subject matter. For every (closed, convex) subset of $\mathbb{R}^n$, say C, we denote by $\psi_C$ its *indicator function*, i.e. $\psi_C(x)=0$ if $x\in C$ and $+\infty$ otherwise. The *subdifferential* $\partial\psi_C(x)$ is known to equal the normal cone to C at point x (empty if $x\notin C$).

PROPOSITION 5.1 *A smooth motion, with initial data* $q(t_0)$ *belonging to* $\Phi$, *is a solution of* (5.4), (5.5) *if and only if the velocity function* u *associated with* q *through* (2.1) *satisfies Lebesgue-a.e. in* I *the differential inclusion*

$$-A(q)\,u' + K(t,q,u) \in \partial\psi_{V(q)}(u). \qquad (5.6)$$

PROOF. For every t such that (5.6) holds, the right-hand side is nonempty, hence $u(t)\in V(q(t))$. Since, by assumption, u is a (locally absolutely) continuous function, (2.1) entails that u(t) equals, for every $t\in \operatorname{int} I$, the (two-sided) derivative of $t\to q(t)$ Through Proposition 2.4, one concludes that, if (5.6) is verified Lebesgue-a.e., then (5.5) holds. Furthermore, (5.6) implies (5.4) because $\partial\psi_{V(q)}(u)$ is essentially a subset of N(q), the polar cone of V(q).

Conversely, suppose that the function $t\to q(t)$ satisfies (5.5). Since u is continuous, Proposition 2.2 shows that u(t), for every $t\in \operatorname{int} I$, belongs to $V(q(t))\cap -V(q(t))$ which is the linear subspace of $\mathbb{R}^n$ orthogonal to N(q(t)). Therefore, if (5.4) holds Lebesgue-a.e., then for Lebesgue-almost every t, the values of u and $-A\,u'+K$ are orthogonal and respectively belong to the pair of mutually polar cones V(q(t)) and N(q(t)); consequently they are *conjugate points* relative to this pair, i.e. (5.6) holds. ∎

REMARK 5.2 At the present stage, where the motion smoothness, i.e. the local absolute continuity of u, is a priori assumed, the same symmetry between past and future is observed as in the classical case of bilaterally constrained frictionless systems. In particular, for the differential inclusion (5.6) to take care automatically of condition $q \in \Phi$, it is enough that $q(\tau)$ belongs to $\Phi$ at some $\tau \in I$, nonnecessarily the initial instant. Also an equivalence similar to what is stated in the above Proposition may symmetrically be established, with (5.6) replaced by

$$-A(q)\,u' + K(t,q,u) \in -\,\partial\psi_{-V(q)}(u). \tag{5.7}$$

Similarly to (5.6), this implies the orthogonality of the elements u and $-Au'+K$ of $\mathbb{R}^n$. From such an orthogonality, the same *power equation* may be derived as in the traditional case of frictionless time-independent bilateral constraints:

$$\frac{d}{dt}\mathcal{E}_c = F.u\,. \tag{5.8}$$

REMARK 5.3 As an introduction to forthcoming Sections, let us indicate how the formulation (5.6) directly suggests a procedure of time discretization for computing approximately the motion consequent to some initial data

$$q(t_0) = q_0, \qquad \text{given in } \Phi \tag{5.9}$$

$$u(t_0) = u_0, \qquad \text{given in } V(q_0). \tag{5.10}$$

Let $(t_I,t_F)$ be a time-step (here I is understood as referring to "initial" and F to "final"), with length $h=t_F-t_I$ and midpoint $t_M=t_I+\frac{1}{2}h$ . From the approximate values $q_I$, $u_I$ of the functions q and u at $t_I$, one has to compute $q_F$ and $u_F$ , corresponding to $t_F$.

Using $(u_F-u_I)/h$ as an approximant of $u'$, one discretizes the differential inclusion (5.6) into

$$-A(q_M)\,(u_F-u_I)/h + K(t_M,q_M,u_I) \in \partial\psi_M(u_F). \tag{5.11}$$

Here $q_M=q_I+\frac{1}{2}hu_I$ is a midpoint approximant of q; by $\psi_M$ is denoted the indicator function of $V(q_M)$. Inserting $u_F$ as an approximant of u in the right-hand side tends to qualify this discretization scheme as "implicit". However, the smoothness of the given function K allows one to replace in it u by $u_I$, so the procedure may be said "semi-implicit".

We shall come back later to more general algorithms of the same sort; let us only show here how (5.11) uniquely determines $u_F$. Suppose, for simplicity, that $A(q_M)$ reduces to the unit matrix. This actually entails no loss of generality: in the line of Remark 2.6, it amounts to make of the tangent space, at the point $q_M$ of the position manifold, a Euclidean linear space, with scalar product defined through $A(q_M)$, and to take an orthonormal base in this space (more detail on the practical use of this trick may be found in [15]). Then (5.11) becomes

$$-(u_F-u_I) + hK \in \partial\psi_M(u_F).$$

The multiplication of both members of (5.11) by the positive number h has not altered the right-hand side, because $\partial\psi_M(u_F)$ is a cone. In view of the classical characterization of the *proximal point* to $u_I$+hK in the closed convex subset $V(q_M)$ of the Euclidean linear space, this is equivalent to

$$u_F = \text{prox}(\, u_I + hK,\, V(q_M)). \qquad (5.12)$$

Observe that $u_I$+ hK is nothing but the value that $u_F$ would take in the case $J(q_M)=\emptyset$, i.e. the value that the discretization of Lagrange equations would yield in the absence of superimposed unilateral constraints.

After determining $u_F$, one finishes the computation step by calculating $q_F = q_M + \frac{1}{2} h\, u_F$.

# 6. PERCUSSIONS AND FRICTIONLESS SHOCKS

The preceding Section was restricted to motions a priori assumed smooth. For such a motion, there may exist in particular a time interval during which one or more of the contacts persist, say the contact expressed by $f_\alpha = 0$. Call $t_c$ the end of this interval and suppose that a nonzero interval follows, over which $f_\alpha < 0$. In other words, as soon as t exceeds $t_c$, the set $J(q(t))$ ceases to contain $\alpha$; thus the cone $V(q(t))$ suddenly increases. This involves no contradiction with the assumed (locally absolute) continuity of u, nor with the inclusion $u \in V \cap -V$, resulting from Prop.2.2. Certain motions of a unilateral pendulum provide familiar examples of this.

Computationally, there is no difficulty in approximating a motion showing such a smooth contact break, through the algorithm of Remark 5.3. It only happens that, from a certain time-step to the next, the dimension of $V(q_M)$ suddenly increases, without producing any notable irregularity in the sequence of the calculated values of u.

Imagine, on the contrary, that an interval of smooth motion ends at some instant $t_s$ *with the occurence of new contacts*, i.e. $J(q)$ suddenly increases. In view of Prop.2.2, the continuity of u at $t_s$ would require of the left-limit $u^-(t_s)$, an element of $-V(q(t_s))$, to belong also to $V(q(t_s))$. This would mean that the new contacts are attained tangentially, an event which cannot be expected in general. So, a discontinuity of u at $t_s$ has to be contemplated. This is called a *shock* and, to deal with it, Classical Mechanics provides the concept of *percussion*, that we are to review in a few words.

Assume that $t_s$ is followed by another interval of smooth motion. It will be understood that, because of slight deformability in the system parts, the velocity change is not rigorously instantaneous, but takes place on a "very short" time interval, say $(t_s, t_s+\theta)$, over which the differential equations of smooth dynamics supposedly hold. In view of the steep velocity change, these equations are expected to involve "very large" values of the contact forces. By equalling the integrals on $(t_s, t_s+\theta)$ of both members of the differential equations, one obtains the *momentum change formula*. This is a balance equation, through which the net velocity change is related to the time integral, say $\Pi$, of the function $t \to r(t) \in \mathbb{R}^n$ which, in the notations of Sec.4, represents the contact forces. Compared to it, the term K yields a negligible integral, because $\theta$ is "very small"; for the same reason, the variations of q on the interval are neglected. Once obtained, the balance formula is inserted into the former setting of instantaneous shock; the element $\Pi$ of $\mathbb{R}^n$ is then said to make *the components of the contact percussions*. This procedure may be seen as an early example of a *multiple scaling* ; it permits to calculate the after-shock velocity $u^+(t_s)$, as far as sufficient information is available about contact percussions.

Usually, the above reasoning is applied under the assumption of frictionless contacts. Then it seems natural to admit, similarly to (5.2), that, for every $\alpha \in J(q_s)$, the contribution to $\Pi$ of the corresponding contact, say $\Pi^\alpha$, has the following form

$$\exists M_\alpha \geq 0 \;:\; \Pi^\alpha = - M_\alpha \nabla f_\alpha (q(t_s)). \tag{6.1}$$

The argument commonly proposed in support to this assertion about $\Pi^\alpha$ is that, in (5.2), the vector $\nabla f_\alpha(q(t))$ should remain nearly constant during the very short time interval $(t_s, t_s+\theta)$, because the variations of q are very small; thus $M_\alpha$ would simply equal the integral of the nonnegative real function

$t \to \mu_\alpha(t)$. In the author's opinion, this conclusion cannot be accepted without further discussion, though physical situations certainly exist in which (6.1) accurately agrees with reality. In fact, the use of the equations of regular Dynamics rests on the smoothing effect of a certain amount of deformation in the system parts. The very meaning of the parametrization $q$ may then be questioned. Furthermore, the resulting alterations of the functions $f_\alpha$, however small in amplitude and possibly concentrated in the vicinity of the point $q(t_s)$, are prone to generate nonnegligible variations of the vectors $\nabla f_\alpha(q(t))$ in the course of the interval $(t_s, t_s+\theta)$. A similar discussion could also be conducted on equations (4.7) and (5.1): the deformation of the contacting bodies, however small, may appreciably alter the vector $\mathcal{N}_q^\alpha$ in $\mathcal{E}_3$ as well as the mapping $\mathcal{G}_q^{\alpha *}: \mathcal{E}_3 \to \mathbb{R}^n$.

At all events, (6.1) does not bring enough information about $\Pi$ to determine $u^+(t_s)$ completely, even in the simplest case where $J(q(t_s))$ consists of a single element. Classically, a shock (in a scleronomic system) is said *elastic* if it preserves the total kinetic energy; this additional assertion permits, in the case of a single contact, to determine $u^+(t_s)$ unambiguously. This may be given a geometric form by using in $\mathbb{R}^n$ the kinetic metric, i.e. the Euclidean metric defined by the matrix $A(q(t_s))$. Then, if $J(q(t_s))=\{\alpha\}$, one finds that $u^+(t_s)$ equals the mirror image of $u^-(t_s)$ relative to the tangent plane at the hypersurface $f_\alpha=0$.

But one can hardly justify energy conservation by any physical argument. In fact the deformation induced by the impact is expected to propagate dynamically all through the various parts of the system and possibly also through the external connected bodies. Even if the (very slightly) deformable materials of which the system is built may be asserted perfectly elastic, the various parts usually remain, after a bounce, in a state of vibration which, in the energy balance drawn at macroscopic level, amounts to dissipation.

In short, predicting accurately the outcome of a shock requires some higher order information, unavailable in usual situations. The pertinence of the model of an elastic shock has to be discussed in each particular application. The same is true for the other sort of shock we shall present in Sec.8, which however offers the advantage of better formal consistency and easier numerical handling.

## 7. NONSMOOTH DYNAMICS

We now intend to insert the description of shocks into a generalized formulation of the dynamics of the investigated system, which does not require the local absolute continuity of the velocity function $u: I \to \mathbb{R}^n$. This function will only be assumed to have *locally bounded variation*, i.e. to have bounded variation on every compact subinterval of $I$ ; notation: $u \in lbv(I, \mathbb{R}^n)$. Classically, with such a function, an $\mathbb{R}^n$-valued measure on the interval $I$ is associated, that we shall call the *differential measure* of $u$ and denote by $du$. A characteristic property of this measure is that, for every compact subinterval $[\sigma,\tau] \subset \operatorname{int} I$, one has

$$\int_{[\sigma,\tau]} du = u^+(\tau) - u^-(\sigma). \tag{7.1}$$

In fact, the bounded variation assumption secures the existence of the one-sided limits of $u$ at any point of $\operatorname{int} I$. Equality (7.1) remains valid for $\sigma = t_0$ provided that, as already proposed in Sec.2, we agree on the convention

$u^-(t_0)=u(t_0)$. The symmetrical convention may also be used for $u^+$ at the possible right end of I.

By making $\sigma=\tau$, one sees in particular that, if u is discontinuous at point $\tau$, then the measure du is expected to possess an *atom* at this point, with value equal to the total jump of u.

It is clear on (7.1) that du depends on the function u only through $u^+$ and $u^-$. *The values that* u *may take at its discontinuity points are immaterial. Neither have these values any effect on the expression* (2.1) *of* q *since the set of the discontinuity points of* u *is countable, hence Lebesgue-negligible.*

The reader may refer to [7] as a monograph on the lbv functions of a real interval and their differential measures, with values in a Banach space X. For a vast class of Banach spaces, in particular for $X = \mathbb{R}^n$, there comes out that, if u is locally absolutely continuous, the measure du possesses, relatively to the Lebesgue measure on I, here denoted by dt, a *density function*, say $u'_t \in \mathcal{L}^1_{loc}(I, dt; \mathbb{R}^n)$. Of course, the latter is defined up to the addition of a dt-negligible function. This is commonly expressed by saying that the $\mathbb{R}^n$-valued measure du equals *the product of the real measure* dt *by the* $\mathbb{R}^n$-*valued, locally* dt-*integrable, function* $u'_t$; notation $du=u'_t\,dt$. Also du is said *locally* dt-*continuous*. Conversely, if du has such a form, the function $u \in lbv(I,\mathbb{R}^n)$, possibly after correcting the unessential values it takes on a countable set, is locally absolutely continuous. Furthermore, for Lebesgue-almost every t, the value $u'_t(t)$ of the density function makes the (two-sided) derivative of u.

Throughout these lectures, we shall comply with the widespread usage of affecting the character $\mathcal{L}$ to non-Hausdorff spaces consisting of functions defined everywhere, while L will refer to the corresponding Hausdorff spaces of equivalence classes.

In this setting, let us come back to Lagrange equations. If u is locally absolutely continuous, with q related to it through (2.1), the notation used in Sec.4 becomes

$$A_{ij}(q)\, u'^{j}_{t} + (A_{ij,k}(q) - \tfrac{1}{2} A_{jk,i}(q))\, u^j u^k = c_i. \tag{7.2}$$

The right-hand side refers to the covariant components of the totality of the forces acting on the system, including the reactions of possible contacts. For these differential equations to make sense, the n functions $t \to c_i$ have to be elements of $\mathcal{L}^1_{loc}(I, dt; \mathbb{R})$ ; so each of the n equations (7.2) may equivalently be expressed as an *equality of measures* on the interval I,

$$A_{ij}(q)\, du^j + (A_{ij,k}(q) - \tfrac{1}{2} A_{jk,i}(q))\, u^j u^k\, dt = c_i\, dt. \tag{7.3}$$

One readily checks that the functions of t, by which the measures $du^j$ or dt, on the left-hand side, are multiplied, have the local integrability properties required in order that the products make sense.

Now, this new writing keeps meaningful for general $u \in lbv(I, \mathbb{R}^n)$, and allows one to replace the terms $c_i\, dt$, on the right-hand side by some real measures $dC_i$ , said to be the covariant components of the *total impulsion* dC experienced by the system. These will equal the sum of the dt-continuous measures $F_i(t,q,u)\,dt$, describing given forces, and of the covariant components $dR_i$ of the *contact impulsion* dR, an $\mathbb{R}^n$-valued measure on I. For instance, in the case of an isolated collision occuring at instant $t_s$ , as investigated in Sec.6, the measure dR would involve an atom with mass $\Pi$, placed at point $t_s$ ; this should be added to the dt-continuous measure r dt, expressing the contact actions in the course of possible episodes of smooth motion with persistent contact.

As before, we shall denote by K the known $\mathbb{R}^n$-valued function with components

$$K_i(t,q,u) = F_i(t,q,u) - (A_{ij,k}(q) - \frac{1}{2} A_{jk,i}(q))\, u^j u^k .$$

*Then the following equality of* $\mathbb{R}^n$*- valued measures on the interval* I *will be adopted as governing the dynamics of possibly nonsmooth motions*:

$$A(q)\, du - K(t,q,u)\, dt = dR . \tag{7.4}$$

The connection between such an extension of Lagrange equations and the principles of Classical Dynamics is discussed with more precision in [9].

# 8. CONTACT SOFTNESS

Let us consider again the differential inclusion (5.6), which has been found to govern the assumedly smooth motions of the system. This inclusion equivalently means that the expression $t \to r(t) \in \mathbb{R}^n$ of the covariant components of the contact forces makes an element of $\mathcal{L}^1_{loc}(I, dt, \mathbb{R}^n)$ which verifies, for Lebesgue-almost every t,

$$- r(t) \in \partial\psi_{V(q(t))}(u(t)). \tag{8.1}$$

In the context of Sec.5, q(t) belongs to $\Phi$ for every t; therefore, due to Prop. 2.2 and to the continuity of u, one has $u(t) \in V(q(t))$ for every t. Hence, the right-hand side of (8.1) is nonempty for every t (it contains at least the zero of $\mathbb{R}^n$). This enables one, by altering the function r on a Lebesgue-negligible subset of I, to make (8.1) hold *everywhere* in I.

In the language of Sec.7, the $\mathbb{R}^n$-valued measure $dR = r\,dt$ constitutes, for the above case, the total *contact impulsion*. We now propose to adapt (8.1) to the more general setting of Nonsmooth Dynamics.

The Lebesgue measure thus loses its preeminence and we shall definitely give up using the expression "almost everywhere". As observed in Sec.7, if u is discontinuous, only $u^+$ and $u^-$ have relevance to the motion; so, one has to make a choice about what is to replace u in the right-hand side of (8.1).

DEFINITION 8.1 *The set of superimposed constraints is said* frictionless and soft *if the total contact impulsion admits a representation* $dR = R'_\mu d\mu$, *where* $d\mu$ *denotes a nonnegative real measure on* I *and* $R'_\mu$ *an element of* $\mathcal{L}^1_{loc}(I, d\mu; \mathbb{R}^n)$ *such that, for* every t *in* I,

$$- R'_\mu(t) \in \partial\psi_{V(q(t))}(u^+(t)). \qquad (8.2)$$

This first implies that the right-hand side is nonempty, hence $u^+(t) \in V(q(t))$ for every t. Therefore, if the initial data satisfy $q(t_0) \in \Phi$, (8.2) will secure, thanks to Prop.2.4, that $q(t) \in \Phi$ for every $t \in I$.

The following Proposition shows that the concept introduced by Definition 8.1 does not actually depend on the peculiar choice of $d\mu$.

PROPOSITION 8.2 *Inclusion* (8.2) *holds for every* $t \in I$ *if and only if the same is true after replacing* $d\mu$ *by another nonnegative real measure relative to which* dR *possesses a density function.*

PROOF Suppose that (8.2) holds for every t in I and denote by $d\nu$ another nonnegative real measure such that $R'_\nu$ exists. If we put $d\sigma = d\mu + d\nu$, the Radon-Nikodym theorem ensures the existence of the density functions $\mu'_\sigma$ and $\nu'_\sigma$, nonnegative elements of $\mathcal{L}^\infty(I, d\sigma; \mathbb{R})$. Then $dR = R'_\mu \mu'_\sigma d\sigma = R'_\nu \nu'_\sigma d\sigma$; so the $\mathbb{R}^n$-valued functions $R'_\mu \mu'_\sigma$ and $R'_\nu \nu'_\sigma$ are equal, except possibly in some $d\sigma$-negligible (hence $d\nu$-negligible) subset S of I. The subset $N=\{t \in I : \nu'_\sigma(t)=0\}$ is $d\nu$-negligible. Outside the union $S \cup N$, the above implies $R'_\nu = R'_\mu \mu'_\sigma / \nu'_\sigma$, with $\mu'_\sigma / \nu'_\sigma \geq 0$; then the expected inclusion holds, since the right-hand member of (8.2) is a cone. For $t \in S \cup N$, this (closed, convex) cone, being nonempty, contains at least the zero of $\mathbb{R}^n$. After replacing by zero the values that the function $R'_\nu$, as formerly defined, may take in $S \cup N$, one obtains the asserted conclusion, with $d\mu$ and $d\nu$ playing symmetric roles. ∎

If the superimposed constraints agree with Definition 8.1, the elimination of dR between (8.2) and the equation (7.4) of Nonsmooth Dynamics yields the following characterization of the possible motions

$$- A(q)\, u'_\mu + K(t,q,u)\, t'_\mu \in \partial\psi_{V(q(t))}(u^+(t)), \tag{8.3}$$

required to hold for *every* $t \in I$. Here $d\mu$ may equivalently be replaced by any nonnegative real measure, relative to which du and dt possess density functions. The existence of such measures is a priori secured by the Radon-Nikodym theorem; one may take, for instance, $d\mu = |du| + dt$, where $|du|$ denotes the nonnegative real measure *modulus* (or *variation measure*) [6][7] of the vector measure du. Since K is continuous and since the discontinuity set of u is dt-negligible, it does no matter to replace u, on the left-hand side, by $u^+$ or $u^-$.

The indifference of (8.3) regarding the choice of $d\mu$ (this could be checked directly, through the same reasoning as in the proof of Prop. 8.2), suggests to strip down the writing into

$$- A(q)\, du + K(t,q,u)\, dt \in \partial\psi_{V(q(t))}(u^{+}(t)). \qquad (8.4)$$

This may be called a *measure differential inclusion*. The existence of solutions to initial value problems governed by conditions of this sort and their possible uniqueness have so far been studied only in some special cases [11][12][13][31].

As before, one observes that, provided that $q(t_0) \in \Phi$, inclusion (8.3) entails $q(t) \in \Phi$ for every $t \in I$.

REMARK 8.3 Similarly to what precedes, the simple assumption of no-friction, without reference to "softness", as it has been expressed for the case of smooth motions in (5.3), may be adapted to Nonsmooth Dynamics. It will merely consist in stating that the contact impulsion dR possesses, relative to some $d\mu \geqslant 0$, a density function which satisfies for every $t \in I$

$$- R'_{\mu}(t) \in N(q(t)). \qquad (8.5)$$

Here again, the fact that the right-hand member is a cone entails that such an assumption does not actually depend on the peculiar choice of $d\mu$.

Since the right-hand member of (8.2) is contained in $N(q(t))$, (8.5) constitutes a *weaker assumption* than what expresses Definition 8.1, i.e. the set of the superimposed constraints may be frictionless without being soft.

REMARK 8.4 Classically, a subdifferential relation such as (8.2) admits some alternative formulations. Here, since the polar cone of $V(q)$ equals $N(q)$, (8.2) is found equivalent to asserting, for every t,

$$u^+ \in V(q) \ , \quad - R'_\mu \in N(q) \ , \quad u^+ . R'_\mu = 0 \tag{8.6}$$

# 9. SOFT COLLISIONS

In the setting of the above Section, the following characterizes the possible *velocity jumps*.

PROPOSITION 9.1 *Let* $\tau \in I$ *, different from the possible right end of this interval. For any motion satisfying* (8.3)*, one has*

$$u^+(\tau) = \text{prox}\,(u^-(\tau), V(q(\tau)), \tag{9.1}$$

*where the proximation is understood in the sense of the* kinetic *metric, i.e. the Euclidean metric defined in* $\mathbb{R}^n$ *by the matrix* $A(q(\tau))$

PROOF In view of (7.1),

$$u^+(\tau) - u^-(\tau) = \int_{\{\tau\}} du = \int_{\{\tau\}} u'_\mu d\mu = \mu_\tau\, u'_\mu(\tau),$$

where $\mu_\tau \geq 0$ denotes the integral of $d\mu$ over the singleton $\{\tau\}$. If motion agrees with (8.3), one has

$$- A(q(\tau))\, u'_\mu(\tau) + K(\tau, q(\tau), u(\tau))\, t'_\mu(\tau) \ \in \partial\psi_{V(q(\tau))}(u^+(\tau)).$$

Now $\mu_\tau t'_\mu(\tau)=0$, since the Lebesgue measure dt has no atom. Then, after multiplying both members of the above by $\mu_\tau$, one obtains

$$- A(q(\tau))(u^+(\tau) - u^-(\tau)) \in \partial\psi_{V(q(\tau))}(u^+(\tau)).$$

If one uses in $\mathbb{R}^n$ a base orthonormal relatively to the metric in view, $A(q(\tau))$ becomes the unit matrix, reducing this to the classical characteristic property of proximal points. ■

This Proposition, which, under the convention $u^-(t_0)=u(t_0)$, also holds for $\tau=t_0$, shows that u exhibits a nonzero jump at point $\tau$ if and only if $u^-(\tau)\notin V(q(\tau))$, i.e. a *nontangential impact* occurs at this instant.

Assume in addition that $\tau>\tau_0$, so $u^-(\tau)\in -V(q(\tau))$ in view of Prop.2.2. Then, the condition for nonzero velocity jump becomes

$$u^-(\tau)\in V(q(\tau))\cap -V(q(\tau)).$$

The right-hand member is a linear subspace of $\mathbb{R}^n$. If $J(q(\tau))=\emptyset$, i.e. $q(\tau)\in \operatorname{int}\Phi$, this is the whole of $\mathbb{R}^n$. If $J(q(\tau))$ consists of a singleton, say $\{\alpha\}$, the subspace is the vector hyperplane tangent at $q(\tau)$ to hypersurface $f_\alpha$=const.. For larger $J(q(\tau))$, the point $q(\tau)$ lies on what may be called, in the wide sense, an *edge* and the said subspace (possibly reduced to $\{0\}$) is declared tangent to this edge.

Also for $\tau>\tau_0$, one observes that $u^-(\tau)$, being an element of $-V(q(t))$, cannot belong to the interior of $V(q(\tau))$; thus (9.1) yields that $u^+$ actually lies on the boundary of this polyhedral cone.

REMARK 9.2 Here is another consequence of Prop.9.1. Suppose that, on some open subinterval I' of I, the motion evolves in such a way that the set-valued function $t\to J(q(t))$ never increases. In other words, during this time interval, some of the contacts may get loose, but no collision occurs. Let $\tau\in I'$ and let $\alpha\in J(q(\tau))$. Then, for every $t\in I'$ such that $t\leq\tau$, one has $f_\alpha(q(t))=0$, an equality which, through the chain rule, entails

$$u^-(\tau).\nabla f_\alpha(q(\tau)) = 0.$$

Hence $u^-(\tau)\in V(q(\tau))$, so Prop.9.1 shows that no velocity jump occurs at instant $\tau$.

In the traditional treatment of unilateral constraints, the latter is taken for granted: one accepts to enter percussions into the analysis only at instants where geometry makes them unavoidable. This agrees with the heuristic maxim of the "minimal singularity", but does not result from any explicit mechanical assumption. Here is a familiar example demonstrating this method deficiency.

Suppose an object performing a sliding motion in the contact of a table (or simply at rest on it). If an operator hits the table with a hammer, the object is commonly observed to jump. So the table has imparted an impulse to the contacting object, without itself exhibiting any motion at the macroscopic observation scale. In contrast, the assumption that unilateral constraints are frictionless and *soft* rules out such an active behaviour of boundaries.

Incidentally, the replacement of softness, in frictionless unilateral constraints, by the quite different assumption of *energy conservation* would also permit a deductive treatment of the above situation. It will be shown in Sec.10 that energy loss, in velocity jumps, should on the contrary be expected when frictionless soft constraints are present.

REMARK 9.3 Equation (9.1) expresses that, in the considered motion, all velocity jumps are of the sort that the author has previoulsy called *standard inelastic shocks* [14][15]. These were proposed as a generalization of the shocks which, in the case of a system involving a single constraint inequality, say $f(q)\leqslant 0$, are traditionally called "inelastic" or also "soft". In fact, if $f(q(t_s))=0$, the tangent cone to the feasible region $\Phi$ of $\mathbb{R}^n$ at point $q_s=q(t_s)$ is simply the half-space $V(q_s) = \{v\in\mathbb{R}^n: v.\nabla f(q_s)\leqslant 0\}$. Since the left-side velocity $u^-(t_s)$ must belong to $-V(q_s)$ (at least if one supposes $t_s>t_0$), equ.(9.1) yields in this special case that $u^+(t_s)$ equals *the orthogo-*

*nal projection of* $u^-(t_s)$ *to the vector hyperplane tangent at* $q_s$ *to hypersurface* $f=0$. More information on general standard inelastic shocks may be found in [15] ; incidentally, the *contact percussion* receives an extremal characterization, dual to (9.1).

# 10. ENERGY BALANCE

PROPOSITION 10.1 *For every motion satisfying* (8.3) *the function* $t \to \mathcal{E}_c$ *belongs to* $\mathrm{lbv}(I,\mathbb{R})$. *In the sense of the ordering of real measures on* $I$, *one has*

$$d\mathcal{E}_c \leqslant F.\,dq, \tag{10.1}$$

*with equality if and only if* $u$ *has no jump in* $I$.

PROOF For the traditional case of smooth motions with frictionless (time-independent) superimposed constraints, the *power equation*

$$\frac{d}{dt}\mathcal{E}_c = F_i(t,q,u)\,u^i$$

is easily derived from Lagrange equations, with left-hand side developed in the form (4.3). For a motion governed by (7.4), with the contact impulsion $dR=R'_\mu d\mu$ satisfying (8.2), there is only to retrace the same calculation, under the replacement of some steps, based on the rules of usual Differential Calculus, by what follows.

1° If $u^i$ and $u^j$ belong to $\mathrm{lbv}(I,\mathbb{R})$, the same holds for the product $u^iu^j$ and its differential measure is given [7] [32] by

$$d(u^iu^j) = u^{i+}\,du^j + u^{j-}\,du^i.$$

The products of measures by functions, which appear on the right-hand side, make sense because the functions $t \to u^{i+}(t)$ and $t \to u^{j-}$ belong to $\mathrm{lbv}(I,\mathbb{R})$; so they are locally integrable relative to any real measure. Hence, in view of the symmetry of A,

$$A_{ij}(q)\, d(u^i u^j) = A_{ij}(q)(u^{i+} + u^{i-})\, du^j.$$

Furthermore, since $t \to q(t)$ is locally absolutely continuous, with $t \to u(t)$ as derivative, the differential measure of $t \to A_{ij}(q(t))$ equals

$$dA_{ij} = A_{ij,k}\, u^k\, dt$$

and one has

$$d(A_{ij} u^i u^j) = u^i u^j\, dA_{ij} + A_{ij}\, d(u^i u^j).$$

2° The real measure $A_{ij}(q)(u^{i+} - u^{i-})du^j$ is nonnegative; it vanishes if and only if $u$ is continuous on $I$. In fact, this measure consists of a countable and locally summable collection of point measures located at the jump instants of $u$. Let $t_s$ denote one of these instants; under the notations $u^+(t_s)=u_s^+$, $u^-(t_s)=u_s^-$, $q(t_s)=q_s$, the mass of the corresponding point measure equals

$$A_{ij}(q_s)(u_s^{i+} - u_s^{i-})\,(u_s^{j+} - u_s^{j-}).$$

Since $A_{ij}(q)$ is, for every $q$, a positive definite matrix, this real number is nonnegative; it vanishes if and only if $u_s^+ - u_s^- = 0$.

3° Equality $u.r=0$, a consequence of the no-friction assumption in the case of smooth motions, is replaced at present by $u^+.R'_\mu=0$, a fact observed in (8.6). ∎

REMARK 10.2 A more general concept than frictionless soft constraints is obtained by inserting into the right-hand side of (8.2), instead of $u^+$, some weighted mean

$$u_\delta = \frac{1}{2}(1+\delta)\, u^+ + \frac{1}{2}(1-\delta)\, u^-, \qquad (10.2)$$

where $\delta$ is a chosen real number, here supposed independent of $t$, for simplicity. Since $u_\delta(t)=u(t)$, except at the jump points of $u$, which make a

Lebesgue-negligible subset of I, the law of constraint

$$- R'_{\mu}(t) \in \partial\psi_{V(q(t))}(u_{\delta}(t)) \tag{10.3}$$

implies, exactly like (8.2), that condition $q(t)\in\Phi$ is satisfied for every $t$ as soon as it holds for $t=t_0$.

For $\delta>0$ (even larger than 1), the law of constraint (10.3) entails the same as what has been stated in Prop.10.1. Choosing $\delta<0$ would yield the reverse inequality, physically unacceptable (unless the possible bounces are artificially enhanced, as in some electric billiard games).

If $\delta=0$, i.e. $u_\delta$ equals the *midpoint* of $u^-$ and $u^+$, one finds *equality* $d\mathcal{E}_c=F.dq$ , expressing the same *energy conservation* as in smooth motions. In that sense, the constraint law (10.3) with $\delta=0$ may be said "elastic".

We suggest to call $\delta$ the *dissipation index* of the constraint law (10.3).

# 11. TIME DISCRETIZATION ALGORITHM

The principle of such an algorithm has been introduced, for smooth motions, in Remark 5.3. The main observation we now have to make is that the same numerical technique applies in the framework of Nonsmooth Dynamics, provided that the superimposed frictionless unilateral constraints are assumed soft.

The expression in (7.4) of the system dynamics, in term of measures, directly suggests time discretization: the rule will be to equal some approximants of the respective integrals of both members over each subinterval of I determined by the discretization nodes.

Let $(t_I,t_F)$ be one of these intervals (here I refers to "initial" an F to "final"), with length $h=t_F-t_I$ possibly variable from one step to another. From the approximants $q_I$ and $u_I$ of q and u at the beginning of the interval, one has to compute some approximants $q_F$ and $u_F$ , assigned to instant $t_F$ and which, in turn, will be used as initial values in the next step.

*Stage 1.* Calculate $t_M=t_I+\frac{1}{2}h$ and some *midpoint approximants*

$$q_M=q_I+\frac{1}{2}hu_I \in \mathbb{R}^n, \qquad A_M=A(q_M) \in \mathbb{R}^{n\times n}, \qquad K_M=K(t_M,q_M,u_I) \in \mathbb{R}^n.$$

Then

$$u_L=u_I + h\, A_M^{-1}K_M$$

is the value that the discretized equations of Dynamics would yield for $u_F$ in the absence of contact force (here one may read the subscript L as referring to "loose").

If $q_M \in \operatorname{int}\Phi$ (i.e. all $f_\alpha(q_M)$ are strictly negative) or $u_L \in V(q_M)$, one considers that contact forces have no effect on the calculated step; so one makes $u_F=u_L$ and goes to Stage 3 (this decision is actually a trivial case of Stage 2 below).

*Stage 2.* On the left-hand side of (8.4), let us replace $A(q)$ and $K(t,q,u)$ by $A_M$ and $K_M$ . Then, an approximant of the integral of this left-hand side over $(t_I,t_F)$ is $-A_M(u_F-u_L)$ . Concerning the right-hand side, it will be considered that the set $V(q(t))$ keeps, throughout the interval, the constant value $V(q_M)$, whose indicator function will be denoted by $\psi_M$. Furthermore, let us take $u_F$ as an approximant -one may rather say a *simulation*- of $u^+$. This yields as a simulation of (8.4) on the said interval

$$-A_M(u_F-u_L) \in \partial\psi_M(u_F)$$

If the matrix $A_M$ is used in order to define a Euclidean metric on $\mathbb{R}^n$, this characterizes $u_F$ as the *proximal point* to $u_L$ in the closed convex set $V(q_M)$. Therefore, computing $u_F$ is a Quadratic Programming problem: *to minimize on* $V(q_M)$ *the real function* $x \to (x-u_L).A_M(x-u_L)$. Recall that $V(q_M)$ is a polyhedral cone, the intersection of a collection of half-spaces determined by the index set $J(q_M)$. These correspond to the values of $\alpha$ such that, in the *test position* $q_M$, the inequality $f_\alpha \leq 0$ holds as an equality or is *violated*. If their number is not too large, the proximal point will be constructed algebraically; otherwise some of the classical algorithms of Quadratic Programming will have to be applied.

*Stage 3.* One terminates the computation step by

$$q_F = q_M + \frac{1}{2} h u_F .$$

REMARK 11.1 In [24] are reported some computer experiments with methods of the above sort. These methods prove to be stable. The finer the time discretization is, the better the computed motion complies with the inequalities $f_\alpha \leq 0$. In that respect, some improvement may be achieved by evaluating, in Stage 2, the cone $V(q)$ at another point than $q_M$. A good choice appears to be the point $q_I + h\dot{u}_I$.

If these methods are applied to the calculation of mechanisms, which in reality always involve some imperfectly known friction, no great precision can anyway be expected in predicting the motion. From that viewpoint, a moderately fine time discretization will be enough. In contrast, the violation of the constraint inequalities must sharply be kept in check. A very effective way of doing it consists in completing each time step by a stage of *linear correction of the possible violation*. Let us explain it in the simple case where the position $q_F$, calculated in Stage 3, violates only one of the constraint inequalities. In other words, the real number $\varphi = f(q_F)$ is found strictly positive, with $f$ denoting one of the functions $f_\alpha$. A plausible correction of this violation (naturally assumed "small") would be to replace $q_F$ by its proximal point, say $q_C$, in the region $f \leqslant 0$. Proximity here should be understood in the sense of some Euclidean metric on $\mathbb{R}^n$; the most justified choice is to rely on the kinetic metric, defined by the matrix $A(q)$. To save computation, the latter will be evaluated at a point where it has already been calculated in the current step. Using an affine approximation of the function $f$, one obtains $0=f(q_C)=\varphi+(q_C-q_F).\nabla f+..$ . Here the gradient $\nabla f$ should be evaluated at some neighbour point. This precisely must have been done at the Stage 2 of the current step, where also the vector $A^{-1}\nabla f$ has been needed. Defining $q_C$ as the proximal point to $q_F$ in the region $f \leqslant 0$, relative to the metric in view, means that the vector $q_C-q_F$ is parallel to $A^{-1}\nabla f(q_C)$. Since the latter is estimated to be close to the calculated value of $A^{-1}\nabla f$, this finally yields the approximate formula

$$q_C = q_F - \varphi\,(\nabla f.A^{-1}\nabla f)^{-1}\, A^{-1}\nabla f + \dots .$$

The same techniques of violation control applies to the numerical methods presented in the sequel, for unilateral constraints with friction.

REMARK 11.2 Also in [24], a numerical procedure is developed for a system with a frictionless unilateral constraint expressed by a single inequality $f \leqslant 0$, assuming that the dissipation index, as we have defined it in Sec.10, equals 1. In other words, the possible collisions are *elastic bounces*. In contrast with the excellent stability of the preceding method, a careful check of the energy balance of each time-step here is needed, in order to prevent divergent oscillations when the algorithm is applied to the computation of a motion with assumedly persistent contact.

# 12. COULOMB FRICTION

This lectures are meant to provide only an introduction to the treatment of friction. So we shall restrict ourselves to a system involving *a single unilateral constraint* with geometric condition $f(q)\leqslant 0$ (there is presented in [30] a rather usual case, where a system with multiple possible contacts may be reduced, through decomposition, to this simple setting).

For every q such that $f(q)\geqslant 0$, the cone $V(q)$ equals the half-space $\{v\in\mathbb{R}^n: v.\nabla f(q)\leqslant 0\}$, with boundary

$$T(q) = \{v\in\mathbb{R}^n: v.\nabla f(q)=0\};$$

the latter is the vector hyperplane tangent at point q to the hypersurface f=const. drawn through this point.

Let a motion satisfy $f(q(t))\leqslant 0$ for every $t\in I$. If the velocity function u is continuous at some $\tau>t_0$, then, in view of Prop.2.2, $u(\tau)$ belongs to $V(q(\tau))\cap -V(q(\tau))$; this set equals $T(q(\tau))$ if $f(q(\tau))\geqslant 0$ and otherwise the whole of $\mathbb{R}^n$. The same is true for $\tau=t_0$, as far as the initial data have the meaning we agreed to give them in Sect.2, namely $u_0$ equals the left-limit $u^-(t_0)$ in a motion taking place before $t_0$, with the unilateral constraint already in effect.

To fix the ideas, suppose, as in Sec.3, that equality $f(q)=0$ expresses that in the position q, some part $\mathcal{B}_1$ of the system touches the unmoving external obstacle $\mathcal{B}_0$. Then, under the previous notations,

$$\mathcal{U} = \mathcal{V}(\mathcal{M}_1(q),q,u) = G_q u$$

is the velocity of $\mathcal{B}_1$ relative to $\mathcal{B}_0$ at the contact point.

Denoting again by $\mathcal{N}_q$ the common normal unit vector to the contacting bodies, directed toward $\mathcal{B}_1$, we assume $G_q^*\mathcal{N}_q \neq 0$ (see Remark 3.2). Then the assertion $u\in T(q)$ is equivalent to $\mathcal{U}$ belonging to $\mathcal{T}(q)$, the linear subspace of

$\mathcal{E}_3$ orthogonal to $\mathcal{N}_q$, i.e. the common tangent vector plane at the contact point.

The above is the situation in which, traditionally, Coulomb's law of dry friction is formulated. This law is a relation between the *sliding velocity* $\mathcal{U}\in\mathcal{T}(q)$ and the contact force $\mathcal{R}\in\mathcal{E}_3$ experienced by $\mathcal{B}_1$

For brevity, let us write $\mathcal{N}$ for $\mathcal{N}_q$ and $\mathcal{T}$ for $\mathcal{T}(q)$. The familiar formulation rests on the decomposition of $\mathcal{R}$ into

$$\mathcal{R} = \mathcal{R}_T + \rho\,\mathcal{N}, \quad \text{with } \mathcal{R}_T\in\mathcal{T} \text{ and } \rho\geqslant 0, \tag{12.1}$$

and consists of two well known separate assertions concerning the respective cases $\mathcal{U}=0$ and $\mathcal{U}\neq 0$ in $\mathcal{T}$.

In some of the author's early papers [20] [21], it has been observed that, as far as the normal component $\rho$ is treated as known, this pair of assertions is equivalent to a relation, between the elements $\mathcal{R}_T$ and $\mathcal{U}$ of the linear space $\mathcal{T}$, which derives from a "pseudo-potential". Furthermore, this formulation readily extends to the description of possibly *anisotropic friction*, as a relation of the form

$$-\mathcal{U}\in\partial\psi_{\mathcal{D}}(\mathcal{R}_T), \tag{12.2}$$

with $\mathcal{D}=\rho\mathcal{D}_1$. By $\mathcal{D}_1$ is denoted a given closed convex subset of $\mathcal{T}$, containing the origin; in the traditional case of isotropic friction, $\mathcal{D}_1$ equals the disk centered at the origin, with radius equal to the *friction coefficient*, say $\gamma$.

*We shall restrict ourselves in these lectures to the case of* bounded friction, *i.e. the set $\mathcal{D}_1$ is bounded*. To take $\mathcal{D}_1$ unbounded would provide a way of including in our approach the situation traditionally called a *nonholonomic constraint*. But this would cause some complications in further statements.

Through the standard calculation rules of Convex Analysis, (12.2) may equivalently be written as

$$-\mathcal{R}_T \in \partial\varphi(u), \tag{12.3}$$

where the *dissipation function* $\varphi$ equals the support function of the set $-\mathcal{D}_1$; in particular, for isotropic friction $\varphi = \gamma\|.\|$.

Elementary applications, where $\rho$ in fact is known, may be found in [22]. Furthermore, having to treat $\rho$ as known does not prevent using this pseudo-potential formulation (or an equivalent variational inequality which expresses a "principle of maximal dissipation") in the proof of existence of solutions to dynamical problems. See [33], where the normal components of the contact forces become the primary unknowns in some functional analytic arguments

In the present lectures, we choose to formulate the same law under a *conical* equivalent form [24], avoiding the decomposition (12.1). The *friction cone* $C$ at the contact point is introduced, a closed convex conical subset of the linear space $\mathcal{E}_3$ (recall that, speaking of a cone in a linear space, one understands that it has vertex at the origin). In traditional isotropic friction, this is a cone of revolution about $\mathcal{N}$; *generally, $C$ equals the cone generated in $\mathcal{E}_3$ by the set $\mathcal{D}_1 + \mathcal{N}$*. So $C$ contains $\mathcal{N}$ and lies entirely on the corresponding side of $\mathcal{T}$.

PROPOSITION 12.1 *The pair of relations* (12.1), (12.2) *is equivalent to*

$$-\mathcal{U} \in \mathrm{proj}_{\mathcal{T}}\ \partial\psi_C(\mathcal{R}). \tag{12.4}$$

PROOF Assume that (12.4) holds; hence $-\mathcal{U} \in \mathcal{T}$. Denoting by $[\mathcal{N}]$ the linear subspace generated in $\mathcal{E}_3$ by $\mathcal{N}$, one has

$$\exists \mathcal{V} \in \partial\psi_C(\mathcal{R}),\ \exists \mathcal{W} \in [\mathcal{N}] :\ -\mathcal{U} = \mathcal{V} + \mathcal{W}. \tag{12.5}$$

Let us decompose $\mathcal{R}$ in the form (12.1); necessarily $\rho \geq 0$, since (12.4) implies that $\partial\psi_C(\mathcal{R})$ is nonempty, thus $\mathcal{R}\in C$. By construction, $\mathcal{R}$ belongs to the affine plane $\mathcal{A}=\mathcal{T}+\rho\mathcal{N}$. Therefore $\partial\psi_{\mathcal{A}}(\mathcal{R})=[\mathcal{N}]$ and (12.5) means that

$$-\mathcal{U}\in\partial\psi_{\mathcal{A}}(\mathcal{R})+\partial\psi_C(\mathcal{R})\subset\partial\psi_{\mathcal{A}\cap C}(\mathcal{R}). \tag{12.6}$$

In view of the definition of $\mathcal{D}$, one has $\mathcal{A}\cap C=\mathcal{D}+\rho\mathcal{N}$, thus, using a translation in the evaluation of subdiffentials,

$$\partial\psi_{\mathcal{A}\cap C}(\mathcal{R})=\partial\psi_{\mathcal{D}}(\mathcal{R}-\rho\mathcal{N})=\partial\psi_{\mathcal{D}}(\mathcal{R}_T).$$

Then (12.6) entails that (12.2) holds in the sense of the Euclidean autoduality of $\mathcal{E}_3$. Since $\mathcal{U}$ and $\mathcal{R}_T$ are elements of the linear subspace $\mathcal{T}$, the same is true relatively to the Euclidean autoduality of this subspace.

Conversely, let us assume that $\mathcal{U}\in\mathcal{T}$ and that (12.2) holds in the sense of the Euclidean autoduality of $\mathcal{T}$, with $\mathcal{R}_T$ defined by (12.1) (observe that $\rho\geq 0$ is stated at this place). Then (12.2) is true also in the sense of the autoduality of $\mathcal{E}_3$ and, using translations as above, one concludes

$$-\mathcal{U}\in\partial\psi_{\mathcal{A}\cap C}(\mathcal{R}).$$

Here we need to know whether the inclusion on the right side of (12.6) actually holds as an equality of sets. If $\rho>0$, this equality results from a known calculation rule for the subdifferential of a sum of l.s.c. proper convex functions in finite-dimensional spaces: in fact there exists a point in the relative interior of $C=\operatorname{dom}\psi_C$ where $\psi_{\mathcal{A}}$ takes a finite value (see [34], Theorem 23.8). In that case, by going from (12.6) backward to (12.5), one establishes (12.4).

It is only for $\rho=0$ that the assumption of *bounded friction*, made once for all in the preceding, has to be used. If $\rho=0$, the set $\mathcal{D}$ reduces to $\{0\}$, so (12.2) simply consists in the assertion:" $\mathcal{U}$ arbitrary in $\mathcal{T}$ and $\mathcal{R}_T=0$ "; one has to establish that such is also the meaning of (12.5) in this special situation. For $\mathcal{R}=0$, the subdifferential $\partial\psi_C(\mathcal{R})$ consists of the *polar cone* $C°$ of $C$. Because the section of $C$ by the plane $\mathcal{A}$ (constructed for instance with $\rho=1$) is compact, $C°$ contains the vector $-\mathcal{N}$ in its interior (see e.g.[35], parag.8.7). Consequently, the projection of $C°$ to $\mathcal{T}$ equals the whole of this subspace; this completes the proof. ■

In turn (12.4) may be transformed as follows [25]:

PROPOSITION 12.2 *Define in $\mathcal{E}_3$ the extended real function* $\theta$

$$\mathcal{U}\to\theta(\mathcal{U})=\frac{1}{2}\|\mathcal{U}\|^2+\psi_{\mathcal{T}}(\mathcal{U}). \qquad (12.7)$$

*Then* (12.4) *is equivalent to*

$$0\in\partial\psi_C(\mathcal{R})+\partial\theta(\mathcal{U}). \qquad (12.8)$$

PROOF Since $\frac{1}{2}\|.\|^2$ is a smooth function, with gradient mapping equal to identity, one has for every $\mathcal{U}$ in $\mathcal{E}_3$

$$\partial\theta(\mathcal{U})=\mathcal{U}+\partial\psi_{\mathcal{T}}(\mathcal{U}).$$

Now, $\partial\psi_{\mathcal{T}}(\mathcal{U})$ equals $[\mathcal{N}]$ if $\mathcal{U}\in\mathcal{T}$ and, otherwise, is empty. Therefore, (12.8) is equivalent to

$$\mathcal{U}\in\mathcal{T} \quad \text{and} \quad 0\in\partial\psi_C(\mathcal{R})+\mathcal{U}+[\mathcal{N}],$$

which is precisely (12.4). ■

Incidentally, observe that resistance laws involving a pair of subdiffential mappings, as in (12.8), have a wider interest than describing contact friction in three-dimensional space. For instance, in Plasticity, this form may be used in formulating constitutive laws for which the flow rule is not "associated" with the yield criterion.

Recall that $\mathcal{T}$ and $C$, in what precedes, depend on the position q of the system, under the assumption f(q)=0. Let us make the same writing meaningful *also for positions which do not involve contact.*

To this end, we shall agree that, when f(q)<0, the cone $C(q)$ reduces to {0} and that $\mathcal{T}(q)=\mathcal{E}_3$. Then the relations (12.4) or (12.8) simply express that $\mathcal{R}=0$, with $\mathcal{U}$ arbitrary in $\mathbb{R}^n$.

Furthermore, in what concerns evolution problems under condition f⩽0, it is immaterial to choose any (adequately smooth) extension of the multifunction $q \rightarrow C(q) \subset \mathcal{E}_3$ to the case f(q)>0. Similarly, the linear mapping $\mathcal{G}_q: \mathbb{R}^n \rightarrow \mathcal{E}_3$ will be extended to such q , as well as the normal unit $\mathcal{N}_q$, with attention to preserving (3.2). The reason for such extensions lies in numerical methods, where a certain amount of violation of the desired inequality f⩽0 has naturally to be faced.

The set of the values of $\mathcal{R} \in \mathcal{E}_3$ that (12.4) or (12.8) make correspond with each $\mathcal{U} \in \mathcal{E}_3$ (actually the empty set if $\mathcal{U} \notin \mathcal{T}$) is a cone, since the multiplication of $\mathcal{R}$ by any strictly positive number leaves $\partial\psi_C(\mathcal{R})$ invariant. Like in preceding Sections, this fact will now prove essential, as we come to formulating Nonsmooth Dynamics in terms of measure differential inclusions.

In the course of a smooth motion, the contact force $\mathcal{R}$ is a function of time that we may denote by $\mathcal{P}'_t$. This in fact is the density, relative to the Lebesgue measure dt, of the *local contact impulsion* $d\mathcal{P}$, an $\mathcal{E}_3$-valued measure on the time interval I. For nonsmooth motions, $d\mathcal{P}$ can no more be expected to possess a density relative to dt, but in any case this measure may be represented in the form $\mathcal{P}'_\mu d\mu$, where $d\mu$ is a nonnegative real measure on I and $\mathcal{P}'_\mu \in \mathcal{L}^1_{loc}(I,\ d\mu;\ \mathcal{E}_3)$. The $\mathbb{R}^n$-valued contact impulsion dR, as introduced in Sect.7 for insertion into the measure equation of Nonsmooth Dynamics (7.4), has the form $dR = R'_\mu d\mu$, and, similarly to (4.7),

$$R'_\mu(t) = G^*_{q(t)}\ \mathcal{P}'_\mu(t) \qquad (12.9)$$

holds for every t.

As far as Coulomb's law is accepted for the description of dry friction, one naturally admits that, in possible nonsmooth motions, the density $\mathcal{P}'_\mu(t)$ of the local contact impulsion will be related, for every t, to the sliding velocity $\mathcal{U}$ through the same relation as $\mathcal{R}$ is in (12.4), or equivalently in (12.8). This, at least, raises no discussion when $t \to \mathcal{U}$ is continuous. At instants of velocity jumps, *we decide that the same relation will hold with* $\mathcal{U}$ *replaced by its right-limit*, namely $\mathcal{U}^+ = G_q u^+$, since the linear mapping $G_q: \mathbb{R}^n \to \mathcal{E}_3$ continuously depends on q. This assumption entails that $\mathcal{U}^+(t)$ belongs to $\mathcal{T}(q(t))$ for every t (an immaterial assertion when $f(q(t))<0$, since it has been agreed that $\mathcal{T}(q(t)) = \mathcal{E}_3$ in this event). In view of Proposition 3.1, this is equivalent to $u^+(t) \in T(q(t))$, a property which, in the frictionless case of Sect.9, has been identified as characterizing the "softness" of unilateral constraints. We thus are induced to put the following definition.

DEFINITION 12.3 *The unilateral constraint investigated above is said* soft with Coulomb friction *if, in any motion with l.b.v. velocity function, the contact impulsion has the form* $d\mathcal{P} = \mathcal{P}'_\mu d\mu$, *with* $\mathcal{P}'_\mu \in \mathcal{L}^1_{loc}(I,\ d\mu\ ;\mathcal{E}_3)$ *verifying for every* t

$$-\mathcal{U}^+(t) \in \operatorname{proj}_{\mathcal{T}(q(t))} \partial\psi_{C(q(t))}(\ \mathcal{P}'_\mu(t)), \qquad (12.10)$$

*(recall that* $\mathcal{T}(q) = \mathcal{E}_3$ *and* $C(q) = \{0\}$ *when* $f(q)<0$ *) or the equivalent form*

*given to a relation of this sort by Prop. 12.2.*

Because this relation is conical with regard to the element $\mathcal{P}'_\mu(t)$, the reasoning already used in the proof of Prop.8.2 shows that the choice of the nonnegative real measure $d\mu$ is immaterial, as long as $d\mathcal{P}$ possesses, relative to it, a density function.

REMARK 12.4 Since we have agreed to extend the definitions of $\mathcal{T}(q)$ and $C(q)$ to positions such that $f(q)>0$, (12.10) makes sense also in that case. But, as previously observed, this relation implies that $u^+(t)$ belongs to $T(q(t))$, hence to $V(q(t))$. In view of Prop.2.4, this secures that, *provided the initial data satisfy* $f(q(t_0))\leqslant 0$, *inequality* $f(q(t))\leqslant 0$ *will hold throughout* I.

REMARK 12.5 The validity of (12.10) for $t=t_0$ calls for some comments. In beginning this section, we recalled the meaning given to the initial condition $u(t_0)=u_0$ of an evolution problem. It is understood that $u_0$ equals the left-limit $u^-(t_0)$ in some anterior motion, during which the unilateral constraint was already in effect. In particular, one may have $f(q_0)=0$ and $u_0$ interior to the half-space $-V(q_0)$; this implies that a collision takes place at instant $t_0$. Then softness, as expressed by (12.10), makes that $u^+(t_0)$ belongs to the linear space $T(q_0)=V(q_0)\cap -V(q_0)$. But the case $f(q_0)=0$ with $u_0$ interior to $V(q_0)$, i.e. initial velocity implying contact break, is excluded from the present study. This will cause no great inconvenience in practice.

REMARK 12.6 Put the notation $\mathcal{G}_q(V(q))=\mathcal{V}(q)$; this is the closed half-space of $\mathcal{E}_3$ lying on the same side as $\mathcal{N}_q$ with respect to $\mathcal{T}(q)$. The set $\partial\psi_C(\mathcal{P}'_\mu)$ is contained in $C°$, the polar cone of $C$, which in turn is contained in the half-space $-\mathcal{V}(q)$. Then, in (12.10), the operation $\mathrm{proj}_{\mathcal{T}(q)}$ might equivalently be replaced by the proximation mapping to $\mathcal{V}(q)$. This would be numerically inconvenient, but is liable to improve the consistency of some further developments.

REMARK 12.7 In applications, describing dry friction through Coulomb's law can only provide a rather crude approximation. However, as this law retains the essential features of the phenomenon, its use at the stage of a first study is extremely valuable in numerous situations. As soon as this law has been adopted, under its traditional form, there is little doubt that its generalization (12.10) can also be accepted for every motion in which the velocity function is continuous, even in the absence of local absolute continuity. We shall discuss later its use in the case of a *velocity jump*.

# 13. TWO-DIMENSIONAL CONTACT

In the same setting as in the above Section, we now make the following additional assumption. However three-dimensional the contact may physically be, we shall suppose that, for every q such that f(q)=0 (and also for f(q)⩾0, after the extension we have agreed to make) *the range* $G_q(\mathbb{R}^n)$ *of* $G_q$ *reduces to a two-dimensional subspace* $\mathcal{W}_q$ *of* $\mathcal{E}_3$. Such is the case, for instance, if the part $\mathcal{B}_1$ of the system is astrained, by the primitive constraints, to only perform motions parallel to a fixed plane. The condition found in Remark 3.2, in order that $G_q^*\mathcal{N}_q \neq 0$ will be supposed fullfilled, i.e. $\mathcal{W}_q$ *and the tangent plane* $\mathcal{T}_q$ *are distinct*: let us denote by $\mathit{1}$ a unit vector of their intersection.

Due to the expression (4.7) of the covariant components of the contact force $\mathcal{R}$ (or, in Nonsmooth Dynamics, the covariant components of the density $\mathcal{P}'_\mu(t)$ of contact impulsion), the dynamically significant information concerning this vector is entirely conveyed through its equivalence class modulo ker $\mathcal{G}_q^*$. This kernel equals the subspace of $\mathcal{E}_3$ orthogonal to $\mathcal{W}_q$, consequently, we may in the sequel replace $\mathcal{R}$ by its orthogonal projection to $\mathcal{W}_q$, also called $\mathcal{R}$, by abuse of notation.

We shall come back, in Sec.17, to a discussion of what in general becomes the three-dimensional law of friction under such a geometric two-dimensional reduction. Let us restrict ourselves at present to the usual case where the result is simply the familiar *two-dimensional version of Coulomb's law*. The considerations of the preceding section might readily be adapted to this case. Here we shall rather choose to express the two-dimensional Coulomb law in the following alternative form. For more readibility, the subscript q will momentarily be omitted.

*There exist in the vector plane* $\mathcal{W}$ *two half-lines* $\mathcal{D}_+$ and $\mathcal{D}_-$, *emanating from the origin and lying, with respect to* $\mathcal{T}$, *on the same side as* $\mathcal{N}$. The convex cone $C$ now equals their convex hull. In smooth motions, the sliding velocity is essentially an element of $\mathcal{T}\cap\mathcal{W}$, say $\mathcal{U}= s\mathcal{I}$ with $s\in\mathbb{R}$. *Coulomb's law consists of the three implications*

$$s>0 \Rightarrow \mathcal{R}\in\mathcal{D}_+ \tag{13.1}$$

$$s<0 \Rightarrow \mathcal{R}\in\mathcal{D}_- \tag{13.2}$$

$$s=0 \Rightarrow \mathcal{R}\in C \tag{13.3}$$

The angles that $\mathcal{D}_+$ and $\mathcal{D}_-$ make with $\mathcal{I}$ have respective tangent equal to $-1/\gamma_+$ and $1/\gamma_-$, where the positive real numbers $\gamma_+$ and $\gamma_-$ respectively are the familiar *friction coefficients* corresponding to positive and negative sliding.

All these elements are defined for such q that $f(q)=0$; as before, we shall imagine an extension of them to every q such that $f(q)\geqslant 0$ (at least in a neighborhood of hypersurface $f=0$).

The vector plane $\mathcal{W}$ depends on $q$ ; it will prove convenient to send it onto a fixed copy of $\mathbb{R}^2$, called the *calculation plane*, equipped with the usual base consisting of vectors $\mathbf{i} = (1,0)$ and $\mathbf{j} = (0,1)$. To this end, a regular linear mapping $\mathcal{F}_q : \mathcal{W}_q \to \mathbb{R}^2$ will be constructed, in such a way that $\mathcal{F}_q \mathcal{I}$ is a positive vector of the first axis. Denote by $G_q : \mathbb{R}^n \to \mathbb{R}^2$ the product mapping $\mathcal{F}_q \mathcal{G}_q$. Assume that the real function f is at least $C^2$ ; then it will be possible to choose $\mathcal{F}_q$ in order that $G_q$ depend on $q$ in a $C^1$ way.

Let us mean by $\mathcal{F}_q^*$ the transpose of $\mathcal{F}_q$, in the sense of the natural scalar products of $\mathbb{R}^2$ and $\mathcal{W}$, and denote by $D_+$ and $D_-$ the images of $\mathcal{D}_+$ and $\mathcal{D}_-$ under $\mathcal{F}_q^{*-1}$.

There would remain enough arbitrariness in the choice of $\mathcal{F}_q$ for making these images equal two fixed half-lines of $\mathbb{R}^2$, e.g. the half-lines generated by $(-1,1)$ and $(1,1)$. The drawback is that the two friction coefficients $f_+$ and $f_-$ would then have to depend smoothly on $q$. So we shall not use this trick in the sequel and only assume that *the convex hull, say* C, *of* $D_+$ *and* $D_-$ *contains* $\mathbf{j}$ *in its interior.*

Under these notations, the friction law (13.1) to (13.3) may equivalently be formulated as a relation of the same form between $R = \mathcal{F}_q^{*-1}\mathcal{R}$ and $U = \mathcal{F}_q \mathcal{U} = G_q u = \sigma \mathbf{i}$. Due to the assumptions made, $\sigma$ is a real number of the same sign as s or vanishing with it. Henceforth, there only is to replace $s$, $\mathcal{D}_+$, $\mathcal{D}_-$, $C$ respectively by $\sigma$, $D_+$, $D_-$, C .

In order to express the dynamics of the considered system, one has to bring together the above reaction law and the Lagrange equations. The covariant components of the contact force make the element r of $\mathbb{R}^n$, related to R through

$$r = \mathcal{G}_q^* \mathcal{R} = \mathcal{G}_q^* \mathcal{F}_q^* R = G_q^* R \,. \qquad (13.4)$$

After having so restated the two-dimensional law of Coulomb, we now propose to extend it to Nonsmooth Dynamics, in the same line as in Sec. 12. Recall that, in such a context, velocity functions are significant only through their one-sided limits. Instead of u, $\mathcal{U}$ and U, the *right-limits* $u^+$, $\mathcal{U}^+ = \mathcal{G}_q u^+$ and $U^+ = G_q u^+$ are introduced into the above writing. The contact force $\mathcal{R}$ is replaced, for every t, by the value $\mathcal{P}'_\mu(t)$ of the density function of the three-dimensional contact impulsion, or equivalently by the orthogonal projection of $\mathcal{P}'_\mu(t)$ to $\mathcal{W}(q(t))$. When coming to the use of the calculation plane, the function $P'_\mu = \mathcal{F}_q^{*-1} \mathcal{P}'_\mu$ is considered.

The dynamics of the system is now expressed by the Lagrange equation in $\mathbb{R}^n$

$$A(q)\, u'_\mu - K(t,q,u)\, t'_\mu = R'_\mu \,, \qquad (13.5)$$

to be joined with

$$R'_\mu = G_q^* P'_\mu \qquad (13.6)$$

$$\exists \sigma \in \mathbb{R}: \quad U^+ = \sigma\, \mathbf{i} \qquad (13.7)$$

$$\sigma > 0 \Rightarrow P'_\mu \in D_+ \qquad (13.8)$$

$$\sigma < 0 \Rightarrow P'_\mu \in D_- \qquad (13.9)$$

$$\sigma = 0 \Rightarrow P'_\mu \in C \,, \qquad (13.10)$$

for every t such that $f(q(t)) \geqslant 0$.

On the contrary, when $f(q(t)) < 0$ , then $R'_\mu = 0$.

## 14. VELOCITY JUMPS IN FRICTIONAL DYNAMICS

The formulation given above, for the Nonsmooth Dynamics of a system with single possible contact and two-dimensional Coulomb friction, will now be used in discussing the event of a *velocity jump*.

This means that, at some instant $t_s$, the $\mathbb{R}^n$-valued measure du possesses an atom, the mass of which equals

$$u^+(t_s) - u^-(t_s) = u'_\mu(t_s)\,\mu_s .$$

Here, the positive number $\mu_s$ is the mass of the atom that the measure $d\mu$ should possess at point $t_s$.

Recalling that the Lebesgue measure dt has no atom, one derives from (13.5) that, at $t=t_s$,

$$u^+ - u^- = A_q^{-1}\, G_q^*\, P'_\mu \mu_s \tag{14.1}$$

If $f(q(t_s))<0$, i.e. no contact, one has $P'_\mu(t_s)=0$, so no jump of u can occur. We therefore shall assume $f(q(t_s))\geqslant 0$ (equivalently $f(q(t_s))=0$) and denote simply by G the value of $G_q$ at $q=q(t_s)$. Then, for $t=t_s$, one has $R'_\mu=G^*P'_\mu$. Put $P'_\mu(t_s)\mu_s=P$ and apply the linear mapping $G:\mathbb{R}^n\to\mathbb{R}^2$ to both members of (14.1); this yields

$$U^+ - U^- = HP, \tag{14.2}$$

*where* H *denotes the symmetric* 2×2 *positive definite matrix* $GA^{-1}G^*$.

This has to be joined with two-dimensional Coulomb law, expressed as in Sec. 13 by

$$\exists\sigma\in\mathbb{R}: \quad U^+ = \sigma\,\mathbf{1} \tag{14.3}$$

$$\sigma>0 \Rightarrow P\in D_+ \tag{14.4}$$

$$\sigma<0 \Rightarrow P\in D_- \tag{14.5}$$

$$\sigma=0 \Rightarrow P\in C. \tag{14.6}$$

We are to discuss how, starting with given $u^-$, the system of conditions (14.1) to (14.6) allows one to determine $u^+$.

Combining (14.2) and (14.3), one obtains

$$P = -H^{-1}U^- + \sigma H^{-1}\mathbf{1}, \tag{14.7}$$

which expresses that $P$ belongs to the line $\Delta$ of $\mathbb{R}^2$ drawn through the point $-H^{-1}U^-$, with $H^{-1}\mathbf{i}$ as directing vector. Easy calculation (for instance by observing that $\Delta$ is orthogonal to $H\mathbf{j}$) yields that this line intersects the second axis at point

$$S = -(\mathbf{j}.U^-)\mathbf{j}/(\mathbf{j}.H\mathbf{j}). \tag{14.8}$$

Due to the way we have chosen the mapping $\mathcal{F}$, the real number $\mathbf{j}.U^- = \mathcal{F}^*\mathbf{j}.\mathcal{U}^-$ has the same sign as $\mathcal{N}.\mathcal{U}^- = G_q^*\mathcal{N}.u^-$ or vanishes with it. Through Prop.3.1, one concludes that the coordinate of $S$ on the $\mathbf{j}$ axis is positive, negative or zero if and only if the same is true for $u^-.\nabla f(q(t_s))$.

*First case* : impact.

This is the event where $f(q(t_s))=0$, with $u^-(t_s).\nabla f(q(t_s))>0$. Consequently $f(q(t))<0$ on some left-neighborood of $t_s$, i.e. $t_s$ is an instant of *nontangential collision*. The point $S$ lies in the interior of $C$, so the line $\Delta$ intersects $C$ and does not pass through the origin. Observe that $\mathbf{i}.H^{-1}\mathbf{i}>0$, i.e. the vector $H^{-1}\mathbf{i}$ directs $\Delta$ from left to right. Therefore, the point $P$, as expressed in (14.7), lies on the right of $H^{-1}U^-$ if $\sigma>0$ and on the left in the reverse case. By comparison with (14.4) to (14.6), one concludes that the formulated set of conditions is satisfied if and only if $P$ equals *the nearest point to* $H^{-1}U^-$ *in the intersection* $\Delta \cap C$.

So the problem of determining $u^+$ possesses a *unique solution* in that case.

*Second case* : sliding.

We now assume that $f(q(t_s))=0$, with $u^-(t_s).\nabla f(q(t_s))=0$ (equivalently $u^-(t_s).\nabla f(q(t_s))\leqslant 0$, since the anterior motion is assumed to agree with $f\leqslant 0$).

Then $S=0$, i.e. the line $\Delta$ passes through the origin.

- If $\Delta \cap C$ consists only of the origin, the formulated set of condition admits $P=0$, hence $u^+ = u^-$, as unique solution: *no velocity jump occurs.*
- If $\Delta$ intersects also the interior of C, one has to determine whether it is possible for P to lie in this interior. In view of conditions (14.4) to (14.6), this requires $U^+=0$. It is in fact a solution if $-H^{-1}U^- \in \text{int}\, C$: then the sliding suddenly sticks. Concurrently $P=0$, with $u^+ = u^-$, i.e. no velocity jump, *is also a solution* in this case.
- If $\Delta$ contains one of the half-lines $D_+$ or $D_-$, a value of P on this half-line meets the requirements provided the resulting value of $U^+ = U^- + HP = \sigma \mathbf{i}$ agrees with conditions (14.4) to (14.6). For instance, imagine $D^+ \subset \Delta$: these conditions require $\sigma \geqslant 0$. This can happen only if $-H^{-1}U^- \in D_+$ and then *every value of* P *belonging to the line segment* $[-H^{-1}U^-, 0]$ *is a solution.* The corresponding values of $U^+$ cover the line segment $[0, U^-]$.

REMARK 14.1 The latter is the most interesting item of the discussion. It shows that, contrary to the frictionless case, an episode of smooth motion with persistent contact may end with a velocity jump *without any collision occurring.* This is a dynamical analogue to the locking effect, well known in the statics of mechanisms with dry friction.

The earliest reference we know of, where such a possibility is asserted, is a note by L. Lecornu [26]. At the time, a controversy has been opened by P. Painlevé [36], with the observation that, in systems involving Coulomb friction, some initial value problems could have no solution, or also several solutions. In addition, the behaviour of the system depended on its constants in a discontinuous way. To Painlevé, and later to E. Delassus [37][38], these findings seemed in contradiction with the very bases of Physics. In the subsequent years, different opinions were sustained by such authors as F. Klein [39], R.v.Mises [40], G. Hamel [41] or L. Prandtl [42]. Even after H. Beghin [43][44] had clearly demonstrated that the incriminated findings actually agreed with common observation, some suspicion remained in the scientific community that Coulomb's law could be intrinsically illogical.

Today, one is accustomed to meet multiple solutions or the absence of solution to physical problems, usually ascribing these facts to the nature of the treated information, without opening any discussion about determinism in Physics. One is also familiar with discontinuous behaviour.

Dynamic locking, that we propose to call a *frictional catastrophe*, is commonly observed in practice. The example of the chattering motion of a piece of chalk driven at an angle against a blackboard, so that a dotted line is drawn, was already put forward by E. Delassus [37]. A model of this phenomenon is presented in [30], with some drawings generated by a computer using the time discretization procedure of Sect.15 below. This displays an instance of a "stick-slip" motion. Depending on the system constants, frictional catastrophes and intermittent contact breaking may occur or not. No attempt so far has been made at comparing this model, quantitatively, with experimental measurements.

A very simple example of frictional catastrophe is presented in Sect.15, as a demonstration of the ability of our numerical technique to handle nonsmooth solutions.

REMARK 14.2 From the mathematical standpoint, we think that little has to be retained of the early discussions on the subject. At the time, differential equations were implicitely understood in the sense of the elementary theory:

solutions should possess some *regulated functions* as their derivatives of the highest order involved. In other words, it was admitted that the acceleration $q''$ possessed a right-limit and a left-limit at every t. Certain of these limits played an essential role in the discussions by Painlevé and his followers Unfortunately, in usual instances where an interval of smooth motion precedes the catastrophic instant $t_s$, one finds that the norm of $q''$ actually tends to infinity when t tends to $t_s$ from the left. The concept of a measure differential inclusion, on which the present lectures are based, provides a more synthetic view, since it allows one to express Dynamics on the whole interval I, including $t_s$, and does not rely on the existence of one-side accelerations.

REMARK 14.3 There remains to discuss whether the velocity jumps agreeing with the constraint law (12.10) are physically realistic. We have already stressed that, even in the absence of friction, predicting safely the outcome of a shock would require some high order of information, actually unavailable in engineering situations. Things naturally become worse if friction is entered into account. What do we know about the Physics of high pressure friction, during the "very short" interval of time on which the velocity change takes place? Already for the frictionless case, we have in Sect.6 been reluctant in accepting the invariance of the direction of $\nabla f(q)$ during this interval. The latter invariance, if admitted, makes of the no-friction assumption a time-independent linear condition imposed on the contact force at every instant; therefrom, the normality of the contact percussion is inferred, by integration on the interval. On the contrary, Coulomb law imposes on contact force a nonlinear condition. Even if one assumes this condition independent of time, it cannot be expected in general to commute with integration. The contact percussion can safely be asserted to verify (12.10) only as far as the sliding velocity is sure to remain zero or to keep a constant direction for the short duration of the investigated shock. It is of course in the case of two-dimensional friction that the latter event proves the easiest to discuss.

The reader may find in [2] an attempt at analysing frictional impact, in the line formerly suggested by G. Darboux. This consists in investigating the variation of the velocity as a function of some "micro-time", relatively to which the system position remains a constant. Even so, the conclusion is subject to some assumption about the shock end which seems difficult to justify.

In conclusion, the concept of a frictional and soft contact, as involved in Definition 12.3, only generates a special model of frictional shock, with the advantage of good theoretical consistency. As we shall see in the next Section, the corresponding motions are also very tractable numerically. Physical situations to which this model is relevant very probably exist, but experimentation is still needed to safely identify them in practice.

# 15. ALGORITHM FOR TWO-DIMENSIONAL FRICTION

We now present a time-discretization procedure for computing a motion under the conditions of Sec.13. The notations are the same as in the frictionless case, exposed in Sec.11.

*Stage 1.* Calculate $t_M=t_I+\frac{1}{2}h$ , the midpoint approximants

$$q_M=q_I+\frac{1}{2}hu_I \in \mathbb{R}^n , \qquad A_M=A(q_M) \in \mathbb{R}^{n\times n} , \qquad K_M=K(t_M,q_M,u_I) \in \mathbb{R}^n$$

and the "loose velocity"

$$u_L=u_I + h\, A_M^{-1} K_M .$$

As in Sec.13, a single constraint inequality, say $f \leq 0$, is taken into account.

If $f(q_M) < 0$ or $u_L.\nabla f(q_M) \leq 0$, then make $u_F = u_L$ and go to Stage 3.

*Stage 2.* If, on the contrary, $f(q_M) \geq 0$ and $u_L.\nabla f(q_M > 0$, contact is estimated to have effect on the considered time-step. One constructs a discrete analog to the measure differential inclusion of Dynamics

$$du = A^{-1}(q)\,K(t,q,u)\,dt + A^{-1}(q)\,dR$$

by equalling some approximate values of the integrals on $(t_I,t_F)$ of the respective members, namely

$$u_F - u_I = hA_M^{-1}\,K_M + A_M^{-1}R. \tag{15.1}$$

Let $G_M$ denote the value taken at $q=q_M$ by the linear mapping $G_q: \mathbb{R}^n \to \mathbb{R}^2$ defined in Sec.13. Then an approximate version of (13.7) reads

$$R = G_M^* P,$$

where P, an element of the calculation plane, is an approximant of the total contact impulsion on the interval $(t_I,t_F)$. Putting $U_I = G_M u_I$ and $U_F = G_M u_F$, one derives from (15.1) that

$$U_F - U_I = hG_M A_M^{-1}\,K_M + H_M P. \tag{15.2}$$

Here, similarly to what has been done in Sec.13, one denotes by $H_M$ the symmetric positive definite 2×2 matrix $G_M A_M^{-1} G_M^*$.

According to the decision made in Sec.11, of considering $u_F$ as a simulation of $u^+$, the softness condition (13.7) will, in the present discretization procedure, be transcribed as

$$\exists \sigma \in \mathbb{R}: \qquad U_F = \sigma\,\mathbf{i}. \tag{15.3}$$

Discretizing the two-dimensional Coulomb law consists in relating P to $U_F$ by the system of implications

$$\sigma > 0 \quad \Rightarrow \quad P \in D_+$$

$$\sigma < 0 \quad \Rightarrow \quad P \in D_-$$

$$\sigma = 0 \quad \Rightarrow \quad P \in C.$$

In the case where the friction coefficients depend on q, the elements $D_+$, $D_-$ and C will be evaluated at $q = q_M$.

The determination of P and $u_F$ from this set of conditions is similar to what has been done in one of the cases of shock investigated in Sec.14. If one puts $U_L = G_M u_L$, (15.2) becomes

$$U_F = U_L + H_M P. \tag{15.4}$$

By combination with (15.3), this yields

$$\exists \sigma \in \mathbb{R} : \quad P = -H_M^{-1} U_L + \sigma H_M^{-1} \mathbf{i}.$$

This expresses that P belongs to the line Δ of $\mathbb{R}^2$, drawn through the known point $-H_M^{-1}U_L$, with known directing vector $H_M^{-1}\mathbf{i}$. This line intersects the second axis of $\mathbb{R}^2$ at point

$$S_M = -(\mathbf{j}.U_L)\mathbf{j}/(\mathbf{j}.H_M\mathbf{j}).$$

A similar expression was discussed in (14.8). Since the present computation stage is developed under the assumption $u_L.\nabla f(q_M) > 0$, one finds in the same way as in Sec.14 that $\mathbf{j}.U_L < 0$, so $S_M$ is sure to be interior to C. Observe in addition that P lies in Δ on the right or on the left of $-H_M^{-1}U_L$ according to the sign of σ. One concludes to the existence of a unique solution P, characterized as *the nearest point to* $-H_M^{-1}U_L$ *in the intersection* $\Delta \cap C$.

For computation it is more convenient to formulate the same as follows:

- If $-H_M^{-1}U_L \in C$, then $P = -H_M^{-1}U_L$.
- Otherwise $-H_M^{-1}U_L.\mathbf{i} < 0 \Rightarrow P = \Delta \cap D_+$
  $-H_M^{-1}U_L.\mathbf{i} > 0 \Rightarrow P = \Delta \cap D_-$.

After P is calculated, one derives $u_F$ from (15.1), namely

$$u_F = u_L + A_M^{-1} G_M^* P.$$

*Stage 3.* The computation step finishes with $q_F = q_M + \frac{1}{2}hu_F$.

REMARK 15.1 A more intuitive description of the above discretization procedure may be found in [24]. Instead of relying on the calculation plane, it uses the image of $C$ under $\mathcal{G}_q^{*-1}$, a two-dimensional cone in $\mathbb{R}^n$. This makes the comparison with the frictionless case clearer, but numerically is less effective. The calculation plane is also useful at the stage of deriving the inequalities needed in the study of existence and regularity of solutions.

REMARK 15.2 The case $-H_M^{-1}U_L \in C$ yields $U_F=0$, i.e. zero sliding velocity (the discussion here is simpler than that of the similar geometric construction made in Sec.14, because $S_M$ is certainly interior to C).

In that connection, the algorithm works very well to compute a motion involving the event which, in Remark 14.1, we have called a *frictional catastrophe*. Now, we have just seen that each computation step is deterministic, i.e. it yields a unique pair $q_F$, $u_F$. This contrasts with the conclusions of Sec.14, showing multiple possible outcomes for such a catastrophe: all the points of a line segment are solutions in what concerns the contact percussion P and similarly in what concerns $u^+(t_s)$ or $U^+(t_s)$ (the latter may take any value between zero and $U^-(t_s)$).

In fact, the algorithm is able to approximate any of these solutions. As soon as the successive discretization intervals are chosen, a sequence of values of $u_F$ is unambiguously generated. This sequence is smooth, except for a jump in one of the intervals, said catastrophic. Before this jump, the computation of the motion, from given initial data $q(t_0)$, $u(t_0)$, yields consistent results, for arbitrarily fine discretizations. But the value of the jump obtained in a catastrophic interval depends on the ratio in which the exact instant divides this interval. So, calculations made with different discretization meshes may yield different results after the catastrophe.

EXAMPLE 15.3 A round-tipped rigid body $\mathcal{B}_1$ performs motions parallel to a vertical plane. It is submitted to gravity and confined by a horizontal fixed boundary $\mathcal{B}_0$, with friction coefficient equal to 0.5. Initial conditions are those of contact, with negative angular velocity and sliding velocity directed to the right.

On Figure 15.1, the computer has drawn the profile of $\mathcal{B}_1$ for every third step of the time-discretization (numbers refer to these steps). After an epi-

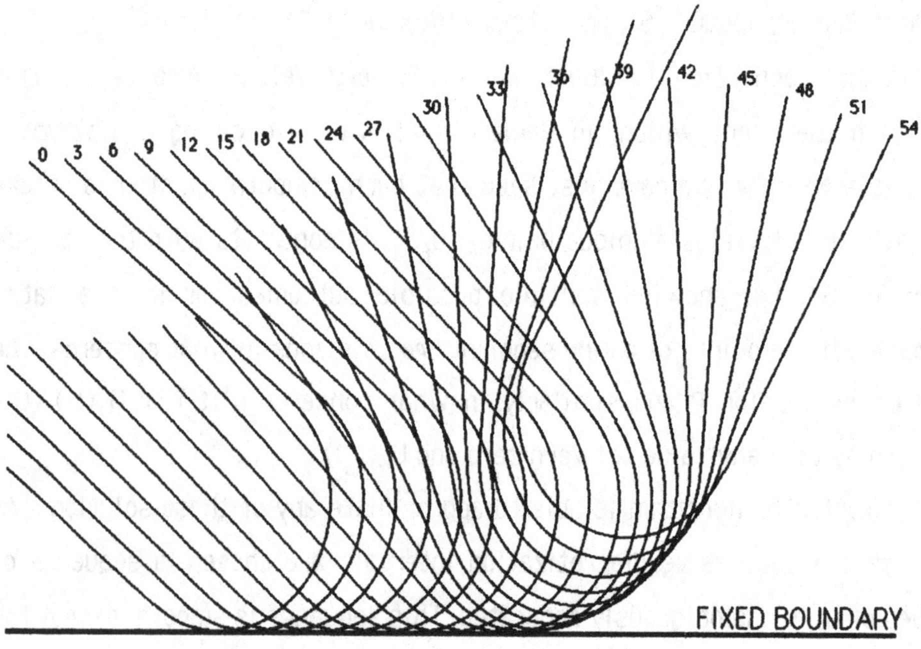


Figure 15.1

sode of persistent contact with sliding of constant direction, a catastrophe occurs. The horizontal component of the velocity of the lowest particle of $\mathcal{B}_1$ (this equals the sliding velocity in the case of contact) presents a sudden drop. At the same instant, contact is broken, with zero normal velocity. In the process, a percussion is imparted to $\mathcal{B}_1$ from $\mathcal{B}_0$, making the negative angular velocity increase in magnitude.

Though discretization is rather rough, the corrective procedure of the possible constraint violation, described in Remark 11.1, has *not* been used. Drawing is however found to comply very well with the unilateral constraint.

On Figure 15.2, the horizontal component of the velocity of the lowest particle is plotted versus time. All curves correspond to the same initial conditions as above, but are computed with finer discretization. In order to display the multiple possible outcomes of the catastrophe, computation has

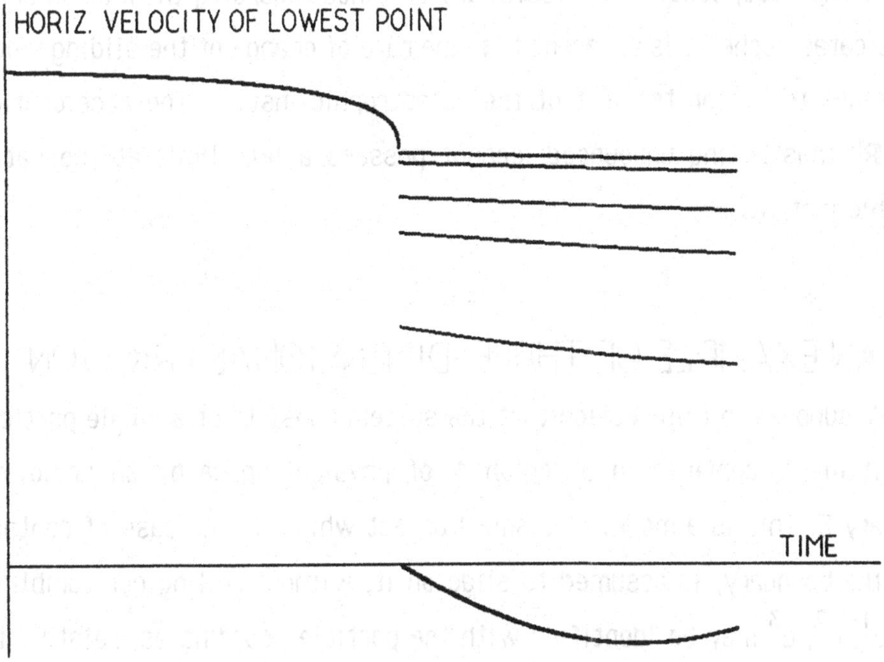


Figure 15.2

been repeated, each time with uniform time-mesh, but successively using different values of the step-length, namely

$$h = 0.00046 - 0.0001 * \mathrm{RND}.$$

Here RND denotes the built-in random sequence of the computer, with values in $[0,1[$. The *maximal catastrophe*, i.e. sliding velocity dropping to zero, is repeatedly obtained, more than two times out of three on an average, and followed with a well defined contact-free motion. The other curves show various sliding velocity drops of smaller amplitudes. More extensive experimentation has confirmed that each value of the sliding velocity drop corresponds to a well defined consequent motion. Statistically, the values of the drop are not uniformly distributed: frequency is found maximal in the vicinity of zero drop.

As expected, consistent results are obtained regarding the motion prior to the catastrophe. It is apparent that the rate of change of the sliding velocity tends to $-\infty$ on the left of the catastrophic instant. The acceleration $q'': I \to \mathbb{R}^n$ thus beeing unbounded, cannot possess a left-limit at the catastrophic instant.

## 16. AN EXAMPLE OF THREE-DIMENSIONAL FRICTION

We suppose in this section that the system consists of a single particle of unit mass, confined in a region $\Phi$ of physical space by an unmoving boundary $\Sigma$. This is a model of a small object which, in the case of contact with the boundary, is assumed to slide on it, without rolling nor tumbling. Then $q^1$, $q^2$, $q^3$ may be identified with the particle coordinates, relative to some inertial orthonormal Cartesian frame and $A(q)$ consists, for every $q$, in the unit matrix.

In addition to the possible frictional reaction of the boundary, the particle is submitted to a force given as a smooth function F of t, q, u. The case of a boundary with prescribed motion may be reduced to this one, through changing the reference frame; there only is to include in F the fictitious forces, thanks to which the new reference frame may be treated as inertial.

The particle dynamics is expressed by this equality of $\mathcal{E}_3$-valued measures on the time-interval I

$$du = dR + F(t, q, u)\, dt, \tag{16.1}$$

i.e. after representing vector measures by density functions relative to some nonnegative real measure $d\mu$,

$$u'_\mu(t) = R'_\mu(t) + F(t, q, u)\, t'_\mu(t),$$

an equality to be satisfied for every $t \in I$.

In the present case, the mapping $\mathcal{G}_q$ reduces to identity for every q. Then, using the law of frictional contact in the form (12.8), one obtains the measure differential inclusion

$$0 \in \partial\psi_{C(q)}(u'_\mu - F(t, q, u)\, t'_\mu) + \partial\theta_q(u^+). \tag{16.2}$$

The feasible region $\Phi$ of $\mathbb{R}^3$ is defined as before by a single inequality $f(q) \leq 0$. For $f(q)=0$, C(q) denotes the friction cone at the point q of the boundary. Again, let us agree to extend its definition, in a smooth arbitrary way, to the values of q such that $f(q) \geq 0$. In addition, C(q) is interpreted as reducing to the zero of $\mathbb{R}^3$ when $f(q)<0$. With every q such that $f(q) \geq 0$, the vector plane T(q) also is associated, orthogonal to $\nabla f(q)$. For $f(q)<0$, we agree to understand T(q) as consisting of the whole of $\mathbb{R}^3$. The extended-real function $\theta_q$, as defined in (12.7), equals $\frac{1}{2}\|.\|^2 + \psi_{T(q)}$.

In particular, at every t such that $f(q(t))<0$, one has $\partial\theta_{q(t)}(x) = \{x\}$ for every $x \in \mathbb{R}^3$, while the value of $\partial\psi_{C(q(t))}(x)$ equals $\mathbb{R}^3$ for $x=0$ and $\emptyset$

otherwise. Therefore (16.2) reduces in that case to the differential equation $u'_t - F(t,q,u) = 0$ of contact-free motion.

Recall that a condition such as (16.2) implies that $q(t) \in \Phi$ for every t, as soon as this is assumed to hold for $t=t_0$.

Some cases of existence, for the solutions to the coresponding initial value problem, are investigated in [13].

Here we shall only present a time-discretization method for their approximate computation [25]. With the same notations as in previous sections, each time-step runs as follows.

*Stage 1.* Calculate $t_M = t_I + \frac{1}{2}h$, the midpoint approximant $q_M = q_I + \frac{1}{2}hu_I$, the force estimation $F_M = F(t_M, q_M, u_I)$ and the "loose velocity" $u_L = u_I + h F_M$.

*Stage 2.*

- If $f(q_M) < 0$ or $u_L . \nabla f(q_M) \leqslant 0$, then $u_F = u_L$.
- Otherwise, $u_F$ is determined by a semi-implicit discretization of (16.2).

In view of the positive homogeneity of the multifunction $\partial\psi_C$, this is

$$0 \in \partial\psi_C(u_F - u_L) + \partial\theta(u_F),$$

whith the cone C and the function $\theta$ evaluated at point $q_M$. Using the definition of $\theta$, one gives to this inclusion the form $0 \in \partial\psi_D(u_F) + u_F$, where D denotes the set $T(q_M) \cap (C(q_M) + u_L)$. Through elementary Convex Analysis, this means that $u_F$ equals the proximal point to the origin in this set, with regard to the usual Euclidean metric of $\mathbb{R}^3$. For the traditional, isotropic, Coulomb law, D is a disk, so the proximal point is specially easy to calculate.

*Stage 3.* Calculate $q_F = q_M + \frac{1}{2}h u_F$.

REMARK 16.1 Even if one assumes isotropic friction, computing the motion of systems involving three-dimensional friction is not in general as simple as above, due to the role of the mapping $\mathcal{G}_q$. The determination of $u_F$, at each time-step, usually is a non convex problem, which may possess several solutions.

EXAMPLE 16.2 The numerical technique presented here has been applied [24] to the motion of a particle P submitted to gravity and confined by a plane boundary with prescribed motion. This particle represents an object which, in the case of contact, may slide on the boundary, without rolling nor tumbling. The plane boundary may be the ground surface, in the course of an earthquake, or also a vibrating table. Motions of the following sort are common in industrial conveyors.

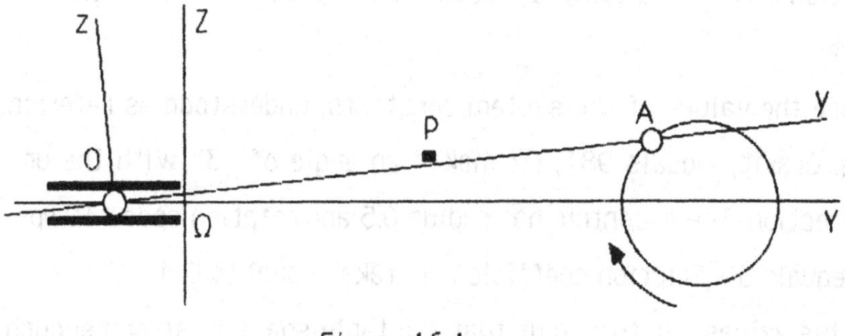


Figure 16.1

The vibrating table is assumed to have the motion of the shaft in a *crank and shaft* mechanism. Let orthonormal axes Oxyz be attached to the table, with Oxy in its surface. Axes Oy and Oz move in fixed plane ΩYZ. The point O is guided along a segment of the line ΩY. The point A (0, a, 0) of the table is astrained, by an eccentric, to describe, at constant velocity, a circle in the plane ΩYZ, centered on ΩY. Therefore, the plane ΩXY is the mean position of the vibrating table.

In order to produce a clear pattern of trajectories, the whole machine is set at a slant: the plane ΩZX is vertical, but ΩZ is not in line with gravity. Hence, ΩX is the direction of steepest descent in the plane ΩXY and determines the general trend of the motion of P.

A computer program, using the numerical technique described above, draws the projections to Oxy an to Oyz of the trajectories of P *relative to the table*. No experiment has so far been conducted for comparison with reality.

At the initial instant, P is left on the table with zero relative velocity. Subsequent trajectories are drawn for several choices of this initial position, at various distances of Ox. Motions taking place sufficiently far from Ox involve intermittent contact break; the loops then observed on the Oyz projection correspond to the parabolic motion that P have, when referred to fixed axes.

Here are the values of the system constants, understood as referring to c.g.s. units. Gravity equals 981; ΩZ makes an angle of 13° with the upward vertical direction. The eccentric has radius 0.5 and rotation speed 10 rps. The length OA equals 50. Friction coefficient is taken equal to 0.4.

With this values, it turns out that the table shake is strong enough for leaving no place where P could rest without sliding. The whole pattern of trajectories admits Ox as an axis of symmetry. Depending on the direction in which the eccentric rotates, this line is a locus of attraction or of repulsion, a fact which could be asserted from qualitative reasoning. More inexpected is the existence of other lines of attraction or of repulsion (they exchange their roles when rotation is reversed), parallel to this one. Such a "quantic" effect appears to be connected with the number of flappings that the table performs while the particle runs through each episode of contact-free motion. Actually, the farther they lie from Ox, the more confuse these attraction loci appear, due to the chaotic behaviour that trajectories then have.

Figure 16.2 shows the trajectories of P, consequent to 15 initial positions equally spaced from y=5 to y=320; the eccentric rotates in the reverse direction to that indicated on Fig. 16.1.

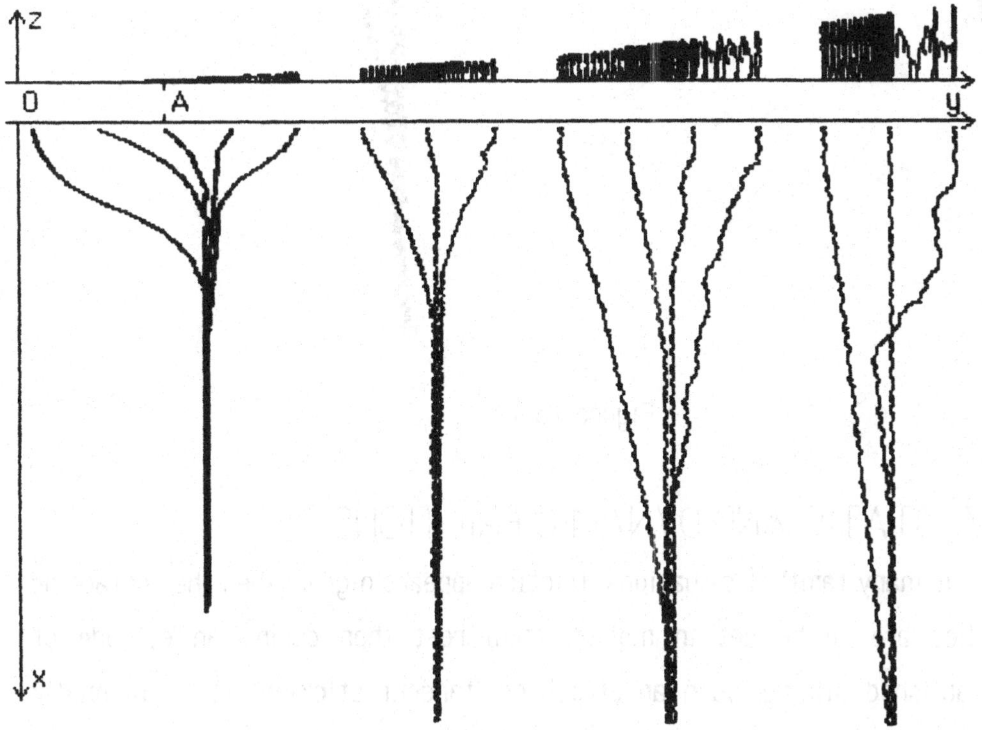


Figure 16.2

Fig.16.3 displays a larger scale drawing, corresponding to the same direction of rotation as on Fig.16.1. The trajectories correspond to 7 initial positions equally spaced from y=5 to y=150.

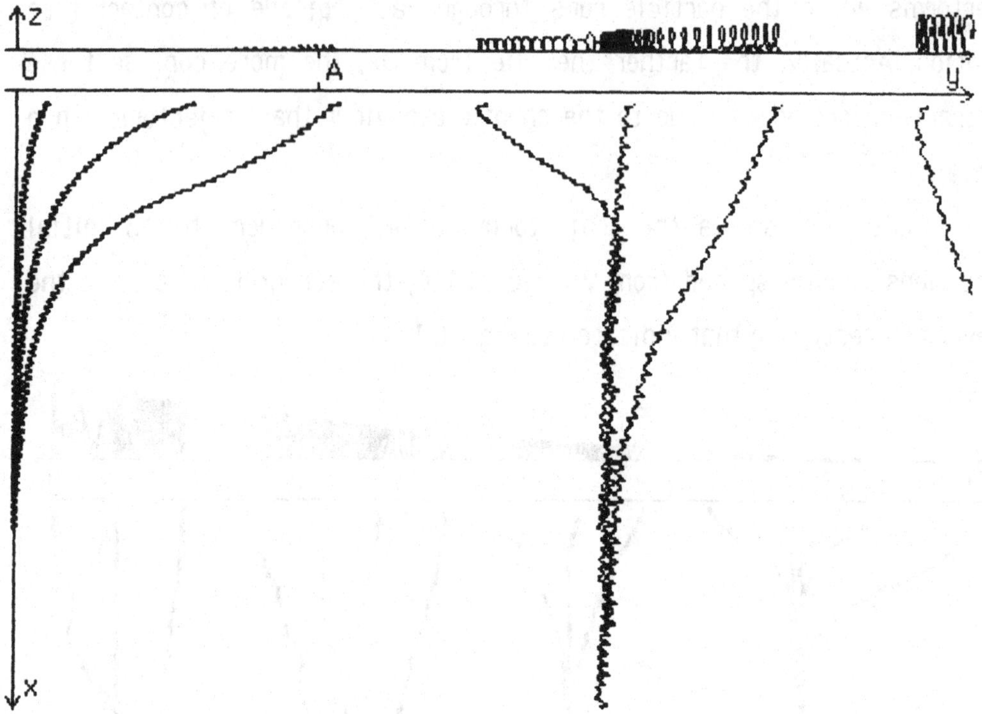


Figure 16.3

## 17. STATIC AND DYNAMIC FRICTIONS

In many familiar situations, friction appears higher when the contacting bodies are to be set in motion from rest than during an episode of established sliding. Such an effect of "tangent sticking" is traditionally accounted for by introducing a larger value for the *static* friction coefficient, i.e. relative to zero sliding velocity, than for the *dynamic* one.

The numerical techniques proposed in the foregoing sections handle this refinement without difficulty. There only is, in each step of time discretization, to make the cone $C$ depend on the sliding status in the antecedent step. For instance, in the algorithm described in Sec.15, this antecedent step involves sliding or no sliding, according to $-H_M^{-1}U_L$ belonging to $C$ or not. This precisely makes one of the branching conditions one anyway had to consider.

At the stage of the analytical formulation, one has to make the cone $C$ depend not only on q, but also on $\mathcal{U}$ or, if the possibility of a velocity jump is considered, on $\mathcal{U}^-$. At first glance, this seems to reduce the advantage of the formulations of friction presented in the foregoing. But the example below tends to demonstrate that, far from being an heterogeneous addition to the previous theorization, such a refinement actually proves inherent in the subject matter. In fact, this example shows that, even if one starts from a law of friction with single coefficient, it may happen that the logical derivation of consequences eventually makes some contact force appear to obey a Coulomb law with coefficient depending on sliding velocity.

Let us consider again the situation of Sec.13, namely, in a position q of the system, with f(q)=0, the range $\mathcal{G}_q(\mathbb{R}^n)$ is assumed to reduce to a two-dimensional subspace $\mathcal{W}$ of $\mathcal{E}_3$, different from the common tangent plane $\mathcal{T}$ to the contacting bodies. In a motion with continuous velocity, it is assumed that the contact force $\mathcal{R} \in \mathcal{E}_3$ is related to $\mathcal{U}$ through Coulomb's law, expressed as before in the form

$$-\mathcal{U} \in \operatorname{proj}_{\mathcal{T}} \partial\psi_C(\mathcal{R}). \qquad (17.1)$$

As already observed, $\mathcal{R}$ pertains to the equations of mechanics only through its equivalence class modulo ker $\mathcal{G}_q^*$, a natural representative of which is

$$\mathcal{R}^* = \operatorname{proj}_{\mathcal{W}} \mathcal{R}.$$

Furthermore, since $\mathcal{U}$ essentially belongs to the tangent vector plane $\mathcal{T}$, it has the form $\mathcal{U} = s\mathcal{I}$, where $\mathcal{I}$ denotes a unit vector of the line $\mathcal{T} \cap \mathcal{W}$, and $s \in \mathbb{R}$. We are going to show that the resulting relation between s and $\mathcal{R}^*$ may have a more general form than it has been assumed in Sect. 13.

Let us consider only, for simplicity, the traditional case where $C$ is a *cone of revolution* about $\mathcal{N}$. Clearly, for all s>0, the set of the values of $\mathcal{R}$ that relation (17.1) associates with $\mathcal{U}=s\mathcal{I}$ consists of a certain boundary half-line, say $\mathcal{H}^+$, of the cone. With all s<0 is associated the half-line $\mathcal{H}^-$, symmetric to $\mathcal{H}^+$ relative to the cone axis. Finally, to s=0, correspond all the points of $C$. The orthogonal projections of $\mathcal{H}^+$ and $\mathcal{H}^-$ to $\mathcal{W}$ are two half-lines, say $\mathcal{D}^+$ and $\mathcal{D}^-$, lying on the same side of $\mathcal{T}\cap\mathcal{W}$. The expected relation between s and $\mathcal{R}^*$ is then expressed by the three implications

$$s>0 \Rightarrow \mathcal{R}^* \in \mathcal{D}^+$$

$$s<0 \Rightarrow \mathcal{R}^* \in \mathcal{D}^-$$

$$s=0 \Rightarrow \mathcal{R}^* \in \mathrm{proj}_{\mathcal{W}} C.$$

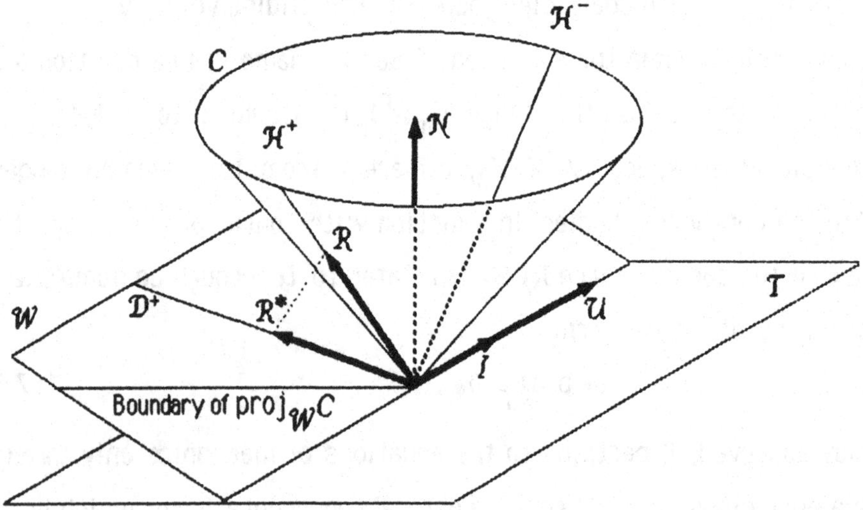


Figure 17.1

Depending on the span of the cone $C$ and on the angle that $\mathcal{W}$ makes with $\mathcal{N}$, the orthogonal projection of $C$ to $\mathcal{W}$ may equal the whole of $\mathcal{W}$ or some angular region $C^*$ of this two-dimensional space.

In the latter case, $C^*$ contains $\mathcal{D}^+$ and $\mathcal{D}^-$ but has no reason in general to possess them as its edges: this means that the relation found between $\mathcal{U}$ and $\mathcal{R}^*$ is equivalent to *a two-dimensional Coulomb law with static coefficient larger than the dynamic one*. The equality of these coefficients, namely the simple case studied in Sec. 13, is however achieved if $\mathcal{W}$ is orthogonal to $\mathcal{T}$.

If $\mathrm{proj}_{\mathcal{W}} C = \mathcal{W}$, every value of $\mathcal{R}^*$ in $\mathcal{W}$ is associated with $\mathcal{U}=0$, possibly making with $\mathcal{N}$ an angle larger than $\pi/2$. In statical problems, this could be described as a *wedging* effect.

The above discussion provides an example of the interaction between the constraints of the system and the frictional effects at possible contact points. The treatment of systems involving *several contacts with Coulomb friction*, a question left aside in these lectures, leads in general to similar situations.

## REFERENCES

1. Delassus, E.: Mémoire sur la théorie des liaisons finies unilatérales, Ann. Sci. Ecole Norm. Sup., 34 (1917), 95-179.
2. Pérès, J.: Mécanique Générale, Masson, Paris 1953.
3. Moreau, J.J.: Les liaisons unilatérales et le principe de Gauss, C.R. Acad. Sci. Paris, 256 (1963), 871-874.
4. Moreau, J.J.: Quadratic programming in mechanics: dynamics of one-sided constraints, SIAM J. Control, 4 (1966), 153-158.
5. Dunford, N. and J.T. Schwartz: Linear Operators, Part I: General Theory, Interscience Pub. Inc., New York 1957.
6. Dinculeanu, N.: Vector Measures, Pergamon, London, New York 1967.
7. Moreau, J.J.: Bounded variation in time, in: Topics in Nonsmooth Mechanics (Ed. J.J. Moreau, P.D. Panagiotopoulos and G. Strang), Birkhäuser, to appear.
8. Pandit, S.G. and S.G. Deo: Differential Systems Involving Impulses,

Lecture Notes in Math., vol. 954, Springer-Verlag, Berlin, Heidelberg, New York 1982.
9. Moreau, J.J.: Une formulation de la dynamique classique, C.R. Acad. Sci. Paris, Sér.II, 304 (1987), 191-194.
10. Panagiotopoulos, P.D.: Inequality Problems in Mechanics and Applications, Birkhäuser, Boston, Basel, Stuttgart 1985.
11. Monteiro Marques, M.D.P.: Chocs inélastiques standards: un résultat d'existence, Travaux du Séminaire d'Analyse Convexe, Univ. des Sci. et Techniques du Languedoc, vol.15, exposé n° 4, Montpellier, 1985.
12. Monteiro Marques, M.D.P.: Rafle par un convexe semi-continu inférieurement, d'intérieur non vide, en dimension finie, C.R. Acad. Sci. Paris, Sér.I, 299 (1984), 307-310.
13. Monteiro Marques, M.D.P.: Inclusões Diferenciais e Choques Inelásticos, Doctoral Dissertation, Faculty of Sciences, University of Lisbon, 1988.
14. Moreau, J.J.: Liaisons unilatérales sans frottement et chocs inélastiques, C.R. Acad. Sci. Paris, Sér.II,296 (1983),1473-1476.
15. Moreau, J.J.: Standard inelastic shocks and the dynamics of unilateral constraints, in: Unilateral Problems in Structural Analysis (Ed. G. Del Piero and F. Maceri), CISM Courses and Lectures No.288, Springer-Verlag, Wien, New York 1985.
16. Schatzman, M.: A class of nonlinear differential equations of second order in time, J. Nonlinear Analysis, Theory, Methods and Appl.,2 (1978),355-373.
17. Buttazzo, G. and D. Percivale: On the approximation of the elastic bounce problem on Riemannian manifolds, J. Diff. Equations, 47 (1983), 227-245.
18. Carriero, M. and E. Pascali: Uniqueness of the one-dimensional bounce problem as a generic property in $L^1([0,T];R)$, Bolletino U.M.I.(6) 1-A (1982), 87-91.
19. Percivale, D.: Uniqueness in the elastic bounce problem, J. Diff. Equations, 56(1985), 206-215.
20. Moreau, J.J.: Sur les lois de frottement, de plasticité et de viscosité, C.R. Acad. Sci. Paris, Sér.A , 271 (1970), 608-611.
21. Moreau, J.J.: On unilateral constraints, friction and plasticity, in: New Variational Techniques in Mathematical Physics (Ed. G. Capriz and G. Stampacchia), CIME 2° ciclo 1973, Edizioni Cremonese, Roma, 1974, 173-322.
22. Moreau, J.J.: Application of convex analysis to some problems of dry friction, in: Trends in Applications of Pure Mathematics to Mechanics,

vol.2 (Ed. H. Zorski), Pitman Pub. Ltd., London 1979, 263-280.

23. Duvaut, G. and J.L. Lions: Les Inéquations en Mécanique et en Physique, Dunod, Paris 1972.
24. Moreau, J.J.: Dynamique de systèmes à liaisons unilatérales avec frottement sec éventuel; essais numériques, Note Technique 85-1, Lab. de Mécanique Générale des Milieux Continus, Univ. des Sci. et Techniques du Languedoc, Montpellier, 1985.
25. Moreau, J.J.: Une formulation du contact à frottement sec; application au calcul numérique, C.R. Acad. Sci. Paris, Sér.II,302 (1986), 799-801.
26. Lecornu, L.: Sur la loi de Coulomb, C.R. Acad. Sci. Paris, 140 (1905), 847-848.
27. Oden, J.T. and J.A.C. Martins: Models and computational methods for dynamic friction phenomena, Computer Methods in Appl. Mech. and Engng., 52 (1985), 527-634.
28. Jean, M. and G. Touzot: Implementation of unilateral contact and dry friction in computer codes dealing with large deformations problems, to appear in: Numerical Methods in Mechanics of Contacts Involving Friction, J. de Mécanique théor. et appl., Special issue, 1988.
29. Abadie, J.: On the Kuhn-Tucker theorem, in: Nonlinear Programming (Ed. J.Abadie), North-Holland Pub. Co., Amsterdam 1967, 19-36.
30. Jean, M. and J.J. Moreau: Dynamics in the presence of unilateral contacts and dry friction; a numerical approach, in: Unilateral Problems in Structural Analysis 2" (Ed. G. Del Piero and F. Maceri), CISM Courses and Lectures No 304, Springer-Verlag, Wien 1987, 151-196.
31. Moreau, J.J.: Evolution problem associated with a moving convex set in a Hilbert space, J. Diff. Equations, 26 (1977), 347-374.
32. Moreau, J.J.: Sur les mesures différentielles de fonctions vectorielles et certains problèmes d'évolution, C.R. Acad. Sci. Paris, Sér.A, 282 (1976), 837-840.
33. Jean, M. and E. Pratt: A system of rigid bodies with dry friction, Int. J. Engng. Sci., 23 (1985), 497-513.
34. Rockafellar, R.T.: Convex Analysis, Princeton Univ. Press, Princeton 1970.
35. Moreau, J.J.: Fonctionnelles Convexes, Lecture Notes, Séminaire sur les Equations aux Dérivées Partielles, Collège de France, Paris 1967.
36. Painlevé, P.: Sur les lois du frottement de glissement, C.R. Acad. Sci. Paris, 121 (1895), 112-115. Same title, ibid. 141 (1905), 401-405 and 141 (1905), 546-552.
37. Delassus, E.: Considérations sur le frottement de glissement, Nouv. Ann.

de Math. (4ème série), 20 (1920), 485-496.
38. Delassus, E.: Sur les lois du frottement de glissement, Bull. Soc. Math. France, 51 (1923), 22-33.
39. Klein, F.: Zu Painlevés Kritik des Coulombschen Reibungsgesetze, Zeitsch. Math. Phys., 58 (1910), 186-191.
40. Mises, R.v.: Zur Kritik der Reibungsgesetze, ibid., 191-195.
41. Hamel, G.: Bemerkungen zu den vorstehenden Aufsätzen der Herren F. Klein und R. v. Mises, ibid., 195-196.
42. Prandtl, L.: Bemerkungen zu den Aufsätzen der Herren F. Klein, R. v. Mises und G. Hamel, ibid., 196-197.
43. Beghin, H.: Sur certains problèmes de frottement, Nouv. Ann. de Math., 2 (1923-24), 305-312.
44. Beghin, H.: Sur l'indétermination de certains problèmes de frottement, Nouv. Ann. de Math., 3 (1924-25), 343-347.

# NONCONVEX SUPERPOTENTIALS AND HEMIVARIATIONAL INEQUALITIES. QUASIDIFFERENTIABILITY IN MECHANICS

**P.D. Panagiotopoulos**
**Aristotle University, Thessaloniki, Greece**
**R.W.T.H., Aachen, FRG**

ABSTRACT

These lectures concern the study of mechanical problems involving nonconvex and possibly nondifferentiable energy functions, the superpotentials. First superpotentials connected with the notion of generalized gradient are considered, then the V-superpotentials are defined and finally the quasidifferential. After the defintion and formulation of several classes of mechanical problems involving superpotentials we study the questions of existence and approximation of the solution of a variational-hemivariational inequality resulting in the delamination theory of von Kármán laminated plates. Then a general semicoercive hemivariational inequality is studied and necessary and sufficient conditions are derived. The last section is devoted to the new notion of quasidifferentiability and its application to the study of mechanical problems.

## 1. INTRODUCTION

The present Chapter deals mainly with the study of mechanical problems expressed in terms of nonconvex superpotentials and generalized gradients. These problems have as variational formulations hemivariational inequalities, expressions of the principle of virtual work or power in inequality form. We mention here, for instance, problems involving nonmonotone, possibly multivalued laws between stresses and strains or reactions and displacements, or generally speaking, between generalized forces and velocities. A further aim of the present Chapter is the discussion of some other types of variational formulations which result from V-superpotentials and from quasidifferentials. All the aforementioned problems are expressed in terms of nonconvex energy functions and their treatment differs inherently from that of problems having as variational formulations variational inequalities.

The study of variational inequalities started in 1963 with the works of G. Fichera [1],[2] and of J.L. Lions with G. Stampacchia [3]. An important achievement of J.J. Moreau [4] in 1968 was the introduction of the notion of (convex) superpotential and his connection with the theory of variational inequalities. By means of Moreau's superpotential large classes of complicated, yet unsolved problems in mechanics and engineering, could now be formulated and solved correctly, which could not be achieved with the methods of classical bilateral mechanics. Because the variational inequalities are expressions of the principle of virtual work or power in inequality form all these problems are also called inequality problems.

Until 1981 all the inequality problems studied were expressed in terms of variational inequalities and included convex energy functions (superpotentials). But the convexity of an energy function implies the monotonicity, e.g. of the underlying stress- strain relation. Thus, only monotone, possibly multivalued conditions between generalized forces and velocities could be considered. In order to overcome the constraint of monotonicity the notion of nonconvex superpotential was introduced by P.D. Panagiotopoulos [5],[6] by using the generalized gradient of F.H. Clarke. Thus a new type of variational expressions in inequality form was obtained, the hemivariational inequalities which are no longer connected with monotonicity. With respect to static hemivariational inequalities we have substationarity "principles" for the potential and the complementary energy and nonuniqueness of the solution, instead of the minimum "principles" and the uniqueness of the solution as happens in the case of convex superpotentials. For the formulation and treatment of variational inequality problems we refer the reader to [2],[7],[8],[9],[10],[11],[12] and for the hemivariational inequalities to [6],[12] both for the mechanical and the mathematical aspects of the theory.

In this Chapter several types of hemivariational inequalities are studied. The mechanical problems chosen are representative and form the appropriate substratum upon which the mathematical theory is developed. In Sec. 2 we give some definitions and notations for the reader's convenience and we derive formally for certain mechanical problems some hemivariational and variational-hemivariational inequalities. Also the numerical treatment of the arising problems is discussed, together with the V-superpotentials and the basic questions of stability and unloading. Sec. 3 concerns the mathematical study of laminated von Kármán plates subjected to a nonmonotone interlaminar law and to monotone boundary conditions, i.e. we deal with variational-hemivariational inequalities. In Sec. 4 a general semicoercive hemivariational inequality is studied and necessary and sufficient conditions for the solution of the problem have been derived. Finally in Sec. 5 the notion of quasidifferentiability is introduced into mechanics and certain applications are given. It is worth noting that all the mathematical methods developed here for the study of the coercive and semicoercive hemivariational inequalities which arise in concrete mechanical problems can be extended without many modifications to a more abstract mathematical setting.

Closing this introduction we would like to point out once again the inherent differences between "Smooth Mechanics", based on the notion of classical potentials, and "Nonsmooth Mechanics" which is concerned with the study of the nonsmooth and/or nonfinite, convex or nonconvex superpotentials and the related mechanical problems.

## 2. MATHEMATICAL PRELIMINARIES. FORMULATION OF MECHANICAL PROBLEMS

### 2.1 Mathematical Notions and Definitions. A Survey

Let X be a locally convex Hausdorff topological vector space, X' its dual space and $\langle x',x\rangle$ the duality pairing. The reader who is not familiar with this terminology could perceive X as a classical Hilbert space (then $X=X'$ and $\langle x',x\rangle$ is the scalar product of the Hilbert space), or more simply as the classical n-dimensional Euclidean space $\mathbb{R}^n$ (then $\langle x',x\rangle = x_i'x_i$ $i=1,2,\dots,n$). A functional $f:X\to(-\infty,+\infty]$ with $f\not\equiv\infty$ is called proper. Moreover f is lower semicontinuous (l.s.c) on X if and only if the set $\mathrm{epi} f = \{(x,\lambda)\,|\,f(x)\leqq\lambda,\ \lambda\in\mathbb{R}\}$ is closed in $X\times\mathbb{R}$. We also recall that f is Lipschitzian (locally) at x if a neighborhood U of x exists on which f is finite and

$$|f(x_1)-f(x_2)| \leqq c p(x_1-x_2) \quad \forall x_1,x_2\in U , \tag{2.1}$$

where c is a positive constant depending on U and p is a continuous seminorm on X. If (2.1) holds at every $x\in A\subset X$ then f is called (locally) Lipschitzian on A. For instance f is Lipschitzian at x, if it is continuously differentiable at x, or convex (resp. concave) and finite at x or a linear combination of Lipschitzian functions at x. We denote by $f^o(x,y)$ the directional differential in the sense of F.H. Clarke at x in the direction y which is defined for f Lipschitzian at x by the expression (see e.g. [13])

$$f^o(x,y) = \limsup_{\substack{\lambda\to 0_+ \\ h\to 0}} \frac{f(x+h+\lambda y)-f(x+h)}{\lambda} . \tag{2.2}$$

Then the generalized gradient of f at x for f(x) finite is by definition

$$\bar{\partial} f(x) = \{x'\in X' \,|\, f^o(x,x_1-x) \geq \langle x',x_1-x\rangle \quad \forall x_1\in X\} . \tag{2.3}$$

It should be noted that if f is convex then $\bar{\partial} f(x)$ coincides with the subdifferential $\partial f(x)$ defined for f(x) finite by the expression

$$\partial f(x) = \{x'\in X' \,|\, f(x_1)-f(x) \geq \langle x',x_1-x\rangle \quad \forall x_1\in X\} . \tag{2.4}$$

If f is continuously differentiable at x then $\bar{\partial} f(x)=\{\text{grad} f(x)\}$.
Note that in the mathematical literature the term subdifferential and the symbol $\partial f$ are used both for the subdifferential (2.4) of convex analysis and the generalized gradient (2.3) for which here the notation $\bar{\partial} f$ is used. The generalized gradient can be defined for any functional $f:X\to[-\infty,+\infty]$. In this case $f^o(x,y)$ should be replaced in (2.3) by the more complicated notion of the upper subdifferential of R.T. Rockafellar [14] $f^\uparrow(x,y)$. Then we have that

$$\bar{\partial} f(x) = \{x'\in X' \,|\, f^\uparrow(x,x_1-x) \geq \langle x',x_1-x\rangle \quad \forall x_1\in X\} . \tag{2.5}$$

Moreover $\bar{\partial} f(x) = \emptyset$ if $f^\uparrow(x,0) = -\infty$, otherwise $\bar{\partial} f(x) \neq \emptyset$. Note that $\bar{\partial} f(x)$ is never empty, if f is Lipschitzian at x, or if f attains a local minimum at x.
For further properties and other equivalent definitions of the generalized gradient we refer the reader, for a short presentation of the results to [12](Ch.4) and for further details and proofs to [13],[14].
We define here anly two other notions. The notion of substationarity and the notion of $\bar{\partial}$-regularity. A point $x_0$ is called a substationarity point of f, if it is a solution of the multivalued equation

$$0 \in \bar{\partial} f(x) \ . \tag{2.6}$$

Every local minimum and every saddle point is a substationarity point. Also a local maximum $x_0$ is a substationarity point if f is Lipschitzian around $x_0$. The substationarity has a more profound equivalent definition in terms of the directions of approximately uniform descent; see in this respect [15] p.102, and [12].

A functional $f:X \to [-\infty,+\infty]$ is called $\bar{\partial}$-regular if

$$f^{\uparrow}(x,y) = \mathcal{F}'(x,y) = \lim_{\lambda \to 0_+} \frac{f(x+\lambda y)-f(x)}{\lambda} \quad \forall y \in X \ . \tag{2.7}$$

Here $\mathcal{F}'(x,y)$ is the one-sided directional Gâteaux-differential of f at x in the direction y defined by the last equality in (2.7). For instance, if f is a convex, or a maximum type function, then f is $\bar{\partial}$-regular. The property of $\bar{\partial}$-regularity is important since it guarantees the additivity of the generalized gradient.

## 2.2 Nonconvex Superpotentials. Generalities

Let $\Sigma$ be a mechanical system characterized by the generalized force vector $f \in F$ and the generalized velocity vector $u \in U$. Here F and U are vector spaces placed in separating duality through the bilinear form $\langle u,f \rangle$ expressing the work produced by f due to u. Let $\Phi:U \to [-\infty,+\infty]$ and suppose that between f and u a relation of the form

$$-f \in \bar{\partial}\Phi(u) \text{ in } \Sigma \tag{2.8}$$

holds. By definition (2.8) is equivalent to the relation

$$\Phi^{\uparrow}(u,v-u) \geq \langle f,v-u \rangle \quad \forall v \in U \tag{2.9}$$

which is called a hemivariational inequality.

i) Suppose now that $\Phi$ is a maximum type function, i.e. $u \mapsto \Phi(u) = \max_i \{\varphi_i(u)\}$, $i=1,\dots,m$, $u=\{u_1,\dots,u_n\}$ and $\varphi_i(\cdot)$ are smooth functions. Let us introduce the sets $A_i = \{u \mid \Phi(u)=\varphi_i(u)\}$. Then (cf. also [15])

$$\bar{\partial}\Phi(u) = \{\text{grad}\varphi_i(u)\} \text{ if } u \in A_i \tag{2.10}$$
$$\bar{\partial}\Phi(u) = \text{co}\{\text{grad}\varphi_i(u), \text{grad}\varphi_j(u)\} \text{ if } u \in A_i \cap A_j \tag{2.11}$$
$$\bar{\partial}\Phi(u) = \text{co}\{\text{grad}\varphi_i(u), \text{grad}\varphi_j(u), \text{grad}\varphi_k(u)\} \text{ if } u \in A_i \cap A_j \cap A_k \ . \tag{2.12}$$

Here "co" denotes the convex hull and thus in (2.11) (resp. (2.12)) $\bar{\partial}\Phi(u)$ is the line segment (resp. the triangle) connecting the ends of the two (resp. the three) gradients.

ii) Suppose that $\Phi$ is the indicator $I_C$ of a closed set $C \subset U = \mathbb{R}^n$, i.e. $\Phi(u) = I_C(u) = \{0 \text{ if } u \in C, \infty \text{ if } u \notin C\}$, and that $u = \{u_1, \ldots, u_n\}$. Then $\bar{\partial}\Phi(u) = N_C(u)$, where $N_C(u)$ denotes the normal cone to C at the point u.

iii) If more specifically $C = \{u \in \mathbb{R}^n : \varphi_i(u) \leq 0 \quad i = 1, \ldots, m\}$ where the functions $\varphi_i$ are smooth then it can be shown (cf.[12] p.146) that

$$N_C(u) = \{f \in F = \mathbb{R}^n \mid f = \sum_{i=1}^{m} \lambda_i \operatorname{grad}\varphi_i(u),\ \lambda_i \geq 0, \varphi_i \leq 0, \lambda_i \varphi_i = 0\} \qquad (2.13)$$

on the assumption that if $u \in$ boundary C a vector y exists such that

$$\langle y, \operatorname{grad}\varphi_i(u) \rangle < 0 \qquad (2.13a)$$

for every i characterizing an active constraint ($\varphi_i(u) = 0$) at u. Roughly speaking property (2.13) implies the validity of Lagrange multiplier type formulas for the nonconvex problems.

We may make use of (2.13) in the study of the equilibrium problem of a material point which is constrained to remain in C. Then we can easily verify that if R is the reaction force, then a constraint of the form $-R \in N_C(u)$ holds and thus, if $F_o$ is the external force acting on the material point, we shall have

$$F = -R \in N_C(u) = \bar{\partial} I_C(u), \qquad (2.14)$$

where u denotes the displacement, or equivalently

$$I_C^{\uparrow}(u, v-u) \geq \langle F, v-u \rangle \quad \forall v \in U. \qquad (2.15)$$

But $I_C^{\uparrow}(u,v) = I_{T_C(u)}(v)$ [14], where $T_C(u)$ denotes the tangent cone to C at the point u and thus (2.15) is equivalent to

$$\langle F, v-u \rangle \leq 0 \quad \forall v \in T_C(u) + u. \qquad (2.16)$$

If the variations v-u are sufficiently small, as is assumed in the classical formulation of the principle of virtual work and if at u the admissible set C is tangentially regular (e.g., if C has not a reentrant corner at u, cf. also [15] p.41) we obtain the same inequality as in (2.16) holding for every $v \in C$, i.e. we find the inequality of the classical principle of virtual work [16]. However, (2.16) is the exact and general formulation of this principle for the problem under consideration.

### 2.3 Nonmonotone Unilateral Contact and Friction Conditions

We consider a deformable body $\Omega$ and let $\Gamma$ be its boundary. $\Omega$ is referred to a fixed orthogonal Cartesian system $Ox_1x_2x_3$. $\Gamma$ is decomposed into three nonoverlaping parts $\Gamma_F$, $\Gamma_U$ and $\Gamma_S$. On $\Gamma_F$ the boundary forces $S=\{S_i\}$ are given, i.e.

$$S_i = F_i, \quad F_i = F_i(x) \;, \tag{2.17}$$

and on $\Gamma_U$ the displacements are prescribed

$$u_i = U_i, \quad U_i = U_i(x) \;, \tag{2.18}$$

where $S_i=\sigma_{ij}n_j$ (i,j=1,2,3 - summation convention), $\sigma=\{\sigma_{ij}\}$ is the stress tensor and $n=\{n_i\}$ is the outward unit normal vector to $\Gamma$. We assume that on $\Gamma_S$ S (and u) are decomposed into normal and tangential components $S_N$ and $S_T$, and $u_N$ and $u_T$ respectively. The assumption of given tangential displacements or forces on $\Gamma_S$ say

$$S_{T_i} = C_{T_i}, \quad C_{T_i} = C_{T_i}(x) \tag{2.19}$$

can be combined with a nonlinear relation between $S_N$ and $u_N$. Thus we may write on $\Gamma_S$ a relation of the form (Fig. 1a)

$$\begin{aligned} &\text{if } u_N < 0 \quad \text{then} \quad S_N = 0 \\ &\text{if } u_N \geq 0 \quad \text{then} \quad S_N + k(u_N) = 0 \;, \end{aligned} \tag{2.20}$$

where $k(\cdot)$ is a nonmonotone function. (2.20) describes the unilateral contact with a granular support (rock, concrete, soil). Moreover the graph of k may include some jumps which describe locking and local crushing effects. If the contact surface exhibits a small resistance in tension, e.g. due to an adhesive material, we can write the condition (Fig. 1b)

$$\begin{aligned} &\text{if } u_N > u_{N_0} \quad \text{then} \quad S_N + k(u_N) = 0 \\ &\text{if } u_N = u_{N_0} \quad \text{then} \quad 0 \leq S_N \leq k(u_{N_0}) \\ &\text{if } u_N < u_{N_0} \quad \text{then} \quad S_N = 0 \quad . \end{aligned} \tag{2.21}$$

The dotted line in Fig. 1b is more realistic, since we avoid the ideally brittle behaviour. All the aforementioned laws can be put in the form

$$-S_N \in \bar{\partial} j_N(u_N) \ , \tag{2.22}$$

where $j_N(\cdot)$ results by integration. For instance, in the case of Fig.1a

$$j_N(u_N) = \int_0^{u_{N+}} k(\zeta)d\zeta, \tag{2.23}$$

where $u_{N+} = \sup(0, u_N)$.

In the tangential direction the previous normal laws may be combined with a general tangential nonmonotone relation of the form

$$-S_T \in \bar{\partial} j_T(u_T) \tag{2.24}$$

in order to describe frictional phenomena. For $S_N = C_N$, given, we depict in Fig. 1c the friction law of Coulomb. In Fig. 1d to Fig. 1f some other nonmonotone, possibly multivalued, friction laws are depicted. All of them hold for $\Omega \in \mathbb{R}^2$ ; in the case $\Omega \in \mathbb{R}^3$ the additional assumption that the vectors $-S_T$ and $u_T$ are collinear has to be made. The law of Coulomb can be written in the form (2.24) with $j_T(u_T) = \mu |C_N| |u_T|$ ; $\mu$ is the coefficient of friction. Due to the convexity of the superpotential $\bar{\partial}$ can be replaced by the subdifferential $\partial$. The laws of Figs. 1d - 1f are realistic generalizations of Coulomb's law. The graphs of $(-S_T, u_T)$ diagrams may include vertical jumps describing local cracking and crushing of the contact surface asperities. Worthy of note is the existing analogy between Scanlon's diagram for reinforced concrete in tension (see e.g. [17] and Fig. 1g) and the sawtooth friction diagrams describing the modification of the mechanical properties of the contact surface during the frictional contact.

In Fig. 1h the friction diagram between reinforcement and concrete is depicted. Here $[u_T]$ denotes the relative tangential displacement between the two structural elements in contact. All the above diagrams and their twodimensional extensions can be put in the form (2.24) or if $S_N$ is not prescribed in the form

$$-S_T \in \bar{\partial} j_T(u_T; S_N) \; . \tag{2.25}$$

The nonmonotone unilateral contact boundary condition coupled with nonmonotone friction has the form

$$\begin{aligned} -S_N \in \bar{\partial} j_N(u_N) \text{ and if } S_N = 0 \text{ then } S_T = 0 \\ \text{if } S_N < 0 \text{ then } -S_T \in \bar{\partial} j_T(u_T; S_T) . \end{aligned} \tag{2.26}$$

Relation (2.24) includes [18] as a special case the anisotropic friction. Let us give further a method for the derivation of general nonmonotone friction laws, which are not necessarily onedimensional, or which do not assume the collinearity of $-S_T$ and $u_T$. Let $K(S_N)$ be a closed nonconvex subset of the $S_T$-space which may depend on $S_N$. For the present we assume no relation between $S_N$ and $S_T$. Then we consider the law (generalized gradient with respect to $S_T$)

$$-u_T \in \bar{\partial} I_K(S_T; S_N) = N_K(S_T; S_N) \tag{2.27}$$

which implies (cf. (2.14 - 2.16)) that for $S_T \in K(S_N)$

$$\langle -u_T, S_T^* - S_T \rangle \leqq 0 \qquad \forall S_T^* \in T_{K(S_N)}(S_T) + S_T \; . \tag{2.28}$$

This last inequality recalls Hill's "principle" in holonomic elastoplasticity; it coincides with this principle if $K(S_N)$ is tangentially regular at $S_T$ and the variation $S_T^* - S_T$ is appropriately small. The tangential regularity is guaranteed, e.g., if $K(S_N)$ does not have a reentrant corner at $S_T$. Note also that (2.27) implies that $\langle u_T, S_T \rangle$ is substationary with respect to $K(S_N)$ at $S_T$.
Suppose now that $K(S_N) = \{S_T \mid f_i(S_T; S_N) \leqq 0 \quad i=1,\dots,m\}$ where $f_i$ are continuously differentiable functions of $S_T$ defining the boundary between sliding friction and adhesive friction. Moreover, let a condition similar to (2.13a) be valid. Thus we may write the nonmonotone friction law in the form

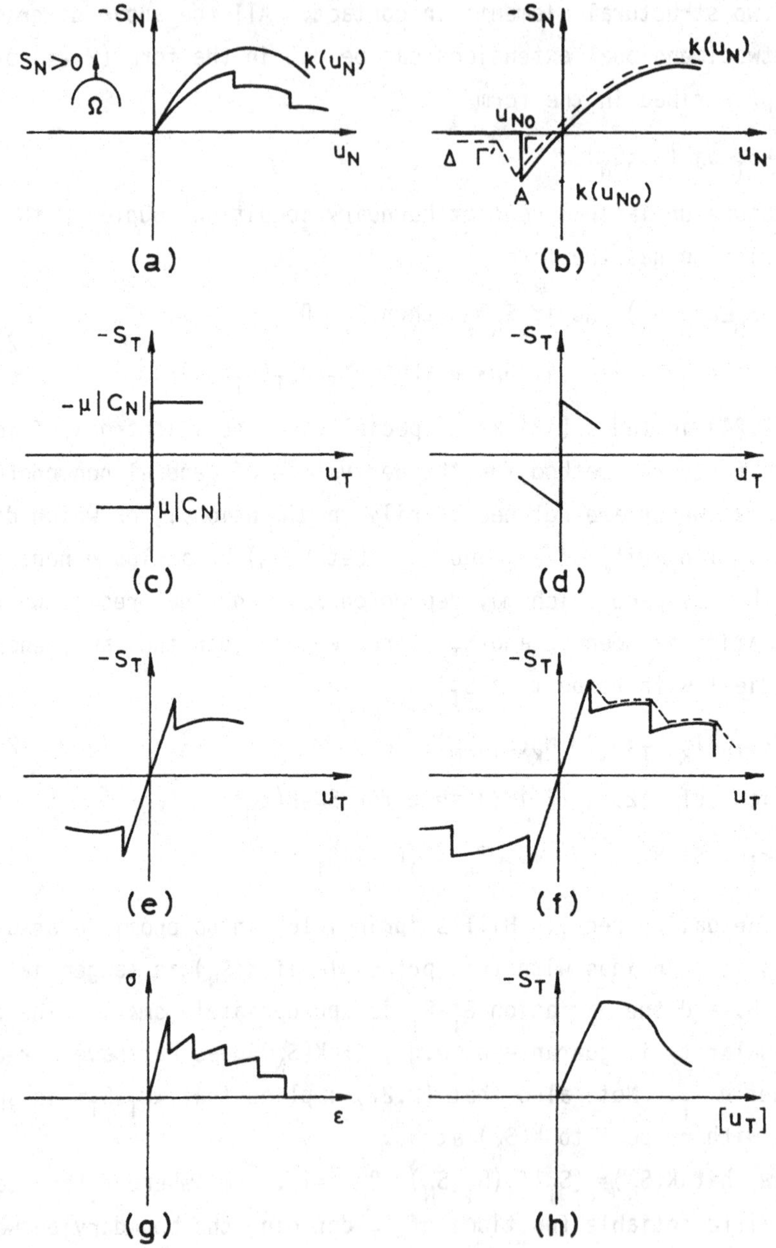


Fig. 1, Nonconvex Superpotential Laws

$$-u_T = \sum_{i=1}^{m} \lambda_i \frac{\partial f_i(S_T;S_N)}{\partial S_T}, \quad \lambda_i \geq 0, \quad f_i(S_T;S_N) \leq 0, \quad \lambda_i f_i(S_T;S_N) = 0. \quad (2.29)$$

With respect to the general friction law (2.25) we can adapt the idea of nonlocal effects caused by the asperities of the contact surface [19]. In this case $S_N$ should be replaced by $Q(S_N)$, where Q is an appropriate smoothing operator.

The previous static friction laws are of importance for static and certain quasistatic problems. Due to the similarity of this theory to the deformation theory of plasticity (holonomic theory of plasticity), the existing criticism and comparisons between this theory and the flow theory of plasticity applies also to all the static friction laws. For dynamic problems we have to consider analogous relations between $S_T$ and $v_T$, where $v_T$ is the tangential velocity. In this case we may write the same laws as before by replacing $u_T$ by $v_T$. We can explain the nonconvexity of the boundary between the sliding and adhesive friction by means of analogous arguments to those of Green and Naghdi [20] in plasticity; thus a star-shaped friction boundary results. Moreover we can justify a nonconvex superpotential law of the form

$$-v_T \in \bar{\partial} W(S_T;S_N) \qquad (2.30)$$

if we assume that the elastic properties of the asperities change during the tangential deformation. The justification we shall present is analogous to the one given in nonconvex plasticity [12]. Let $S_T$ be the actual tangential stress vector assumed to be on the friction surface (Fig.2a) and let $S_T^*$ be on or inside the friction surface; we denote by K the admissible set for the $S_T$'s. An external action changes the $S_T^*$ into $S_T$ and let an additional action produce an additional displacement increment $-du_T$. Let $dS_T$ be the tangential stress vector increment which does not need to be tangential to the friction surface, if the latter changes with frictional deformation (analogously to the hardening effects in plasticity). Further the external action releases $dS_T$ and the tangential vector returns to the initial value $S_T^*$. We assume that for the stress path ABΓΔEA the work produced consists of one conservative

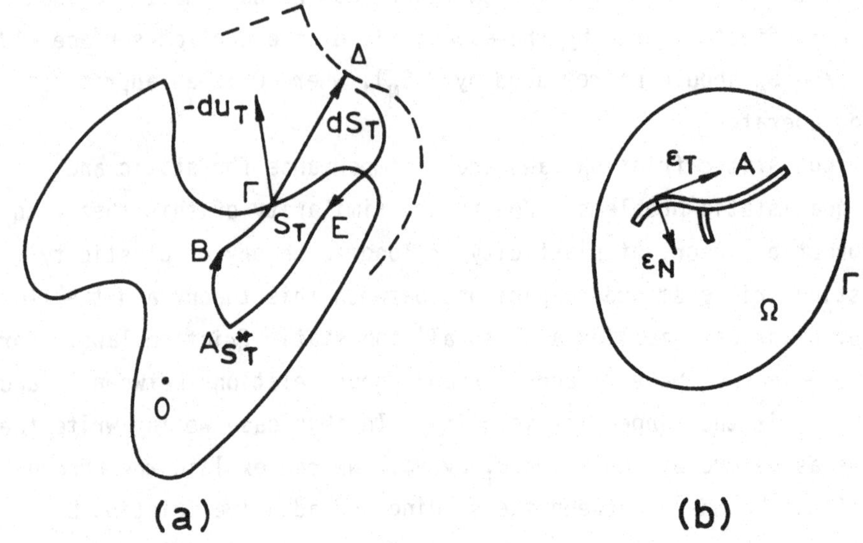


Fig. 2, Nonconvex Friction Laws. Interfaces

part, which does not depend on the path, and a nonconservative part. The first results from the elastic deformation of the asperities, whereas the second from irreversible local fractures causing a change of the elastic properties of the asperities. Let us assume that the nonrecoverable work $\tilde{w}$ is a function of $S_T, S_T^*$, of the path followed and of $S_N$. Let also $\tilde{w}=0$ if $S_T^*=S_T$. We assume that the work produced over ΑΒΓΔΕΑ is positive, i.e.

$$-\langle(S_T-S_T^*),du_T\rangle-\langle dS_T,du_T\rangle+\tilde{w}(S_T^*,S_T,\text{path},S_N)>0 . \tag{2.31}$$

Setting $S_T=S_T^*$, implies $-\langle dS_T,du_T\rangle>0$ and by making this expression as small as possible we get the inequality

$$-\langle S_T-S_T^*,du_T\rangle+\tilde{w}(S_T^*,S_T,\text{path},S_N)\geq 0 \quad \forall S_T^*\in K . \tag{2.32}$$

Now we assume the existence of a function $W:\mathbb{R}^3\rightarrow[-\infty,+\infty]$ such that

$$W^{\uparrow}(S_T,(S_T^*-S_T);S_N) = \begin{cases} \widetilde{W}(S_T,S_T^*,\text{path},S_N) & \text{if } S_T^*\in K \\ \infty & \text{otherwise} \end{cases} \qquad (2.33)$$

and thus (2.32) implies that

$$-du_T \in \partial W(S_T;S_N) \qquad (2.34)$$

or, if one replaces $du_T$ by the velocity $v_T$, (2.30) is obtained. Many dissipation mechanisms giving rise to nonmonotone stress-strain relations can be explained by the generalized (nonconvex) hypothesis of normal dissipation after an appropriate choice of hidden variables. This hypothesis, first developed in [12](p.157), is a direct generalization of the hypothesis of normal dissipation of B. Halphen and Nguyen Q. Son [21], see also [22], based on convexity and subdifferentiability. By means of this new hypothesis we can extend the theory of Rice of the viscoplastic potential, the convex theory of plasticity and viscoplasticity to the case of lack of convexity. For nonconvex locking materials we refer to [6].

## 2.4 Interface Laws, Adhesive Joints and Composite Materials

Let us assume that in a body $\Omega$ there is a surface A along which debonding and slip of the material is possible. The surface A may be a rock joint, a fault or an interface between different structural components, e.g. between the adjacent laminae in laminated structures, or between a fiber and the matrix in a fibre reinforced structure or between reinforcement and concrete etc. Much research work has been done on the simulation of the behaviour of an interface and on the influence of the interface on the equilibrium of the whole structure. Here we propose a model based on nonconvex superpotentials. We assume that the interface can be simulated by joint elements having two strain components: the normal strain $\varepsilon_N$ and the shearing strain $\varepsilon_T$ (Fig. 2b). Let $\sigma_N$ and $\sigma_T$ be the corresponding stresses. Then we may assume between $\sigma_N$ and $\varepsilon_N$, and between $\sigma_T$ and $\varepsilon_T$ the same laws as those of Fig. 1, where, now, $-S_N$ and $-S_T$ are replaced by $\sigma_N$ and $\sigma_T$ respectively (note the sign difference between boundary tractions and (internal) stresses). This treatment assumes A as a part of a given body $\Omega$ with different material behaviour. There is also another approach which is in use especially for laminated structures. We consider that each side of the interface belongs to a different body and then we assume on the interface contact and friction laws between $-S_N$ (resp. $-S_T$) and the relative normal (resp. tangential) displacement of the two bodies $[u_N]$(resp. $[u_T]$) expressed in terms of nonconvex superpotentials as in (2.22) and in (2.25). Of course uncoupling of the normal and the tangential action might be questionned; more general superpotential laws in the form

$$-\begin{bmatrix} S_N \\ \overline{S_T} \end{bmatrix} \in \bar{\partial}\zeta\left(\begin{bmatrix} u_N \\ u_T \end{bmatrix}\right) \tag{2.35}$$

could be considered. At this point we would like to note that diagrams of the form depicted in Fig. 1b simulate in the normal and in the tangential direction the delamination effect, as well as the interlaminar slip effect in laminated structures with adhesive material between the laminae and lead to the development of a technical theory for these structures [23].

If we want to study the behaviour of a structure under a given loading we may consider stress-strain diagrams of each element defined in the whole length of the strain axis (Fig. 3a) [24],[25]. Even, if only the final values $\varepsilon_0$ and $\varepsilon_0'$ are known, we may render the respective law complete by assuming a brittle (vertical lines) or a semibrittle behaviour beyond $\varepsilon_0$ and $\varepsilon_0'$. Accordingly, a complete law [25] has as a main advantage the consideration of several nonlinearities like cracking, crushing, delamination, slip etc. using a static or a quasi-static method. This approach may be applied to composite material structures, because it provides an estimation of the behaviour of the structure through purely static methods, as is the case in several engineering theories, e.g., the calculation of reinforced concrete structures etc. But this is possible only using superpotentials, since the sawtooth form is predominant in the stress-strain curves obtained experimentally in composite materials. The diagram of Fig. 3b describes the progressive, almost ideally brittle failure of the fibres in a fibre reinforced material [26]. In the peaks of the diagram fibre failure happens; the jumps (which are complete) correspond to the load transfer to the rest of the fibres. The upward parts are due to the strength recovery after the redistribution of the stresses between fibres and matrix. Slow pull-off tests of reinforced specimens of laminated structures give the diagram of Fig. 3c [27]. For other diagrams of this kind and their interpretation the reader is referred to [28],[29],[30].

It is obvious that all the aforementioned onedimensional phenomenological laws can be written through a nonconvex superpotential in the form

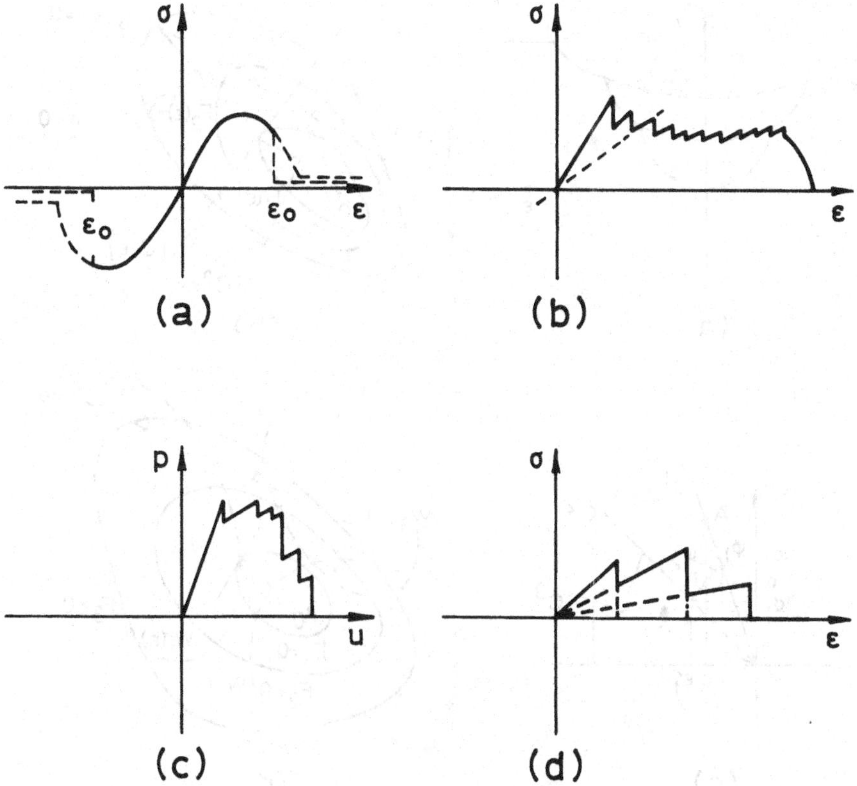


Fig. 3, Complete Diagrams. Behaviour of Fibre Reinforced Materials (Phenomenological Diagrams)

$\sigma \in \bar{\partial} w(\varepsilon)$, where w denotes the area between the graph and the horizontal axis on which the strains are measured.

Generalizations of the above onedimensional examples for threedimensional continua can be constructed in analogy to the ideally plastic and locking materials. Let us e.g. construct a threedimensional analogon to the onedimensional stress-strain diagram of Fig. 3d. One method would be to consider four elastic energy structures which intersect each other and form reentrant corners causing the stress jumps. Moreover the last energy surface should guarantee that $\sigma=0$ after the last jump (cf.Fig.4a).

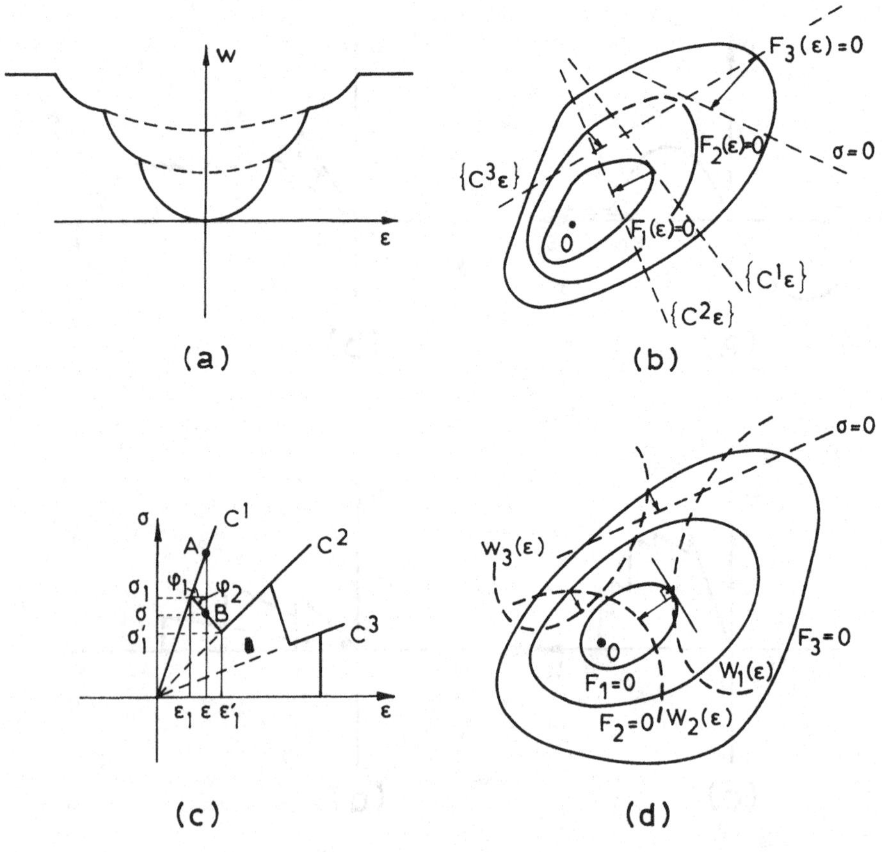


Fig. 4, Construction of Superpotentials for Composite Materials for 3D-Bodies

However, a more general definition of the superpotential of a composite material is the following: Let us define in the space $\mathbb{R}^6$ of the symmetric strain tensors $\varepsilon=\{\varepsilon_{ij}\}$ certain surfaces $F_i(\varepsilon)=0$ $i=1,2,\ldots,m$. We assume that the $F_i$'s are continuously differentiable. Moreover $K_1=\{\varepsilon|F_1(\varepsilon)\leq 0\}$ is a closed set. If $\varepsilon\in \operatorname{int} K_1=\{\varepsilon|F_1(\varepsilon)<0\}$ then we have a linear elastic stress-strain behaviour governed by the elasticity tensor $C^1=\{C^1_{ijhk}\}$. If $\varepsilon$ satisfies $F_1(\varepsilon)=0$ then we have a brittle fracture (vertical segment in Fig. 3d) and we assume that stress

$$\sigma=\{\sigma_{ij}\}=\{C^1_{ijhk}\varepsilon_{hk}\}+\lambda_1 \operatorname{grad}F_1(\varepsilon) \quad \lambda_1\in[-a_1,0] \tag{2.36}$$

and $a_1\geq 0$ satisfies the relation that

$$\{C^1_{ijhk}\varepsilon_{hk}\}-a_1 \operatorname{grad}F_1(\varepsilon)=\{C^2_{ijhk}\varepsilon_{hk}\}\ , \tag{2.37}$$

where $C^2=\{C^2_{ijhk}\}$ is the elasticity tensor of the subsequent linear elastic region. This region lies within the admissible strain set $K_2=\{\varepsilon|F_2(\varepsilon)\leq 0\}$. For $\varepsilon\in\{\varepsilon|F_1(\varepsilon)>0\ F_2(\varepsilon)<0\}$ we have that $\sigma_{ij}=C^2_{ijhk}\varepsilon_{hk}$ and for $\varepsilon$ lying on the boundary of $K_2$ we can again assume a law of the form (2.36). We continue in this way until the last brittle fracture surface, after which the stress becomes zero, i.e. the right-hand-side of (2.37) becomes zero (Fig. 4b). The aforementioned procedure can be modified to include the case of semibrittle fractures (Fig. 4c). We note that the segment AB in Fig. 4c is equal to $(\varepsilon-P_{K_1}\varepsilon)(\operatorname{tg}\phi_1+\operatorname{tg}\phi_2)$, where $P_{K_1}(\varepsilon)$ denotes the projection of $\varepsilon$ onto $K_1=\{\varepsilon|\varepsilon\leq\varepsilon_1\}$; in our case $P_{K_1}(\varepsilon)=\varepsilon_1$. Therefore we may generalize the law of Fig. 4c in three dimensions by writing that for $F_1(\varepsilon)\geq 0$ we have

$$\sigma=\{\sigma_{ij}\}=\{C^1_{ijhk}\varepsilon_{hk}\}-\mu_1\{\varepsilon_{ij}-(P_{K_1}\varepsilon)_{ij}\} \quad \mu_1>0\ . \tag{2.38}$$

Moreover for $F_1(\varepsilon)<0$ we have a linear elastic behaviour governed by the Hooke's tensor $C^1$. Relation (2.38) holds for all $\{\varepsilon_{ij}\}$ which satisfy the inequality

$$\{C^1_{ijhk}\varepsilon_{hk}\}-\mu\{\varepsilon_{ij}-(P_{K_1}\varepsilon)_{ij}\}\geq\{C^2_{ijhk}\varepsilon_{hk}\}\ . \tag{2.39}$$

Let us denote their set by $S^1$. Let $\{\varepsilon'_{1ij}\}$ be the value for which (2.39) is satisfied as equality. This procedure continues until the final surface $\sigma=0$ is reached. Here we may assume convexity and closedness of $K_1$, $K_2$ etc. in order to guarantee the uniqueness of the projection.

Note that in case of semibrittle fracture we may characterize the fracture by means of closed (and for simplicity convex) sets defined in the stress space by the inequalities $\bar{F}_1(\sigma)\leq 0$ $\bar{F}_2(\sigma)\leq 0$ etc. Indeed we have from Fig. 4c that $\varepsilon=\sigma\cot g\,\phi_1+(P_K\sigma-\sigma)(\cot g\phi_2+\cot g\phi_1)$ where $P_K\sigma$ is now the projection of $\sigma$ onto the set $\sigma\leqq\sigma_1$. The generalization is obvious (cf. also [12] p.97) and reads

$$\text{if } \bar{F}_1(\sigma)<0 \quad \text{then} \quad \varepsilon_{ij}=c^1_{ijhk}\sigma_{hk} \tag{2.40}$$

$$\text{if } \bar{F}_1(\sigma)\geq 0 \quad \text{then} \quad \varepsilon_{ij}=c^1_{ijhk}\sigma_{hk}-\bar{\mu}_1[\sigma_{ij}-(P_{\bar{K}_1}\sigma)_{ij}] \quad \bar{\mu}>0. \tag{2.41}$$

Here $c^1$ is the inverse tensor of $C^1$, $\bar{K}_1$ is the set $\{\sigma|\bar{F}_1(\sigma)\leq 0\}$. Relation (2.41) holds until the value $\{\sigma'_{1ij}\}$ is attained satisfying the relation

$$c^2_{ijhk}\sigma'_{1hk}=c^1_{ijhk}\,\sigma'_{1hk}-\bar{\mu}_1[\sigma'_{1ij}-(P_{\bar{K}_1}\sigma'_1)_{ij}]\ . \tag{2.42}$$

This scheme continues until the final surface $\sigma=0$ is attained. It is also worth noting that the linear elasticity laws can be replaced by nonlinear elasticity laws of the form $\sigma=\frac{\partial w}{\partial\varepsilon}$, or more generally by the monotone law $\sigma\in\partial w(\varepsilon)$, where w is a convex, l.s.c and proper functional and $\partial$ denotes the subdifferential (Fig. 4d). All the aforementioned threedimensional generalizations of the sawtooth onedimensional laws for composite materials, due to their formulation, can be put in the general forms

$$\sigma\in\bar{\partial}w(\varepsilon), \quad \text{or } \varepsilon\in\bar{\partial}\tilde{w}(\sigma)\ , \tag{2.43}$$

where w or $\tilde{w}$ are nonconvex superpotentials and $\bar{\partial}$ denotes the generalized gradient. The form of w is given in the case of (2.38) by the expression

$$w(\varepsilon)=\begin{cases}\frac{1}{2}\varepsilon_{ij}C^1_{ijhk}\varepsilon_{hk}-\mu_1\|\varepsilon-P_{K_1}\varepsilon\|^2\ (=A_1) \text{ if } \varepsilon\in S^1\\ A_1-\frac{1}{2}\varepsilon'_{1ij}C^2_{ijhk}\varepsilon'_{1hk}+\frac{1}{2}\varepsilon_{ij}C^2_{ijhk}\varepsilon_{hk}-\mu_2\|\varepsilon-P_{K_2}\varepsilon\|^2 \text{ if } \varepsilon\notin S^1, \varepsilon\in S^2\\ \text{etc.}\end{cases} \tag{2.44}$$

In the expressions, which we have formulated, $F_i$ and $\overline{F}_i$ are very general. We can, e.g., assume for them analogous forms to the ones for ideally locking and ideally plastic materials. In the case of brittle fractures we may assume that the mutual positions of the hyperplanes, defined by the elasticity tensors $C_i$, and of the gradients to the surfaces $F_i(\varepsilon)=0$ always give bounded stresses for every strain; then the nonconvex superpotential w is locally Lipschitz continuous. A deeper discussion would impose further restrictions on the forms of $\overline{F}_i$, $\overline{F}_1$, $C^1$ etc.; e.g. if one wanted to assure the feasibility of the previous geometric constructions (i.e. nonvoid intersections of hypersurfaces, finite $a_i$ in (2.37) etc.) and that $\overline{\partial}w(\varepsilon)$ (resp. $\overline{\partial}\tilde{w}(\sigma)$) is nonempty and bounded for each $\varepsilon$ (resp. $\sigma$).

All the above considerations hold for the case of loading. Obviously the unloading along an elastic path could be included in the same model, but then we have to work with stress and strain increments (cf. Sec.2.9).

## 2.5 Derivation of Hemivariational Inequalities and of Variational-Hemivariational Inequalities

In this Section we shall derive formally some hemivariational inequalities. We choose a mechanical problem, the problem of adhesive contacts of elastic bodies and for a given external loading we ask for the possible equilibrium positions. This is, for instance, the case of laminated structures; we have the delamination problem between the laminae, where a priori it is not known which parts between the laminae remain in contact and which not. If the diagram of Fig. 1b holds with $u_N$ replaced by $[u_N]$, we do not know a priori which parts of the adhesive have $(-S_N,[u_N])$ on AB, on AΓ (or AΓ'), or on ΓΔ (or Γ'Δ). Analogous is the situation concerning the interlaminar slip, with the difference that there is a coupling between delamination and slip which we shall discuss in Sec. 2.9.

Let $\Omega_m$ m=1,2,...,l be a set of deformable bodies possibly with different elasticity properties, with the boundaries $\Gamma_m$ m=1,2,...,l assumed to be appropriately regular. Let $x=\{x_i\}$ i=1,2,3 be a point of $\mathbb{R}^3$ and let $\sigma^{(m)}=\sigma^{(m)}_{ij}$ and $\varepsilon^{(m)}=\varepsilon^{(m)}_{ij}$ i,j=1,2,3 be the stress and strain tensors of the m-body. We denote by $f^{(m)}=\{f_i^{(m)}\}$ and $u^{(m)}=\{u_i^{(m)}\}$ the volume force and the displacement vector in each body. If $n^{(m)}=\{n_i^{(m)}\}$ is the outward unit normal vector to $\Gamma^{(m)}$, the boundary force on $\Gamma^{(m)}$ is $S_i^{(m)}= =\sigma_{ij}^{(m)} n_{ij}^{(m)}$ (summation convention). Let $S_N^{(m)}$ and $S_T^{(m)}$ be the normal and tangential components of it respectively. The corresponding displacement components are $u_N^{(m)}$ and $u_T^{(m)}$. The boundary $\Gamma^{(m)}$ is divided into three non-overlaping parts $\Gamma_U^{(m)}$, $\Gamma_F^{(m)}$ and $\Gamma_S^{(m)}$. On $\Gamma_U^{(m)}$ the displacements are given, i.e.,

$$u_i^{(m)} = U_i^{(m)} \quad \text{on} \quad \Gamma_U^{(m)} \tag{2.45}$$

on $\Gamma_F^{(m)}$ the forces are prescribed, i.e.,

$$S_i^{(m)} = F_i^{(m)} \quad \text{on} \quad \Gamma_F^{(m)} \tag{2.46}$$

and on $\Gamma_S^{(m)}$ - which corresponds to the interface of structure m with other substructures - nonmonotone interface conditions hold describing slip and delamination effects. We write in the general case $\Omega^{(m)} \subset \mathbb{R}^3$ the interface conditions in the form

$$-S_N^{(m)} \in \bar{\partial} j_{N(m)}(S^{(m)};[u_N^{(m)}]) \tag{2.47}$$

$$-S_T^{(m)} \in \bar{\partial} j_{T(m)}(S^{(m)};[u_T^{(m)}]) \tag{2.48}$$

in the normal and in the tangential direction to the interface. The superpotentials $j_N$ and $j_T$ are assumed to be functions of the interlayer gap $[u_N]$ and slip $[u_T]$ (locally Lipschitz continuous) respectively and of the interface stresses S (e.g. in the case of Coulomb's friction), which is also a function of u. Here, however, we assume that (2.47) and (2.48) are uncoupled, i.e. that $S^{(m)}$ is considered as having a given value, or that $j_{N(m)}$ and $j_{T(m)}$ do not depend on $S^{(m)}$. Then (2.47,48) are equivalent to the inequalities

$$j_{N(m)}^{o}([u_N^{(m)}],v-[u_N^{(m)}]) \geq -S_N^{(m)}(v-[u_N^{(m)}]) \quad \forall v \in R \tag{2.49}$$

$$j_{T(m)}^{o}([u_T^{(m)}],v-[u_T^{(m)}]) \geq -S_{T_i}^{(m)}(v_i-[u_T^{(m)}]) \quad \forall v_i \in R \quad i=1,2,3 . \tag{2.50}$$

In the framework of small strains and linear elastic behaviour for

$\Omega^{(m)}$ m=1,2,...,l we can write the relations

$$\sigma^{(m)}_{ij,j} + f^{(m)}_i = 0 \,, \tag{2.51}$$

$$\varepsilon^{(m)}_{ij} = \frac{1}{2}(u^{(m)}_{i,j} + u^{(m)}_{j,i}) = \varepsilon_{ij}(u^{(m)}) \,, \tag{2.52}$$

$$\sigma^{(m)}_{ij} = C^{(m)}_{ijhk}\varepsilon^{(m)}_{hk} \,. \tag{2.53}$$

The comma denotes the differentiation and Hooke's tensor $C^{(m)}=\{C^{(m)}_{ijhk}\}$ satisfies the well-known symmetry and ellipticity conditions. We write the principle of virtual work for every body $\Omega^{(m)}$ in the form

$$\int_{\Omega^{(m)}}\sigma^{(m)}_{ij}\varepsilon^{(m)}_{ij}(v^{(m)}-u^{(m)})d\Omega = \int_{\Omega^{(m)}}f^{(m)}_i(v^{(m)}_i-u^{(m)}_i)d\Omega +$$

$$+\int_{\Gamma^{(m)}_F}F^{(m)}_i(v^{(m)}_i-u^{(m)}_i)d\Gamma + \int_{\Gamma^{(m)}_S}\left[S^{(m)}_N(v^{(m)}_N-u^{(m)}_N) + S^{(m)}_{T_i}(v^{(m)}_{T_i}-u^{(m)}_{T_i})\right]d\Gamma$$

$$\forall v\in U^{(m)}_{ad} \,, \tag{2.54}$$

where $U^{(m)}_{ad}$ is the kinematically admissible set of $\Omega^{(m)}$, i.e.

$$U^{(m)}_{ad} = \{v^{(m)}\,|\,v^{(m)} = v^{(m)}_i, v^{(m)}_i\in U(\Omega^{(m)}), v^{(m)}_i = U^{(m)}_i \text{ on } \Gamma^{(m)}_U\} \,. \tag{2.55}$$

Here $U(\Omega^{(m)})$ denotes a space of functions defined on $\Omega^{(m)}$. Adding with respect to m all the expressions (2.54) and taking into account the interconnection of the bodies yields a relation of the form

$$\sum_{m=1}^{\ell}\int_{\Omega^{(m)}}\sigma^{(m)}_{ij}\varepsilon_{ij}(v^{(m)}-u^{(m)})d\Omega = \sum_{m=1}^{\ell}[\int_{\Omega^{(m)}}f^{(m)}_i(v^{(m)}_i-u^{(m)}_i)d\Omega +$$

$$+ \int_{\Gamma^{(m)}_F}F^{(m)}_i(v^{(m)}_i-u^{(m)}_i)d\Gamma] + \sum_{q=1}^{k}[\int_{\Gamma^{(q)}}S^{(q)}_N([v^{(q)}_N]-[u^{(q)}_N]d\Gamma +$$

$$+ \int_{\Gamma^{(q)}}S^{(q)}_{T_i}([v^{(q)}_{T_i}]-[u^{(q)}_{T_i}])d\Gamma] \quad \forall v\in U_{ad} \,, \tag{2.56}$$

where $U_{ad} = \bigcup_{m=0}^{\ell} U^{(m)}_{ad}$ .

In (2.56) we introduce the integrals along the joints $\Gamma_q$, q=1,...,k. The new enumeration of the $\Gamma_S^{(m)}$-boundaries has the advantage that finally the energy of each joint appears. Care should be taken of the fact that the variation of the energy of each constraint of the form (2.47) and (2.48) must appear only once in the last terms of (2.56). Further we introduce the elastic energy of the m-structure

$$a(u^{(m)},v^{(m)}) = \int_{\Omega^{(m)}} C^{(m)}_{ijhk}\varepsilon_{ij}(u^{(m)})\varepsilon_{hk}(v^{(m)})d\Omega \tag{2.57}$$

and by taking into account (2.49),(2.50) and (2.57) we obtain from (2.56) the following hemivariational inequality:
Find $u\in U_{ad}$ such as to satisfy

$$\sum_{m=1}^{\ell} a(u^{(m)},v^{(m)}-u^{(m)}) + \sum_{q=1}^{k} [\int_{\Gamma^{(q)}} [j^{o}_{N(q)}([u_N^{(q)}],[v_N^{(q)}]-[u_N^{(q)}]) +$$

$$+ j^{o}_{T(q)}([u_T^{(q)}],[v_T^{(q)}]-[u_T^{(q)}])]d\Gamma \geq \sum_{m=1}^{\ell} [\int_{\Omega^{(m)}} f_i^{(m)}(v_i^{(m)}-u_i^{(m)})d\Omega +$$

$$+ \int_{\Gamma_F^{(m)}} F_i^{(m)}(v_i^{(m)}-u_i^{(m)})d\Gamma] \quad \forall v\in U_{ad} . \tag{2.58}$$

This hemivariational inequality is the expression of the principle of virtual work in its inequality form for the structure under consideration.
To check in which sense a solution of (2.58) satisfies (2.51), the boundary conditions on $\Gamma_F^{(m)}$ m=1,...,l and the interface relations (2.47),(2.48) we must make the functional setting of the problem more precise. So we assume that $f_i^{(m)}\in L^2(\Omega^{(m)})$, $F^{(m)}\in L^2(\Gamma_F^{(m)})$, $C^{(m)}_{ijhk}\in L^\infty(\Omega^{(m)})$, $u_i^{(m)},v_i^{(m)}\in H^1(\Omega^{(m)})$ (classical Sobolev space).
Then $u_N^{(m)}$, $u_{T_i}^{(m)}\in H^{1/2}(\Gamma^{(m)})$ and $S_N^{(m)}$, $S_{T_i}^{(m)}\in H^{-1/2}(\Gamma^{(m)})$. We set in (2.58) $v_i^{(m)}-u_i^{(m)} = \pm\varphi_i^{(m)}$ where $\varphi_i^{(m)}$ belongs to the space of infinitely differentiable functions with compact support in $\Omega^{(m)}$, $\mathcal{D}(\Omega^{(m)})$. Then from (2.58) by setting $v_i^{(m)}-u_i^{(m)} = \pm\varphi_i^{(m)}$ for m=n and $v_i^{(m)}-u_i^{(m)} = 0$ for m≠n we obtain

$$a(u^{(n)},\phi^{(n)}) = \int_{\Omega^{(n)}} f_i^{(n)}\phi_i^{(n)}d\Omega \tag{2.59}$$

since $\varphi_i^{(n)}=0$ on $\Gamma^{(n)}$. (2.59) implies that (2.51) holds on $\Omega^{(n)}$ in the sence of distributions over $\Omega^{(n)}$. This procedure is repeated for n=1,2,...,1. Now applying the Green-Gauss theorem to each body we obtain the equality

$$a(u^{(m)},v^{(m)}-u^{(m)}) = \int_{\Omega^{(m)}} f_i^{(m)}(v_i^{(m)}-u_i^{(m)})d\Omega + \int_{\Gamma_F^{(m)}} S_i^{(m)}(v_i^{(m)}-u_i^{(m)})d\Gamma +$$
$$+ \int_{\Gamma_S^{(m)}} [S_N^{(m)}(v_N^{(m)}-u_N^{(m)}) + S_{T_i}^{(m)}(v_{T_i}^{(m)}-u_{T_i}^{(m)})]d\Gamma . \tag{2.60}$$

More correctly in (2.60) we should have instead of $\int_{\Gamma_F^{(m)}}$ and $\int_{\Gamma_S^{(m)}}$ the duality pairing <.,.> between $H^{1/2}(\Gamma)$ and $H^{-1/2}(\Gamma)$. From (2.60) and (2.58) we derive the inequality

$$\sum_{q=1}^{k} [\int_{\Gamma_q}[j^o_{N(q)}([u_N^{(q)}],[v_N^{(q)}]-[u_N^{(q)}]) + j^o_{T(q)}([u_T^{(q)}],[v_T^{(q)}]-$$
$$-[u_T^{(q)}])]d\Gamma] + \sum_{m=1}^{\ell} <S_i^{(m)}-F_i^{(m)},v_i^{(m)}-u_i^{(m)}>_{\Gamma_F^{(m)}} + \sum_{q=1}^{k}\{<S_N^{(q)},[v_N^{(q)}]-$$
$$-[u_N^{(q)}]>_{\Gamma_q} + <S_{T_i}^{(q)},[v_{T_i}^{(q)}]-[u_{T_i}^{(q)}]>_{\Gamma_q}\} \geq 0 \quad \forall v\in U_{ad} . \tag{2.61}$$

If in (2.61) we consider that on $\Gamma_F^{(m)}$, $v_i^{(m)}-u_i^{(m)} = \pm r_i^{(m)} \in H^{1/2}(\Gamma^{(m)})$ for m=n, and that $v_i^{(m)}-u_i^{(m)}=0$ for m≠n on $\Gamma_F^{(m)}$ and on $\Gamma_q$ for every q, we obtain $S_i^{(n)}=F_i^{(n)}$ as an equality in $H^{-1/2}(\Gamma^{(n)})$; this can be shown for every n. From (2.61) by setting $[v_N^{(q)}]-[u_N^{(q)}]=r_N^{(q)}$ on $\Gamma_q$ for q=n and the same difference =0 for q≠n, and setting $[v_T^{(q)}]-[u_T^{(q)}]=0$ on $\Gamma_q$ for every q we obtain

$$\int_{\Gamma_n} j^o_{N(n)}([u_N^{(n)}],r_N^{(n)})d\Gamma \geq -<S_N^{(n)},r_N^{(n)}>_{\Gamma_n} \quad \forall r_N^{(n)}\in H^{1/2}(\Gamma) \tag{2.62}$$

which constitutes a "weak" formulation of (2.47) on $H^{-1/2}(\Gamma)\times H^{1/2}(\Gamma)$. Analogously we obtain from (2.61) a weak form of (2.48).

It is noteworthy that the interface superpotential does not need to be a Lipschitz continuous function. Then Clarke's directional differentials $j_N^o(.,.)$ and $j_T^o(.,.)$ must be replaced in the previous hemivariational inequalities by $j_N^{\uparrow}(.,.)$ and $j_T^{\uparrow}(.,.)$. Analogously we proceed in dynamic problems on the assumption of small displacements. Then $f_i^{(m)}$ has to be replaced by $f_i^{(m)} - \rho^{(m)} \frac{\partial^2 u_i^{(m)}}{\partial t^2}$, where $\rho^{(m)}$ is the density of the m-body, and initial conditions for the displacements $u_i^{(m)}$ and the velocities $\partial u_i^{(m)}/\partial t$ have to be considered. The resulting hemivariational inequality is analogous to (2.58) and expresses the d'Alembert's principle in inequality form. In the dynamic case the holonomic interface relations (2.47) and (2.48) may be replaced by the relations

$$-S_N^{(m)} \in \bar{\partial} j_{N(m)}(S^{(m)};\ [\frac{\partial u_N^{(m)}}{\partial t}]), \tag{2.63}$$

$$-S_T^{(m)} \in \bar{\partial} j_{T(m)}(S^{(m)};\ [\frac{\partial u_T^{(m)}}{\partial t}]). \tag{2.64}$$

In this case we write again (2.54) but by considering instead of displacement variations, velocity variations. Thus we are led to a hemivariational inequality as the (2.58) but now having instead of $v^{(m)} -u^{(m)}$ and $[v_N^{(m)}]-[u_N^{(m)}]$ the variations $v^{(m)}- \frac{\partial u^{(m)}}{\partial t}$ and $[v_N^{(m)}]-[\frac{\partial u_N^{(m)}}{\partial t}]$. Suppose now that the substructures $\Omega^{(m)}$ $m=1,\ldots,l$ obey a general nonmonotone law of the form $\sigma^{(m)} \in \bar{\partial} W_{(m)}(\varepsilon)$, where $W_{(m)}$ is an extended real-valued function nonconvex and noneverywhere differentiable (cf. e.g. (2.43)). Then the variational expression of the problem is the same as (2.58) but now the term $\sum_{m=1}^{\ell} a(u^{(m)},v^{(m)}-u^{(m)})$ has to be replaced by $\sum_{m=1}^{\ell} \int_{\Omega^{(m)}} w_{(m)}^{\uparrow}(\varepsilon(u^{(m)}),\varepsilon(v^{(m)}-u^{(m)}))d\Omega$. If the $w_{(m)}$'s are convex superpotentials, i.e. are convex, l.s.c and proper functionals, then we define

$$W_{(m)}(\varepsilon) = \begin{cases} \int_{\Omega^{(m)}} w_{(m)}(\varepsilon)d\Omega & \text{if} \quad w_m(\cdot)\in L^1(\Omega^{(m)}) \\ \infty & \text{otherwise} \end{cases} \tag{2.65}$$

and we are led to a variational formulation analogous to (2.58); now the elastic energy function a(.,.) is replaced by the difference $\sum_{m=1}^{\ell}[W_m(\varepsilon(v^{(m)})) - W_m(\varepsilon(u^{(m)}))]$. This variational form is called variational-hemivariational inequality and is the expression of the principle of virtual work for the considered problem.

In all the above variational formulations we can assume that one or more of the $\Omega_i$'s are rigid. Thus we can incorporate into our model boundary conditions of the form (2.22) and (2.24).

## 2.6 Substationarity of the Potential and of the Complementary Energy

Let us consider the following substationarity problem:

Find $u\in U_{ad}$ such that the potential energy of the structure

$$\Pi(v) = \frac{1}{2}\sum_{m=1}^{\ell} a(v^{(m)},v^{(m)}) + \sum_{q=1}^{k}\int_{\Gamma_q}[j_{N_{(q)}}([v_N^{(q)}])+j_{T_{(q)}}([v_T^{(q)}])]d\Gamma -$$

$$- \sum_{m=1}^{\ell}\int_{\Omega^{(m)}} f_i^{(m)}v_i^{(m)}d\Omega - \sum_{m=1}^{\ell}\int_{\Gamma^{(m)}} F_i^{(m)}v_i^{(m)}d\Gamma \tag{2.66}$$

is substationary at v=u, where $v\in U_{ad}$. In other words $u\in U_{ad}$ is a solution of the inclusion

$$0\in \bar{\partial}\Pi(u) \quad \text{for} \quad v\in U_{ad} \,. \tag{2.67}$$

The following proposition holds.

Proposition 2.1: Suppose that the superpotentials $\xi\to j_{N(m)}(\xi)$ and $\xi\to j_{T(m)}(\xi)$ are locally Lipschitz and $\bar{\partial}$-regular for m=1,...,l. Then every solution of the substationarity problem is a solution of the hemivariational inequality (2.58) and conversely.

Proof: In the functional framework introduced in the previous section we can write (2.67) in the equivalent form

$$\ell \in \bar{\partial}\Pi_1(u) \quad \text{for} \quad v \in U_{ad} \tag{2.68}$$

where

$$(1,v) = \sum_{m=1}^{\ell} \int_{\Omega^{(m)}} f_i^{(m)} v_i^{(m)} d\Omega + \sum_{m=1}^{\ell} \int_{\Gamma^{(m)}} F_i^{(m)} v_i^{(m)} d\Gamma \tag{2.69}$$

and

$$\Pi(v) = \Pi_1(v) - (1,v) \; . \tag{2.70}$$

Now let us compute directly $\bar{\partial}\Pi_1(u)$ by using the definition (2.2) of $\tilde{\Pi}^o(u,v-u)$. Note that $A(u^{(m)}) = 1/2a(u^{(m)},u^{(m)})$ is $\bar{\partial}$-regular and

$$A^o(u^{(m)},v^{(m)}) = a(u^{(m)},v^{(m)}) \; . \tag{2.71}$$

Moreover let us denote by $J_{N(q)}$ the integral

$$J_{N(q)}([u_N^{(q)}]) = \int_\Gamma j_{N(q)}([u_N^{(q)}]) d\Gamma \; , \tag{2.72}$$

which may be either finite or infinite.
We shall show first that,

$$J^o_{N(q)}([u_N^{(q)}],[v_N^{(q)}]) = \int_\Gamma j^o_{N(q)}([u_N^{(q)}],[v_N^{(q)}]) d\Gamma \; . \tag{2.73}$$

In this proof we omit N,(q) for the sake of simplicity.
Let us denote by $g_{\lambda,h}$ the difference quotient

$$g_{\lambda,h}(u,v) = \frac{j(u+h+\lambda v) - j(u+h)}{\lambda} \; . \tag{2.74}$$

But $\xi \to j(\xi)$ is locally Lipschitz and thus

$$|g_{\lambda,h}(u,v)| \leq c|v| \; , \tag{2.75}$$

where c depends on the neighborhood of $(u+h)(x)$. We note that $u,v \in L^2(\Gamma)$ and that $\xi \to j(\xi)$ is continuous. Thus $x \to g_{\lambda,h}(u(x),v(x))$ is measurable. Accordingly we apply Fatou's lemma - for not necessarily integrable functions, see [31]p.152 - and we have (changing the signs) that

$$\int_\Gamma \limsup_{\substack{\lambda\to 0_+ \\ h\to 0}} (g_{\lambda,h} - c|v|) d\Gamma \geq \limsup_{\substack{\lambda\to 0_+ \\ h\to 0}} \int_\Gamma (g_{\lambda,h} - c|v|) d\Gamma \; . \tag{2.76}$$

Here c is a function of the neighborhood of $(u+h)(x)$ but disappears from both sides: Indeed we assume in (2.76) that $h\to 0$ taking values in a denumberable dense subset of $L^2(\Gamma)$, since this space is separable.Let us denote these functions by $h_i$ $i=1,\ldots,n\ldots$ and we have respective constants $c_i$ depending on the neighborhood of $(u+h_i)(x)$. Defining $\bar{c}>\limsup_{n\to\infty} c_i$ leads to the existence of M such that $n>M$ implies $c_n<\bar{c}$. Thus we can put in (2.76) a c independent of h namely $c=\max\{\bar{c},c_1,\ldots,c_M\}$. From (2.76) we obtain that

$$\int_\Gamma j^0(u,v)d\Gamma \geq J^0(u,v)\ , \tag{2.77}$$

where the integrals are either finite or infinite. We have also using the definition of limsup,Fatou's lemma,the $\bar{\partial}$-regularity of j, and (2.77), that

$$J^0(u,v)\geq \liminf_{\lambda\to 0_+} \frac{J(u+\lambda v)-J(u)}{\lambda} \geq \int_\Gamma \liminf_{\lambda\to 0_+} \frac{j(u+\lambda v)-j(u)}{\lambda}d\Gamma =$$

$$= \int_\Gamma \lim_{\lambda\to 0_+} \frac{j(u+\lambda v)-j(u)}{\lambda}d\Gamma = \int_\Gamma j'(u,v)d\Gamma = \int_\Gamma j^0(u,v)d\Gamma \geq J^0(u,v). \tag{2.78}$$

Accordingly we have shown (2.73). At this point it is worth noting that if j satisfies an inequality of the form

$$|j(u)-j(v)| \leq c(x)|u(x)-v(x)| \tag{2.79}$$

with $c(x)\in L^2(\Gamma)$, then $j(u(\cdot))\in L^1(\Gamma)$ and thus J is finite and Lipschitzian.(1)

From (2.71),(2.72) and the same equality for $J_T^0(.,.)$ we obtain the hemivariational inequality (2.58). Now we shall show the converse, i.e. that any solution of (2.58) is a solution of the substationarity problem (2.68). We show first that $J_N$ and $J_T$ are $\bar{\partial}$-regular: Indeed we have as in (2.78) due to Fatou's lemma(2)

(1) The finiteness of J may result from the properties of u on $\Gamma$, e.g. if $u\in C^0(\Gamma)$ as it happens for $\Gamma$ regular in plate theory $(H^2(\Omega)\subset C^0(\bar{\Omega})\subset C^0(\Gamma))$. Then $x\to j(u(x))$ is Riemann integrable.

(2) For the proof of the first equality in (2.81) the dominated convergence theorem is applied by some authors. The proof here seems to be more general. Moreover in (2.76) is omitted in the literature the proof that c is a function of a neighborhood of u and not of u+h.

$$J^o(u,v) \geq \liminf_{\lambda\to 0_+} \frac{J(u+\lambda v)-J(u)}{\lambda} \geq \int_\Gamma \liminf_{\lambda\to 0_+} \frac{j(u+\lambda v)-j(u)}{\lambda} =$$

$$= \int_\Gamma \tilde{j}'(u,v)d\Gamma = \int_\Gamma \limsup_{\lambda\to 0_+} \frac{j(u+\lambda v)-j(u)}{\lambda} \geq \limsup_{\lambda\to 0_+} \frac{J(u+\lambda v)-J(u)}{\lambda} \geq$$

$$\geq \liminf_{\lambda\to 0_+} \frac{J(u+\lambda v)-J(u)}{\lambda} . \tag{2.80}$$

Thus

$$\tilde{J}'(u,v) = \int_\Gamma \tilde{j}'(u,v)d\Gamma = \int_\Gamma j^o(u,v)d\Gamma = J^o(u,v) . \tag{2.81}$$

The $\bar{\partial}$-regularity of the functions of $\Pi_1$ implies that (the proof follows the same steps indicated in (2.78) and (2.80) but now for finite sums [1])

$$\Pi_1^o(u,v-u) = \tilde{\Pi}'(u,v-u) \tag{2.82}$$

and therefore (2.58) is equivalent, due to (2.82),(2.81) and the $\bar{\partial}$-regularity of $A(u^{(m)})$, to

$$\Pi_1^o(u,v-u) \geq (l,v-u) \quad \forall u \in U_{ad} , \tag{2.83}$$

which implies (2.68) q.e.d.

Note that if instead of (2.47) and (2.48) relations of the form

$$-[u_N^{(m)}] \in \bar{\partial} j^*_{N(m)}(S_N^{(m)}) \tag{2.84}$$

$$-[u_T^{(m)}] \in \bar{\partial} j^*_{T(m)}(S_T^{(m)}) \tag{2.85}$$

hold, then (2.49) and (2.50) are replaced by the inequalities

$$j^{*o}_{N(m)}(S_N^{(m)},T_N-S_N^{(m)}) \geq -[u_N^{(m)}](T_N-S_N^{(m)}) \quad \forall T_N \in \mathbb{R} , \tag{2.86}$$

and

---

(1) Instead of Fatou's lemma the inequality

$$\limsup_{y\to x}(f+g)(y) \leq \limsup_{y\to x} f(y) + \limsup_{y\to x} g(y)$$

is applied and the analogous one for liminf.

$$j^{*o}_{T(m)}(S^{(m)}_T, T_T - S^{(m)}_T) \geq -[u^{(m)}_{Ti}](T_{Ti} - S^{(m)}_{Ti}) \quad \forall T_{Ti} \in \mathbb{R} \quad i=1,2,3 \ . \tag{2.87}$$

For the problem defined by (2.45),(2.46),(2.51-53), and (2.84),(2.85) we can formulate a hemivariational inequality by considering stress variations $\tau^{(m)}_{ij} - \sigma^{(m)}_{ij}$ and the statically admissible set

$$\Sigma^{(m)}_{ad} = \{\tau^{(m)} | \tau^{(m)} = \{\tau^{(m)}_{ij}\}, \tau^{(m)}_{ij} \in \Sigma(\Omega^{(m)}), \tau^{(m)}_{ij,j} + f^{(m)}_i = 0 \text{ in } \Omega^{(m)},$$
$$T^{(m)}_i = F^{(m)}_i \text{ on } \Gamma^{(m)}_F\} \tag{2.88}$$

and by formulating the principle of complementary virtual work for the m-body. We obtain for the whole structure through addition that

$$\sum_{m=1}^{\ell} \int_{\Omega^{(m)}} \varepsilon^{(m)}_{ij} (\tau^{(m)}_{ij} - \sigma^{(m)}_{ij}) d\Omega = \sum_{m=1}^{\ell} \int_{\Gamma^{(m)}_U} U^{(m)}_i (T^{(m)}_i - S^{(m)}_i) d\Gamma +$$
$$+ \sum_{q=1}^{k} [\int_{\Gamma_q} [u^{(q)}_N](T^{(q)}_N - S^{(q)}_N) + \int_{\Gamma_q} [u^{(q)}_{Ti}](T^{(q)}_{Ti} - S^{(q)}_{Ti})] d\Gamma$$
$$\forall \tau \in \Sigma_{ad} \ . \tag{2.89}$$

Here $T_i = \tau_{ij} n_j$, $T_N$ and $T_T$ correspond to $S_N$ and $S_T$, $\Sigma_{ad} = \bigcup_{m=0}^{\ell} \Sigma^{(m)}_{ad}$, and $\tau = \{\tau^{(1)}, \ldots, \tau^{(\ell)}\}$. Moreover let us introduce the elastic energy of the m-structure expressed in terms of the stresses

$$A(\sigma^{(m)}, \sigma^{(m)}) = \int_{\Omega^{(m)}} c^{(m)}_{ijhk} \sigma^{(m)}_{ij} \sigma^{(m)}_{hk} d\Omega \ , \tag{2.90}$$

where $c^{(m)} = \{c^{(m)}_{ijhk}\}$ is the inverse tensor of $C^{(m)}$. From (2.86),(2.87), (2.89) and (2.90) we obtain the following hemivariational inequality which expresses the principle of complementary virtual work for the whole structure: Find $\sigma \in \Sigma_{ad}$ such as to satisfy

$$\sum_{m=1}^{\ell} A(\sigma^{(m)},\tau^{(m)}-\sigma^{(m)}) + \sum_{q=1}^{k} \int_{\Gamma_q} [j^{*o}_{N(q)}(S_N^{(q)},T_N^{(q)}-S_N^{(q)}) +$$

$$+ j^{*o}_{T(q)}(S_T^{(q)},T_T^{(q)}-S_T^{(q)})]d\Gamma \geq \sum_{m=1}^{\ell} \int_{\Gamma_U^{(m)}} U_i^{(m)}(T_i^{(m)}-S_i^{(m)})d\Gamma$$

$$\forall \tau \in \Sigma_{ad} \,. \tag{2.91}$$

Let $\Pi^*$ be the complementary energy of the structure

$$\Pi^*(\tau) = \frac{1}{2} \sum_{m=1}^{\ell} A(\tau^{(m)},\tau^{(m)}) + \sum_{q=1} \int_{\Gamma_q} [j^*_{N(q)}(T_N^{(q)}) + j^*_{T(q)}(T_T^{(q)})]d\Gamma -$$

$$- \sum_{m=1}^{\ell} \int_{\Gamma_U^{(m)}} U_i^{(m)} T_i^{(m)} d\Gamma \,. \tag{2.92}$$

Then the following substationarity problem is formulated:
Find $\sigma \in \Sigma_{ad}$ such that

$$0 \in \bar{\partial}\Pi^*(\tau) \quad \text{for} \quad \tau \in \Sigma_{ad} \,. \tag{2.93}$$

The relation between the solution of the hemivariational inequality (2.91) and the solution of the substationarity problem (2.93) obeys a proposition analogous to Prop. 2.1.

If each of the substructures $\Omega^{(m)}$ obeys a general nonmonotone law of the form $\varepsilon^{(m)} \in \bar{\partial} W^*_{(m)}(\sigma^{(m)})$ then the variational formulation of the problem results from (2.91) by replacing the quadratic terms by $\sum_{m=1}^{\ell} \int_{\Omega^{(m)}} W^{*\uparrow}_{(m)}(\sigma^{(m)},\tau^{(m)}-\sigma^{(m)})d\Omega$. If moreover, monotone laws $\sigma^{(m)} \in \partial W_{(m)}(\varepsilon^{(m)})$ hold, where the $w_{(m)}$'s are convex, l.s.c and proper functionals, then the conjugate functionals $w^c_{(m)}$ permit to write the above relation in the form $\varepsilon^{(m)} \in \partial w^c_{(m)}(\sigma^{(m)})$. Then we define

$$W^c_{(m)}(\sigma) = \begin{cases} \int_{\Omega^{(m)}} w^c_{(m)}(\sigma)d\Omega & \text{if} \quad w^c_m(\cdot)\ L^1(\Omega^{(m)}) \\ \infty & \text{otherwise} \end{cases} \tag{2.94}$$

and we are led to a variational formulation analogous to (2.91) but with the elastic energy replaced by $\sum_{m=1}^{\ell} [W^c_{(m)}(\tau^{(m)}) - W^c_{(m)}(\sigma^{(m)})]$, which is a variational-hemivariational inequality.

## 2.7 Applications of the Nonconvex Superpotentials to Reinforced Concrete and Cracked Structures

Until now we have given general examples showing the range of applicability of nonconvex superpotentials. Here we shall focus our attention on the reinforced concrete structures, where several nonmonotone and sawtooth stress-strain relations can be experimentally verified.

i) It is known that in the behaviour of reinforced concrete structures the phenomenon of concrete confinement plays an important role, especially in the case of large displacements and inelastic stress states [32]. Various analytical models based on experiments have been proposed to predict the concrete behaviour when it is confined by rectangular ties. The confinement of the concrete is characterized by zones of tensile stresses in the direction of confining forces. These zones appear both in the levels of ties and along the longitudinal reinforcement. The cross-section inside this area is the section of the effectively confined concrete (the shaded areas in Fig. 5). The boundary ABC of the effectively confined concrete is not a priori known and in the literature it is assumed to be a parabola. Using the theory of hemivariational inequalities we can determine this free boundary; the resulting free BVP has a great similarity with analogous problems in hydraulics as, e.g. the determination of the free surface in an earth dam, where the theory of variational inequalities plays an important role [9]. In order to construct an adequate model for the present problem it is necessary to have a "complete" law for the structural element, i.e. a stress-strain diagram which will be able to describe both the cracking and the crushing of the concrete, the behaviour of the steel and the behaviour of the interface between steel and concrete with the arising contact and sliding effects. Thus a diagram analogous to the one of Fig. 3a can be adopted for the concrete, where the two jumps correspond to the cracking and the crushing effects. The interface be-

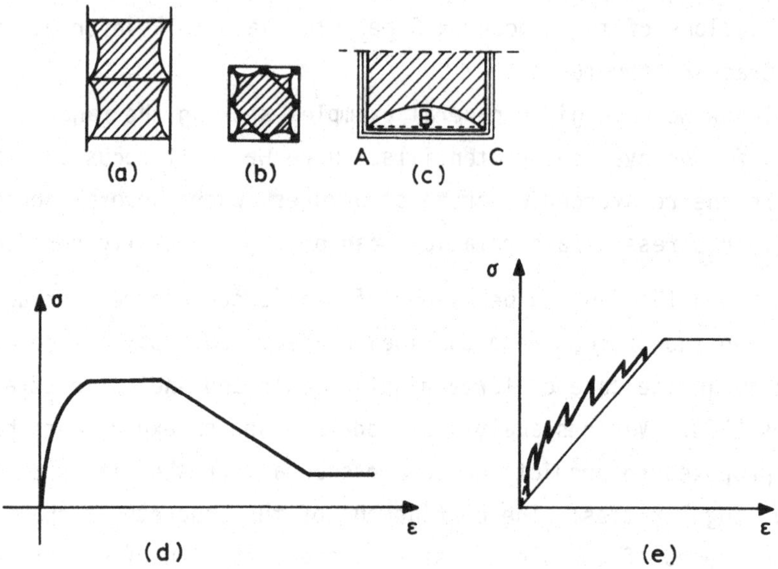


Fig. 5, On the Behaviour of a Reinforced Concrete Element

haviour can be described e.g. by the diagrams of Figs. 1b and 1e, whereas for the steel an appropriate [33] inelastic stress-strain diagram should be considered. From all the above it is apparent that the global behaviour of a reinforced concrete structure is described by a hemivariational inequality analogous to (2.58) or to (2.91), with the quadratic energy terms replaced by terms involving upper subderivatives (cf. Secs. 2.5, 2.6).

For the practical calculations concerning the concrete confinement the experimentally verified nonmonotone law of Fig. 5d is adopted. Obviously this law implies a nonmonotone moment-rotation diagram. Let us denote further by $\Pi_R$ the potential energy of the reinforcement and by $\Pi_{R/C}$ the interaction energy between reinforcement and concrete (e.g. representing sliding, lack of adhesion etc). On the assumption that $\Pi_R$ and $\Pi_{R/C}$ are convex the dynamic behaviour of the structure is governed by the hemivariational-variational inequality

$$\sum_{i=1}^{n} w_i^o(\varepsilon_i(u),\varepsilon_i(v)-\varepsilon_i(u))+\Pi_R(v)-\Pi_R(u)+\Pi_{R/C}(v)-\Pi_{R/C}(u) \geq$$

$$\geq (p-M\frac{d^2u}{dt^2}-c\frac{du}{dt})^T(v-u) \quad \forall v\in X_d . \tag{2.95}$$

Here $w_i$ is the energy of the i-element of the structure i=1,...,n, M (resp. C) is the mass (the damping) matrix and $X_d$ is the kinematically admissible set. If $\Pi_R$ and/or $\Pi_{R/C}$ are nonconvex then instead of the differences $\Pi_R(v)-\Pi_R(u)$ etc. the terms $\Pi_R^o(u,v-u)$ etc. appear.

ii) The theory of hemivariational inequalities combined with the "smeared-fracture" concept [17] permits the calculation of structures at the cracked stage and the determination of the regions where fracture (cracking or crushing) has occured. The structure (Fig. 6a) is considered to be discretized through a finite element scheme. At an integration point say $P_\rho$ (Fig. 6b) the material of the proportionate subdomain of the considered element is assumed to be either intact or crushed or cracked with an assumed pattern of fracture (e.g. equidistant parallel cracks in one or two given directions). The action of the fractured area is described by means of one or two fictitious springs at the integration points located normally to the given directions of the cracks and simulating the action of the cracks around the integration point under consideration (Fig. 6b).

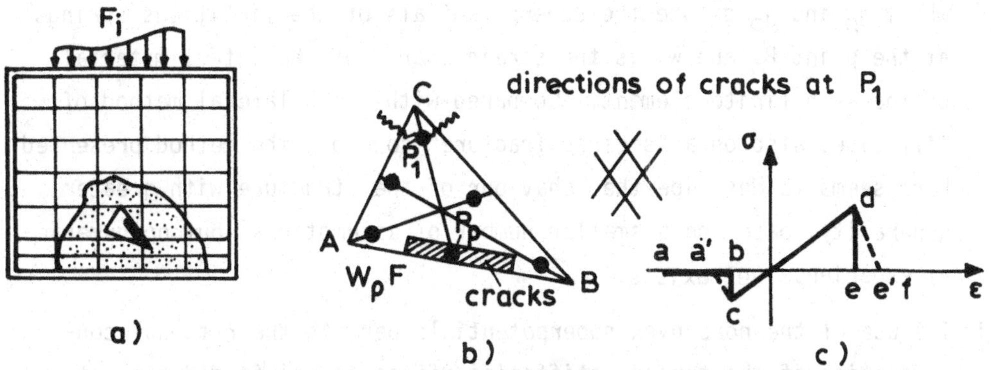


Fig. 6, On the Determination of Cracked Zones in a Structure

Note that for each crack direction we may introduce both a normal and a tangential spring in order to study the tangential interface effects (e.g. friction) at the crack. The directions of the fictitious springs may be different at every integration point but they indicate the directions of all the cracks in the area of the proportionate subdomain of the element. Thus, if $w_\rho$ is the weight coefficient of the integration point $P_\rho$ for the numerical integration and F is the area of the considered element ABC then the cracked or crushed area around $P_\rho$ is $w_\rho F$ (Fig. 6b). The aforementioned springs have a stress-strain law like the one depicted in Fig. 6c whose values are determined by the ultimate strength in tension and compression of the material. Further nonlinearities of the stress-strain law of the material may be incorporated into the classical finite element analysis of the structure. Due to the consideration of the law of Fig. 6c we can determine the areas which are cracked ($\sigma=0$ $\varepsilon\in[e,f)$), crushed ($\sigma=0$ $\varepsilon\in(a,b]$), intact ($(\sigma,\varepsilon)$ is on cd) or they are at an intermediate stage ($(\sigma,\varepsilon)$ on de or de' and on cb or ca'). The problem is described by the hemivariational inequality (cf. (2.95))

$$\sum_{i=1}^{n} w_i^o(\varepsilon_i(u),\varepsilon_i(v-u))+\sum_{\rho=1}^{n'}(j_\rho^o(u_\rho,v_\rho-u_\rho)+\tilde{j}_\rho^o(u_\rho,v_\rho-u_\rho))+$$
$$+\Pi_R(v)-\Pi_R(u)+\Pi_{R/C}^o(u,v-u)\geq p^T(v-u) \quad \forall v\in X_\rho \tag{2.96}$$

where $j_\rho$ and $\tilde{j}_\rho$ denote the superpotentials of the fictitious springs at the point $P_\rho$ and $w_i$ is the strain energy of the intact material of the i-th finite element. Compared with the bilateral method of [17] based also on a "smeared-fracture" concept, the method presented here seems to describe the behaviour of the structure with greater generality, based on a smaller number of assumptions, but no numerical experience yet exists.

iii) The use of the nonconvex superpotentials permits the rational consideration of the tension stiffening effect in reinforced concrete structures, i.e. of the effect representing the capacity of the intact concrete between adjacent cracks to sustain tensile stresses. For the study of this effect stress-strain laws of having the form of Fig. 1g (resp. Fig. 5e) are proposed for the concrete in tension (resp. the tension steel) [17], which obviously can be written in terms of nonconvex superpotentials (cf. e.g. (2.44)).

## 2.8 On the Properties of Solids Containing Voids or Cracks

The aim of the present section is to show the mode of inheritance of certain ultimate strength properties from the microscopic to the macroscopic level. Let us consider in the framework of small strain theory a nonlinear elastic solid characterized by the stress energy density function $\sigma \mapsto w(\sigma)$ assumed to be locally Lipschitz and $\bar{\partial}$-regular. Also let us consider a macroscopic volume element V of this solid which includes certain plane microcracks and traction-free voids. We denote by $V_M$ (resp. $V_v$) the region occupied by the material (resp. by the voids) and let $A_c$ be the planes of the cracks included into $V_M$. We denote here by $\Sigma$ and E the macroscopic stress and strains; generally, capital letters denote macroscopic quantities whereas minuscules denote local pointwise (called here in an abuse of the language "microscopic") quantities. The element V is loaded on its boundary $\Gamma$ by a uniform traction $\Sigma_{ij}n_j$ where $n=\{n_i\}$ is the local outward unit normal and $\Sigma$ is independent of the position. This loading causes at a microscopic level the strains $\varepsilon$, the stresses $\sigma$ and the displacements u. According to [35] the macroscopic strain is given by the formula

$$E_{ij} = \frac{1}{2V}\int_\Gamma (u_i n_j + u_j n_i)d\Gamma \ . \tag{2.97}$$

Assume now that

$$\varepsilon \in \bar{\partial} w(\sigma) \ . \tag{2.98}$$

If $\Sigma^*-\Sigma$ is a variation of $\Sigma$ we have, applying the principle of virtual work, that

$$E_{ij}(\Sigma^*_{ij}-\Sigma_{ij}) = V^{-1}\int_\Gamma u_i n_j(\Sigma^*_{ij}-\Sigma_{ij})d\Gamma = V^{-1}\int_\Gamma u_i n_j(\sigma^*_{ij}-\sigma_{ij})d\Gamma =$$

$$= V^{-1}\int_{V_M} \varepsilon_{ij}(\sigma^*_{ij}-\sigma_{ij})dV \leq V^{-1}\int_{V_M} w^o(\sigma,\sigma^*-\sigma)dV \ . \qquad (2.99)$$

Let us introduce the "macroscopic" superpotential

$$\Sigma \to W(\Sigma) = \frac{1}{V}\int_{V_M} w(\sigma)dV \ . \qquad (2.100)$$

Due to the $\bar{\partial}$-regularity of the locally Lipschitz microscopic superpotential $\sigma\to w(\sigma)$ we can show (see the proofs in Sec. 2.6) that

$$W^o(\Sigma,\Sigma^*-\Sigma) = \frac{1}{V}\int_{V_M} w^o(\sigma,\sigma^*-\sigma)dV \ . \qquad (2.101)$$

Thus (2.99) and (2.101) imply that

$$E\in\bar{\partial}W(\Sigma) \ , \qquad (2.102)$$

where $W$ is again $\bar{\partial}$-regular. Thus (2.98), (2.100) and (2.102) indicate how the properties of the $\varepsilon\to\sigma$ relation are inherited in the macroscopic $E$-$\Sigma$ relation. For instance, if $\varepsilon$-$\sigma$ has sawtooth parts simulating e.g. the friction on the crack planes, then (2.100) "transmits" this properties to the macroscopic level. The same happens obviously with the ultimate strength properties of the microscopic level: a $\varepsilon$-$\sigma$ law of the form of Fig. 6c is inherited by the $E$-$\Sigma$ level in the sense of (2.100) (consider e.g. a onedimensional homogeneous macroscopic element). Analogously we may argue for the relation between microscopic and macroscopic yield frunctions. As shown already (cf. [35] and the references given there), even if the local yield surface is independent of the mean stress, the macroscopic yield surface for a material containing voids and cracks will generally depend on $\Sigma_{ii}$, will be convex, and nonsmooth. Repeating the proof of (2.30) in Sec. 2.2 we can show that the convexity is no longer guaranteed if the state of voids and cracks changes with the elastic deformation (cf. [12] p.154) thus causing a change of the elastic properties of the macroscopic volume element. We leave this proof as an exercise for the reader.

## 2.9 V-superpotentials, Unloading, Numerical Methods and Miscellanea

i) The subdifferential $\partial$ and the generalized gradient $\bar{\partial}$ are the operators which give rise to variational and hemivariational inequalities when acting on superpotentials. The arising mechanical laws are multivalued and with respect to (2.3) and (2.4) $f(x_1)-f(x)$ and $f^0(x,x_1-x)$ express the virtual power (or work) of the constraint (or the law) for a generalized velocity x and for a virtual generalized velocity $x_1-x$. Let us put ourselves in the framework of (2.8). On the mechanical system $\Sigma$ a multivalued mechanical law of the form

$$-f\in W(u) \tag{2.103}$$

is introduced, which is by definition equivalent to the expression

$$W(u)=\{f\in F \mid G(u,v-u,\{B\},\{\beta\},\text{path},\text{history},\text{time derivatives},\ldots)\geqq$$
$$\geqq \langle f,v-u\rangle \quad \forall v\in U\}. \tag{2.104}$$

G is an appropriately defined function expressing the virtual power (or work) of the system. It may be of a nonlocal nature depending on the observable variables {B} and on the hidden variables {β}, on the "path" of the variation v-u, on the history of the mechanical system etc. The law (2.103) is called V-superpotential (virtual power superpotential) law and may serve for the following "Hypothesis of Dissipation".

(H) for every real thermodynamic process a V-superpotential W can be determined such that the "flux" $u\in U$ associated with the "force" $f\in F$ satisfies (2.103).

This hypothesis is a generalization of the hypothesis of normal dissipation of [21]. We can easily verify that if

$$G(u,-u,\{B\},\{\beta\},\text{path},\text{history},\ldots)\leqq 0 \tag{2.105}$$

then the inequality $f_iu_i\geqq 0$ expressing the second principle of thermodynamics is satisfied. If furthermore we assume that

$$G(u,v,\ldots)>G(u,o,\ldots) \quad \forall v\in U,\ v\neq 0 \tag{2.106}$$

then we can show that, if at u

$$G(u,o,\dots) \geqq 0 \tag{2.107}$$

then there corresponds to u at least one force f such as to satisfy (2.103) and conversely. Indeed if (2.103) holds we obtain from (2.104) for v=u (2.107). Conversely from (2.107) and (2.106) we obtain that $W(u)\neq\emptyset$, since F and U are finite dimensional spaces. A special case of the V-superpotentials are the superpotentials connected with Ioffe's fans [36]. Through the use of V-superpotentials we may define generalized viscoplastic and viscoelastic materials by following the method of [12] p.158. Defining also "normal" cones through the indicator we can adapt the ideas of [37] to this more general framework in order to obtain more realistic friction laws (cf. also [18]).

ii) Until this point we have not considered the case of unloading in the stress-strain relations which we have introduced. We recall that all these relations are generally multivalued and nonmonotone. In [38] a method was developed for the calculation of a structure for a given unloading path and the method of the "macroincrements" was developed. Here we shall present a more general method which takes into account all the possibilities of loading and/or unloading (not only along a given unloading path), i.e. it describes completely the phenomenon. This is possible by using appropriately defined multifunctions involving $\varepsilon,\sigma,\dot{\varepsilon}$ and $\dot{\sigma}$. Here the dot means the time partial derivatives (assumptions of small strains and displacements). Let us assume that the unloading is linear and that the modulus of elasticity changes with the strain. In Fig. 7 the onedimensional case is depicted or an equivalent stress-strain uniaxial law. At each point along the softening branch AB we pose the question whether $\dot{\varepsilon}=\{\dot{\varepsilon}_{ij}\}$ is positive, i.e. that we remain on AB, or negative and then the elastic unloading paths AC, A'C' etc. should be realized. If the stress-strain of an element are on OAB at the end of a load increment, then for the next load increment we can write that

$$\sigma\epsilon\ \bar{\partial}\varphi(\varepsilon) \text{ if } \varepsilon<\varepsilon_0, \text{ or if } \varepsilon\geqq\varepsilon_0 \text{ and } \dot{\varepsilon}\geqq 0 \tag{2.108}$$

$$\sigma= C(\varepsilon)\varepsilon \text{ if } \varepsilon\geqq\varepsilon_0 \text{ and } \dot{\varepsilon}< 0 . \tag{2.109}$$

If the stress and strain of an element are, e.g., on A'C' then

$$\sigma= C(\varepsilon')\varepsilon \tag{2.110}$$

where $\varepsilon'$ denotes the strain at A'. We may easily conclude that (2.108-110) can be put in the form

$$\sigma\in\bar{\partial}\varphi(\varepsilon)\psi(\varepsilon) + [\bar{\partial}\varphi(\varepsilon)\chi_+ + C(\varepsilon)(\varepsilon)\chi_-]q(\varepsilon,\sigma)(1-\psi(\varepsilon)) + \\ + C(\varepsilon')(\varepsilon)(1-q(\varepsilon,\sigma))(1-\psi(\varepsilon)) , \tag{2.111}$$

where $\psi(\varepsilon)=\{1 \text{ for } \varepsilon<\varepsilon_0,\ 0 \text{ for } \varepsilon\geqq\varepsilon_0\}$ and $\varepsilon_0$ is the strain for which irreversible strains appear, $\chi=\frac{\dot{\varepsilon}}{|\dot{\varepsilon}|}$ and $\chi_+$ and $\chi_-$ are the positive and the negative parts of $\chi$ respectively, $q(\varepsilon,\sigma)=\{1 \text{ if } (\varepsilon,\sigma)\in E,\ 0 \text{ otherwise}\}$ with $E=\{(\varepsilon,\sigma)\,|\,\sigma\in\bar{\partial}\varphi(\varepsilon)\}$. Note that in (2.111) $C(\varepsilon')$ is assumed as known and corresponds to $(\varepsilon,\sigma)\notin E$. The aforementioned thoughts hold for any type of nonmonotone stress-strain law with unloading, i.e. uniaxial or multiaxial laws. Thus, for instance, if (2.44) holds then the corresponding $\sigma,\varepsilon$ and $\dot{\varepsilon}$ satisfy an inclusion similar to (2.111) with the difference that $\psi(\varepsilon)=\{1 \text{ if } \varepsilon\in\text{elastic region},\ 0 \text{ otherwise}\}$. Note that a characterization of the elastic and plastic regions in terms of $\varepsilon$, $\sigma$ or both is necessary for the formulation of any loading-unloading law. The general form of such a law is a differential inclusion of the form

$$0\in F(\varepsilon,\sigma,\dot{\varepsilon}) , \tag{2.112}$$

where F is an appropriately defined multifunction. For instance, the classical elastoplastic law is a special case of (2.112). Let us know $\sigma$ and $\varepsilon$ for a given load p and let a new load $\dot{p}\delta t$ be imposed. Now for the total load $p+\dot{p}\delta t$ the stresses and strains become $\sigma+\dot{\sigma}\delta t$ and $\varepsilon+\dot{\varepsilon}\delta t$ and we may write generally an incremental relation of the form

$$\dot{\sigma}\in F_1(\varepsilon,\sigma,\dot{\varepsilon}) . \tag{2.113}$$

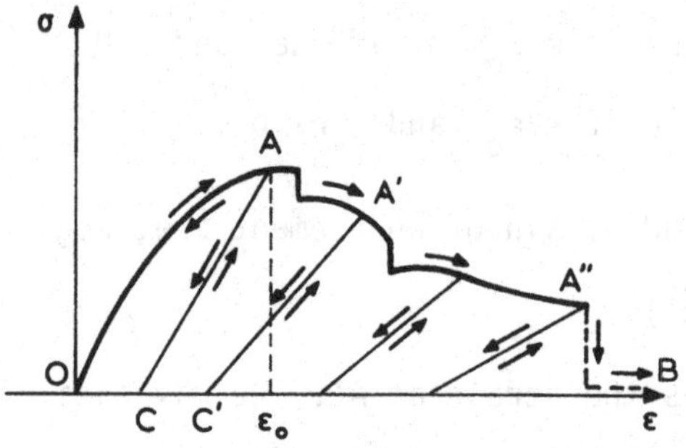


Fig. 7, Nonmonotone Law with Unloading

For (2.112) the B.V.P will be defined by the following relations in the framework of small strains and small displacements:
Find in $\Omega x(0,T)$ a displacement field u which satisfies

$$\rho\ddot{u}_i=\sigma_{ij,j}+f_i,\ \varepsilon_{ij}=\frac{1}{2}(u_{i,j}+u_{j,i}),\ 0\in F(\varepsilon,\sigma,\dot{\varepsilon}) \tag{2.114}$$

together with the initial conditions $u=u_0$, $\dot{u}=u_1$, $\sigma=\sigma_0$ and the boundary conditions e.g. given displacements on $\Gamma_U$ and given tractions on $\Gamma_F$ ($\Gamma=\Gamma_F\cup\Gamma_U\cup\Gamma_\phi$, mes$\Gamma_\phi=0$, $\Gamma_F$ and $\Gamma_U$ nonoverlapping). For (2.113) the corresponding B.V.P (e.g. the quasistatic one) reads:
Find in $\Omega x(0T)$ a displacement increment $\dot{u}$ such as to satisfy

$$\dot{\sigma}_{ij,j}+\dot{p}_i=0,\ \dot{\varepsilon}_{ij}=\frac{1}{2}(\dot{u}_{i,j}+u_{j,i}),\dot{\sigma}\in F_1(\varepsilon,\sigma,\dot{\varepsilon}) \tag{2.115}$$

together with the initial conditions $u=u_0$ and $\sigma=\sigma_0$ and the boundary condition (e.g. as before). Both BVPs are still open; it seems until now that some polynomial growth assumptions for some selections of the multifunctions play an important role in the discussion of the existence of the solution.

iii) Several methods have been proposed until now for the numerical treatment of the hemivariational inequalities. Let us begin for the moment with the monotone case and with a very special discrete problem, the elastic cable structures for small displacements (see e.g. [39],[40]):
The arising system of equalities and inequalities is equivalent to a quadratic programming (Q.P.) problem (cf. also [12] p.350).

The initial complete system of equalities and inequalities - and under the term "complete" we mean that we have to consider both the static and the kinematic inequalities as well as the complementarity relations - are the Kuhn-Tucker conditions for the Q.P. problem, which, if the matrix of the quadratic form is positive definite, are necessary and sufficient for the existence of the solution. The numerical treatment of such a cable-structure under a given loading seems to be difficult, because we do not know a priori which cables are slack and which not, and this thought might lead someone to the use of a combinatorial-like method. However things are more simple! Indeed the inequality constrained Q.P. problem when numerically solved by an appropriate Q.P. algorithm "decides" which cables are slack and which have tensile forces. For a positive definite matrix the solution is unique. Moreover any classical method (incremental-iterative, trial and error etc.) can be applied; its solution will be solution of the problem if it satisfies (cf. [41]) the complete set of Kuhn-Tucker conditions i.e. all the statical and kinematical equations and inequalities together with the complementarity conditions (stating that a cable will be either slack or will have a tensile force i.e. $s_i v_i=0$, where $s_i \geq 0$ is the cable force and $v_i \geq 0$ is the negative slackness of it). Any attempts for obtaining the solution through trial and error methods which do not control finally all the Kuhn-Tucker conditions may lead to a wrong result (cf. [12] Ch.10). Now let us assume that the final value of the loading is attained through a sequence of load increments $\dot{p}$. Suppose then that by using any classical solution scheme we find a displacement, a strain and a stress field. Obviously we have to check whether this "solution" satisfies the Kuhn-Tucker conditions and then we may conclude that it is the solution of the problem due to the uniqueness of it. However the classical schemes of incremental methods (cf. [42]) may not be able to give a result due to the horizontal and possibly vertical part of the graphs of the cable law (e.g. if the law $s_i \geq 0$, $v_i \geq 0$, $s_i v_i=0$ is considered).

Analogous questions arise in all problems expressed in terms of variational inequalities (monotone case) i.e. involving convex superpotentials. Again we may apply a minimization procedure, or any other classical incremental interative method always with the restrictions cited before: the final result must satisfy all the static and kinematic relations of the problem and the incremental method may fail to work for horizontal or vertical parts of the arising $\varepsilon$-$\sigma$ material laws. For these parts the use of Q.P. algorithms is advisable in order to overcome this difficulty (cf. e.g. [40],[41]). Moreover, in the case of large displacements or lack of uniqueness of the solution [43], one must be careful in the application of an incremental iterative method. The increments must be small enough to avoid the omission of equilibrium configurations during the incremental procedure. Indeed, because of stability questions, there is an inherent risk of jumps from one solution path to another (cf. e.g. the bifurcation mode of a simple cantilever beam). For nonconvex superpotentials and hemivariational inequalities we have analogous but more difficult questions (cf. also [44]). Let us consider first the purely static problem: The total load p is given and we want to determine all possible positions of equilibrium. The problem is formulated as a substationarity problem and every local minimum of the potential energy corresponds to an equilibrium configuration. There are several numerical difficulties concerning the determination of a local minimum of a nonconvex generally nondifferentiable function. We mention here as a possible tool the bundle method of Lemarechal-Strodiot [45] which is now at the stage of experimentation concerning its applicability to this problem. Analogously to the case of cable structures, and, let us say, with respect to the diagram of Fig. 1g, one could make some combinatorial attempts to find on which part of Fig. 1g the solution can appear. But the substationarity property of the solution gives the answer and no need of combinatorial thoughts exists. Another method related to the previous one is the regularization method whose convergence has been already discussed by the author in several problems ([46] ÷ [48]). Let us describe the method with respect to the hemivariational inequality (2.58). We assume that superpotentials $j_{N(q)}$ and $j_{T(q)}$ are approximated by the smooth functions $j_{N\varepsilon}$ and $j_{T\varepsilon}$ depending on $\varepsilon$ (we omit (q)), such that, as $\varepsilon \to 0$ $j_{N\varepsilon}$ (resp. $j_{T\varepsilon}$) tends to $j_N$ (resp. $j_T$). Any heuristic method for the regularization can be used: For instance the graphs of the corresponding stress-strain laws can be appropriately smoothened and vertical lines should be replaced by inclined lines. Due to the smoothness of $j_{N\varepsilon}$ and $j_{T\varepsilon}$, the derivatives $j_N^o(\xi,z)$, $j_T^o(\xi,z)$ are replaced by $\frac{dj_N(\xi)}{d\xi} z$ and $\frac{dj_T(\xi)}{d\xi} z$ and thus we are led to consider the following problem: Find $u_\varepsilon^{(m)}$, $m=1,2,\ldots,\ell$, such as to satisfy the variational equality

$$\sum_{m=1}^{\ell} a(u_\varepsilon^{(m)},v^{(m)}) + \sum_{q=1}^{k} \Big\{ \int_{\Gamma_q} \Big\{\Big[\frac{dj_{N\varepsilon(q)}}{du_N}([u_{N\varepsilon}^{(q)}])[v_N^{(q)}] + \frac{dj_{T\varepsilon(q)}}{du_T}([u_{T\varepsilon}^{(q)}])[v_T^{(q)}]\Big\} d\Gamma\Big\} - \sum_{m=1}^{\ell} \Big[\int_{\Omega^{(m)}} f_i^{(m)} v_i^{(m)} d\Omega + \int_{\Gamma_F^{(m)}} F_i^{(m)} v_i^{(m)} d\Gamma\Big] = 0$$

$$\forall v \in U_{ad}\ . \tag{2.116}$$

By means of discretization, (2.116) is equivalent to a system of semilinear algebraic equations depending on $\varepsilon$. The solution of this system "tends" as $\varepsilon \to 0$ to the solution of the initial problem. The system of nonlinear algebraic equations has the general form

$$Ku + B_\varepsilon(Tu) = p \tag{2.117}$$

where u (resp. p) is the total displacement (resp. load) vector and K is the stiffness matrix. Here T is a transformation matrix such that Tu gives the appropriate relative displacements and $B_\varepsilon(\cdot)$ is the vector of the nonlinear terms depending on $\varepsilon$. Noteworthy is the sparsity of the nonlinear terms which in some cases only slightly influences the monotonicity of the problem. In this case the classical Newton-Raphson algorithm has given acceptable results. However, it remains still an open question the testing of other types of algorithms like the fixed point algorithms and the homotopy algorithms. At this point let us make a parenthesis and let us examine what happens if the superpotentials depend on $S^{(q)}$ which itself is an unknown function of u. Then the foregoing methods cannot be used. This happens, for instance in the case of real Coulomb's friction. For such problems the following algorithm is proposed, which seems to be reasonable from the standpoint of mechanics, on the assumption that in (2.58) $j_N$ depends on $S_T$ and $j_T$ depends on $S_N$. We split the problem into the following two subproblems: In the first (resp. the second) we assume that $S_T^{(q)} = C_{T1}^{(q)}$,(resp.$S_N^{(q)}=C_{N2}^{(q)}$),where $C_{T1}^{(q)}$ (resp. $C_{N2}^{(q)}$) is given and we solve (2.58) with $\overset{o}{j}_{T(q)} = 0$ (resp. $\overset{o}{j}_{N(q)} = 0$) and with the terms $\int_{\Gamma_q} C_{T1i}^{(q)}(v_{Ti}^{(q)}-u_{Ti}^{(q)})d\Gamma$ (resp. $\int_{\Gamma_q} C_{N2}^{(q)}(v_{N2}^{(q)}-u_{N2}^{(q)})d\Gamma$) $q=1,\dots k$, added in the right-hand side of it. Obviously, the first problem is the normal contact problem and the second is the tangential contact problem. From the first problem we obtain a value for $S_N^{(q)}$, say $C_{N2}^{(q)}$, which is used for the solution of the second problem. From it we obtain a value of $S_T^{(q)}$, say $C_{T3}^{(q)}$, which is used for the solution of the first problem, and so on, until the difference of the displacements of two consecutive steps becomes appropriately small. The proposed algorithm recalls the fixed point algorithm of [49] proved in [50], and the decomposition techniques in multilevel optimization [51].

Let us come back now to the numerical implementation of hemivariational inequalities with strong nonmonotonicities. In this case the microspring or sawtooth approximation of the decreasing branches is proposed by Koltsakis [52]. Then the Newton-Raphson algorithm is applied. This idea is roughly described in Fig. 8a. The dotted

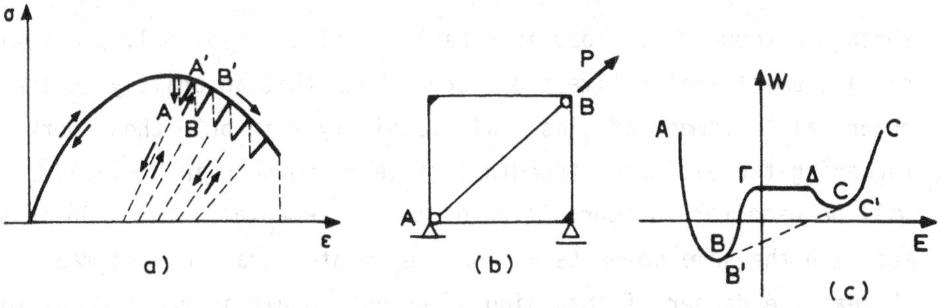


Fig. 8, Microspring Approximation of Decreasing Branches and the Question of Stability

vertical parts of the graph denote local jumps and the decreasing branch is "attained" by small monotone increasing "stress recovery" parts AA',BB'. The physical explanation of this approximation is obvious and is based on the idea of gradual failure of the material on the decreasing branch. Small numerical examples solved by this method are given in [23] and in [53]. The method is fully examined in [52] and has certain advantages concerning its ability to describe the localization of the failure. However there are considerable numerical difficulties in obtaining numerically all the solutions of the problem. At this point we would like to remark that a structure may have a position of overall equilibrium in which one ore some elements of the structure are in unstable equilibrium. For instance the framework of Fig. 8b with strong linear elastic vertical and horizontal beams is always in equilibrium under the loading P even if the diagonal AB has a stress-strain law derived by the potential function of Fig. 8c. Note that a simple rod-element having this strain energy would have as a position of stable equilibrium the point B and as other possible position of equilibrium the point C and all the points between Γ and Δ (ΓΔ is an horizontal line segment). Indeed all these positions of equilibrium are substationarity points. It is also noteworthy that a convexification of the energy would eliminate several positions of equilibrium (dotted line in Fig. 8c). Let us now assume that the final value of the loading is attained through a sequence of load increments. Let us then apply a classical incremental-iterative procedure. Note that in analogy to the numerical treatment of classical plasticity with or without work-hardening the ultimate strength surface introduced in Sec. 2.4 will be used for the correction of the incremental paths. On this approach the same comments as for the monotone case can be made, whereas the danger of obtaining a "wrong" result is greater due to the lack of monotonicity of the problem. It is well-known (see e.g. [44]) from the numerical work on nonmonotone laws with the classical incremental-iterative methods that severe numerical difficulties appear especially if one wants to take into account all the causes of structural failure in a finite element scheme, for instance complete vertical jumps in a stress-strain diagram for which the classical incremental-iterative schemes may have difficulties. Analogously to the large displacement theory of cable structures where an incremental-iterative procedure is combined with a Q.P. problem [42],[43], we propose here combination of the incremental-iterative procedures with a substationarity problem which can take into consideration irregularities like jumps; then the numerical solution within the increment in which the substationarity problem is considered is obtained from a local optimization problem or from a regularization procedure. Closing this Section we would like to comment that the numerical efforts concerning the hemivariational inequalities are still at the stage of experimentation.

## 3. STUDY OF VARIATIONAL HEMIVARIATIONAL INEQUALITIES

### 3.1 Delamination Effects of Laminated von Kårmån plates. Formulation of the Problem

In this Section we show the general method for the proof of existence and approximation of the solution of a problem connected with nonconvex superpotentials. This is achieved by studying a mechanical problem: the delamination effect for laminated plates undergoing large displacements (von Kårmån plates) is studied. Recall that the delamination, which is mainly due to the normal interlayer stresses [54], is one of the main causes of strength-degradation of composite plates. Here the mechanical behaviour of the interlayer binding material, together with the possibility of local cracking and crushing effects, are described by a nonmonotone, possibly multivalued law between the interlaminar bonding forces with the corresponding relative displacements. This law can be derived from a nonconvex superpotential and leads to a hemivariational inequality. Here in order to study a larger and more realistic class of problems we assume at the boundary of the von Kármán plate monotone boundary conditions which include e.g. the Signorini-Fichera boundary condition, the plastic hinge condition, etc. (see e.g. [7],[12]). These conditions are derived through subdifferentiation from a convex generally nondifferentiable and nonfinite functional, the convex superpotential of Moreau.

Unilateral problems for a single plate have been studied for monotone boundary conditions in [7] and [12] and for nonmonotone boundary conditions in [47]. A laminated-plate theory concerning Kirchhoff plates (small displacements - no stretching) has been presented in [55],[56]. We refer also at this point the work of Naniewicz and Woźniak [57],[58],[59] on the hemivariational inequalities and their applications to the debonding processes in layered composites.

Let us consider a laminated plate consisting of two laminae and the binding material between them (Fig. 9). In the undeformed state the middle surface of each lamina, say the j one, occupies an open, bounded and connected subset $\Omega_j$ of $\mathbb{R}^2$. $\Omega_i$ is referred to a fixed Cartesian right-handed coordinate system $0x_1x_2x_3$. The boundary of the j-th lamina $\Gamma_j$ is assumed to be a Lipschitz boundary $C^{0,1}$. The interlaminar binding material occupies a subset $\Omega'$ such that $\Omega' \subset \Omega_1 \cap \Omega_2$ and $\bar{\Omega}' \cap \Omega_1 = \emptyset$, $\bar{\Omega}' \cap \Omega_2 = \emptyset$. Let us denote by $\zeta^{(j)}(x)$ the vertical deflection of the point $x \in \Omega_j$ of the j-th lamina, by $f^{(j)} = (0,0,f_3^{(j)}(x))$ the distributed vertical load on the j-th lamina, and by $u^{(j)} = \{u_1^{(j)}, u_2^{(j)}\}$ the in-plane displacements of the j-th lamina. The j-th lamina has constant thickness $h_j$, while the interlaminar binding layer has constant thickness h. For each lamina the von Kármán plate theory is adopted, i.e. each lamina is regarded as a thin plate undergoing large deflections. Thus we have the following system of differential equations:

$$K_j\Delta\Delta\zeta^{(j)}-h_j(\sigma^{(j)}_{\alpha\beta}\zeta^{(j)}_{,\beta})_{,\alpha} = f^{(j)} \quad \text{in } \Omega_j , \tag{3.1}$$

$$\sigma^{(j)}_{\alpha\beta,\beta} = 0 \quad \text{in } \Omega_j , \tag{3.2}$$

$$\text{and} \quad \sigma^{(j)}_{\alpha\beta} = C^{(j)}_{\alpha\beta\gamma\delta}(\varepsilon^{(j)}_{\gamma\delta}(u^{(j)})+\tfrac{1}{2}\zeta^{(j)}_{,\gamma}\zeta^{(j)}_{,\delta}) \quad \text{in } \Omega_j . \tag{3.3}$$

The subscripts $\alpha,\beta,\gamma,\delta = 1,2$ indicate the coordinate directions; $\{\sigma^{(j)}_{\alpha\beta}\}$, $\{\varepsilon^{(j)}_{\alpha\beta}\}$ and $\{C^{(j)}_{\alpha\beta\gamma\delta}\}$ are the stress, strain and elasticity tensors in the plane of the j-lamina. The components of $C^{(j)}$ are assumed to be elements of $L^{\infty}(\Omega_j)$ and they satisfy the well-known symmetry and ellipticity properties of the Hooke's tensor. Moreover, $K_j = Eh_j^3/12(1-\nu^2)$ is the bending rigidity of the j-th plate with E Young's modulus and $\nu$ Poisson's ratio. Here we consider isotropic homogeneous plates of constant thicknesses. However, in place of (3.1-4) we could assume the appropriate equations for orthotropic or anisotropic plates in bending. The mathematical treatment is the same. In order to model the action of the interlaminar normal stresses $\sigma_{33}$, $f^{(j)}$ is splitted into $\bar{f}^{(j)}$, which describes the interaction of the two plates, and $\bar{\bar{f}}^{(j)}\in L^2(\Omega_j)$, which represents the external loading of the j-th plate:

$$f^{(j)} = \bar{f}^{(j)} + \bar{\bar{f}}^{(j)} \quad \text{in } \Omega_j, \quad j=1,2 . \tag{3.4}$$

If f denotes the stress in the interlaminar binding layer, we have that

$$f = \bar{f}^{(1)} = -\bar{f}^{(2)} \quad \text{in } \Omega' . \tag{3.5}$$

Then a phenomenological law connecting f with the relative deflection of the plates (see Fig. 9b)

$$[\zeta] = \zeta^{(1)} - \zeta^{(2)} \tag{3.6}$$

is introduced. Let f be a multivalued nonmonotone function $\hat{\beta}$ (cf. Fig. 10a), of $[\zeta]$ i.e.

$$-f\in\hat{\beta}([\zeta]) \quad \text{in } \Omega' . \tag{3.7}$$

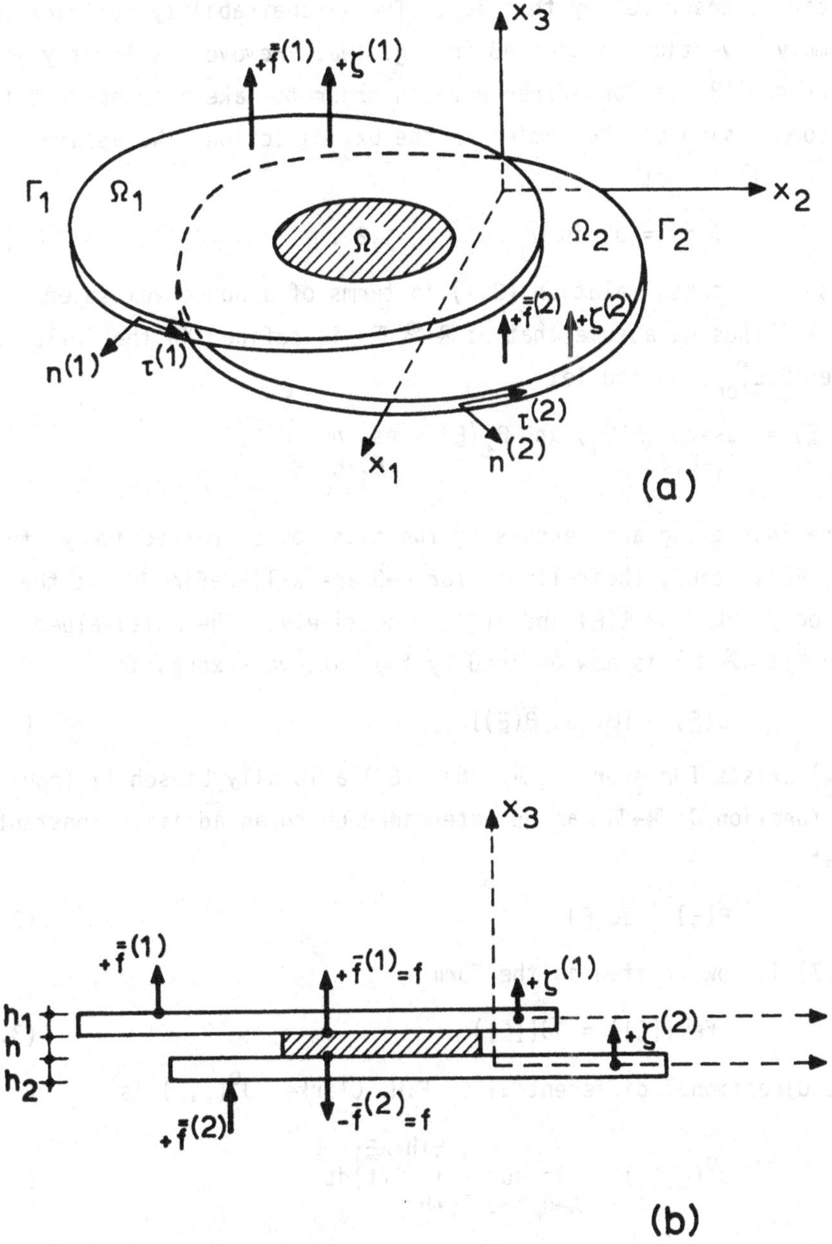


Fig. 9, The Laminated Plate Geometry

Note that cracking as well as crushing effects of brittle or semi-brittle nature can be described by this law. The impenetrability restriction would imply a vertical branch AB in Fig. 10a. However, a lightly inclined branch AB' is considered here in order to take into account the slight compression of the lamina in the $Ox_3$-direction. We assume further for $\bar{f}^{(j)}$ that

$$\bar{f}^{(j)} = 0 \quad \text{in } \Omega_j - \Omega', \quad j=1,2 . \tag{3.8}$$

Now we shall express relation (3.7) in terms of a nonconvex superpotential. Thus we assume that $\hat{\beta}: \mathbb{R} \to \mathcal{P}(\mathbb{R})$ is defined in the following way. Let $\beta \in L^{\infty}_{loc}(\mathbb{R})$ and let

$$\bar{\beta}_{\varepsilon}(\xi) = \operatorname*{esssup}_{|\xi_1-\xi| \leq \varepsilon} \beta(\xi_1) \text{ and } \underline{\beta}_{\varepsilon}(\xi) = \operatorname*{essinf}_{|\xi_1-\xi| \leq \varepsilon} \beta(\xi_1) . \tag{3.9}$$

These are increasing and decreasing functions of $\varepsilon$, respectively, for any $\varepsilon>0$, $\xi \in \mathbb{R}$; thus, their limits for $\varepsilon \to 0$ are well-defined; let these limits be denoted by $\bar{\beta}(\xi)$ and $\underline{\beta}(\xi)$ respectively. The multivalued function $\hat{\beta}: \mathbb{R} \to \mathcal{P}(\mathbb{R})$ is now defined by the interval-expression

$$\hat{\beta}(\xi) = [\underline{\beta}(\xi), \bar{\beta}(\xi)] . \tag{3.10}$$

If $\beta(\xi_{\pm 0})$ exists for every $\xi \in \mathbb{R}$, then [60] a locally Lipschitz (nonconvex) function $J: \mathbb{R} \to \mathbb{R}$ can be determined up to an additive constant, such that

$$\hat{\beta}(\xi) = \bar{\partial} J(\xi) . \tag{3.11}$$

Thus (3.7) is now written in the form

$$-f \in \hat{\beta}([\zeta]) = \bar{\partial} J([\zeta]) . \tag{3.12}$$

Here the directional differential of F.H. Clarke, $J^0(.,.)$ is

$$J^0(\xi,\xi_1) = \limsup_{\lambda \to 0_+,\, h \to 0} \frac{1}{\lambda} \int_{\xi+h}^{\xi+h+\lambda\xi_1} \beta(t)\,dt . \tag{3.13}$$

Moreover, we assume that the following boundary conditions hold on the subset $\tilde{\Gamma}_j$ of the plate boundaries:

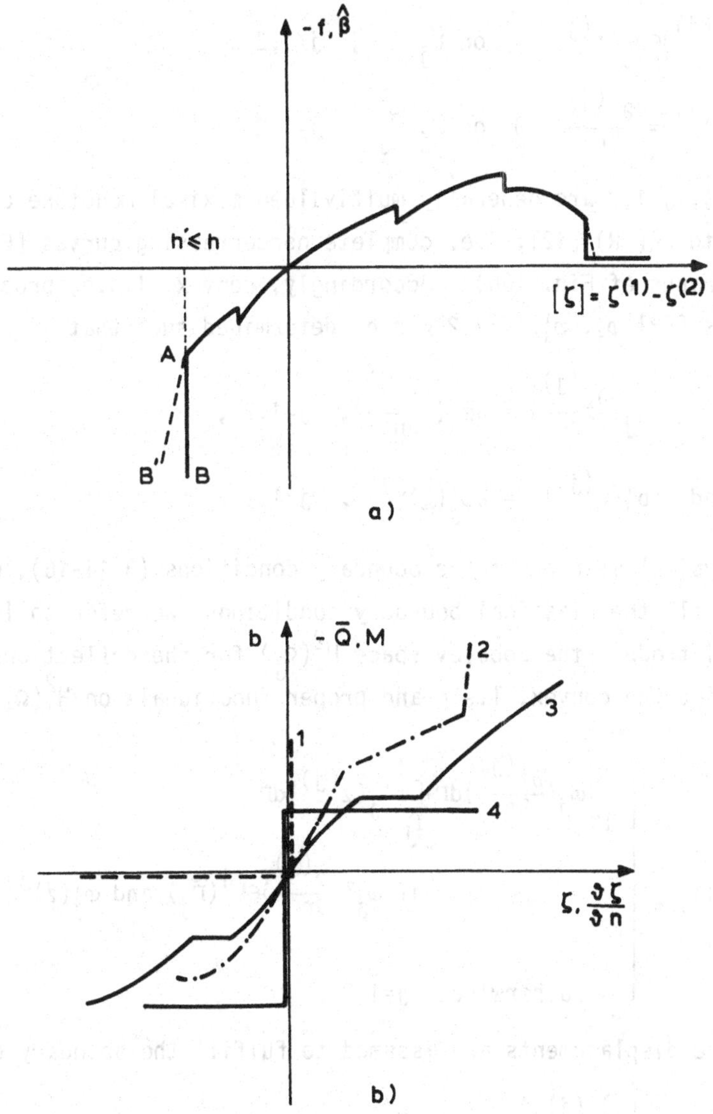


Fig. 10, The Binding Behaviour of the Interlaminar Adhesive (a) and Subdifferential Boundary Conditions (b)

$$M_j(\zeta^{(j)}) \in b_j(\frac{\partial \zeta^{(j)}}{\partial n} \quad \text{on } \tilde{\Gamma}_j \quad , \quad j=1,2 \; , \tag{3.14}$$

$$-Q_j(\zeta^{(j)}) \in b_j'(\zeta^{(j)} \quad \text{on } \tilde{\Gamma}_j \quad , \quad j=1,2 \; , \tag{3.15}$$

$$\text{and } \zeta^{(j)} = \frac{\partial \zeta^{(j)}}{\partial n} = 0 \quad \text{on } \Gamma_j - \tilde{\Gamma}_j \; , \quad j=1,2 \; . \tag{3.16}$$

Here $b_j$, $b_j'$, j=1,2 are generally multivalued maximal monotone operators from $\mathbb{R}$ into $\mathcal{P}(\mathbb{R})$ [12], i.e. complete nondecreasing curves [61] (see also the graphs of Fig. 10b). Accordingly, convex, l.s.c, proper functionals [62] $\varphi_j$, $\varphi_j'$, j=1,2 can be determined such that

$$b_j(\frac{\partial \zeta^{(j)}}{\partial n}) = \partial \varphi_j(\frac{\partial \zeta^{(j)}}{\partial n}), \quad j=1,2 \; , \tag{3.17a}$$

$$\text{and} \quad b_j'(\zeta^{(j)}) = \partial \varphi_j'(\zeta^{(j)}) \; , \quad j=1,2 \; . \tag{3.17b}$$

For the physical meaning of the boundary conditions (3.14-16), which include also all the classical boundary conditions, we refer to [7],[12]. Futher we introduce the Sobolev space $H^2(\Omega_j)$ for the deflections $\zeta^{(j)}$ and we define the convex, l.s.c and proper functionals on $H^2(\Omega_j)$, j=1,2:

$$\Phi_j(z^{(j)}) = \begin{cases} \int_{\tilde{\Gamma}_j} \varphi_j(\frac{\partial z^{(j)}}{\partial n}) d\Gamma + \int_{\tilde{\Gamma}_j} \varphi'_j(z^{(j)}) d\Gamma, \\ \qquad \text{if } \varphi_j(\frac{\partial z^{(j)}}{\partial n}) \in L^1(\tilde{\Gamma}_j) \text{ and } \varphi_j'(z^{(j)}) \; L^1(\Gamma_j) \\ \infty \quad \text{otherwise}, \quad j=1,2 \; . \end{cases} \tag{3.18}$$

The in-plane displacements are assumed to fulfill the boundary conditions

$$\sigma_{\alpha\beta}^{(j)} n_\beta^{(j)} = 0 \quad \text{on } \Gamma_j \; . \tag{3.19}$$

For the derivation of the variational formulation of the problem we assume first sufficiently regular functions. Then multiplying (3.1) by $z^{(j)} - \zeta^{(j)}$, integrating and applying the Green-Gauss theorem, implies the expressions:

$$\alpha(\zeta^{(j)},z^{(j)}-\zeta^{(j)})+\int_{\Omega_j} h_j\sigma_{\alpha\beta}^{(j)}\zeta_{,\alpha}^{(j)}(z^{(j)}-\zeta^{(j)})_{,\beta}d\Omega =$$

$$= \int_{\Gamma_j} h_j\sigma_{\alpha\beta}^{(j)}\zeta_{,\beta}^{(j)}n_\alpha^{(j)}(z^{(j)}-\zeta^{(j)})d\Gamma+\int_{\Omega_j}\bar{\bar{f}}^{(j)}(z^{(j)}-\zeta^{(j)})d\Omega+$$

$$+\int_{\Gamma_j}\bar{Q}_j(\zeta^{(j)})(z^{(j)}-\zeta^{(j)})d\Gamma-\int_{\Gamma_j}M_j(\zeta^{(j)})\frac{\partial(z^{(j)}-\zeta^{(j)})}{\partial n^{(j)}}d\Gamma ,$$

$$j=1,2, \quad \alpha,\beta=1,2 . \tag{3.20}$$

Here $n^{(j)}$ denotes the outward normal unit vector to $\Gamma_j$,

$$\alpha(\zeta,z)= K\int_\Omega[(1-\nu)\zeta_{,\alpha\beta}z_{,\alpha\beta}+\nu\Delta\zeta\Delta z]d\Omega, \quad \alpha,\beta=1,2, \ 0<\nu<0,5 , \tag{3.21}$$

$$M(\zeta) = -K[\nu\Delta\zeta+(1-\nu)(2n_1n_2\zeta_{,12}+n_1^2\zeta_{,11}+n_2^2\zeta_{,22}] , \tag{3.22}$$

$$\bar{Q}(\zeta) = -K[\frac{\partial\Delta\zeta}{\partial n}+ (1-\nu)\frac{\partial}{\partial\tau}[n_1n_2(\zeta_{,22}-\zeta_{,11})+(n_1^2-n_2^2)\zeta_{,12}]] . \tag{3.23}$$

Here $\tau$ is the unit vector tangential to $\Gamma$, such that $n,\tau$ and the $0x_3$-axis form a right-handed system. In (3.22-23) M and $\bar{Q}$ are the bending moment and the total shearing force ($\bar{Q}= Q-\partial M/\partial\tau$) respectively on the boundary $\Gamma$. Analogously we get from (3.2) the expression

$$\int_{\Omega_j}\sigma_{\alpha\beta}^{(j)}\varepsilon_{\alpha\beta}^{(j)}(v^{(j)}-u^{(j)})d\Omega=\int_{\Gamma_j}\sigma_{\alpha\beta}^{(j)}n_\beta^{(j)}(v_\alpha^{(j)}-u_\alpha^{(j)})d\Gamma,$$

$$j=1,2, \quad \alpha,\beta=1,2 . \tag{3.24}$$

The following notations are introduced:

$$R(m,k) = \int_\Omega C_{\alpha\beta\gamma\delta}m_{\alpha\beta}k_{\gamma\delta}d\Omega, \quad \alpha,\beta,\gamma,\delta=1,2 \tag{3.25}$$

and

$$P(\zeta,z) = \{\zeta_{,\alpha}z_{,\beta}\} , \ P(\zeta,\zeta) = P(\zeta) , \tag{3.26}$$

where $m= \{m_{\alpha\beta}\}$ and $k= \{k_{\alpha\beta}\}$, $\alpha,\beta=1,2$. We assume also that $u^{(j)},v^{(j)}\in[H^1(\Omega_j)]^2$ and that $\zeta^{(j)},z^{(j)}\in Z_j$, where

$$Z_j = \{z | z\ H^2(\Omega_j),\ z=0 \text{ on } \Gamma_j-\tilde{\Gamma}_j\ ,$$
$$\frac{\partial\zeta}{\partial n} = 0 \text{ a.e. on } \Gamma_j-\tilde{\Gamma}_j\}\ . \tag{3.27}$$

Taking into account (3.25, 3.26), (3.20) and (3.24), the boundary conditions (3.14-16) and (3.19), the inequalities defining the subdifferentials $\partial$ in (3.17a,b), and (3.18), as well as the interface conditions (3.5),(3.8),(3.12) and the definition of the generalized gradient $\bar{\partial}$ we obtain from (3.20) for j=1 and j=2 and addition and from (3.24) the following problem $\tilde{\mathcal{P}}$, in which due to the aforementioned addition the relative deflection $[\zeta]$ appears.

Problem $\tilde{\mathcal{P}}$: Find $u^{(j)}\in[H^1(\Omega_j)]^2$ and $\zeta^{(j)}\in Z_j$, j=1,2, such as to satisfy the variational-hemivariational inequality

$$\sum_{j=1}^{2} \alpha_j(\zeta^{(j)},z^{(j)}-\zeta^{(j)}) + \sum_{j=1}^{2} h_j R(\varepsilon(u^{(j)})+\tfrac{1}{2}P(\zeta^{(j)}),P(\zeta^{(j)},z^{(j)}-\zeta^{(j)}))+$$
$$+ \int_{\Omega'} J^0([\zeta],[z]-[\zeta])d\Omega + \sum_{j=1}^{2} \{\Phi_j(z^{(j)}) - \Phi_j(\zeta^{(j)})\} \geqq$$
$$\geqq \sum_{j=1}^{2} \int_{\Omega_j} \bar{\bar{f}}^{(j)}(z^{(j)}-\zeta^{(j)})d\Omega,\quad \forall z^{(1)}\in Z_1,\ \forall z^{(2)}\in Z_2\ , \tag{3.28}$$

and the variational equalities

$$R(\varepsilon(u^{(j)})+\tfrac{1}{2}P(\zeta^{(j)}),\ \varepsilon(v^{(j)}-u^{(j)})) = 0\ ,$$
$$\forall v^{(j)}\in[H^1(\Omega_j)]^2,\ j=1,2\ . \tag{3.29}$$

Inequality (3.28) is called a variational-hemivariational inequality, because of the convex (resp. nonconvex) terms $\Phi(z)-\Phi(\zeta)$ (resp. $J^0(.,.)$) as in the theory of variational (resp. hemivariational) inequalities, and together with (3.29) is the expression of the principle of virtual work.

It is obvious that the same method is applicable to the case of r-laminae. Then in (3.28) the summation $\sum_{j=1}^{2}$ is replaced by $\sum_{j=1}^{r}$, the term

$\int_{\Omega'} J^o(.,.)d\Omega$ is replaced by the term $\sum_{m=1}^{m'} \int_{\Omega'_{(m)}} J^o_m(.,.)d\Omega$, where $m'=r-1$ is the total number of interfaces, and (3.28) must hold for very $z^{(j)}\in Z_j$, $j=1,...,r$.

Several other types of boundary conditions may be considered. The only condition which should be fulfilled is that the classical boundary conditions must ensure the coerciveness of the quadratic form of bending $\alpha_j(\zeta^{(j)},\zeta^{(j)})$, i.e. they must prevent any rigid-plate deflection. This is guaranteed, if, e.g., a plate is partially clamped or simply supported on a curved boundary part. Moreover the boundary conditions (3.19) can be replaced by $u_1=u_2=0$ or any other combination of classical boundary conditions which, together with the boundary conditions in bending, will guarantee the vanishing of the term $\int_{\Gamma_j} h_j\sigma^{(j)}_{\alpha\beta}\zeta^{(j)}_{,\beta}n^{(j)}_\alpha(z^{(j)}-\zeta^{(j)})d\Gamma$ in (3.20). For nonhomogeneous classical boundary conditions we obtain through an appropriate translation analogous variational problems as in the homogeneous case. In the general model problem $\tilde{P}'$, which will be studied further, the classical boundary conditions of each plate in bending define the kinematically admissible sets $Z_j$ which are assumed to be subspaces $H^2(\Omega)$ on which the bilinear form $\alpha_j(\zeta^{(j)},\zeta^{(j)})$ are coercive. Our study is mainly based on [63].

Problem $\tilde{P}'$: Find $\zeta^{(j)}\in Z_j$ and $u^{(j)}\in[H^1(\Omega_j)]$, $j=1,...,r$, such as to satisfy the variational-hemivariational inequality

$$\sum_{j=1}^{r} \alpha_j(\zeta^{(j)},z^{(j)}-\zeta^{(j)}) + \sum_{j=1}^{r} h_j R(\varepsilon(u^{(j)}) + \tfrac{1}{2}P(\zeta^{(j)}),P(\zeta^{(j)},z^{(j)}-\zeta^{(j)})) +$$
$$+ \sum_{m=1}^{m'} \int_{\Omega'_m} J^o_m([\zeta]^{(m)},[z]^{(m)}-[\zeta]^{(m)})d\Omega + \sum_{j=1}^{r} \{\Phi_j(z^{(j)}) - \Phi_j(\zeta^{(j)})\} \geq$$
$$\geq \sum_{j=1}^{r} \int_{\Omega_j} \bar{\bar{f}}^{(j)}(z^{(j)}-\zeta^{(j)})d\Omega, \quad \forall z^{(j)}\in Z_j,\ j=1,...,r. \qquad (3.30)$$

and the variational equalities

$$R(\varepsilon(u^{(j)})+\tfrac{1}{2}P(\zeta^{(j)}),\ \varepsilon(v^{(j)}-u^{(j)}))=0,$$
$$\forall v^{(j)}\in[H^1(\Omega_j)]^2,\ j=1,\dots,r. \tag{3.31}$$

3.2 On the Existence of the Solution. Convergence of the Galerkin Scheme

$R(.,.)$ is a continuous, symmetric, coercive bilinear form on $[L^2(\Omega)]^4$, and $P:[H^2(\Omega)]^2\to[L^2(\Omega)]^4$ of (3.26), is a completely continuous operator [64]. Thus (3.31) implies by the Lax-Milgram theorem that to every deflection $\zeta^{(j)}\in Z_j$, $j=1,2,\dots,r$ there corresponds a plane displacement $u^{(j)}(\zeta^{(j)})\in[H^2(\Omega_j)]^2$. Indeed by Korn's inequality [2] $R(\varepsilon(u),\varepsilon(v))$ is a bilinear coercive form on the quotient space $[H^1(\Omega)]^2/\bar{R}$, where $\bar{R}$ is the space of in-plane rigid-plate displacements defined by

$\bar{R}=\{\bar{r}\,|\,\bar{r}\in[H^1(\Omega)]^2,\ \bar{r}_1=a_1+bx_2,\ \bar{r}_2=a_2-bx_2,\ a_1,\ a_2,\ b\in\mathbb{R}\}$.

Thus (3.31) implies that $\varepsilon^{(j)}(u^{(j)}(\zeta^{(j)})):Z_j\to[L^2(\Omega_j)]^4$ is uniquely determined and is a completely continuous quadratic function of $\zeta^{(j)}$, $j=1,2,\dots,r$, since $\varepsilon^{(j)}(u^{(j)}(\zeta^{(j)}))$ is a linear continuous function of $P(\zeta^{(j)})$. Then the completely continuous, quadratic function $G_j:Z_j\to\to[L^2(\Omega_j)]^4$ defined by:

$$\zeta^{(j)}\to G_j(\zeta^{(j)})=\varepsilon^{(j)}(u^{(j)}(\zeta^{(j)}))+\tfrac{1}{2}P(\zeta^{(j)}) \tag{3.32}$$

is introduced, which satisfy the equations

$$R(G_j(\zeta^{(j)}),\ \varepsilon^{(j)}(u^{(j)}(\zeta^{(j)})))=0\ . \tag{3.33}$$

Let us now introduce the operators $A_j:Z_j\to Z_j'$ and $C_j:Z_j\to Z_j'$ (where $Z_j'$ is the dual space to $Z_j$) such that

$$\alpha(\zeta^{(j)},z^{(j)})=\langle A_j\zeta^{(j)},z^{(j)}\rangle \tag{3.34}$$

and

$$h_jR(G_j(\zeta^{(j)}),P(\zeta^{(j)},z^{(j)}))=\langle C_j(\zeta^{(j)}),z^{(j)}\rangle\ . \tag{3.35}$$

$A_j$ is a continuous linear monotone operator, $C_j$ is a completely continuous operator, and $\langle.,.\rangle$ denotes the duality pairing between $Z_j$ and $Z_j'$.

Moreover

$$<C_j(\zeta^{(j)}),\zeta^{(j)}> = h_j R(G_j(\zeta^{(j)}),2G_j(\zeta^{(j)})) \geq 0,$$

$$\forall\zeta^{(j)} \in Z_j, \quad j=1,2,\ldots,r \,. \tag{3.36}$$

From the coercivity of $\alpha_j(\zeta^{(j)},\zeta^{(j)})$ on $Z_j$ we obtain

$$<(A_j+C_j)(\zeta^{(j)}),\zeta^{(j)}> = <T_j(\zeta^{(j)},\zeta^{(j)}> \geq c\|\zeta^{(j)}\|^2,$$

$$\forall\zeta^{(j)} \in Z_j, \quad j=1,2,\ldots,r, \; c \text{ const.} > 0, \tag{3.37}$$

where $\|\cdot\|$ denotes the $Z_j$-norm.

Thus Problem $\tilde{\mathcal{P}}'$ takes the following form (here $(.,.)$ is the $L^2$-product):

Problem P: Find $\zeta^{(j)} \in Z_j$, $j=1,2,\ldots,r$ such as they satisfy the variational-hemivariational inequality

$$\sum_{j=1}^{r} <T_j(\zeta^{(j)}),z^{(j)}-\zeta^{(j)}> + \sum_{m=1}^{m'} \int_{\Omega'_m} J^o_m([\zeta]^{(m)},[z]^{(m)}-[\zeta]^{(m)})d\Omega +$$

$$+ \sum_{j=1}^{r} \{\Phi_j(z^{(j)})-\Phi_j(\zeta^{(j)})\} \geq \sum_{j=1}^{r} (\bar{\bar{f}}^{(j)},z^{(j)}-\zeta^{(j)}),$$

$$\forall z^{(j)} \in Z_j \,. \tag{3.38}$$

The basic assumption for the existence and the approximation of the solution of our problem, concerns the interface law: For some $\xi \in \mathbb{R}$

$$\operatorname*{ess\,sup}_{(-\infty,-\xi)} \beta^{(m)}(\xi) \leq \operatorname*{ess\,inf}_{(+\xi,+\infty)} \beta^{(m)}(\xi) \;, \quad m=1,2,\ldots,m' \,, \tag{3.39}$$

i.e. that the nonmonotone $\beta^{(m)}$ increases "ultimately" to the right. Obviously one can assume by performing an appropriate translation of the coordinate axis, that for some $\xi \in \mathbb{R}$

$$\operatorname*{ess\,sup}_{(-\infty,-\xi)} \beta^{(m)}(\xi) \leq 0 \leq \operatorname*{ess\,inf}_{(+\xi,+\infty)} \beta^{(m)}(\xi), \quad m=1,2,\ldots,m' \,. \tag{3.39a}$$

Two cases will be separately treated: The "differentiable" and the "nondifferentiable". In the first the $\mathrm{grad}\Phi_j$ exists everywhere for every j, in the second this does not occur.

a) The "differentiable" case: Relation (3.38) of P is equivalently written in the form

$$\sum_{j=1}^{r} \langle T_j(\zeta^{(j)}), z^{(j)}-\zeta^{(j)}\rangle + \sum_{m=1}^{m'} \int_{\Omega'_m} J_m^o([\zeta]^{(m)}, [z]^{(m)}-[\zeta]^{(m)})d\Omega +$$
$$+ \sum_{j=1}^{r} \langle \mathrm{grad}\Phi_j(\zeta^{(j)}), z^{(j)}-\zeta^{(j)}\rangle \geqq \sum_{j=1}^{r} (\bar{\bar{f}}^{(j)}, z^{(j)}-\zeta^{(j)}),$$
$$\forall z^{(j)} \in Z_j \; . \tag{3.40}$$

Indeed from (3.38) by setting $z^{(j)}=\zeta^{(j)}+\lambda(\tilde{z}^{(j)}-\zeta^{(j)})$, $\lambda\in(0,1)$, $\lambda\to 0_+$ we obtain (3.40). (Consider that $\xi\to J^o([\zeta],\xi)$ is positively homogeneous and the definition of $\mathrm{grad}\Phi_j$). Conversely (3.40) implies (3.38) because of the convexity inequality

$$\Phi(z^{(j)}) - \Phi(\zeta^{(j)}) \geqq \langle \mathrm{grad}\Phi_j(\zeta^{(j)}), z^{(j)}-\zeta^{(j)}\rangle, \quad \forall z^{(j)} \in Z_j \; . \tag{3.41}$$

Now the regularized problem $P_\varepsilon$ is defined:
Let p be a mollifier ($p\in C_c^\infty(-1,+1)$, $p \geqq 0$ with $\int_{-\infty}^{+\infty} p(\xi)d\xi = 1$), and let

$$\beta_\varepsilon = p_\varepsilon * \beta, \; \varepsilon > 0 \; , \tag{3.42}$$

where $p_\varepsilon(\xi) = (1/\varepsilon)p(\xi/\varepsilon)$ and $*$ denotes the convolution product.

Problem $P_\varepsilon$: Find $\zeta_\varepsilon^{(j)} \in Z_j$, j=1,...,r, such as to satisfy the variational equality

$$\sum_{j=1}^{r} \langle T_j(\zeta_\varepsilon^{(j)}), z^{(j)}\rangle + \sum_{j=1}^{r} \langle \mathrm{grad}\Phi_j(\zeta_\varepsilon^{(j)}), z^{(j)}\rangle +$$
$$+ \sum_{m=1}^{m'} \int_{\Omega'_m} \beta_\varepsilon^{(m)}([\zeta_\varepsilon]^{(m)})[z]^{(m)}d\Omega = \sum_{j=1}^{r} (\bar{\bar{f}}^{(j)}, z^{(j)}) \; ,$$
$$\forall z^{(j)} \in Z_j \; . \tag{3.43}$$

Further a Galerkin basis of $Z_j$, j=1,...,r is introduced. Let $Z_{jn}$ be the corresponding n-dimensional subspace of $Z_j$. The corresponding finite dimensional Problem $P_{\varepsilon n}$ reads:

Problem $P_{\varepsilon n}$: Find $\zeta_{\varepsilon n}^{(j)} \in Z_{jn}$, $j=1,2,\ldots,r$ such that

$$\sum_{j=1}^{r} \langle T_j(\zeta_{\varepsilon n}^{(j)}), z^{(j)}\rangle + \sum_{j=1}^{r} \langle \operatorname{grad}\Phi_j(\zeta_{\varepsilon n}^{(j)}), z^{(j)}\rangle +$$

$$+ \sum_{m=1}^{m'} \int_{\Omega'_m} \beta_\varepsilon^{(m)}([\zeta_{\varepsilon n}]^{(m)})[z]^{(m)} d\Omega = \sum_{j=1}^{r} (\bar{\bar{f}}^{(j)}, z^{(j)}),$$

$$\forall z^{(j)} \in Z_{jn}, \quad j=1,\ldots,r . \tag{3.44}$$

We make now the simplifying assumption

$$\operatorname{grad}\Phi_j(0) = 0 , \quad j=1,\ldots,r . \tag{3.45}$$

Proposition 3.1: Let assumptions (3.39a) and (3.45) hold. Then for $\bar{\bar{f}}^{(j)} \in L^2(\Omega_j)$, $j=1,\ldots,r$ problem P has at least one solution.

Proof: (3.44) is written in the form

$$\langle \Lambda(\tilde{\zeta}_{\varepsilon n}), \tilde{z}\rangle = 0, \text{ for } \tilde{z} = \{z^{(j)}\}, \quad \forall z^{(j)} \in Z_{jn}, \quad j=1,2,\ldots,r . \tag{3.46}$$

We verify easily that an inequality analogous to (3.39a) can be written for $\beta_\varepsilon$ (now only with inf and sup). As a result of it we may determine for every m, $\rho_1^{(m)}>0$ and $\rho_2^{(m)}>0$ such that $\beta_\varepsilon^{(m)}(\xi)\geq 0$ if $\xi>\rho_1^{(m)}$, $\beta_\varepsilon^{(m)}(\xi)\leq 0$ if $\xi<-\rho_1^{(m)}$ and $|\beta_\varepsilon^{(m)}(\xi)|\leq\rho_2^{(m)}$ if $|\xi|\leq\rho_1^{(m)}$ and thus

$$\int_{\Omega'_m} \beta_\varepsilon^{(m)}([\zeta_{\varepsilon n}]^{(m)})[\zeta_{\varepsilon n}]^{(m)} d\Omega = \int_{|[\zeta_{\varepsilon n}]^{(m)}|>\rho_1^{(m)}} \ldots d\Omega + \int_{|[\zeta_{\varepsilon n}]^{(m)}|\leq\rho_1^{(m)}} \ldots d\Omega \geq$$

$$\geq 0 - \rho_1^{(m)}\rho_2^{(m)} \operatorname{mes}\Omega'_m . \tag{3.47}$$

The monotonicity of $\operatorname{grad}\Phi_j$ and (3.45) imply that[1]

$$\langle \operatorname{grad}\Phi_j(\zeta_{\varepsilon n}^{(j)}) - \operatorname{grad}\Phi_j(0), \zeta_{\varepsilon n}^{(j)} - 0\rangle \geq 0, \quad \forall \zeta_{\varepsilon n}^{(j)} \in Z_{jn} . \tag{3.48}$$

[1] (3.45) could be omitted. Then we shall use the fact that there is $\zeta_0$ such that $\operatorname{grad}\Phi_j(\zeta_0)$ is finite. Again (3.48) with $\zeta_0$ instead of zero would be used.

From (3.44) using (3.37),(3.47) and (3.48) we obtain that

$$<\Lambda(\zeta_{\varepsilon n}),\zeta_{\varepsilon n}> \geq \sum_{j=1}^{r} c_j \|\zeta_{\varepsilon n}^{(j)}\|^2_{Z_j} + 0 - \sum_{m=1}^{m'} \rho_1^{(m)}\rho_2^{(m)} \mathrm{mes}\Omega'_m -$$

$$- \sum_{j=1}^{r} c'_j \|\zeta_{\varepsilon n}^{(j)}\|_{Z_j}, \quad c_j, c'_j \text{ const.}>0, \quad j=1,\ldots,r \ . \tag{3.49}$$

Due to (3.49) we can apply Brouwer's fixed point theorem and thus (3.46) has a solution $\zeta_{\varepsilon n}$ with $\|\zeta_{\varepsilon n}^{(j)}\| \leq c_j$, $j=1,2,\ldots,r$. Now we may determine subsequences, again denoted by $\{\zeta_{\varepsilon n}^{(j)}\}$, $j=1,\ldots,r$, such that for $\varepsilon\to 0, n\to\infty$

$$\zeta_{\varepsilon n}^{(j)} \to \zeta^{(j)} \text{ weakly in } Z_j, \ j=1,\ldots,r \ , \tag{3.50}$$

and because of the compact imbeddings $H^2(\Omega_j) \subset L^2(\Omega_j)$, $j=1,\ldots,r$

$$\zeta_{\varepsilon n}^{(j)} \to \zeta^{(j)} \text{ strongly in } L^2(\Omega_j), \ j=1,\ldots,r \ . \tag{3.51}$$

Accordingly

$$\zeta_{\varepsilon n}^{(j)} \to \zeta^{(j)} \text{ a.e. in } \Omega_j \ , \quad j=1,\ldots,r \ . \tag{3.52}$$

Further we may omit the indices and (m) if no ambiguity occurs. Now we prove that $\beta_\varepsilon^{(m)}([\zeta_{\varepsilon n}]^{(m)})$ is weakly precompact in $L^1(\Omega'_m)$, $m=1,\ldots,m'$. Equivalently, due to the Dunford-Pettis theorem (e.g. [62] p.239) we must show that for each $\mu_m>0$ a $\delta_m(\mu_m)>0$ can be determined such that for $\omega_m \subset \Omega'_m$ with $\mathrm{mes}\,\omega_m<\delta_m$

$$\int_{\omega_m} |\beta_\varepsilon^{(m)}([\zeta_{\varepsilon n}]^{(m)})| d\Omega_m < \mu_m \ . \tag{3.53}$$

The inequality

$$\xi_0 |\beta_\varepsilon(\xi)| \leq |\beta_\varepsilon(\xi)\xi| + \xi_0 \sup_{|\xi|\leq\xi_0} |\beta_\varepsilon(\xi)| \tag{3.54}$$

implies that

$$\int_\omega |\beta_\varepsilon([\zeta_{\varepsilon n}])| d\Omega \leq \frac{1}{\xi_0} \int_\omega |\beta_\varepsilon([\zeta_{\varepsilon n}])[\zeta_{\varepsilon n}]| d\Omega +$$

$$+ \int_\omega \sup_{|[\zeta_{\varepsilon n}(x)]|\leq\xi_0} |\beta_\varepsilon([\zeta_{\varepsilon n}])| d\Omega \ . \tag{3.55}$$

But

$$\int_{\Omega'}|\beta_\varepsilon([\zeta_{\varepsilon n}])[\zeta_{\varepsilon n}]|d\Omega = \int_{|[\zeta_{\varepsilon n}(x)]|>\rho_1}|\ldots|d\Omega + \int_{|[\zeta_{\varepsilon n}(x)]|\leqq\rho_1}|\ldots|d\Omega =$$

$$= \int_{|[\zeta_{\varepsilon n}(x)]|>\rho_1}|\ldots|d\Omega - \int_{|[\zeta_{\varepsilon n}(x)]|\leqq\rho_1}|\ldots|d\Omega + 2\int_{|[\zeta_{\varepsilon n}(x)]|\leqq\rho_1}|\ldots|d\Omega \leqq$$

$$\leqq \int_{|[\zeta_{\varepsilon n}(x)]|>\rho_1}|\ldots|d\Omega + \int_{|[\zeta_{\varepsilon n}(x)]|\leqq\rho_1}\ldots d\Omega + 2\int_{|[\zeta_{\varepsilon n}(x)]|\leqq\rho_1}|\ldots|d\Omega =$$

$$= \int_{\Omega'}\beta_\varepsilon([\zeta_{\varepsilon n}])[\zeta_{\varepsilon n}]d\Omega + 2\int_{|[\zeta_{\varepsilon n}(x)]|\leqq\rho_1}|\beta_\varepsilon([\zeta_{\varepsilon n}])[\zeta_{\varepsilon n}]|d\Omega . \tag{3.56}$$

Now let us set $z^{(1)}=\tilde{z}^{(1)}\in Z_{1n}$ in (3.44) such that $\tilde{z}^{(1)}=z^{(2)}$ on $\Omega_1'$, $z^{(2)}=\tilde{z}^{(2)}\in Z_{2n}$ such that $\tilde{z}^{(2)}=z^{(3)}$ on $\Omega_2',\ldots,$ $z^{(j-1)}=\tilde{z}^{(j-1)}\in Z_{(j-1)n}$ such that $\tilde{z}^{(j-1)}=\zeta_{\varepsilon n}^{(j)}$ on $\Omega_{j-1}'$, $z^{(j)}=\zeta_{\varepsilon n}^{(j)}$ on $\Omega_j$, $z^{(j+1)}=\zeta_{\varepsilon n}^{(j+1)}$ on $\Omega_{j+1}$, $z^{(j+2)}=\tilde{z}^{(j+2)}\in Z_{(j+2)n}$ such that $\tilde{z}^{(j+2)}=\zeta_{\varepsilon n}^{(j+1)}$ on $\Omega_{j+1}'$ and finally $z^{(r)}=\tilde{z}^{(r)}\in Z_{rn}$ such that $\tilde{z}^{(r)}=z^{(r-1)}$ on $\Omega_{r-1}'$. Thus, using (3.56) and (3.48) we obtain

$$\int_{\Omega_m'}|\beta_\varepsilon^{(m)}([\zeta_{\varepsilon n}]^{(m)})[\zeta_{\varepsilon n}]^{(m)}|d\Omega \leqq \sum_{j=1}^{r}(\bar{\bar{f}}^{(j)},\zeta_{\varepsilon n}^{(j)}) -$$

$$- \sum_{j=1}^{r}\langle T_j(\zeta_{\varepsilon n}^{(j)}),\zeta_{\varepsilon n}^{(j)}\rangle - \sum_{j=1}^{r}\langle \mathrm{grad}\Phi_j(\zeta_{\varepsilon n}^{(j)}),\zeta_{\varepsilon n}^{(j)}\rangle +$$

$$+ 2\int_{|[\zeta_{\varepsilon n}(x)]^{(m)}|\leqq\rho_1^{(m)}}|\beta_\varepsilon^{(m)}([\zeta_{\varepsilon n}]^{(m)})[\zeta_{\varepsilon n}]^{(m)}|d\Omega \leqq$$

$$\leqq c + 2\rho_1^{(m)}\rho_2^{(m)}\mathrm{mes}\Omega_m' - \sum_{j=1}^{r}\langle \mathrm{grad}\Phi_j(\zeta_{\varepsilon n}^{(j)})-\mathrm{grad}\Phi_j(0),\zeta_{\varepsilon n}^{(j)}-0\rangle \leqq$$

$$\leqq c + 2\rho_1^{(m)}\rho_2^{(m)}\mathrm{mes}\Omega_m' , \quad \forall m=1,2,\ldots,m' . \tag{3.57}$$

The definition of the mollifier and (3.42) imply easily the estimate

$$\sup_{|\xi|\leq\xi_0} |\beta_\varepsilon(\zeta)| \leq \operatorname*{ess\,sup}_{|\xi|\leq\xi_0+1} |\beta(\xi)| \ . \tag{3.58}$$

Choosing $\xi_0$, such that for all $\varepsilon$ and $n$

$$\frac{1}{\xi_0}\int_\omega |\beta_\varepsilon([\zeta_{\varepsilon n}])[\zeta_{\varepsilon n}]|\, d\Omega' \leq \frac{1}{\xi_0}(c+2\rho_1\rho_2 \operatorname{mes}\Omega') \leq \frac{\mu}{2} \ , \tag{3.59}$$

and $\delta$ such that for $\operatorname{mes}\omega<\delta$

$$\operatorname*{ess\,sup}_{|\xi|\leq\xi_0+1} |\beta(\xi)| \leq \frac{\mu}{2\delta} \ , \tag{3.60}$$

we obtain that

$$\int_\omega \sup_{|[\zeta_{\varepsilon n}(x)]|\leq\xi_0} |\beta_\varepsilon([\zeta_{\varepsilon n}])|\, d\Omega \leq$$

$$\leq \operatorname*{ess\,sup}_{|[\zeta_{\varepsilon n}(x)]|\leq\xi_0+1} |\beta([\zeta_\varepsilon])| \operatorname{mes}\omega \leq \frac{\mu}{2\delta}\delta = \frac{\mu}{2} \ . \tag{3.61}$$

From (3.55),(3.57),(3.59) and (3.61),(3.53) results i.e. the weak precompactness of $\beta_\varepsilon^{(m)}([\zeta_{\varepsilon n}]^{(m)})$, $m=1,2,\ldots,m'$ in $L^1(\Omega'_m)$. Thus, as $\varepsilon\to 0$ and $n\to\infty$, a subsequence again denoted by $\{\beta_\varepsilon^{(m)}([\zeta_{\varepsilon n}]^{(m)})\}$ can be determined such that

$$\beta_\varepsilon^{(m)}([\zeta_{\varepsilon n}]^{(m)}) \to \chi^{(m)} \text{ weakly in } L^1(\Omega'_m),\ m=1,2,\ldots,m' \ . \tag{3.62}$$

For the present specific plate problem we have that $H^2(\Omega)\subset C^0(\bar\Omega)\subset L^\infty(\Omega)$ (imbeddings). Moreover $\beta_\varepsilon$ is a $C^\infty$-function and thus (3.62) results immediately without using the Dunford-Pettis theorem and the estimates (3.53)÷(3.61) based on [65]. However, this procedure is necessary in most other types of nonmonotone nonlinearity and in any other type of functional setting. Indeed in any hemivariational inequality formulated in the $H^1$-space the above procedure is necessary because $H^1(\Omega)\not\subset C^0(\bar\Omega)$. The same holds for any type of nonlinearity term $\beta(\cdot)$ for which the previous $C^0$-imbedding does not hold: For instance, let us consider in a plate problem a nonmonotone boundary condition $M\in\hat{\beta}(\frac{\partial\zeta}{\partial n})$; then $\frac{\partial\zeta}{\partial n}\in H^{1/2}(\Gamma)$ but $H^{1/2}(\Gamma)\not\subset C^0(\Gamma)$. On the contrary, if in a plate problem a boundary condition $-Q\in\hat{\beta}(\zeta)$ is considered, we have the imbeddings $H^2(\Omega)\subset C^0(\bar{\Omega})\subset C^0(\Gamma)\subset L^\infty(\Gamma)$ and again we can omit the general procedure of the estimates (3.53)÷(3.61).

From (3.44) we find using special variations that

$$|\langle \mathrm{grad}\Phi_j(\zeta^{(j)}_{\varepsilon n}),z^{(j)}\rangle| \leq c\,\|z^{(j)}\|_{Z'_j}\ , \quad \forall z^{(j)}\in Z_j,\ j=1,2,\ldots,r\ , \tag{3.63}$$

i.e.

$$\|\mathrm{grad}\Phi_j(\zeta^{(j)}_{\varepsilon n})\|_{Z'_j} \leq c, \quad j=1,2,\ldots,r\ . \tag{3.64}$$

Thus, there is a subsequence again denoted by $\{\mathrm{grad}\Phi_j(\zeta^{(j)}_{\varepsilon n})\}$ such that, as $\varepsilon\to 0$, $n\to\infty$

$$\mathrm{grad}\Phi_j(\zeta^{(j)}_{\varepsilon n})\to \Psi^{(j)} \text{ weakly in } Z'_j\ , \quad j=1,2,\ldots,r\ . \tag{3.65}$$

From (3.44),(3.62) and (3.65) we obtain passing to the limit $n\to\infty$, $\varepsilon\to 0$ that

$$\sum_{j=1}^{r}\langle T_j(\zeta^{(j)}),z^{(j)}\rangle+\sum_{j=1}^{r}\langle\Psi^{(j)},z^{(j)}\rangle+\sum_{m=1}^{m'}\int_{\Omega'_m}\chi^{(m)}[z]^{(m)}d\Omega =$$

$$=\sum_{j=1}^{r}(\bar{\bar{f}}^{(j)},z^{(j)}),\quad \forall z^{(j)}\in Z_j,\quad j=1,2,\ldots,r\ , \tag{3.66}$$

where we have used the fact that

$$[z]^{(m)}\in H^2(\Omega'_m)\subset L^\infty(\Omega'_m) \text{ for } \Omega'_m\subset \mathbb{R}^2,\quad \forall m=1,\ldots,m'\ . \tag{3.67}$$

At this point it is worth noting that, if (3.67) did not hold, then we would choose appropriately the Galerkin basis. For instance for an analogous problem formulated in a V-space with $V\not\subset C^0(\bar{\Omega})\subset L^\infty(\Omega)$ we would consider a Galerkin basis in the space $V\cap L^\infty(\Omega)$. Then it is necessary that $V\cap L^\infty(\Omega)$ be dense in V for the V-norm (cf. Sec.4 and [12] p.272).

To complete the proof we will show that

$$\Psi^{(j)}=\mathrm{grad}\Phi_j(\zeta^{(j)}) \text{ in } Z'_j\ , \quad j=1,2,\ldots,r\ , \tag{3.68}$$

and that

$$\chi^{(m)}\in\hat{\beta}^{(m)}([\zeta]^{(m)}),\quad m=1,2,\dots,m'\ . \tag{3.69}$$

To prove (3.68) first the nonnegative expression

$$X_n = \sum_{j=1}^{r} \langle \operatorname{grad}\Phi_j(\zeta_{\varepsilon n}^{(j)})-\operatorname{grad}\Phi_j(\vartheta_j),\zeta_{\varepsilon n}^{(j)}-\vartheta_j\rangle \geqq 0\ ,$$

$$\forall\vartheta_j\in Z_j,\quad j=1,2,\dots,r \tag{3.70}$$

is considered, which by means of (3.44) implies that

$$X_n = -\sum_{j=1}^{r}\langle T_j(\zeta_{\varepsilon n}^{(j)}),\zeta_{\varepsilon n}^{(j)}\rangle - \sum_{m=1}^{m'}\int_{\Omega'_m}\beta_\varepsilon^{(m)}([\zeta_{\varepsilon n}]^{(m)})[\zeta_{\varepsilon n}]^{(m)}d\Omega +$$

$$+\sum_{j=1}^{r}(\bar{\bar{f}}^{(j)},\zeta_{\varepsilon n}^{(j)}) - \sum_{j=1}^{r}\langle \operatorname{grad}\Phi_j(\zeta_{\varepsilon n}^{(j)}),\vartheta_j\rangle - \sum_{j=1}^{r}\langle \operatorname{grad}\Phi_j(\vartheta_j),\zeta_{\varepsilon n}^{(j)}-\vartheta_j\rangle \geqq 0,$$

$$\forall\vartheta_j\in Z_j,\quad j=1,2,\dots,r\ . \tag{3.71}$$

Furthermore as $\varepsilon\to 0$, $n\to\infty$

$$\lim\int_{\Omega'}\beta_\varepsilon^{(m)}([\zeta_{\varepsilon n}]^{(m)})[\zeta_{\varepsilon n}]^{(m)}d\Omega = \int_{\Omega'}\chi^{(m)}[\zeta]^{(m)}d\Omega,\quad m=1,2,\dots,m'\ . \tag{3.72}$$

To prove it we form the difference

$$\int_{\Omega'}\{\beta_\varepsilon([\zeta_{\varepsilon n}])[\zeta_{\varepsilon n}] - \chi[\zeta]\}d\Omega = \int_{\Omega'}\beta_\varepsilon([\zeta_{\varepsilon n}])([\zeta_{\varepsilon n}] - [\zeta])d\Omega +$$

$$+\int_{\Omega'}[\zeta](\beta_\varepsilon([\zeta_{\varepsilon n}]) - \chi)d\Omega \quad = \quad A+B\ . \tag{3.73}$$

Since $H^2(\Omega')$ is compactly imbedded into $L^\infty(\Omega')$, we have that $\lim B=0$ due to (3.62). Also it implies that

$$\|\beta_\varepsilon([\zeta_{\varepsilon n}])\|_{L^1(\Omega')} < c\ . \tag{3.74}$$

and thus

$$|A| \leq \|\beta_\varepsilon([\zeta_{\varepsilon n}])\|_{L^1(\Omega')}\,\|[\zeta_{\varepsilon n}] - [\zeta]\|_{L^\infty(\Omega')}\ , \tag{3.75}$$

which yields $\lim A = 0$. The proof of relations of the type (3.72) is necessary only in the case of variational hemivariational inequalities in order to treat the monotone nonlinearity, i.e. to show (3.68). Their proof relies on the properties of the function space considered. Thus for instance in a $H^1$-space ($H^1(\Omega)\not\subset C^0(\bar{\Omega})$) the proof of a relation (3.72) is not possible. From (3.71), using (3.72), we obtain for $\varepsilon\to 0$ $n\to\infty$

$$0 \leq \limsup X_n \leq \sum_{j=1}^{r} <T_j(\zeta^{(j)}),\zeta^{(j)}> - \sum_{m=1}^{m'} \int_{\Omega'_m} \chi^{(m)}[\zeta]^{(m)} d\Omega + $$

$$+ \sum_{j=1}^{r} (\bar{\bar{f}}^{(j)},\zeta^{(j)}) - \sum_{j=1}^{r} <\Psi^{(j)},\vartheta_j> - \sum_{j=1}^{r} <\mathrm{grad}\Phi_j(\vartheta_j),\zeta^{(j)}-\vartheta_j> . \quad (3.76)$$

(3.66) and (3.76) imply

$$\sum_{j=1}^{r} \{<\Psi^{(j)},\zeta^{(j)}-\vartheta_j> - <\mathrm{grad}\Phi_j(\vartheta_j),\zeta^{(j)}-\vartheta_j>\} \geq 0, \ \forall\vartheta_j\in Z_j, \ j=1,\dots,r , \quad (3.77)$$

which by Minty's well-known monotonicity argument (cf. e.g. [7] p.55) yields that

$$\Psi^{(j)} = \mathrm{grad}\Phi_j(\zeta^{(j)}) \text{ in } Z'_j , \quad j=1,2,\dots,r . \quad (3.78)$$

To show (3.69) we proceed as follows [66]: (3.52) implies by Egoroff's theorem, that for every $\alpha>0$, we can determine $\omega_j$, $j=1,\dots,r$ with $\mathrm{mes}\,\omega_j<\alpha$ such that

$$\zeta^{(j)}_{\varepsilon n} \to \zeta^{(j)} \text{ uniformly in } \Omega_j-\omega_j , \quad j=1,2,\dots,r , \quad (3.79)$$

with $\zeta^{(j)}\in L^\infty(\Omega_j-\omega_j)$. Due to this uniform convergence, for every $\alpha>0$ a $\omega$ with $\mathrm{mes}\,\omega<\alpha$ can be found such that for any $\mu>0$ and for $\varepsilon<\varepsilon_0<\mu/2$ and $n>n_0>2/\mu$

$$|[\zeta_{\varepsilon n}] - [\zeta]| < \frac{\mu}{2} , \quad \forall x\in\Omega'-\omega . \quad (3.80)$$

Owing to (3.42) we obtain easily the inequality

$$\beta_\varepsilon(\xi) = (p_\varepsilon * \beta)(\xi) = \int_{-\varepsilon}^{+\varepsilon} \beta(\xi-t)p_\varepsilon(t)dt \leq \operatorname*{ess\,sup}_{|t|\leq\varepsilon} \beta(\xi-t) , \quad (3.81a)$$

and analogously

$$\operatorname*{ess\,inf}_{|t|\le\varepsilon} \beta(\xi-t) \le \beta_\varepsilon(\xi) \;. \tag{3.81b}$$

From (3.80) and (3.81a,b) we have that

$$\beta_\varepsilon([\zeta_{\varepsilon n}]) \le \operatorname*{ess\,sup}_{|[\zeta_{\varepsilon n}]-\xi|\le\varepsilon} \beta(\xi) \le \operatorname*{ess\,sup}_{|[\zeta_{\varepsilon n}]-\xi|<\mu/2} \beta(\xi) \le$$

$$\le \operatorname*{ess\,sup}_{|[\zeta]-\xi|<\mu} \beta(\xi) = \overline{\beta}_\mu([\zeta]) \tag{3.82}$$

and

$$\underline{\beta}_\mu([\zeta]) \le \operatorname*{ess\,inf}_{|[\zeta]-\xi|<\mu} \beta(\xi) \le \beta_\varepsilon([\zeta_{\varepsilon n}]) \;, \tag{3.83}$$

respectively. Due to (3.82),(3.83) we obtain by choosing $e\geqq 0$ a.e. in $\Omega'-\omega$ with $e\in L^\infty(\Omega'-\omega)$ that

$$\int_{\Omega'-\omega} \underline{\beta}_\mu([\zeta])e\,d\Omega \le \int_{\Omega'-\omega} \beta_\varepsilon([\zeta_{\varepsilon n}])e\,d\Omega \le \int_{\Omega'-\omega} \overline{\beta}_\mu([\zeta])e\,d\Omega \tag{3.84}$$

which as $\varepsilon\to 0$ and $n\to\infty$ becomes

$$\int_{\Omega'-\omega} \underline{\beta}_\mu([\zeta])e\,d\Omega \le \int_{\Omega'-\omega} \chi e\,d\Omega \le \int_{\Omega'-\omega} \overline{\beta}_\mu([\zeta])e\,d\Omega \;. \tag{3.85}$$

Now passing to the limit $\mu\to 0$ we have by Lebesque's theorem

$$\int_{\Omega'-\omega} \underline{\beta}([\zeta])e\,d\Omega \le \int_{\Omega'-\omega} \chi e\,d\Omega \le \int_{\Omega'-\omega} \overline{\beta}([\zeta])e\,d\Omega \;, \tag{3.86}$$

and since $e\geqq 0$ is arbitrary, (3.86) yields that

$$\chi\in\hat{\beta}([\zeta]) \text{ a.e. in } \Omega'-\omega \;.$$

Now taking $\alpha$ as small as possible, implies (3.69) q.e.d.

b) The "nondifferentiable" case: In this case grad$\Phi_j$ does not exist everywhere for some j. Then we regularize the convex superpotentials $\Phi_j$, by assuming that for every j=1,2,...,r a sequence of convex Gâteaux-differentiable functionals $\{\Phi_{j\rho}\}$ exists, depending on the parameter $\rho$ with the properties:

i) As $\rho\to 0$, $\Phi_{j\rho}(z^{(j)})\to\Phi_j(z^{(j)})$, $\forall z^{(j)}\in Z_j$ . (3.87)

ii) $\operatorname{grad}\Phi_{j\rho}(0)=0$, and (3.88)

iii) if $z_\rho^{(j)}\to z^{(j)}$ weakly in $Z_j$ for $\rho\to 0$, and if

$\Phi_{j\rho}(z_\rho^{(j)})<c$, where c is a constant, then $\liminf_{\rho\to 0}\Phi_{j\rho}(z_\rho^{(j)})\geq\Phi_j(z^{(j)})$,

$$j=1,2,\dots,r\ . \tag{3.89}$$

First the regularized-discretized form of the problem is given:

Problem $P'_{\varepsilon\rho n}$: Find $\zeta_{\varepsilon\rho n}^{(j)}\in Z_{jn}$, $j=1,2,\dots,r$ such as to satisfy the variational equality

$$\sum_{j=1}^{r}<T_j(\zeta_{\varepsilon\rho n}^{(j)}),\ z^{(j)}>+\sum_{m=1}^{m'}\int_{\Omega'}\beta_\varepsilon^{(m)}([\zeta_{\varepsilon\rho n}]^{(m)})[z]^{(m)}d\Omega\ +$$

$$+\sum_{j=1}^{r}<\operatorname{grad}\Phi_{j\rho}(\zeta_{\varepsilon\rho n}^{(j)}),z^{(j)}> = \sum_{j=1}^{r}(\bar{\bar{f}}^{(j)},z^{(j)}),$$

$$\forall z^{(j)}\in Z_{jn}\ ,\quad j=1,2,\dots,r\ . \tag{3.90}$$

For this case we prove the following results:

Proposition 3.2: Suppose that (3.39) (or (3.39a)), (3.87),(3.88) and (3.89) hold. Then Problem P has at least one solution.

Proof: Following the same procedure as in Prop. 3.1 Brouwer's fixed point theorem implies the existence of a solution of $P'_{\varepsilon\rho n}$ such that $||\zeta_{\varepsilon\rho n}^{(j)}||\leq c_j$, $j=1,2,\dots,r$. Thus, as $\varepsilon\to 0$, $n\to\infty$

$$\zeta_{\varepsilon\rho n}^{(j)}\to\zeta_\rho^{(j)}\ \text{weakly in}\ Z_j,\ j=1,2,\dots,r, \tag{3.91}$$

and as $\rho\to 0$

$$\zeta_\rho^{(j)}\to\zeta^{(j)}\ \text{weakly in}\ Z_j,\ j=1,2,\dots,r\ . \tag{3.92}$$

Now, we prove that (3.53) holds for $\{\beta_\varepsilon^{(m)}([\zeta_{\varepsilon\rho n}]^{(m)})\}$ (cf. Prop. 3.1) and thus,

$$\beta_\varepsilon^{(m)}([\zeta_{\varepsilon\rho n}]^{(m)})\to \chi_\rho^{(m)} \text{ weakly in } L^1(\Omega'_m),\ m=1,\dots,m'\ , \tag{3.93}$$

and

$$\chi_\rho^{(m)}\to \chi^{(m)} \text{ weakly in } L^1(\Omega'_m),\ m=1,\dots,m'\ . \tag{3.94}$$

Moreover we get the estimate

$$\|\operatorname{grad}\Phi_{j\rho}(\zeta_{\varepsilon\rho n}^{(j)})\|_{Z'_j}\le c'_j,\quad j=1,2,\dots,r\ , \tag{3.95}$$

where the constants do not depend on $\varepsilon,n,\rho$. Therefore we have for $\varepsilon\to 0$, $n\to\infty$

$$\operatorname{grad}\Phi_{j\rho}(\zeta_{\varepsilon\rho n}^{(j)})\to \Psi_\rho^{(j)} \text{ weakly in } Z'_j\ ,\quad j=1,2,\dots,r\ , \tag{3.96}$$

and for $\rho\to 0$

$$\Psi_\rho^{(j)}\to \Psi^{(j)} \text{ weakly in } Z'_j,\quad j=1,2,\dots,r\ . \tag{3.97}$$

From (3.90) we obtain for $\varepsilon\to 0$, $n\to\infty$

$$\sum_{j=1}^{r}\langle T_j(\zeta_\rho^{(j)}),z^{(j)}\rangle+\sum_{m=1}^{m'}\int_{\Omega'_m}\chi_\rho^{(m)}[z]^{(m)}d\Omega+\sum_{j=1}^{r}\langle\Psi_\rho^{(j)},z^{(j)}\rangle =$$

$$=\sum_{j=1}^{r}(\bar{\bar{f}}^{(j)},z^{(j)}),\quad \forall z^{(j)}\in Z_j,\quad j=1,2,\dots,r\ . \tag{3.98}$$

The relations

$$\Psi_\rho^{(j)}=\operatorname{grad}\Phi_{j\rho}(\zeta_\rho^{(j)}) \text{ in } Z'_j,\quad j=1,2,\dots,r\ , \tag{3.99}$$

are proved as in Prop. 3.1. From (3.98) and (3.99) and the convexity inequalities
$\Phi_{j\rho}(\xi^{(j)})-\Phi_{j\rho}(z^{(j)})\ge\langle\operatorname{grad}\Phi_{j\rho}(z^{(j)}),\xi^{(j)}-z^{(j)}\rangle,\ \forall\xi^{(j)}\in Z_j$ we get

$$\sum_{j=1}^{r}\langle T_j(\zeta_\rho^{(j)}),z^{(j)}-\zeta_\rho^{(j)}\rangle+\sum_{j=1}^{r}\{\Phi_{j\rho}(z^{(j)})-\Phi_{j\rho}(\zeta_\rho^{(j)})\}+$$

$$+\sum_{m=1}^{m'}\int_{\Omega'_m}\chi_\rho^{(m)}([z]^{(m)}-[\zeta_\rho]^{(m)})d\Omega \geqq \sum_{j=1}^{r}(\bar{\bar{f}}^{(j)},z^{(j)}-\zeta_\rho^{(j)}) ,$$

$$\forall z^{(j)}\in Z_j,\quad j=1,2,\ldots,r . \tag{3.100}$$

We let $z^{(j)}$ in (3.100) be such that $\Phi_j(z^{(j)})<\infty$, j=1,2,...,r. Then, due to (3.87) there exist constants $c_j$ such that $\Phi_{j\rho}(z^{(j)})<c_j$, j=1,2,...,r and from (3.100) it results that

$$\Phi_{j\rho}(\zeta_\rho^{(j)})\leqq c'_j ,\quad j=1,2,\ldots,r . \tag{3.101}$$

But (3.101) and (3.91) imply (3.89) for each $\Phi_{j\rho}$. Now let us write (3.100) as

$$\sum_{j=1}^{r}\Phi_{j\rho}(z^{(j)})+\sum_{j=1}^{r}\langle T_j(\zeta_\rho^{(j)}),z^{(j)}\rangle+\sum_{m=1}^{m'}\int_{\Omega'_m}\chi_\rho^{(m)}[z]^{(m)}d\Omega \geqq$$

$$\geqq \sum_{j=1}^{r}\Phi_{j\rho}(\zeta_\rho^{(j)})+\sum_{j=1}^{r}\langle T_j(\zeta_\rho^{(j)}),\zeta_\rho^{(j)}\rangle+\sum_{m=1}^{m'}\int_{\Omega'_m}\chi_\rho^{(m)}[\zeta_\rho]^{(m)}d\Omega +$$

$$+\sum_{j=1}^{r}(\bar{\bar{f}}^{(j)},z^{(j)}-\zeta_\rho^{(j)}),\quad \forall z^{(j)}\in Z_j,\quad j=1,2,\ldots,r . \tag{3.102}$$

For $\rho\to 0$ we obtain

$$\lim_{\rho\to 0}\{\sum_{j=1}^{r}\Phi_{j\rho}(z^{(j)})+\sum_{m=1}^{m'}\int_{\Omega'}\chi_\rho^{(m)}[z]^{(m)}d\Omega\}\geqq$$

$$\geqq \liminf_{\rho\to 0}\{\sum_{j=1}^{r}\Phi_{j\rho}(\zeta_\rho^{(j)})\}+\sum_{j=1}^{r}\langle T_j(\zeta_\rho^{(j)}),\zeta_\rho^{(j)}-z^{(j)}\rangle +$$

$$+\sum_{m=1}^{m'}\int_{\Omega'_m}\chi_\rho^{(m)}[\zeta_\rho]^{(m)}d\Omega\}+\lim_{\rho\to 0}\{\sum_{j=1}^{r}(\bar{\bar{f}}^{(j)},z^{(j)}-\zeta_\rho^{(j)})\} ,$$

$$\forall z^{(j)}\in Z_j,\quad j=1,2,\ldots,r . \tag{3.103}$$

It can be easily verified that due to the functional framework of the problem

$$\lim_{\rho\to 0}\int_{\Omega'}\chi_\rho^{(m)}[\zeta_\rho]^{(m)}d\Omega=\int_{\Omega'}\chi^{(m)}[\zeta]^{(m)}d\Omega,\quad \forall m=1,2,\ldots,m' , \tag{3.104}$$

and that

$$\liminf_{\rho\to 0}\langle T_j(\zeta_\rho^{(j)}),\ \zeta_\rho^{(j)}-z^{(j)}\rangle \geqq \langle T_j(\zeta^{(j)}),\zeta^{(j)}-z^{(j)}\rangle\ ,\ j=1,\ldots,r. \tag{3.105}$$

Relations (3.103),(3.104),(3.105),(3.87) and (3.89) imply that

$$\sum_{j=1}^{r}\{\Phi_j(z^{(j)})-\Phi_j(\zeta^{(j)})\}+\sum_{j=1}^{r}\langle T_j(\zeta^{(j)}),\ z^{(j)}-\zeta^{(j)}\rangle +$$

$$+\sum_{m=1}^{m'}\int_{\Omega'}\chi^{(m)}([z]^{(m)}-[\zeta]^{(m)})d\Omega \geqq \sum_{j=1}^{r}(\bar{\bar{f}}^{(j)},z^{(j)}-\zeta^{(j)})\ ,$$

$$\forall z^{(j)}\in Z_j\ ,\quad j=1,2,\ldots,r\ . \tag{3.106}$$

The proof is completed by proving that

$$\chi^{(m)}\in\hat{\beta}^{(m)}([\zeta]^{(m)})\quad \text{a.e. in } \Omega'_m\ ,\quad m=1,\ldots,m'\ . \tag{3.107}$$

It can be shown exactly as in the differentiable case (relations (3.79)÷(3.86)) q.e.d.

Until now we have shown the weak convergence of the approximate solution in the $H^2$-space and due to the compact imbedding $H^2(\Omega)\subset C^o(\bar{\Omega})$ the strong convergence in the $C^o$-space. From the complete continuity of operator G defined by (3.32) we deduce the strong convergence of the planar stresses in $L^2$. Further we give without proof (see e.g. [63]) a proposition which under some additional assumptions guarantees the strong convergence of the solution $\zeta$ in the $H^2$-space.

Proposition 3.3: Let the assumptions of Prop. 3.2 hold. Moreover, suppose that p,q, $\tilde{q}$ different for each m and j exist and constants c such that

$$|\beta^{(m)}(\xi)| \leq c(1+|\xi|^p), \quad \forall\xi\in\mathbb{R}, \ m=1,\dots,m', \tag{3.108}$$

$$|b_j(\xi)| \leq c(1+|\xi|), \quad \forall\xi\in\mathbb{R}, \tag{3.109}$$

$$|b_{j\rho}(\xi)| \leq c(1+|\xi|), \quad \forall\xi\in\mathbb{R}, \ j=1,\dots,r, \tag{3.110}$$

and

$$|b_j'(\xi)| \leq c(1+|\xi|^q), \quad \forall\xi\in\mathbb{R}, \tag{3.111}$$

$$|b_{j\rho}'(\xi)| \leq c(1+|\xi|^{\tilde{q}}), \quad \forall\xi\in\mathbb{R}, \ j=1,\dots,r. \tag{3.112}$$

Then, for $\varepsilon\to 0$, $n\to\infty$,

$$\zeta_{\varepsilon n}^{(j)}\to\zeta^{(j)} \quad \text{strongly in } Z_j\subset H^2(\Omega_j), \quad j=1,2,\dots,r. \tag{3.113}$$

In the next Section we deal with a semicoercive hemivariational inequality.

## 4. A SEMICOERCIVE HEMIVARIATIONAL INEQUALITY

The method developed in this Section for the study of hemivariational inequalities is general and applies to a general class of hemivariational inequalities. Let us consider for $u,v\in V$ where $V$ is a real Hilbert space, a symmetric continuous bilinear form $a(.,.):V\times V\to\mathbb{R}$. Let $V'$ be the dual space of $V$ and assume that for $\Omega$ open and bounded subset of $\mathbb{R}^n$

$$V\subset L^2(\Omega)\subset V' \tag{4.1}$$

where the injections are continuous. We denote by $(.,.)$ the $L^2$-product and duality pairing, the norm of $V$ by $\|\cdot\|$ and the norm of $L^2(\Omega)$ by $|\cdot|_2$. We recall here that the linear form $(.,.)$ extends uniquely from $V\times L^2(\Omega)$ to $V\times V'$ [67]. Further let us assume that

$$V\subset L^2(\Omega) \text{ is compact} \tag{4.2}$$

and that

$$V\cap L^\infty(\Omega) \text{ is dense in } V \text{ for the } \|\cdot\|\text{-norm}. \tag{4.3}$$

The bilinear form is assumed to have a nonzero kernel, i.e. $\ker a(u,u) = \{q \mid a(q,q) = 0\} \neq \{0\}$, and let

$$\ker a \text{ be finite dimensional} . \tag{4.4}$$

The norm $||v||$ on V is assumed to be equivalent to $|||v||| = p(\tilde{v})+|q|_2$, where $v=\tilde{v}+q$, $q\in\ker a$, $\tilde{v}\in\ker a^\perp$ (i.e. $(\tilde{v},q)=0$ $\forall q\in\ker a$), $p(\tilde{v})$ is a seminorm on V such that $p(v)=p(v+q)$ $\forall v\in V$, $q\in\ker a$ and let

$$a(v,v) \geqq c(p(v))^2 \quad \forall v\in V \quad c \text{ const}> 0 . \tag{4.5}$$

Now the following problem is formulated for $f\in L^2(\Omega)$.

Problem P: Find $u\in V$ such that

$$a(u,v-u)+\int_{\Omega'} j^0(u,v-u)d\Omega \geqq (f,v-u) \quad \forall v\in V . \tag{4.6}$$

Here $\Omega'\subset\Omega$ and j results from $\hat{\beta}\in L^\infty_{loc}(\mathbb{R})$ as in (3.9)÷(3.13). Thus j is a locally Lipschitz function such that

$$\bar{\partial}j(\xi) = \hat{\beta}(\xi) \tag{4.7}$$

where $\hat{\beta}$ is the multivalued function which results from $\beta$ by "filling in" the discontinuities. We denote further by $q_+$ and $q_-$ the positive and the negative parts of q respectively, i.e. $q_+ = \frac{q+|q|}{2}$ and $q_- = \frac{|q|-q}{2}$ .

The following proposition gives a necessary condition for the existence of the solution. Further the notation

$$\beta(-\infty) = \limsup_{\xi\to-\infty} \beta(\xi) \text{ and } \beta(\infty) = \liminf_{\xi\to\infty} \beta(\xi) \tag{4.8}$$

is used.

Proposition 4.1: Let

$$\beta(-\infty) \leqq \beta(\xi) \leqq \beta(\infty) \quad \forall\xi\in\mathbb{R} . \tag{4.9}$$

Then a necessary condition for the existence of a solution $u\in V$ of Problem P is the inequality

$$\int_{\Omega'}[\beta(-\infty)q_+-\beta(\infty)q_-]d\Omega \leqq (f,q) \leqq \int_{\Omega'}[\beta(\infty)q_+-\beta(-\infty)q_-]d\Omega \quad \forall q\in\ker a. \tag{4.10}$$

If (4.9) holds as a strict inequality (with < instead of $\leqq$) the same is the case for the necessary condition (4.10).

Proof: Let us set in (4.6) $v-u=\pm q\in \ker a$, $q\neq 0$. We obtain

$$\int_{\Omega'} j^0(u,\pm q)\,d\Omega \geqq \pm(f,q) \quad \forall q\in \ker a,\ q\neq 0 . \tag{4.11}$$

Then (4.11) is written as

$$\int_{\Omega'} j^0(u,q)\,d\Omega \geqq (f,q) \geqq -\int_{\Omega'} j^0(u,q)\,d\Omega \quad \forall q\in \ker a,\quad q\neq 0 \tag{4.12}$$

because $q\to j^0(u,q)$ is positively homogeneous. From the definition of $j$ and $j^0$ we obtain by means of (4.9) that

$$\int_{\Omega'} j^0(u,q)\,d\Omega \leqq \int_{\Omega'} [\beta(\infty)q_+ - \beta(-\infty)q_-]\,d\Omega \quad \forall q\in \ker a,\quad q\neq 0 \tag{4.13}$$

and analogously for $\int_{\Omega'} j^0(u,-q)\,d\Omega$. Thus (4.10) is shown. The rest of the proposition is obvious q.e.d. Further we give a sufficient condition.

Proposition 4.2: Let

$$\beta(-\infty) < \beta(\infty) . \tag{4.14}$$

Then if

$$\int_{\Omega'} [\beta(-\infty)q_+ - \beta(\infty)q_-]\,d\Omega < (f,q) < \int_{\Omega'} [\beta(\infty)q_+ - \beta(-\infty)q_-]\,d\Omega \quad \forall q\in \ker a,\ q\neq 0 , \tag{4.15}$$

problem P has a solution.

Proof: The proof has some parts similar to the proof of Prop. 3.1. We point out here the new arguments. First a Galerkin basis of $V\cap L^\infty(\Omega)$ is introduced. This is possible due to (4.3). Let $V_n$ be the n-dimensional subspace of $V\cap L^\infty(\Omega)$. We regularize Problem P through the mollifier $p_\varepsilon$. Then the corresponding finite dimensional problem $P_{\varepsilon n}$ is formulated.

Problem $P_{\varepsilon n}$: Find $u_{\varepsilon n}\in V_n$ such that

$$a(u_{\varepsilon n},v) + \int_{\Omega'} \beta_\varepsilon(u_{\varepsilon n})v\,d\Omega = (f,v) \quad \forall v\in V_n . \tag{4.16}$$

Now the finite dimensional variational equality (4.16) is written in the form

$$(\Lambda(u_{\varepsilon n}),v) = 0 \quad \forall v\in V_n . \tag{4.17}$$

Because of (4.14) the estimate (3.47) is obtained for $\beta_\varepsilon$. From (4.17), this estimate, and (4.5) we find that

$$(\Lambda(u_{\varepsilon n}),u_{\varepsilon n})\geq c[p(u_{\varepsilon n})]^2 - c\,|||\tilde{u}_{\varepsilon n}||| - \rho_1\rho_2 \mathrm{mes}\Omega' \quad c\ \mathrm{const}>0. \tag{4.18}$$

Now we will apply Brouwer's theorem to show that (4.16) has at least one solution $u_{\varepsilon n}$ and that $\{||u_{\varepsilon n}||\}$ is bounded. According to this theorem we have to prove that $r>0$ exists such that

$$||u_{\varepsilon n}|| = r \text{ implies } (\Lambda(u_{\varepsilon n}),u_{\varepsilon n})\geq 0 . \tag{4.19}$$

Here it will be shown that a number $M>0$ can be determined such that

$$||u_{\varepsilon n}|| > M \text{ implies } (\Lambda(u_{\varepsilon n})u_{\varepsilon n}) > 0 . \tag{4.20}$$

Then we may take $||u_{\varepsilon n}|| = r > M$. For the proof of (4.20) it is sufficient to prove that

$$(\Lambda(u_{\varepsilon n}),u_{\varepsilon n})\leq 0 \text{ implies } ||u_{\varepsilon n}||\leq c . \tag{4.21}$$

Due to (4.18) we deduce that, if $(\Lambda(u_{\varepsilon n}),u_{\varepsilon n})\leq 0$, then a constant $c>0$ exists such that

$$p(\tilde{u}_{\varepsilon n})\leq c(\sqrt{|q_{\varepsilon n}|_2}+1) \tag{4.22}$$

where $u_{\varepsilon n} = \tilde{u}_{\varepsilon n}+q_{\varepsilon n}$. Accordingly it is sufficient to prove that $(\Lambda(u_{\varepsilon n}),u_{\varepsilon n})\leq 0$ implies $|q_{\varepsilon n}|_2\leq c$, or equivalently, that a number $R>0$ can be determined such that

$$|q_{\varepsilon n}|_2 > R \text{ and } (4.22) \Rightarrow (\Lambda(u_{\varepsilon n}),u_{\varepsilon n}) > 0 . \tag{4.23}$$

This last relation will be shown now. From the definition of $\beta_\varepsilon$ we prove that

$$\beta_\varepsilon(\infty) = \lim_{\xi\to\infty}\beta_\varepsilon(\xi) = \lim_{\xi\to\infty}\int_{-\varepsilon}^{+\varepsilon}\beta(\xi-t)p_\varepsilon(t)dt \geq \lim_{\xi\to\infty}\operatorname{ess\,inf}_{|x-\xi|\leq\varepsilon}\beta(x) \geq$$

$$\geq \lim_{\xi\to\infty}\operatorname{ess\,inf}_{\xi-\varepsilon\leq x<\infty}\beta(x) = \liminf_{x\to\infty}\beta(x) = \beta(\infty) . \tag{4.24}$$

Similarly we show that $\beta_\varepsilon(-\infty)\leq\beta(-\infty)$. Thus (4.15) implies that

$$\int_{\Omega'}[\beta_\varepsilon(-\infty)q_+-\beta_\varepsilon(\infty)q_-]d\Omega<(f,q)<\int_{\Omega'}[\beta_\varepsilon(\infty)q_+-\beta_\varepsilon(-\infty)q_-]d\Omega\ ,$$

$$\forall q\in \ker a\ \ q\neq 0\ . \tag{4.25}$$

Noting that from (4.14) the inequality (3.39) results as a strict inequality, a number $\tilde{M}>\rho_1$ (cf. (3.47)) can be chosen such that for any function $\bar{u}\in V$ with $|\bar{u}(x)|>\tilde{M}$ and sign $\bar{u}(x)=$ sign $q(x)$ for almost every $x\in\Omega'$, we have from (4.25) that for $q>0$, $q=q_+$, $\bar{u}(x)>\tilde{M}$ and thus

$$\int_{\{x|q(x)>0\}}\beta_\varepsilon(\bar{u}(x))q(x)d\Omega-(f,q)>0\ , \tag{4.26}$$

$$-\int_{\{x|q(x)>0\}}\beta_\varepsilon(-\bar{u}(x))q(x)d\Omega+(f,q)>0\ . \tag{4.27}$$

For $q<0$ we have $q=-q_-$, $\bar{u}(x)<-M$ and thus we obtain from (4.25)

$$\int_{\{x|q(x)<0\}}\beta_\varepsilon(\bar{u}(x))q(x)d\Omega-(f,q)>0\ , \tag{4.28}$$

$$-\int_{\{x|q(x)<0\}}\beta_\varepsilon(-\bar{u}(x))q(x)d\Omega+(f,q)>0\ . \tag{4.29}$$

By an appropriate choice of the numbers $\delta\in(0,1]$, $N>1$, $\eta>0$, and $\alpha>0$, and by taking into account that $\beta_\varepsilon(\bar{u}(x))\bar{u}(x)\geq 0$ and that sign $\bar{u}(x)=$ sign $q(x)$ these inequalities imply for every $\bar{u}$ as above, the relations

$$(1-\frac{1}{N})\int_{\{x||q(x)|>\delta\alpha\}}\beta_\varepsilon(\bar{u}(x))q(x)d\Omega-(f,q)>\eta|q|_2 \tag{4.30}$$

$$-(1-\frac{1}{N})\int_{\{x||q(x)|>\delta\alpha\}}\beta_\varepsilon(-\bar{u}(x))q(x)d\Omega+(f,q)>\eta|q|_2 \tag{4.31}$$

as is obvious by taking $\delta\to 0_+$. We write now $u_{\varepsilon n}=\tilde{u}_{\varepsilon n}+q_{\varepsilon n}$ and for N as in (4.30),(4.31) and for $\alpha>\alpha_0=\tilde{M}\,\delta^{-1}(1-\frac{1}{N})^{-1}$ we have

$$\int_{\Omega'}\beta_\varepsilon(u_{\varepsilon n})u_{\varepsilon n}d\Omega=\int_{\substack{|\tilde{u}_{\varepsilon n}(x)|<\frac{\delta\alpha}{N}\\ |q_{\varepsilon n}(x)|>\delta\alpha}}\cdots+\int_{\substack{|\tilde{u}_{\varepsilon n}(x)|\geq\frac{\delta\alpha}{N}\\ |q_{\varepsilon n}(x)|>\delta\alpha}}\cdots+\int_{|q_{\varepsilon n}(x)|\leq\delta\alpha}\cdots\ \geq$$

$$\geq \int_{\substack{|\tilde{u}_{\varepsilon n}(x)|<\frac{\delta\alpha}{N}\\ |q_{\varepsilon n}(x)|>\delta\alpha}} \beta_\varepsilon(u_{\varepsilon n})u_{\varepsilon n}d\Gamma - \rho_1\rho_2 \mathrm{mes}\Omega' \geq$$

$$\geq (1-\frac{1}{N}) \int_{\substack{|\tilde{u}_{\varepsilon n}(x)|<\frac{\delta\alpha}{N}\\ |q_{\varepsilon n}(x)|>\delta\alpha}} q_{\varepsilon n}\beta_\varepsilon(u_{\varepsilon n})d\Omega - \rho_1\rho_2 \mathrm{mes}\Omega' \ . \tag{4.32}$$

Indeed for $|\tilde{u}_{\varepsilon n}(x)|<\delta\alpha/N$ and $|q_{\varepsilon n}(x)|>\delta\alpha$ we have that for $\alpha>\alpha_0$

$$|u_{\varepsilon n}| = |\tilde{u}_{\varepsilon n} + q_{\varepsilon n}| > (1-\frac{1}{N})\delta\alpha > \tilde{M} \tag{4.33}$$

and thus $\beta_\varepsilon(u_{\varepsilon n})u_{\varepsilon n} \geq 0$, and $\beta_\varepsilon(u_{\varepsilon n})q_{\varepsilon n} \geq 0$. Further it can be verified that for $q_{\varepsilon n} > \delta\alpha$

$$u_{\varepsilon n}(x) = \tilde{u}_{\varepsilon n}(x)+q_{\varepsilon n}(x) > -\frac{\delta\alpha}{N} + q_{\varepsilon n}(x) > q_{\varepsilon n}(x)(1-\frac{1}{N}) \tag{4.34}$$

and therefore

$$\beta_\varepsilon(u_{\varepsilon n})u_{\varepsilon n} \geq (1-\frac{1}{N})q_{\varepsilon n}\beta_\varepsilon(u_{\varepsilon n}) \ . \tag{4.35}$$

Similarly for $q_{\varepsilon n}<-\delta\alpha$. Now we get the inequality

$$(\Lambda(u_{\varepsilon n}),u_{\varepsilon n}) \geq (1-\frac{1}{N}) \int_{\substack{|\tilde{u}_{\varepsilon n}(x)|<\frac{\delta\alpha}{N}\\ |q_{\varepsilon n}(x)|>\delta\alpha}} q_{\varepsilon n}\beta_\varepsilon(u_{\varepsilon n})d\Omega \ -$$

$$- \rho_1\rho_2 \mathrm{mes}\Omega' - (f,q_{\varepsilon n}) - (f,\tilde{u}_{\varepsilon n}) \ . \tag{4.36}$$

Using (4.30), we obtain for $\alpha>\alpha_0$ sufficiently large from (4.36), that

$$(\Lambda(u_{\varepsilon n}),u_{\varepsilon n}) > \eta|q_{\varepsilon n}|_2 - \rho_1\rho_2 \mathrm{mes}\Omega' - (f,\tilde{u}_{\varepsilon n}) \ \geq$$

$$\geq \eta|q_{\varepsilon n}|_2 - c_1 - c_2\,||\tilde{u}_{\varepsilon n}|| \geq \eta|q_{\varepsilon n}|_2 - c_1 - c_2'\,|||\tilde{u}_{\varepsilon n}||| =$$

$$= \eta|q_{\varepsilon n}|_2 - c_1 - c_2'\, p\,(\tilde{u}_{\varepsilon n}) \qquad c_1, c_2, c_2' \text{ const} > 0 \ . \tag{4.37}$$

From (4.37) we obtain assuming that $\alpha>\alpha_0$ and that (4.22) holds the final estimate

$$(\Lambda(u_{\varepsilon n}),u_{\varepsilon n})>\eta|q_{\varepsilon n}|_2-c|q_{\varepsilon n}|_2^{1/2}-c', \quad c,c' \text{ const}>0 . \tag{4.38}$$

The right-hand side of (4.38) is positive if we choose $R>0$ such that $|q_{\varepsilon n}|_2>R>\delta\alpha_0(\text{mes}\Omega)^{1/2}$. Thus (4.23) has been proved. Thus by Brouwer's fixed point theorem, problem $P_{\varepsilon n}$ has a solution $u_{\varepsilon n}$ with $||u_{\varepsilon n}||<c$. The rest of the proof is the same as the proof of prop. 3.1: From the boundedness of $u_{\varepsilon n}$ we deduce the weak convergence in $V$ and the strong convergence in $L^2(\Omega)$ because of (4.2), and then using Dunford-Pettis theorem we show as in (3.53)÷(3.61) the weak convergence of $\beta_\varepsilon(u_{\varepsilon n})$ to $\chi$ in $L^1(\Omega')$. Because of (4.3) we pass to the limit $n\to\infty$ $\varepsilon\to 0$ in (4.16) and then by the same procedure as the one indicated by (3.79)÷(3.86) we show that $\chi\in\hat{\beta}(u)$ a.e. in $\Omega'$. q.e.d.

The sufficient condition is similar to the sufficient conditions derived in the Landesman-Lazer theory [68],[69]. Concerning the number of solutions both for the coercive and the semicoercive problem most questions are still open. There are some results along the lines of the theory of semilinear differential equations and more precisely in the area of the contemporary development of the Landesman-Laser theory, which might be extended or directly applied to the theory or hemivariational inequalities. However, the problem of the multiplicity of solutions should be considered as still open.

## 5. APPLICATION OF QUASIDIFFERENTIABILITY TO MECHANICS[(1)]

### 5.1 Generalities and Mathematical Preliminaries

The aim of the present Section is to introduce the notion of quasidifferentiability into Mechanics. Until now we have used nonsmooth energy functions, which, through the subdifferential in the convex case and the generalized gradient in the nonconvex case, give rise to variational or to hemivariational inequalities respectively. Any research

---

(1) The author is indebted to Professor V.F. Demyanov for introducing him to the mathematical theory of quasidifferentiability. This Section is based on an additional Appendix in the Russian translation of the author's book [12].

effort on "nonsmooth mechanics" is closely connected with the careful examination of the generalizations of the classical derivative, especially concerning the possibility of formulating and studying through them mechanical problems involving nonsmooth energy functions. But this is not the only reason why the notion of quasidifferentiability is important for the study of nonsmooth mechanical problems. Let us examine this point more carefully: The subdifferential of convex analysis seems to be a well-defined notion from the standpoint of mechanics; no logical discrepancy appears. The same does not hold for the nonconvex case. Indeed the comparison of the generalized gradient with the derivate container [70] as well as with the quasidifferentiability [71]÷[73] has shown that the generalized gradient "does not contain all possible information on the directional behaviour" of a function at a point. This is caused mainly by the fact that the definition of the generalized gradient $\bar{\partial}f$ is based on the notions of the directional differentials $f^{\uparrow}(.,.)$ and $f^{o}(.,.)$ which are roughly speaking "approximations" of the directional (Gâteaux) differential $\tilde{f}'(.,.)$. On the other hand the "cumulative behaviour of a function" described by the generalized gradient of F.H. Clarke enables us to formulate in Mechanics a wide class of variational formulations in inequality form, the hemivariational inequalities, whenever a mechanical law is expressed in terms of generalized gradients. The generalized gradient $\bar{\partial}f$ leads to hemivariational inequalities because of its definition which is based on the definitions of $f^{\uparrow}(.,.)$ and/or $f^{o}(.,.)$. These definitions have as a main task to achieve for nonconvex nondifferentiable functionals the validity of a relation analogous to the well-known relation

$$\tilde{f}'(x,h) = \max\{\langle x',h\rangle \mid x'\in\partial f(x)\} \quad \forall h\in X \tag{5.1}$$

holding for convex functionals. Indeed, it was proved that a similar relation holds for the directional differential for a maximum type function (see e.g. [73]). Therefore, whenever $\bar{\partial}f(x)\neq\emptyset$ we have for every $y\in X$ the validity of

$$f^{\uparrow}(x,y) = \max\{\langle y,x'\rangle \mid x'\in\bar{\partial}f(x)\} \ , \tag{5.2}$$

if $f^{\uparrow}(x,y)$ is finite for every y. The quasidifferential calculus is appropriately developed with the task to describe the "directional properties" of a function at a point accurately. It leads - in contrast to the calculus of the generalized gradient - to equalities and not to inclusions. Therefore the class of quasidifferentiable functions is a linear space closed with respect to all algebraic operations. All these properties of the quasidifferentiability make this notion very attractive in the theory of optimization. But this notion is equally important in Mechanics for two main reasons: Suppose that in the framework of a mechanical theory a minimum problem is formulated involving non-necessarily differentiable convex or nonconvex functions. For instance, consider the problems of minimum potential energy for a Hencky-body (or an elastic ideally locking body) in unilateral contact with a rigid support, or also a minimum weight problem. In this case it is important, from the numerical point of view, to accurately determine for the function to be minimized the descent directions at given points. This fact reveals the possible advantages of the quasidifferential (cf. [72] p.11). The resulting necessary and sufficient conditions from the theory of quasidifferentiability, when interpreted in terms of mechanical notions, give a more clear description of the behaviour of the mechanical system. Another, more profound reason for the introduction of the quasidifferentiability into Mechanics is the following: When we study a mechanical system we try actually to find relations between the directional variations of certain orders, of the quantities involved in the problem. This general idea already appears in the literature of mechanics in a hidden form concerning differentiable processes (cf. e.g. the zero, first, second gradient mechanical theories [86]).

Before entering into the main part of this Section we would like to point out that especially in the applied mechanics it is believed that always a nonsmooth problem can be approximated by smooth problems. This approach which is very useful in proving existence and general approximation results may lead to erroneous results in the case of a purely numerical procedure because many properties of the original nonsmooth problem can be lost during the smoothening operation. To this context we refer the reader for a very illustrative example to [74] p.7.

Let us define first the notion of quasidifferential introduced in 1979 by V.F. Demyanov, L.N. Polyakova and A.M. Bubinov [76],[77]. Let $f:A\to\mathbb{R}$ where $A$ is an open subset of $\mathbb{R}^n$. The functional $f$ is said to be quasidifferentiable at $x\in A$ if it is directionally differentiable at $x$ (i.e. $\tilde{f}'(x,h)$ exists for every $h\in\mathbb{R}^n$ (cf. (2.7)) and if two convex compact sets $\underline{\partial}'f(x)\subset\mathbb{R}^n$ and $\overline{\partial}'f(x)\subset\mathbb{R}^n$ can be determined such that

$$\tilde{f}'(x,h)=\max_{x_1}\{\langle x_1,h\rangle \mid x_1\in\underline{\partial}'f(x)\}+\min_{x_2}\{\langle x_2,h\rangle \mid x_2\in\overline{\partial}'f(x)\}\quad \forall h\in\mathbb{R}^n . \tag{5.3}$$

(5.3) is equivalently written in the form

$$\tilde{f}'(x,h)=\min_{x_2}\max_{x_1}\{\langle x_1+x_2,h\rangle \mid x_1\in\underline{\partial}'f(x),x_2\in\overline{\partial}'f(x)\}. \tag{5.3a}$$

Here $\langle .,.\rangle$ denotes the $\mathbb{R}^n$-inner product. The pair of sets

$$Df(x)=\{\underline{\partial}'f(x),\overline{\partial}'f(x)\} \tag{5.4}$$

is called a quasidifferential of $f$ at $x$. We call the sets $\underline{\partial}'f(x)$ and $\overline{\partial}'f(x)$ $\partial'$-subdifferential and $\partial'$-superdifferential of $f$ at $x$ respectively and we use the prime in order to indicate that no confusion should be made with the subdifferential $\partial f(x)$ of convex analysis(1). From (5.3) and (5.4) it results that $Df(x)$ is not uniquely determined at $x$. This is not considered a disadvantage. Indeed each form of $Df(x)$ describes equally "well" and "reliably" the directional properties of $f$ at $x$. The above definition can be generalized by replacing $\mathbb{R}^n$ by a locally convex Hausdorff topological vector space $X$. Then $\underline{\partial}'f(x)$ and $\overline{\partial}'(x)$ are $\Sigma(X',X)$-compact subsets of the dual space $X'$ and $\langle .,.\rangle$ denotes the duality pairing between $X$ and $X'$. For further generalizations see [71].

---

(1) In references [71÷77] the prime is omitted as well as in $\partial'$-subdifferential and $\partial'$-superdifferential the symbol $\partial'$.

If f is continuously differentiable on A then f is quasidifferentiable at every $x\in A$ and

$$Df(x)=\{\operatorname{grad}f(x),0\} \quad \text{or} \quad Df(x)=\{0,\operatorname{grad}f(x)\}\ . \tag{5.5}$$

Thus f is both $\partial'$-subdifferentiable and $\partial'$-superdifferentiable at x. If f is convex defined on a convex set A then (5.3) is written in the form (5.1). Accordingly

$$Df(x)=\{\partial f(x),0\}\ . \tag{5.6}$$

Similarly, if f is concave and A is convex then $f_1=-f$ is convex and we obtain that

$$Df(x)=\{0,\tilde{\partial}f(x)\}\ , \tag{5.7}$$

where

$$\tilde{\partial}f(x)=\{x'\in X'\,|\,f(x_1)-f(x)\leq \langle x',x_1-x\rangle\ \forall x_1\in X\} \tag{5.8}$$

is the superdifferential of the concave function f at x [61]. Analogously we may obtain for a $\bar{\partial}$-regular function (see (2.7)) such that $f^{\uparrow}(x,h)\neq\pm\infty$ for every h, that

$$\tilde{f}'(x,h)=f^{\uparrow}(x,h)=\max\{\langle x_1,h\rangle\,|\,x_1\in\bar{\partial}f(x)\} \tag{5.9}$$

and thus

$$Df(x)=\{\bar{\partial}f(x),0\}, \tag{5.10}$$

where $\bar{\partial}f(x)$ is the generalized gradient. For instance, if f is a maximum type function, e.g. $f=\max\{\phi_1,\dots,\phi_m\}$, where $\phi_i=\phi_i(x)$ i=1,...,m, $x\in\mathbb{R}^n$, are smooth functions, then we know that f is $\bar{\partial}$-regular and thus

$$Df(x)=\{\operatorname{co}\{\operatorname{grad}\phi_i(x),\dots,\operatorname{grad}\phi_k(x)\},0\} \tag{5.11}$$

if $x\in\{x\,|\,\phi_i=f,\dots,\phi_k=f\}$, where co denotes the convex hull. This formula can be shown independently of the theory of generalized gradient [73]. Now we consider pairs of sets $\Gamma_i=\{A_i,B_i\}$, i=1,2, where $A_i\subset X'$ and $B_i\subset X'$ and we define their addition by

$$\Gamma_1+\Gamma_2=\{A_1+A_2,\ B_1+B_2\} \tag{5.12}$$

and the multiplication by $\lambda\in\mathbb{R}$ as

$$\lambda\Gamma_i = \begin{cases} \{\lambda A_i, \lambda B_i\} & \text{if } \lambda \geq 0 \\ \{\lambda B_i, \lambda A_i\} & \text{if } \lambda < 0 . \end{cases} \tag{5.13}$$

Using these definitions we can show that, if $f_i$, $i=1,\dots,n$, are quasidifferentiable functions at $x$ then $g = \sum_i a_i f_i$ is also quasidifferentiable at $x (a_i \in \mathbb{R})$ and $Dg = \sum_i a_i Df_i$. Moreover it can be shown that if $f_1$ and $f_2$ are quasidifferentiable at $x$ then $g = f_1 f_2$ has the same property and

$$Dg(x) = f_1(x) Df_2(x) + f_2(x) Df_1(x) \; ; \tag{5.14}$$

moreover for $f_2(x) \neq 0$, $g = f_1/f_2$ is quasidifferentiable at $x$ and

$$Dg(x) = [f_2(x) Df_1(x) - f_1(x) Df_2(x)]/f_2^2(x) . \tag{5.15}$$

The following properties are of importance: With $\phi_i$, $i=1,\dots,n$, $f = \max_i\{\phi_i\}$ is quasidifferentiable at $x$ and for $I(x) = \{i \mid \phi_i(x) = f(x)\}$

$$\underline{\partial}' f(x) = \operatorname{co}\{\underline{\partial}'\phi_k(x) - \sum_{\substack{i \in I(x) \\ i \neq k}} \overline{\partial}'\phi_i(x) \mid k \in I(x)\} \tag{5.16}$$

$$\overline{\partial}' f(x) = \sum_{k \in I(x)} \overline{\partial}'\phi_k(x) . \tag{5.17}$$

Analogous formula holds for a minimum type function $f$. In this case $\overline{\partial}' f(x)$ (resp. $\underline{\partial}' f(x)$) is given by (5.16) (resp. (5.17)) by replacing in the right hand side $\underline{\partial}'$ by $\overline{\partial}'$ and $\overline{\partial}'$ by $\underline{\partial}'$. If $f = f_1 + f_2$ where $f_1$ (resp. $f_2$) is convex (resp. concave) and $f_1(x)$, $f_2(x)$ are finite then $f$ is quasidifferentiable and we have that [78]

$$Df(x) = \{\partial f_1(x), \tilde{\partial} f_2(x)\} . \tag{5.18}$$

Due to (5.3a) for every $x_2 \in \overline{\partial}' f(x)$ (resp. $x_1 \in \underline{\partial}' f(x)$) the function $\max\{\langle q_1, \cdot\rangle \mid q_1 \in x_2 + \underline{\partial}' f(x)\} = r_{x_2}(\cdot)$ (resp. $\min\{\langle q_2, \cdot\rangle \mid q_2 \in x_1 + \overline{\partial}' f(x)\} = r_{x_1}(\cdot)$) is an upper convex (resp. a lower concave) approximation of $f$ at $x$ and the set of functions $\{r_{x_2}(\cdot) \mid x_2 \in \overline{\partial}' f(x)\}$ (resp. $\{r_{x_1}(\cdot) \mid x_1 \in \underline{\partial}' f(x)\}$) represents an exhaustive family of upper convex (resp. lower concave) approximations of $f$ at $x$ [74]. It can be easily shown [73] that if $f$ is quasidifferentiable on X a necessary condition for the existence of a solution of the problem

$$f(x_0) = \min\{f(x) \mid x \in X\} \tag{5.19}$$

resp.

$$f(\tilde{x}_0) = \max\{f(x) \mid x \in X\} \tag{5.20}$$

is

$$-\bar{\partial}' f(x_0) \subset \underline{\partial}' f(x_0) \tag{5.21}$$

resp.

$$-\underline{\partial}' f(\tilde{x}_0) \subset \bar{\partial}' f(\tilde{x}_0) \ . \tag{5.22}$$

A point $x_0$ (resp. x) satisfying (5.21) (resp. (5.22)) is called an inf-stationary (resp. a sup-stationary) point of f on X. It is also verified that if $x_0$ does not satisfy (5.21) and $X = \mathbb{R}^n$ then the direction

$$\xi = -\frac{v_o + w_o}{\|v_o + w_o\|} \ , \ \max_w \min_v \{\|v+w\| \mid w \in \bar{\partial}' f(x_0), v \in \underline{\partial}' f(x_0)\} = \|v_o + w_o\| \tag{5.23}$$

is a direction of steepest descent (not unique) of f at $x_0$, and if $x_0$ does not satisfy (5.22) then the direction

$$\xi = \frac{\tilde{v}_o + \tilde{w}_o}{\|\tilde{v}_o + \tilde{w}_o\|} \ , \ \max_v \min_w \{\|v+w\| \mid w \in \bar{\partial}' f(x_0), v \in \underline{\partial}' f(x_0)\} = \|\tilde{v}_o + \tilde{w}_o\| \tag{5.24}$$

is a direction of steepest ascent (not unique) of f at $x_0$.

## 5.2 Quasidifferentiability in Mechanics

With respect to a mechanical system $\Sigma$ characterized by the triplet $\Sigma = \{U, \langle u,f \rangle, F\}$ any mechanical law of the form (2.7), where $\Phi$ is $\bar{\partial}$-regular can be written in the quasidifferential form

$$\{-f, 0\} \in D\Phi(u) = \{\bar{\partial}\Phi(u), 0\} \ . \tag{5.25}$$

This holds for all theories based on nonconvex superpotentials which are $\bar{\partial}$-regular. This remark enables us to apply the results of quasidifferential calculus for the case of $\bar{\partial}$-regularity, and therefore to obtain sharper "optimality" conditions than with the calculus of generalized gradients [79].

With respect to the same mechanical system we may introduce a more general law of the form

$$\{-\bar{f},-\bar{\bar{f}}\}\in D\Phi(u)=\{\underline{\partial}'\Phi(u),\bar{\partial}'\Phi(u)\} \tag{5.26}$$

where $\bar{f},\bar{\bar{f}}\in F$ and $\Phi$ is a quasidifferentiable function. The greater possibilities offered by (5.26) in comparison with (5.25) are obvious. To explain better (5.26) let us assume, for the sake of simplicity, that $U=\mathbb{R}^n$ and $F=\mathbb{R}^n$. For a bounded subset $A\subset F$ we define the support function

$$s_A(u)=\sup\{\langle u,f\rangle \mid f\in A\}\ . \tag{5.27}$$

Every support function is sublinear (i.e. it is subadditive: $s_A(u_1+u_2)\leq s_A(u_1)+s_A(u_2)$, and positively homogeneous), and every sublinear function h is conversely the support function of a convex compact set which is uniquely determined by the formula

$$A=\{f \mid \langle u,f\rangle \leq h(u)\ \forall u\in U\}\ . \tag{5.28}$$

Accordingly (5.26) can be put by means of the definition (5.3) in the form

$$\tilde{\Phi}'(u,v)=s_{\underline{\partial}'\Phi(u)}(v)-s_{-\bar{\partial}\Phi(u)}(v)\ \ \forall v\in U\ . \tag{5.29}$$

Obviously

$$\underline{\partial}'\Phi(u)\ =\{f \mid \langle f,v\rangle \leq s_{\underline{\partial}'\Phi(u)}(v)\ \forall v\in U\} \tag{5.30}$$

and analogous inequality is valid also for $\bar{\partial}'\Phi(u)$. Thus in (5.29) we have a difference of sublinear functions. Let us denote by Q the class of all convex, compact subsets of F. According to the previous results with every function $\varphi$ which is the difference of two sublinear functions we may associate $A\in Q$ and $B\in Q$ such that

$$\varphi(u)=\max\{\langle u,\bar{f}\rangle \mid \bar{f}\in A\}+\min\{\langle u,\bar{\bar{f}}\rangle \mid \bar{\bar{f}}\in B\}=s_A(u)-s_B(u)\ . \tag{5.31}$$

On $Q\times Q$ we introduce an equivalence relation $\mathcal{R}$ is introduced such that $\{A_1,B_1\}\mathcal{R}\{A_2,B_2\}$ if and only if $A_1-B_2=A_2-B_1$. Further let $\tilde{Q}$ be the quotient space $(Q\times Q)/\mathcal{R}$ and let us denote by [A,B] the equivalence class with representant {A,B}. On $\tilde{Q}$ a vector space structure is imposed by means of (5.12),(5.13). Moreover the norm induced by the Hausdorff distance $d(A,-B)$ turns $\tilde{Q}$ into a normed-space [80]. The vector space $\tilde{\Theta}$ of functions which are the difference of sublinear functions is a normed space for the sup-norm $|||\varphi||| = \sup\{|\varphi(u)|\,|\,u\in U, ||u||=1\}$ and it has been proved [76] that the correspondence $[A,B]\to s_A(.)-s_{-B}(.)$ is an isometry. The above explain the meaning of (5.26). Several mechanical laws of the same form may be added and we shall have for $i=1,\dots,n$

$$\{-\sum_i \bar{f}_i, -\sum_i \bar{\bar{f}}_i\}\in\sum_i D\Phi_i(u) = D\sum_i\Phi_i(u) = \{\underline{\partial}'\sum_i\Phi_i(u), \bar{\partial}'\sum_i\Phi_i(u)\} \;. \tag{5.32}$$

If some of the $\Phi_i$'s in (5.32) are maximum or minimum type functions, as it happens in the case of generalized locking phenomena, then (5.16) and (5.17) permit to write the mechanical law in an interesting multiplier form (due to the appearance of the convex hull). If in (5.32) $\sum_i\Phi_i(u) = \varphi_1(u)+\varphi_2(u)$, where $\varphi_1$ (resp. $\varphi_2$) is convex (resp. concave), then (5.18) implies for $\varphi_1(u)$, $\varphi_2(u)$ finite that

$$\varphi_1(v)-\varphi_1(u) \geq \langle -\sum_i \bar{f}_i, v-u\rangle \quad \forall v\in U \;, \tag{5.33}$$

$$\varphi_2(v)-\varphi_2(u) \leq \langle -\sum_i \bar{\bar{f}}_i, v-u\rangle \quad \forall v\in U \;. \tag{5.34}$$

If the position of equilibrium is characterized by $\sum_i(\bar{f}_1+\bar{\bar{f}}_2) = 0$ then the solution with respect to u of the system of this last equation together with the variational inequalities (5.33,34) supplies the possible positions of equilibrium. Analogously we operate in the general case (5.32) if the position of equilibrium is characterized by a relation (or relations) of the general form $\sum_i(a_i\bar{f}_i+b_i\bar{\bar{f}}_i) = 0$, with $a_i, b_i\in\mathbb{R}$. Finally we can consider mechanical laws of the form

$$\{\bar{u},\bar{\bar{u}}\}\in D\Phi_1(-f) \;. \tag{5.35}$$

Now let us consider a deformable body $\Omega\subset\mathbb{R}^3$ with a fixed boundary $\Gamma$. The principle of virtual work implies ($u_i=0$ on $\Gamma$ $i=1,2,3$)

$$\int_\Omega \sigma_{ij}(u)\varepsilon_{ij}(v-u)d\Omega - \int_\Omega f_i(v_i-u_i)d\Omega = 0 \quad \forall v\in U_{ad} \;, \tag{5.36}$$

where $U_{ad}$ is the kinematically admissible set of the displacements (small strain theory). We assume material laws of the form

$$\sigma_{ij}\varepsilon^*_{ij} = \tilde{w}'(\varepsilon,\varepsilon^*) \quad \forall\varepsilon^* = \{\varepsilon^*_{ij}\} \in \mathbb{R}^6 , \tag{5.37}$$

resp.

$$\sigma_{ij}\varepsilon^*_{ij} \leqq \tilde{w}'(\varepsilon,\varepsilon^*) \quad \forall\varepsilon^* = \{\varepsilon^*_{ij}\} \in \mathbb{R}^6 . \tag{5.38}$$

We are led to the following problem for $f_i \in L^2(\Omega)$.

Problem 1: Find $u \in [\overset{\circ}{H}{}^1(\Omega)]^3$ such as to satisfy the variational equality (resp. the hemivariational inequality)

$$\int_\Omega (\tilde{w}'(\varepsilon(u),\varepsilon(v-u)) - f_i(v_i-u_i))d\Omega = (\text{resp.} \geqq)\ 0 \quad \forall v \in [\overset{\circ}{H}{}^1(\Omega)]^3 . \tag{5.39}$$

(resp.(5.40))

The material law (5.37) describes hyperelastic materials. In the more general law (5.38) the inequality means the appearance of certain dissipation mechanisms. Suppose now that w is quasidifferentiable, i.e. we can determine convex and compact sets $A = \underline{\partial}'w(\varepsilon)$ and $B = \overline{\partial}'w(\varepsilon)$ such that

$$\tilde{w}'(\varepsilon,\varepsilon^*) = \max\{\bar{\tau}_{ij}\varepsilon^*_{ij} \mid \bar{\tau} = \{\tau_{ij}\} \in A\} + \min\{\bar{\bar{\tau}}_{ij}\varepsilon^*_{ij} \mid \bar{\bar{\tau}} = \{\bar{\bar{\tau}}_{ij}\} \in B\} = s_A(\varepsilon^*) - s_{-B}(\varepsilon^*) \quad \forall\varepsilon^* \in \mathbb{R}^6 . \tag{5.41}$$

Using the notation $\bar{\tau}_{ij}\varepsilon^*_{ij} = (\bar{\tau},\varepsilon^*)$ etc, we obtain the relation

$$\tilde{w}'(\varepsilon,\varepsilon^*) \geqq (\bar{\tau},\varepsilon^*) + \min\{(\bar{\bar{\tau}},\varepsilon^*) \mid \bar{\bar{\tau}} \in \overline{\partial}'w(\varepsilon)\} \quad \forall\bar{\tau} \in \underline{\partial}'w(\varepsilon),\ \forall\varepsilon^* \in \mathbb{R}^6 \tag{5.42}$$

and further

$$\exists\bar{\bar{\tau}} \in \overline{\partial}'w(\varepsilon) \text{ such that } \tilde{w}'(\varepsilon,\varepsilon^*) \geqq (\bar{\tau} + \bar{\bar{\tau}},\varepsilon^*) \quad \forall\varepsilon^* \in \mathbb{R}^6 \text{ and } \forall\bar{\tau} \in \underline{\partial}'w(\varepsilon) . \tag{5.43}$$

Therefore (5.39) has the following equivalent formulations.

Problem 2: Find $u \in [\overset{\circ}{H}{}^1(\Omega)]^3$ such as to satisfy

$$\exists\bar{\bar{\tau}}\in\bar{\partial}'w(\varepsilon(u)) \text{ such that } \int_\Omega f_i(v_i-u_i)d\Omega \geq \int_\Omega(\bar{\tau}+\bar{\bar{\tau}},\varepsilon(v-u))d\Omega$$

$$\forall\bar{\tau}\in\underline{\partial}'w(\varepsilon(u)) \text{ and } \forall v\in[\overset{\circ}{H}{}^1(\Omega)]^3 , \qquad (5.44)$$

or equivalently,

Problem 2': Find $u\in[\overset{\circ}{H}{}^1(\Omega)]^3$ such as to satisfy the relation

$$\int_\Omega[\min_{\bar{\bar{\tau}}\in B}\max_{\bar{\tau}\in A}(\bar{\tau}+\bar{\bar{\tau}},\varepsilon(v))-f_iv_i]d\Omega=0$$

$$\forall v\in[\overset{\circ}{H}{}^1(\Omega)]^3 . \qquad (5.45)$$

In the case of (5.40) the equality to zero in (5.45) is replaced by $\geq 0$.

Several materials obey the foregoing material law. For instance materials having as a strain energy density w a finite function which is the sum of a convex and a concave finite part i.e. $w=w_1+w_2$. Then $\underline{\partial}'w(\varepsilon)=\partial w_1(\varepsilon)$ and $\bar{\partial}'w(\varepsilon)=\tilde{\partial}w_2(\varepsilon)$. Analogous is the case of a strain energy which with respect to certain components $\varepsilon_1$ of the strain tensor is convex and with respect to the remaining components $\varepsilon_2$ is concave. Let for instance $\varepsilon\to w(\varepsilon)=w(\varepsilon_1,\varepsilon_2)$ be finite on $E_1\times E_2\subset\mathbb{R}^6$ where $E_1\subset\mathbb{R}^\alpha$, $E_2\subset\mathbb{R}^\beta$, $\alpha+\beta=6$, are open and convex. Then $w(.,\varepsilon_2)$ (resp. $w(\varepsilon_1,.)$) is convex (resp. concave) and the sets $\partial_{\varepsilon_1}w(.,\varepsilon_2)$ - the subdifferential - and $\tilde{\partial}_{\varepsilon_2}w(\varepsilon_1,.)$ - the superdifferential - are nonempty, compact, convex sets. It can be proved [73] p.232) that $\varepsilon\to w(\varepsilon_1,\varepsilon_2)$ is quasidifferentiable on $E_1\times E_2$, and that at $\varepsilon=(\varepsilon_1,\varepsilon_2)$

$$\begin{aligned}&\underline{\partial}'w(\varepsilon_1,\varepsilon_2)=\{\partial_{\varepsilon_1}w(\varepsilon_1,\varepsilon_2),0\}\in\mathbb{R}^\alpha\times\mathbb{R}^\beta \text{ and}\\&\bar{\partial}'w(\varepsilon_1,\varepsilon_2)=\{0,\tilde{\partial}_{\varepsilon_2}w(\varepsilon_1,\varepsilon_2)\}\in\mathbb{R}^\alpha\times\mathbb{R}^\beta\end{aligned} \qquad (5.46)$$

Note that analogous formulations are obtained if we have an elastic body and if the boundary condition

$$-S_iv_i = (\text{or} \leq)\ \tilde{j}'(u,v) \quad \forall v\in\mathbb{R}^3 \qquad (5.47)$$

holds where j is quasidifferentiable.

To give a more concrete example we consider a "total" contact law of the form

$$-S_{T_i} v_{T_i} - S_N u_N = \tilde{j}'(u_T, u_N; v_T, v_N) \tag{5.48}$$

where $u_N \to j(u_T, u_N)$ is concave, ore more generally $\overline{\partial}$-regular, and finite and $u_T \to j(u_T; u_N)$ is convex and finite. For instance, we can take as $u_T \to j(u_T, u_N) = c|u_T|$ - the classical friction potential - and as $u_N \to j(u_T, u_N)$ any $\overline{\partial}$-regular function, e.g. the function resulting from the delaminations law. We can show easily by modifying slightly the proof of [73] p.232, that $(u_T, u_N) \to j(u_T u_N)$ is quasidifferentiable and that conditions analogous to (5.46) hold, with $\tilde{\partial}$ replaced by $-\overline{\partial}$.

The "weakness" of the generalized gradient results roughly speaking from the lack of distinction of the convex and the concave "parts" of the function w under consideration. The influence of the concave (resp. convex) "part" on the directional variations of w is described more accurately by the quasidifferential calculus. For all the minimum problems characterizing positions of equilibrium and containing nonconvex energy functionals a common characteristic is that every local minimum corresponds to a position of equilibrium but not conversely. Indeed in the context of generalized gradients every local minimum is a substationarity point and therefore corresponds to an equilibrium position. Analogous is the situation in the large displacement elasticity [81],[82],[83]. The resulting minimum problems are generally nonconvex and nondifferentiable in the presence of unilateral constraints. The local minima can be checked by the sharp criteria developed by V.F. Demyanov (see e.g. [84]). We shall not elaborate on this idea here and we leave it as an exercise for the reader. We note simply that the resulting interesting expressions of the necessary and sufficient conditions reveal once more the new possibilities offered by the quasidifferentiability. It is true [77] that every closed cone C is associated with a function $\varphi$ which is the difference of two sublinear functions such that $C = \{x \in \mathbb{R}^n \mid \varphi(x) \leq 0\}$.

According to (5.31) we may write that

$$C=\{x\in\mathbb{R}^n \mid \varphi(x)=s_A(x)-s_{-B}(x)\leq 0\}. \tag{5.49}$$

Then the set of all $\varphi$'s satisfying (5.49) is a closed convex cone n[C] in the space $\tilde{\Theta}$ and the set of the corresponding elements [A,B] of $\tilde{Q}$ forms a convex closed cone N[C] in the space $\tilde{Q}$. Now let us consider a closed set $T\subset\mathbb{R}^n$ and $x_0\in T$. Then a vector y is said to be tangent to T at $x_0$ if a sequence $\{x_n\}\in T$ and a sequence $\{t_n\},t_n>0$, with $t_n\to 0_+$ can be determined such that $\lim(x_n-x_0)/t=y$ as $n\to\infty$. All the tangent vectors to T at $x_0$ form the tangent cone $\tau_T(x_0)$ which is closed. Suppose now that $[A,B]\in\tilde{Q}$. We say that [A,B] is quasinormal to T at $x_0$ if for C a closed cone such that $C\subset\tau_T(x_0)$ we have that $[A,B]\in N[C]$. Further let us denote by $n_T(x_0)$ the set of all quasinormals to T at $x_0$. Suppose further that $T=\{x\mid p(x)\leq 0\}$ where p is a quasidifferentiable function and let $x_0$ be such that $p(x_0)=0$. Let us denote by r(.) the function $(s_{\underline{\partial}'p(x_0)}-s_{-\bar{\partial}'p(x_0)})(\cdot)$ and let the nondegeneracy assumption $\overline{\{x\mid r(x)<0\}}=\{x\mid r(x)\leq 0\}$ hold. Then

$$Dp(x_0)=\{\underline{\partial}'p(x_0),\bar{\partial}'p(x_0)\} \tag{5.50}$$

is a quasinormal corresponding to $\tau_T(x_0)$. After this mathematical introduction we define plasticity laws of the form

$$\dot{e}^p=\{\dot{e}^p_1,\dot{e}^p_2\}\in\{\lambda Dp(\sigma)\mid\lambda\geq 0\}. \tag{5.51}$$

This general law includes as a special case the nonassociated plasticity laws for convex yield surfaces.

We shall close this Section with the following interesting observation. Rubinov and Yagubov have shown in [85], that, if $\tilde{f}'(x;h)$ is continuous in h then a pair of star-shaped sets {U,V} exists such that

$$\tilde{f}'(x,h)=\max\{\lambda<0\mid h\in-\lambda V\}+\min\{\lambda>0\mid h\in\lambda U\}. \tag{5.52}$$

We recall here that a closed set $U\subset\mathbb{R}^n$ is called star-shaped if $0\in\operatorname{int}U$

and every ray $\{\lambda x | \lambda \geq 0, x \in U\}$ does not intersect in more than one point the boundary of U. As it is known [20],[87] star-shaped surfaces constitute a very general type of yield surfaces in the theory of plasticity. The expected connection of (5.52) and the calculus developed in [85] with the theory of plasticity and the limit load analysis is still an open problem.

REFERENCES:

[1] Fichera, G.: Problemi elastostatici con vincoli unilaterali: il problema di Signorini con ambigue condizioni al contorno, Mem. Accad. Naz. Lincei, VIII 7 (1964), 91-140.
[2] Fichera, G.: Existence Theorems in Elasticity. In: Encyclopedia of Physics (ed. by S. Flügge) Vol. VIa/2, Springer-Verlag, Berlin 1972.
[3] Lions, J.L. and G. Stampacchia: Variational Inequalities, Comm. pure and applied Math., XX (1967), 493-519.
[4] Moreau, J.J.: La notion de sur-potentiel et les liaisons unilatérales en élastostatique, C.R. Acad. Sc. Paris, 267A (1968), 954-957.
[5] Panagiotopoulos, P.D.: Non-convex Superpotentials in the Sence of F.H. Clarke and Applications, Mech. Res. Comm, 8 (1981), 335-340.
[6] Panagiotopoulos, P.D.: Nonconvex Energy Functions. Hemivariational Inequalities and Substationarity Principles, Acta Mechanica, 42 (1983), 160-183.
[7] Duvaut, G. and J.L. Lions: Les inéquations en Mécanique et en Physique, Dunod, Paris 1972.
[8] Baiocchi, C. and A. Capelo: Variational and Quasivariational Inequalities, J. Wiley, New York 1984.
[9] Kinderlehrer, D. and G. Stampacchia: An Introduction to Variational Inequalities and their Applications, Academic Press, New York 1980.
[10] Friedman, A.: Variational Principles and Free Boundary Problems, J. Wiley, New York 1982.
[11] Rodriguez, J.F.: Obstacle Problems in Mathematical Physics, North Holland, Amsterdam 1987.
[12] Panagiotopoulos, P.D.: Inequality Problems in Mechanics and Applications. Convex and Nonconvex Energy Functions, Birkhäuser Verlag, Boston-Basel 1985 (Rus. Transl. M/R Pub., Moscow 1987).
[13] Clarke, F.H.: Optimization and Nonsmooth Analysis, J. Wiley, New York 1983.
[14] Rockafellar, R.T.: Generalized Directional Derivatives and Subgradients of Non-convex Functions, Can. J. Math., XXXII (1980), 257-280.
[15] Rockafellar, R.T.: La théorie des sous-gradients et ses applications à l'optimization. Fonctions convexes et non-convexes, Les Presses de l'Université de Montréal, Montréal 1979.
[16] Hamel, G.: Theoretische Mechanik, Springer-Verlag, Berlin 1967.

[17] Floegl, H. and H.A. Mang: Tension Stiffening Concept Based on Bond Slip, ASCE, ST 12, 108 (1982), 2681-2701.
[18] Panagiotopoulos, P.D.: Hemivariational Inequalities in Frictional Contact Problems and Applications. In: Mech. of Material Interfaces (ed. by A.P.S. Selvadurai, G.Z. Voyatzis), Elsevier Sc. Publ., Amsterdam 1986.
[19] Pires, E.B. and J.T. Oden: Analysis of Contact Problems with Friction Under Oscillating Loads,Comp. Meth. Appl. Mech. Eng., 39 (1983), 337-362.
[20] Green, A.E. and P.M. Naghdi: A General Theory of an Elastic-Plastic Continuum, Arch. Rat. Mech. Anal., 18 (1965), 251-281.
[21] Halphen, B. and N.Q. Son: Sur les Matériaux Standards Généralisés, J. de Mécanique, 14 (1975), 39-63.
[22] Germain, P.: Cours de Mécanique des milieux continus I, Masson, Paris 1973.
[23] Panagiotopoulos, P.D. and E.K. Koltsakis: Interlayer Slip and Delamination Effect: A Hemivariational Inequality Approach, Trans. of the C.S.M.E., 11 (1987), 43-52.
[24] Baniotopoulos, C.: Analysis of Structures with Complete Stress-strain Laws, Doct. Thesis, Aristotle Univ., Dept. Civil Eng., Thessaloniki 1985.
[25] Panagiotopoulos, P.D. and C. Baniotopoulos: A Hemivariational Inequality and Substationarity Approach to the Interface Problem: Theory and Prospects of Applications, Eng. Anal., 1 (1984), 20-31.
[26] Holister, G.S. and C. Thomas: Fibre Reinforced Materials, Elsevier, London 1966.
[27] Green, A.K. and W.H. Bowyer: The Testing and Analysis of Novel Top-Hat Stiffener Fabrication Methods for Use in GRP Ships, Proc. 1st Int. Conf. on Composite Structures (ed. by I.H. Marshall), Applied Science Publishers, London 1981.
[28] Williams, J.G. and M.D. Rhodes: Effect of Resin on Impact Damage Tolerance of Graphite/Epoxy Laminates, Proc. 6th Int. Conf. on Composite Materials, Testing and Design, ASTM STP 787 (ed. by I.M. Daniel), ASTM, Philadelphia 1982.
[29] Schwartz, M.M.: Composite Materials Handbook, McGraw-Hill, New York 1984.
[30] Baniotopoulos, C.C. and P.D. Panagiotopoulos: A Hemivariational Inequality Approach to the Analysis of Composite Material Structures, Proc. Conf. Composite Structures Patras, Omega Press, London 1987.
[31] Dunford, N. and J. Schwartz: Linear Operators, Part I: General Theory, Interscience Publ., New York 1966.
[32] Sheikh, A. and S.M. Uzumeri: Analytical model for concrete confinement in tied columns, J. Struct. Div. ASCE, 108 (ST12) (1982),2703.
[33] Emori, K. and W.C. Schnobrick: Inelastic behavior of concrete frame-wall structures, J.Struct. Div. ASCE, 107 (ST1) (1981), 145.
[34] Gilbert, J. and R.F. Warner: Tension stiffening in reinforced concrete slabs, J. Struct. Div. ASCE, 104 (ST12) (1978), 1885.
[35] Hutchinson, J.W.: Micromechanics of Damage in Deformation and Fracture, Edition of Solid Mech. Dept., TU Denmark 1987.

[36] Panagiotopoulos, P.D.: Ioffe's Fans and Unilateral Problems: A New Conjecture, Proc. 3nd Conf. Unil. Problems CISM, Springer-Verlag, Wien 1987.
[37] Klarbring, A.: Contact Problems in Linear Elasticity, Linköping Studies in Science and Technology, Dissertation No. 133, Linköping 1985.
[38] Panagiotopoulos, P.D.: Dynamic and Incremental Variational Inequality Principles, Differential Inclusions and their Applications to Co-Existent Phases Problems, Acta Mechanica, 40 (1981), 85-107.
[39] Nitsiotas, G.: Die Berechnung statisch unbestimmter Tragwerke mit einseitigen Bindungen, Ing. Archiv, 41 (1971), 46-60.
[40] Panagiotopoulos, P.D.: A Variational Inequality Approach to the Inelastic Stress-Unilateral Analysis of Cable-Structures, Comp. and Struct., 6 (1976), 133-139.
[41] Mitsopoulou, E.: Unilateral Contact, Dynamic Analysis of Beams by a Time-stepping Quadratic Programming Procedure, Meccanica, 18 (1983), 254-265.
[42] Doudoumis, I.N.: Modelling "infill finite elements" with unilateral Interface conditions and general nonlinear constitutive laws, Doct. Thesis, Aristotle Univ., Dept. Civil Eng., Thessaloniki 1988 (to appear).
[43] Mitsopoulou, E.N. and I.N. Doudoumis: A Contribution to the Analysis of Unilateral Contact Problems with Friction, S.M. Archives, 12 (1987), 165-186.
[44] Crisfield, M.A. and J. Wills: Solution Strategies and Softening Materials, Comp. Meth. Appl. Mech. Eng., 66 (1988), 267-289.
[45] Strodiot, J.J. and V.H. Nguyen: On the Numerical Treatment of the Inclusion $0\in\partial f(x)$. In: Topics in Nonsmooth Mechanics (ed. by J.J. Moreau, P.D. Panagiotopoulos, G. Strang), Birkhäuser Verlag, Basel 1988.
[46] Panagiotopoulos, P.D.: Nonconvex Problems of Semipermeable Media and Related Topics, ZAMM, 65 (1985), 29-36.
[47] Panagiotopoulos, P.D.: Hemivariational Inequalities and Substationarity Principles in the Static Theory of von Kármán Plates, ZAMM, 65 (1985), 219-229.
[48] Panagiotopoulos, P.D. and G. Stavroulakis: A Variational-Hemivariational Inequality Approach to the Laminated Plate Theory under Subdifferential Boundary Conditions, Q. of Appl. Math. (to appear).
[49] Panagiotopoulos, P.D.: A Nonlinear Programming Approach to the Unilateral Contact- and Friction-Boundary Value Problem in the Theory of Elasticity, Ing. Archiv, 44 (1975), 421-432.
[50] Nečas, J., Jarušek, J. and J. Haslinger: On the Solution of the Variational Inequality to the Signorini Problem with Small Friction, Bulletino U.M.I., 17B (1980), 796-811.
[51] Panagiotopoulos, P.D.: Variational Inequalities and Multilevel Optimization Techniques in the Theory of Plasticity, Comp. and Struct., 8 (1978), 649-650.
[52] Koltsakis, E.K.: Doct. Thesis, Aristotle Univ., Dept. Civil Eng., Thessaloniki (to appear).

[53] Panagiotopoulos, P.D. and E.K. Koltsakis: Hemivariational Inequalities for Linear and Nonlinear Elastic Materials, Meccanica, (1987), (to appear).
[54] Jones, R.: Mechanics of composite materials, McGraw Hill, New York 1975.
[55] Stavroulakis, G. and P.D. Panagiotopoulos: Laminated Orthotropic Plates under Subdifferential Boundary Conditions. A Hemivariational Inequality Approach, ZAMM, (to appear).
[56] Panagiotopoulos, P.D. and G. Stavroulakis: A Hemivariational Inequality Approach to the Delamination Effect in the Theory of Layered Plates, Archives of Mechanics, (to appear).
[57] Naniewicz, Z. and C.Z. Woźniak: On the quasi-stationary models of debonding processes in layered composites, Ing. Archiv, (to appear).
[58] Naniewicz, Z.: On some nonmonotone subdifferential boundary conditions in Elastostatics, J. Eng. Math., (to appear).
[59] Naniewicz, Z.: On some Nonconvex Variational Problems related to Hemivariational Inequalities Nonlinear Analysis and Applications, (to appear).
[60] Chang, K.C.: Variational Methods for Non-differentiable Functionals and their Applications to Partial Differential Equations, J. Math. Anal. Appl., 80 (1981), 102-129.
[61] Rockafellar, R.T.: Convex Analysis, Princeton Univ. Press, Princeton 1970.
[62] Ekeland, I. and R. Temam: Convex Analysis and Variational Problems, North-Holland, Amsterdam and American Elsevier, New York 1976.
[63] Panagiotopoulos, P.D. and G.E. Stavroulakis: The Delamination Effect in Laminated v.Kármán Plates under Unilateral Boundary Conditions. A Variational-Hemivariational Inequality Approach, J. of Elasticity, (to appear).
[64] Berger, M.S. and P.C. Fife: Von Kármán's Equations and the Buckling of a Thin Elastic Plate II. Plate with General Edge Conditions, Comm. Pure Appl. Math., XXI (1968), 227-241.
[65] Strauss, W.: On weak solutions of semilinear hyperbolic equations, An. Acad. Brasil. Ci., 42 (1970), 645-651.
[66] Rauch, J.: Discontinuous semilinear differential equations and multiple valued maps. Proc. A.M.S., 64 (1977), 277-282.
[67] Aubin, J.P.: Applied Functional Analysis, J. Wiley, New York 1979.
[68] Landesman, E.M. and A.C. Lazer: Nonlinear Perturbations of Linear Elliptic Boundary Value Problems at Resonance, J. Math. Mech., 19 (1970), 609-623.
[69] McKenna, P.J. and R. Rauch: Strongly Nonlinear Perturbations of Nonnegative Boundary Value Problems with Kernel, J. Diff. eq., 28 (1978), 253-265.
[70] Warga, J.: Derivate Containers, Inverse Functions and Controllability, In: Calculus of Variations and Control Theory (ed. by D.L. Russell), Acad. Press, New York 1976.
[71] Demyanov, V.F. and A.M. Rubinov: On Quasidifferentiable Mappings, Math. Operationsforsch. u.Statist. Ser. Optimization, 14 (1983), 3-21.

[72] Demyanov, V.F. and L.C.W. Dixon (eds): Quasidifferential Calculus, Math. Progr. St., 29 (1986),
[73] Demyanov, V.F. and L.V. Vasil'ev: Nondifferentiable Optimization, Optimization Software Inc., New York 1985.
[74] Demyanov, V.F., Polyakova, L.N. and A.M. Rubinov: Nonsmoothness and Quasidifferentiability, Math. Progr. Study, 29 (1986), 1-19.
[75] Shapiro, A.: On Optimality conditions in quasidifferentiable optimization, SIAM J. Control Opt., 22 (1984), 610-617.
[76] Demyanov, V.F. and A.M. Rubinov: On quasidifferentiable functionals, Soviet Math. Doklady, 21 (1980), 13-17.
[77] Demyanov, V.F., Polyakova, L.N. and A.M. Rubinov: On one generalization of the concept of subdifferential. In: All Union Conf. on Dynamic Control, Abstracts and Reports, Sverdlovsk 1979.
[78] Polyakova, L.N.: On minimizing the sum of a convex function and a concave function, Math. Progr. Study, 29 (1986), 69-73.
[79] Kiwiel, K.C.: A linearization method for minimizing certain quasidifferentiable functions, Math. Progr. Study, 29 (1986), 85-94.
[80] Rådström, H.: An embedding theorem for spaces on convex sets, Proc. AMS, 3 (1952), 165-169.
[81] Ball, J.M.: Convexity Conditions and Existence Theorems in Nonlinear Elasticity, Arch. Rational Mech. Anal., 63 (1977), 337-403.
[82] Ciarlet, P.G. and J. Nečas: Unilateral Problems in Nonlinear Three-Dimensional Elasticity, Arch. Rational Mech. Anal., 87 (1985), 319-338.
[83] Ciarlet, P.G. and J. Nečas: Injectivity and Self-Contact in Nonlinear Elasticity, Arch. Rational Mech. Anal., 97 (1987), 171-188.
[84] Demyanov, V.F.: Quasidifferentiable functions: Necessary conditions and descent directions, Math. Progr. Study, 29 (1986), 20-43.
[85] Rubinov, A.M. and A.A. Yagubov: The space of star-shaped sets and its applications in nonsmooth optimization, Math. Progr. Study, 29 (1986), 176-202.
[86] Germain, P.: La méthode de puissances virtuelles en mécanique des milieux continus, 1ère partie. Théorie du second gradient, J. de Mécanique, 12 (1973), 235-274.
[87] Salençon, J. and A. Tristán-Lopez: Analyse de la stabilité des talus en sols cohérents anisotropes, C.R. Acad. Sc. Paris, 290B (1980), 493-496.

# CONTACT WITH ADHESION

M. Fremond
Laboratoire Central des Ponts et Chaussees
Laboratoire de Modelisation des Materiaux et des Structures du Genie Civil
Paris, France

## 1. INTRODUCTION

Tearing adhesive paper glued onto a support requires a notably greater effort than that needed to overcome its weight. The additional effort is due to the adhesion. We give a description of the phenomena based on continuum mechanics.

The main idea is that usual displacements fields are not sufficient to describe the contact. Two solids in contact can be either strictly bonded (adhesion is present) or simply juxtaposed without any bounding (no adhesion is present). In both situations the displacements are the same but the mechanical situations are different. We therefore introduce a new variable to describe the state of contact : the ratio $\beta$ of active bonds called the intensity of adhesion.

## 2. PRINCIPLE OF VIRTUAL POWER

### 2.1. Kinematic quantities

Let there be a structure, occupying at time t a closed region $\Omega \subset \mathbb{R}^n$ (n = 2 or 3) Along a part $\Gamma$ of its boundary the structure is in contact with a support S which is assumed for the sake of simplicity to be rigid (figure 1).

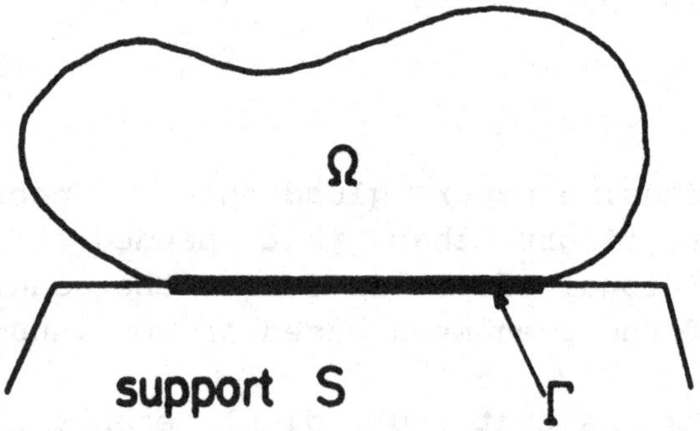


Figure 1
The system formed by $\Omega$ and the part $\Gamma$ of the support.

The mechanical system we consider is the closed domain $\Omega$. Along the part $\Gamma$ of the boundary it comprises both the material points of the support S and those of the structure. The kinematic description on $\Gamma$ is achieved through the velocity field $\vec{u}(x)$ of the points of the structure and the velocity field $\vec{u}_{ext}x)$ of the points of the support. In all other parts of $\Omega$, the velocity $\vec{u}(x)$ describes the kinematic state, (figure 1).

In our opinion these velocities are not sufficient. The formation or destruction of bounds between $\Omega$ and S must be accounted for. Even if the velocities $\vec{u}(x)$ and $\vec{u}_{ext}(x)$ are the same at a point x, the state of bounds between $\Omega$ and S may differ. Thus, a new kinematic variables to describe the microscopic displacements that form or destroy bonds appears to be necessary.

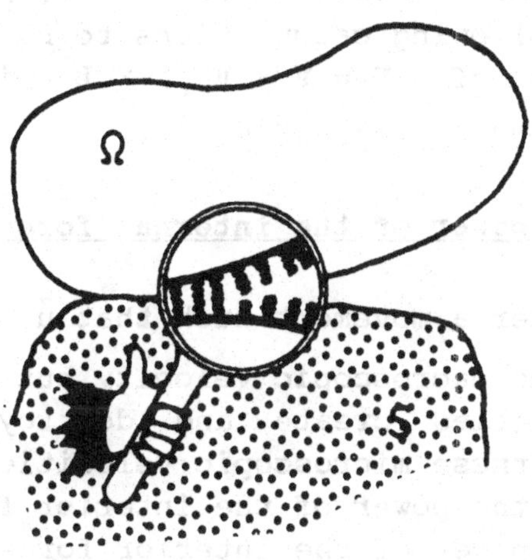


Figure 2
Links ensuring adhesion

To this end we define β, the intensity of adhesion, being the proportion of the active bonds between the structure and the support. If $\beta = 0$, all the bonds are broken, no adhesion exists ; if $\beta = 1$, all the bonds are active, total adhesion is present ; if $0 < \beta < 1$, a part β of the bonds remains active, the remaining bounds are broken, the adhesion is partial. The intensity of adhesion $\beta(x,t)$ depends on the time t and the point x. We assume this point x to be a material point of the structure Ω. Other assumptions are possible. This one is sufficient for our purpose. The velocity $\dot{\beta} = \frac{d\beta}{dt}$ is a macroscopic representation of the microscopic velocities that form and destroy the bonds.

We choose the triplet $(\vec{u}, \vec{u}_{ext}, \dot{\beta})$ to describe the motion of Ω and of its support. One must note that the velocity $\dot{\beta}$ is an element of a linear space although β is not due to the definition of β : $0 \le \beta \le 1$.

We define the linear spaces $\mathcal{V} \times \mathcal{V}_{ext}$ of the velocities of the material points of the system and the linear space $\mathcal{B}$ of the velocities of the adhesion intensities.

We assume that elements of $\mathcal{V} \times \mathcal{V}_{ext} \times \mathcal{B}$ are smooth enough for the following calculations to be coherent. An arbitrary element of $\mathcal{U} = \mathcal{V} \times \mathcal{V}_{ext} \times \mathcal{B}$ is denoted by $d = (\vec{v}, \vec{v}_{ext}, \gamma)$.

## 2.2. Virtual power of the internal forces.

Let us consider a movement such that $\vec{u} = 0$, $\vec{u}_{ext} = 0$ and $\dot{\beta} \neq 0$. There is no macroscopic velocity but in this movement microscopic velocities create and destroy bonds. We think that the power of these microscopic velocities must appear in the expression of the power of the interior forces. The usual expression of the power of the interior forces has the form

$$- \int_{\Omega} \sigma D(\vec{u}) d\Omega - \int_{\Gamma} \vec{R}(\vec{u}_{ext} \vec{u}) d\Gamma,$$

where $\sigma$ denotes the stresses, $D(\vec{u})$ the strain rates $\left(D_{ij}(\vec{u}) = \frac{1}{2}(u_{i,j} + u_{j,i})\right)$ and $\vec{R}$ the internal reaction. It is obvious that this expression does not take into account the power of the microscopic velocities. Thus it should appear in another term having the usual expression of a power : that is a linear function of the velocity. The simpliest choice available is

$$- \int_{\Gamma} F\dot{\beta} \, d\Gamma,$$

where the interior force F has the dimension of a surface work.

The theory that results from this choice is a zero gradient theory in $\beta$ because no spatial derivatives of $\beta$ appears. A first gradient theory would results from the choice

$$- \int_{\Gamma} \{F\dot{\beta} + \vec{J} \operatorname{grad}\dot{\beta}\} d\Gamma,$$

for the power of the microscopic velocities [3].

In order to define the virtual power $\hat{\mathcal{P}}_i$ of the internal forces to a closed domain $\mathcal{D} \subset \Omega$, which is a linear form on the space $\mathcal{U}$, we must fulfill the virtual power axiom [6] :

$$\hat{\mathcal{P}}_i = 0, \text{ for every rigid body velocity field at time t.}$$

To satisfy this axiom we need to define the subset of the rigid body velocities. It is clear that $D(\vec{v}) = 0, \vec{v}_{ext} - \vec{v} = 0$ on $\Gamma$ for such a velocity field. For a rigid body velocity field, the state of the bonds does not change : no bonds are formed or destroyed. It therefore appears reasonable to assume $\gamma = 0$ for a rigid velocity field. Thus a rigid velocity field is defined by

$$D(\vec{v}) = 0 \text{ in } \mathcal{D} \ ; \ \vec{v}_{ext}\vec{v} = 0 \text{ and } \gamma = 0 \text{ in } \mathcal{D} \cap \Gamma.$$

The first two relations are classical. The last one is new. One can point out that it finds strong support from existing experience which emphasises the necessity of some deformation of the structure in order to modify the adherence (that is to say in order to have $\gamma \neq 0$) [9].

Having defined the linear subset of the rigid body velocities, we retain as power of the internal forces to a closed subdomain $\mathcal{D} \subset \Omega$, the expression

$$\forall \mathcal{D} \subset \Omega, \ \forall d \in \mathcal{U}, \quad \hat{\mathcal{P}}_i = - \int_{\mathcal{D}} \sigma D(\vec{v}) d\Omega - \int_{\mathcal{D} \cap \Gamma} \{\vec{R}(\vec{v}_{ext} - \vec{v}) + F\gamma\} d\Gamma,$$

that satisfies obviously the virtuar power axiom.

## 2.3. Virtual power of the external forces.

Within the interior of a domain $\mathcal{D}$ there is the classical volume external force $\vec{f}$ which is a remotely exerted force. On the boundary $\mathcal{D} \cap \Gamma$, one can assume, besides the classical traction $\vec{T}$, that the local action of the outside of $\mathcal{D}$ onto $\mathcal{D} \cap \Gamma$ involves a surface work A. For instance A can represent the work of an electric current which modifies the state of the bonding of two plastic sheets in contact.

The virtual power of the exernal forces is

$$\hat{\mathcal{P}}_e = \int_{\mathcal{D}} \vec{f}\vec{v}\, d\Omega + \int_{\partial\mathcal{D}-\Gamma} \vec{T}\vec{v}\, d\Gamma + \int_{\mathcal{D}\cap\Gamma} \{\vec{T}\vec{v}_{ext} + A\gamma\} d\Gamma.$$

The local external action $\vec{T}$ is applied to the support on the part $\mathcal{D} \cap \Gamma$ because the material points of the support are those which are in contact with the exterior of the system.

**2.4. Principal of virtual power. Equation of equilibrium.**

We restrict ourselves to the quasi-static case for which the principle reads [6]

$$\forall \mathcal{D} \subset \Omega,\ \forall d \in \mathcal{U},\ \hat{\mathcal{P}}_i + \hat{\mathcal{P}}_e = 0.$$

By choosing $\mathcal{D}$ such that $\mathcal{D} \cap \Gamma = \emptyset$, we can first obtain the classical equilibrium equations,

$$\sigma_{ij,j} + f_i = 0 \text{ in } \mathcal{D}, \tag{1}$$

$$\sigma_{ij} n_j = T_i \text{ on } \partial\mathcal{D}-\Gamma, \tag{2}$$

where $n_j$ are the direction cosines of the ourward normal $\vec{n}$ to $\mathcal{D}$. Then by choosing $\mathcal{D}$ such that $\mathcal{D} \cap \Gamma \neq \emptyset$, we obtain the new equations of equilibrium

$$F = A, \tag{3}$$

$$\sigma_{ij} n_j = R_i \text{ and } T_i = R_i, \text{ on } \mathcal{D} \cap \Gamma. \tag{4}$$

In the present zero gradient theory for $\beta$, the internal force F is known as soon as the external force A is known. This situation can be compared to the isostatic situation for beams. The relations (3) and (4) give also physical meaning to the internal forces $\vec{R}$ and F : $\vec{R}$ is the reaction of the support and F is the surface density of work provided to the system.

## 3. ENERGY BALANCE

The internal energy E within a closed domain $\mathcal{D}$ can be decomposed into a volume density $\rho e$ ($\rho$ is the volume density) and two surface densities $\bar{\rho}_\Omega \bar{e}_\Omega$ and $\bar{\rho}_s \bar{e}_s$, where $\bar{\rho}_\Omega$, $\bar{e}_\Omega$ are the surface density and the internal energy of the bonds which remain with the structure ; $\bar{\rho}_s$, $\bar{e}_\Omega$ are the surface density and internal energy of the bonds which remain with the support S. The energy $\bar{\rho}_\Omega$, $\bar{e}_\Omega$ refers to the bonds with macroscopic velocity $\vec{u}$ and the energy $\bar{\rho}_s$, $\bar{e}_s$ to those with macroscopic velocity $\vec{u}_{ext}$.

In accordance with the quasi-static assumption, we neglect the kinetic energy in the energy balance

$$\frac{dE}{dt} = \mathcal{P}_e + \dot{Q} = - \mathcal{P}_i + \dot{Q}, \tag{1}$$

where $\dot{Q}$ is the rate of heat supplied to $\mathcal{D}$ and $\mathcal{P}_e$, $\mathcal{P}_i$ the actual powers of the external and internal forces.

We choose for $\dot{Q}$

$$\dot{Q} = \int_{\mathcal{D}} r \, d\Omega + \int_{\mathcal{D}\cap\Gamma} \bar{r} \, d\Gamma + \int_{\partial\mathcal{D}-\Gamma} (-\vec{q})\vec{n} \, d\Gamma + \int_{\mathcal{D}\cap\Gamma} \omega \, d\Gamma + \int_{\partial(\mathcal{D}\cap\Gamma)} (-\bar{q}\bar{n}) \, d\bar{\gamma},$$

where r is the volume rate of heat production inside $\Omega$, $\bar{r}$ the surface rate of heat production (for instance, the rate of heat produced by a chemical reaction), $\vec{q}$ is the volume heat flux vector in $\Omega$, $\bar{q}$ the surface heat flux vector in $\Gamma$ (we assume that heat can flow through the bonds), $\bar{m}$ is the outward normal to the domain $\mathcal{D} \cap \Gamma$ in $\Gamma$, $\omega$ is the heat which is supplied by the exterior to $\mathcal{D}$ through $\mathcal{D} \cap \Gamma$.

**Note.** Mass conservation on $\Gamma$.

The mass conservation of the bonds which remain with $\Gamma$ reads

$$\frac{\partial \bar{\rho}_\Omega}{\partial t} + \mathrm{div}_\Gamma (\bar{\rho}_\Omega \vec{U}) = 0, \tag{2}$$

where $\mathrm{div}_\Gamma (\bar{\rho}_\Omega \vec{U})$ is the surface divergence.

The mass conservation of the bonds which remain with the support reads

$$\frac{\partial \bar{\rho}_s}{\partial t} + \mathrm{div}_\Gamma (\bar{\rho}_s \vec{U}_{ext}) = 0, \tag{3}$$

Assuming $\Gamma$ to be flat,

$$\mathrm{div}_\Gamma f = f_{\alpha,\alpha},$$

where the $\alpha$'s are the two indices of the orthogonal coordinates in $\Gamma$.

By using relations (1) and (3), it is easy to prove that

$$\frac{d}{dt}\int_{\mathfrak{D}\cap\Gamma} \{\bar{\rho}_\Omega \bar{e}_\Omega + \bar{\rho}_s \bar{e}_s\} d\Omega = \int_{\mathfrak{D}\cap\Gamma} \left\{\bar{\rho}_\Omega \frac{d\bar{e}_\Omega}{dt} + \bar{\rho}_s \frac{d\bar{e}_s}{dt}\right\} d\Omega.$$

Let us recall that the relation

$$\frac{d}{dt}\int_{\mathfrak{D}} \rho e \, d\Omega = \int_{\mathfrak{D}} \rho \frac{de}{dt} d\Omega,$$

is classical. The conservation of energy equation (1) gives

$$\forall \mathcal{D} \subset \Omega, \quad \int_{\mathcal{D}} \rho \frac{de}{dt} d\Omega + \int_{\mathcal{D}\cap\Gamma} \left\{ \bar{\rho}_\Omega \frac{d\bar{e}_\Omega}{dt} + \bar{\rho}_s \frac{d\bar{e}_s}{dt} \right\} d\Gamma$$

$$= \int_{\mathcal{D}} \sigma D(\vec{U}) d\Gamma + \int_{\mathcal{D}\cap\Gamma} \dot{\beta} F + \vec{R}(\vec{u}-\vec{u}_{ext}) d\Gamma + \int_{\mathcal{D}} (r - \text{div } \vec{q}) d\Omega$$

$$+ \int_{\mathcal{D}\cap\Gamma} \{ \bar{r} - \text{div}_\Gamma \bar{q} + (\omega + \vec{q}\vec{n}) \} d\Gamma .$$

By choosing $\mathcal{D}$ such that $\mathcal{D} \cap \Gamma = \emptyset$, we obtain

$$\rho \frac{de}{dt} + \text{div } \vec{q} = \sigma D(\vec{u}) + r, \text{ in } \mathcal{D}, \tag{4}$$

We then obtain

$$\bar{\rho}_\Omega \frac{d\bar{e}_\Omega}{dt} + \bar{\rho}_s \frac{d\bar{e}_s}{dt} + \text{div}_\Gamma \bar{q} = F\dot{\beta} + \vec{R}(\vec{u}-\vec{u}_{ext}) + \bar{r} + (\omega + \vec{q}\vec{n}), \tag{5}$$

This is the energy conservation equation on $\Omega \cap \Gamma$. It can be considered as a boundary condition for the energy conservation equation (4) in $\Omega$, because it gives the heat flux $-\vec{q}\vec{n}$ which is provided from the support to the structure. One can also note that the quantity $(\omega + \vec{q}\vec{n})$ is the difference between the heat $\omega$ that flows into $\Gamma$ from the exterior and the heat $-\vec{q}\vec{n}$ which is provided to the structure and leaves $\Gamma$. The quantity $\text{div}_\Gamma \bar{q}$ accounts for the diffusion of heat through the bonds.

## 4. SECOND PRINCIPLE OF THERMODYNAMICS

We assume again that the entropy $\mathcal{S}$ of a closed domain $\mathcal{D}$ is the sum of volume and surface contributions

$$\mathcal{S} = \int_{\mathcal{D}} \rho s \, d\Omega + \int_{\mathcal{D}\cap\Gamma} \{ \bar{\rho}_\Omega \bar{s}_\Omega + \bar{\rho}_s \bar{s}_s \} d\Gamma .$$

Denoting by T the absolute temperature, the second principle of thermodynamics is

$$\frac{dS}{dt} \geq \int_{\mathcal{D}} \frac{r}{T}\, d\Omega + \int_{\partial\mathcal{D}-\Gamma} - \frac{\vec{q}\vec{n}}{T}\, d\Gamma + \int_{\mathcal{D}\cap\Gamma} \frac{\bar{r}+\omega}{T}\, d\Gamma + \int_{\partial(\mathcal{D}\cap\Gamma)} - \frac{\bar{q}\bar{m}}{T}\, ds,$$

which must be satisfied for any domain $\mathcal{D}$ and any actual evolution. We obtain easily that

$$\rho \frac{ds}{dt} + \operatorname{div} \frac{\vec{q}}{T} - \frac{r}{T} \geq 0, \text{ in } \Omega,$$

$$\bar{\rho}_\Omega \frac{d\bar{s}_\Omega}{dt} + \bar{\rho}_s \frac{d\bar{s}_s}{dt} + \operatorname{div}_\Gamma \left(\frac{\bar{q}}{T}\right) - \frac{\bar{r}+\omega+\vec{q}\vec{n}}{T} \geq 0, \text{ in } \Gamma.$$

Multiplying these equations by T and adding the energy conservation equations (3.4) and (3.5), we obtain

$$\rho\left(T \frac{ds}{dt} - \frac{de}{dt}\right) + \sigma D(\vec{u}) - \frac{\vec{q}\cdot \operatorname{grad} T}{T} \geq 0,$$

$$\bar{\rho}_\Omega \left(T \frac{d\bar{s}_\Omega}{dt} - \frac{d\bar{e}_\Omega}{dt}\right) + \bar{\rho}_s \left(T \frac{d\bar{s}_s}{dt} - \frac{d\bar{e}_s}{dt}\right) +$$

$$+ F\dot{\beta} + \vec{R}(\vec{u}_{ext} - \vec{u}) - \frac{\bar{q}\, \operatorname{grad}_\Gamma T}{T} \geq 0,$$

where $\operatorname{grad}_\Gamma T$ is the surfacic gradient.

Defining $\Psi = e - Ts$, $\bar{\Psi}_\Omega = \bar{e}_\Omega - T\bar{s}_\Omega$, $\bar{\Psi}_s = \bar{e}_s - T\bar{s}_s$ and integrating over a domain $\mathcal{D}$ we obtain the Clausius-Duhem inequality,

$$\forall \mathcal{D}, \quad \int_{\mathcal{D}} \left\{ \rho \frac{d\Psi}{dt} + \rho s \frac{dT}{dt} \right\} d\Omega + \int_{\mathcal{D} \cap \Gamma} \left\{ \bar{\rho}_\Omega \frac{d\bar{\Psi}_\Omega}{dt} + \bar{\rho}_\Omega \bar{s}_\Omega \frac{d^\Omega T}{dt} \right\}$$

$$+ \left\{ \bar{\rho}_s \frac{d\bar{\Psi}_s}{dt} + \bar{\rho}_s \bar{s}_s \frac{d^s T}{dt} \right\} d\Gamma \leq \int_{\mathcal{D}} \left\{ \sigma D(\vec{u}) - \frac{\vec{q} \text{ grad } T}{T} \right\} d\Omega$$

$$+ \int_{\mathcal{D} \cap \Gamma} \left\{ F\dot{\beta} + \vec{R}(\vec{u}_{ext} - \vec{u}) - \frac{\bar{q} \text{ grad}_\Gamma T}{T} \right\} d\Gamma,$$

where the derivative $\frac{d^\Omega T}{dt}$ and $\frac{d^s T}{dt}$ are the material derivatives with respect to the structure and the support :

$$\frac{d_\Omega T}{dt} = \frac{\partial T}{\partial t} + \text{grad } T\, \vec{u}, \qquad \frac{d^s T}{dt} = \frac{\partial T}{\partial t} + \text{grad } T\, \vec{u}_{ext}.$$

The Clausius-Duhem inequality must be satisfied for any domain $\mathcal{D}$ and by any actual evolution ; i.e., for any velocities $\vec{u}$, $\vec{u}_{ext}$, $\dot{\beta}$ and any heat fluxes $\vec{q}$ and $\bar{q}$.

The free energies $\Psi$, $\bar{\Psi}_s$ and $\bar{\Psi}_s$ depend on the temperature T and other quantities $x_i$. Let us define the functions

$$(t,T) \longrightarrow \tilde{\Psi}(t,T) = \Psi(x_i(t),T),$$
$$(t,T) \longrightarrow \tilde{\bar{\Psi}}_\Omega(t,T) = \bar{\Psi}_\Omega(x_i(t),T),$$
$$(t,T) \longrightarrow \tilde{\bar{\Psi}}_s(t,T) = \bar{\Psi}_s(x_i(t),T),$$

and assume the Helmholtz postulate

$$s = -\frac{\partial \Psi}{\partial T}, \qquad \bar{s}_\Omega = -\frac{\partial \bar{\Psi}_\Omega}{\partial T}, \qquad \bar{s}_s = -\frac{\partial \bar{\Psi}_s}{\partial T}.$$

The Clausius-Duhem inequality gives

$$\left.\begin{aligned}&\int_{\mathfrak{D}} \rho \frac{\partial \tilde{\Psi}}{\partial t} d\Omega + \int_{\mathfrak{D}\cap\Gamma} \left\{ \bar{\rho}_{\Omega} \frac{\partial \tilde{\Psi}_{\Omega}}{\partial t} + \bar{\rho}_{s} \frac{\partial \tilde{\Psi}_{s}}{\partial t} \right\} d\Gamma \\ &\leq \int_{\mathfrak{D}} \left\{ \sigma D(\vec{u}) - \vec{q} \frac{\text{grad } T}{\Gamma} \right\} d\Omega \\ &+ \int_{\mathfrak{D}\cap\Gamma} \left\{ F\dot{\beta} + \vec{R}(\vec{u}_{ext} - \vec{u}) - \frac{\vec{q} \text{ grad}_{\Gamma} T}{T} \right\} d\Gamma , \end{aligned}\right\} \quad (1)$$

Let us define the intrinsic dissipation

$$\left.\begin{aligned}\tilde{D}_1 = &\int_{\mathfrak{D}} \left\{ \sigma D(\vec{u}) - \rho \frac{\partial \tilde{\Psi}}{\partial t} \right\} d\Omega + \\ &\int_{\mathfrak{D}\cap\Gamma} \left\{ F\dot{\beta} + \vec{R}(\vec{u}_{ext} - \vec{u}) - \bar{\rho}_{\Omega} \frac{\partial \tilde{\Psi}_{\Omega}}{\partial t} - \bar{\rho}_{s} \frac{\partial \tilde{\Psi}_{s}}{\partial t} \right\} d\Gamma , \end{aligned}\right\} \quad (2)$$

and the thermal dissipation

$$\tilde{D}_2 = \int_{\mathfrak{D}} - \frac{\vec{q} \text{ grad } T}{T} d\Omega + \int_{\mathfrak{D}\cap\Gamma} - \frac{-\bar{q} \text{ grad}_{\Gamma} T}{T} d\Gamma , \quad (3)$$

The Clausius-Duhem inequality takes the form

$$\forall \mathfrak{D},\ \forall \vec{u},\ \vec{u}_{ext},\ \dot{\beta},\ \text{grad } T,\ \text{grad}_{\Gamma} T,\ \tilde{D}_1 + \tilde{D}_2 \geq 0.$$

Following the usual practice, we replace the inequality $\tilde{D}_1 + \tilde{D}_2 \geq 0$ by the two stronger ones

$$\tilde{D}_1 \geq 0 \text{ and } \tilde{D}_2 \geq 0. \quad (4)$$

For the thermal dissipation, we choose as usual the Fourier's laws

$$\vec{q} = - k \text{ grad } T, \ \bar{q} = - \bar{k} \text{ grad}_\Gamma T, \tag{5}$$

where k and $\bar{k}$ are the thermal conductivities of the structure and the bonds. Thus the second inequality of (4) is satisfied.

**Note.**- Let us define the function

$$\Psi_0(t,T) = \int_{\mathcal{D}} \rho\tilde{\Psi}\, d\Omega + \int_{\mathcal{D}\cap\Gamma} \left\{ \bar{\rho}_\Omega \tilde{\Psi}_\Omega + \bar{\rho}_s \tilde{\Psi}_s \right\} d\Gamma .$$

One can prove by using the mass conservation laws (3.1) and (3.2) that

$$\int_{\mathcal{D}} \rho \frac{\partial \tilde{\Psi}}{\partial t}\, d\Omega + \int_{\mathcal{D}\cap\Gamma} \left\{ \bar{\rho}_\Omega \frac{\partial \tilde{\Psi}_\Omega}{\partial t} + \bar{\rho}_s \frac{\partial \tilde{\Psi}_s}{\partial t} \right\} d\Gamma = \frac{\partial \Psi_0}{\partial t} .$$

The quantity $\frac{\partial \Psi_0}{\partial t}$ is the material derivative of $\Psi_0$ where the temperature T is kept to its value T(t) at time t.

Thus the inequality (4) can be written

$\forall \mathcal{D},\ \forall \vec{u},\ \dot{\vec{u}}_{ext},\ \dot{\beta},$

$$0 \le \int_{\mathcal{D}} \{\sigma D(\dot{\vec{u}})\} d\Omega + \int_{\mathcal{D}\cap\Gamma} \left\{ F\dot{\beta} + \vec{R}(\dot{\vec{u}} - \dot{\vec{u}}_{ext}) \right\} d\Gamma - \frac{\partial \Psi_0}{\partial t} = \tilde{D}_1 , \tag{6}$$

## 5. FREE ENERGY WITHIN THE SMALL DEFORMATIONS ASSUMPTION

From now on we suppose the deformations to be small. The quantities $\vec{u}(x,t)$, $\vec{u}_{ext}(x,t)$, will further denote the small displacements.

We assume the free energy to depend on the temperature T, on β and on the small deformations $\varepsilon(\vec{u}) = D(\vec{u})$ and $\vec{u}_{ext} - \vec{u}$.

We assume that if the relative displacement $\vec{u}_{ext}-\vec{u}$ of the support and the structure is not zero, all the bonds are broken. This assumption describes the behaviour of the contact and should be taken into account by the free energy. One can choose an other behaviour allowing interactions at distance in order to describe e.g. a glue film that ruptures progressively as the crack between the support and the structure opens. The behaviour we have chosen implies

$$\beta(\vec{u}_{ext}-\vec{u}) = 0, \tag{1}$$

and

$$(\vec{u}_{ext}-\vec{u})\vec{n} \geq 0, \tag{2}$$

besides

$$0 \leq \beta \leq 1, \tag{3}$$

The last condition results from the definition of $\beta$. The last but one is the non interpenetration condition between the body and the support.

The actual evolutions of the system that satisfy the Clausius-Duhem inequality at every time instant, satisfy also the internal constraints (1), (2) and (3). We can therefore define the free energy at our convenience if the internal constraints are not satisfied, without affecting the Clausius-Duhem inequality. Taking advantage of this, we can include the internal constraints into the free energy by assuming this free energy to take the value $+\infty$ if one of the constraints is not satisfied. As we have said, we do think that the internal constraints are actually parts of the physical or mechanical properties of the system and that the free energy must account for them.

We now suppose, for the sake of simplicity, that the support is rigid : $\vec{u}_{ext} = 0$ and that the structure is fixed along a part $\Gamma_0$ : $\vec{u} = 0$ on $\Gamma_0$ (figure 3).

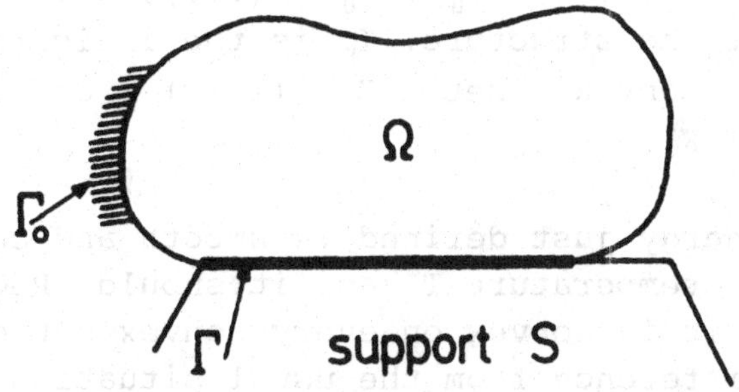


Figure 3
The structure fixed on the part $\Gamma_0$

Let us give some notations :

$$\mathcal{U}_0 = \{\vec{v} \mid \vec{v} \in \mathcal{V} \; ; \; \vec{v} = 0 \text{ on } \Gamma_0\} \times \mathcal{B},$$

$$\mathcal{A}_1(\mathcal{D},\vec{u},\vec{v},T) = \int_{\mathcal{D}} \Big\{ \lambda(T)\varepsilon_{kk}(\vec{u})\varepsilon_{kk}(\vec{v}) + 2\mu(T)\varepsilon_{ij}(\vec{u})\varepsilon_{ij}(\vec{v}) \Big\} d\Omega,$$

$$\mathcal{A}_1(\mathcal{D},\vec{u},\vec{v},T) = \int_{\mathcal{D}} k_2(T)\vec{u}\vec{v} \, d\Gamma,$$

$$K = \{(\vec{v},\gamma) \mid (\vec{v},\gamma) \in \mathcal{U}_0 \; ; \; 0 \le \gamma \le 1; \; \gamma\vec{v} = 0; \; -\vec{v}\vec{n} \ge 0 \text{ on } \Gamma\},$$

where $\lambda(T)$, $\mu(T)$ are the Lamé elastic parameters and $k_2(T)$ a positive elastic rigidity, $(3\lambda+2\mu > 0,\ \mu > 0,\ k_2 > 0)$.

We define the free energy of a domain $\mathcal{D}$ by

$$\Psi(\mathcal{D},\vec{u},\beta,T) = \frac{1}{2}\mathcal{A}_1(\mathcal{D},\vec{u},\vec{u},T) + \frac{1}{2}\mathcal{A}_1(\mathcal{D},-\vec{u},-\vec{u},T)$$

$$- \int_{\mathcal{D}} \Big\{ (3\lambda_0+2\mu_0)\alpha(T-T_0)\varepsilon_{kk}(\vec{u}) + \rho CT \operatorname{Log} T \Big\} d\Omega$$

$$+ \int_{\mathcal{D}\cap\Gamma} \{w(T)(1-\beta)\} d\Gamma + I_K(\vec{u},\beta),$$

where $w(T)$ is the Dupré's energy, $\alpha$ the thermal expansion coefficient, $T_0$ is the reference temperature at which we consider the system ($\lambda_0 = \lambda(T_0)$, $\mu_0 = \mu(T_0)$), C is the thermal heat capacity of the structure, $I_K$ is the indicator function of the non convex set K ($I_K(x) = 0$ if $x \in K$, $I_K(x) = +\infty$ if $x \notin K$).

The free energy just defined is smooth and concave with respect to the temperature T as it should. Regarding the other variables, it is convex on every convex set on which it is finite. The difference from the usual situation is that the function $\Psi$ is not convex with respect to $\beta$ and $\vec{u}$.

In order to calculate the internal energy E let us make the usual assumption that $\lambda(T)$, $\mu(T)$ and $k_2(T)$ are linear function of T. Thus we obtain

$$E = \Psi - T\frac{\partial\Psi}{\partial T} = \int_{\mathcal{D}} \rho CT + (3\lambda_0 + \mu_0)\alpha T_0 \varepsilon_{kk}(\vec{u})\,d\Omega$$

$$+ \int_{\mathcal{D}\cap\Gamma} \left(w(T) - T\frac{dw}{dT}\right)(1-\beta)\,d\Gamma.$$

We can note that the surface energy is equal to zero if the bonds are unaltered ($\beta = 1$). In this situation the contact zone is not distinguished from the structure. It is only when $\beta$ differs from 0 or when some the bonds are brocken that the internal energy of the contact is distinguished from the internal energy of the structure. Of course other assumptions are possible : for instance, one can assume heat capacities $\bar{C}_\Omega$ and $\bar{C}_s$ for the bonds.

**Note.** The function $\Psi_0(T,t)$ previously defined is

$$\Psi_0(T,t) = \Psi(\mathcal{D}, \vec{u}(t), \beta(t), T).$$

One can remark that, due to the small deformations assumption, $\mathcal{D}$ does not depends on the time t.

As we have included the internal constraints in the free energy, the function $\Psi_0$ is not smooth. Thus we must focus on a possible chain rule

$$\frac{\partial \Psi_0}{\partial t} = \frac{\partial \Psi}{\partial \beta} \dot{\beta} + \frac{\partial \Psi}{\partial \vec{u}} \dot{\vec{u}}.$$

We cannot use the subdifferential of a convex function [10] because $\Psi$ is not convex. Thus we define a similar notion the local subdifferential.

## 5.1. Local subdifferential of a non smooth function.

Let $\Phi$ be a mapping from a topological vector space X, in duality with $X^*$, into $\mathbb{R} \cup \{+\infty\}$. Then $\Phi$ is said to be locally subdifferentiable at the point $x \in X$ if it is finite at x and there exists a neighbourhood $\mathcal{W}(x)$ of x and $x^* \in X^*$ such that

$$\forall z \in \mathcal{W}(x),\ \Phi(z) \geq \Phi(x) + < x^*, z-x >,$$

where $< \cdot, \cdot >$ are the duality brackets.

Every $x^*$ satisfying this inequality is called a subgradient of $\Phi$ at x. The set of all the subgradients is the subdifferential of $\Phi$ at x denoted $\bar{\partial}\Phi(x)$.

## 5.2. Properties of the free energy.

The Clausius-Duhem inequality must hold for any subdomain $\mathcal{D} \subset \Omega$. This imposes to $\dfrac{\partial \Psi_0}{\partial t}$ to have the same structure as the first term of (4.6). This implies that the subgradient of $\Psi_0(\Omega)$, that will be calculated, can be restricted to the subdomains of $\Omega$.

This requirement is natural because the behaviour of a material does not depend on the shape of the sample considered.

The elements $d = (\vec{v}, \gamma)$ of $\mathcal{U}_0$ or the deformations and the adhesion intensity are in duality with the internal forces $f = (\sigma, \vec{R}, F)$ by the bilinear form

$$< f,d >_{\mathcal{D}} = \int_{\mathcal{D}} \sigma\varepsilon(\vec{v})d\Omega + \int_{\mathcal{D}\cap\Gamma} \{-\vec{R}\vec{v} + F\gamma\}d\Gamma.$$

We shall show that the function $\Psi(\Omega,u,\beta,T)$ is locally subdifferentiable with respect to the variables $\vec{u}$ and $\beta$ and that among its local subgradients, there exists one whose restriction to $\mathcal{D}$ is a local subgradient of $\Psi(\mathcal{D},u,\beta,T)$.

**PROPOSITION.** The function $\Psi(\Omega,\vec{u},\beta,T)$ is concave and smooth with respect to T and convex with respect to the variables $\vec{u}$, $\beta$ on the convex domains where it is finite. It is locally subdifferentiable with respect to $\vec{u}$ and $\beta$. There exists a subgradient of $\Psi(\Omega,u,\beta,T)$ whose restriction to $\mathcal{D}$ is also a subgradient of $\Psi(\mathcal{D},\vec{u},\beta,T)$, i.e.

$$\forall\mathcal{D} \subset \Omega,\ \forall d \in \mathcal{W}(d_1),\ < f,d-d_1 >_{\mathcal{D}} \le \Psi(\mathcal{D},d,T) - \Psi(\mathcal{D},d_1,T),$$

where $\mathcal{W}(d_1)$ is a neighbourhood of $d_1 = (\vec{u},\beta)$.

**Proof.** The only point not obvious is the local subdifferentiability of $\Psi$. The function $\Psi$ is the sum of a smooth part $\Psi_r$ and a non smooth part $I_K$ ($\Psi = \Psi_r + I_K$). The regular part is convex. We assume that the topology on $\mathcal{U}_0$ is such that $\Psi_r$ is differentiable and that the differential is the classical one defined by the volume density

$$\sigma_r(x) = \lambda(T)\varepsilon_{kk}(\vec{u}(x))1\!\!1 + 2\mu(T)\varepsilon(\vec{u}(x)) - \alpha(3\lambda_0+2\mu_0)(T-T_0)1\!\!1$$

and the surface densities,

$$\vec{R}_r(x) = - k_2(T)\vec{u}(x),$$

$$F_r(x) = - w(T).$$

Let us define $f_r = (\sigma_r,\vec{R}_r,F_r)$ and $d_1 = (\vec{u},\beta)$. Because $\Psi_r$ is convex, we have

$$< f_r, d-d_1 >_{\mathcal{D}} \leq \Psi_r(\mathcal{D},d,T) - \Psi_r(\mathcal{D},d_1,T), \tag{1}$$

for any $d \in \mathcal{U}_0$ and any $\mathcal{D} \subset \Omega$. Moreover, as $I_K \geq 0$ we have

$$\Psi(\mathcal{D},d,T) \geq \Psi_r(\mathcal{D},d,t), \tag{2}$$

For $d_1 = (\vec{u},\beta) \in K$, we have

$$\Psi(\mathcal{D},d_1,T) = \Psi_r(\mathcal{D},d_1,T), \tag{3}$$

Relations (1), (2), (3) give

$$\forall d_1 = (\vec{u},\beta) \in K,$$

there exist $f_r$ such that

$$\forall \mathcal{D} \subset \Omega,\ \forall d \in \mathcal{U}_0,\ < f_r, d-d_1 >_{\mathcal{D}} \leq \Psi(\mathcal{D},d,T) - \Psi(\mathcal{D},d_1,T),$$

which proves that $\Psi(\Omega)$ is locally subdifferential (for this particular subgradient $f_r$ the neighbourhood of $d_1$ is the whole space $\mathcal{U}_0$) and that the restriction of $f_r$ to $\mathcal{D}$ is a subgradient of $\Psi(\mathcal{D})$.

We denotes by $\bar{\partial}\Psi(d_1)$ the set of the all the subgradients of $\Psi(\Omega)$ having the restriction property just mentionned. It is clear that $\bar{\partial}\Psi(d_1)$ does not reduce to the element $f_r$.

## 6. INTRINSIC DISSIPATION. CONSTITUTIVE LAWS.

Let us calculate $\dfrac{\partial \Psi_0}{\partial t}$ which is an element of inequality (4.6).

Let $f \in \bar{\partial}\Psi(d_1)$, $f = (\tau,\vec{G}_1,G_2)$. Assuming $d_1(t) = (\vec{u}(t),\beta(t))$ to be continuous with respect to t, we have

$$d_1(t+\Delta t) \in \mathcal{W}(d_1(t)),$$

for $\Delta t$ small enough. Therefore, since $\Psi_0(\mathcal{D},t,T) = \Psi(\mathcal{D},d_1(t),T)$

$$< f,d_1(t+\Delta t) - d_1(t) >_{\mathcal{D}} \le \Psi_0(\mathcal{D},t+\Delta t,T) - \Psi_0(t,T), \qquad (1)$$

for $\Delta t$ small enough.

Let us divide relation (1) by $\Delta t > 0$ and let $\Delta t$ tend to zero, we obtain

$$\int_{\mathcal{D}} \tau\varepsilon(\vec{u})d\Omega + \int_{\mathcal{D}\cap\Gamma}\{\vec{G}_1(-\vec{u})+G_2\dot{\beta}\}d\Gamma = < f, \frac{dd_1}{dt} > \le \frac{\partial\Psi_0}{\partial t}, \qquad (2)$$

Let us divide relation (1) by $\Delta t < 0$ and let $\Delta t$ tend to zero, we obtain

$$\int_{\mathcal{D}} \tau\varepsilon(\vec{u})d\Omega + \int_{\mathcal{D}\cap\Gamma}\{\vec{G}_1(-\vec{u}) + G_2\dot{\beta}\}d\Gamma = < f, \frac{dd_1}{dt} > \ge \frac{\partial\Psi_0}{\partial t}, \qquad (3)$$

Relations (2), (3) give

$$\int_{\mathcal{D}} \tau\varepsilon(\vec{u})d\Omega + \int_{\mathcal{D}\cap\Gamma}\{\vec{G}_1(-\vec{u}) + G_2\dot{\beta}\}d\Gamma = \frac{\partial\Psi_0}{\partial t}.$$

Therefore we obtain an useful expression for the intrinsic dissipation $\tilde{D}_1$ such that the inequality (4.6) reads

$$\tilde{D}_1 = \int_{\mathcal{D}}(\sigma-\tau)\varepsilon(\vec{u})d\Omega + \int_{\mathcal{D}\cap\Gamma}\{(\vec{R}-G_1)(-\vec{u}) + (G_2-F_2)\dot{\beta}\}d\Gamma \ge 0, \qquad (4)$$

for any $\mathcal{D}$ and any actual velocities $\vec{u}$ and $\dot{\beta}$.

In order to satisfy the inequality (4), we assume that there exists a pseudo-potential of dissipation [7], [11]

$$\Phi(\mathcal{D},\dot{u},\dot{\beta}) = \int_{\mathcal{D}}\varphi(\varepsilon(\vec{u}))d\Omega + \int_{\mathcal{D}\cap\Gamma}\bar{\varphi}(\vec{u},\dot{\beta})d\Gamma, \qquad (5)$$

where $\varphi$ and $\bar{\varphi}$ are convex, positive functions with $\varphi(0) = \bar{\varphi}(0,0) = 0$.

We say that $f^{irr} = (\tau^{irr},\vec{G}_1^{irr},G_2^{irr})$ is an element of the subdifferential $\partial\Phi(\dot{\vec{u}},\dot{\beta})$ if

$$\forall\mathcal{D},\ \forall d \in \mathcal{U}_0,\ < f^{irr}, d-\dot{d}_1 > \ \leq \Phi(\mathcal{D},d) - \Phi(\mathcal{D},\dot{d}_1)$$

with $\dot{d}_1 = (\dot{\vec{u}},\dot{\beta})$.

We choose as constitutive laws

$$f \in \bar{\partial}\Psi(d_1), \tag{6}$$

$$f^{irr} \in \partial\Phi(\dot{d}_1), \tag{7}$$

$$(\sigma,\vec{R},F) = f + f^{irr}. \tag{8}$$

The internal forces are the sum of reversible internal forces and of dissipative (or irreversible) internal forces which depend on the velocities. It is easy to check that relation (4) is satisfied with constitutive laws (6), (7) and (8).

## 7. PROPERTIES OF THE INTERNAL FORCES

The properties of the dissipative forces are classical. We focus on the reversible internal forces and assume the dissipative forces to be zero ($\Phi \equiv 0$). We then have

$$(\sigma,\vec{R},F) \in \bar{\partial}\Psi(\vec{u},\beta).$$

or by denoting $d_1 = (\vec{u},\beta)$,

$$\forall\mathcal{D} \subset \Omega,\ \forall d \in \mathcal{W}(d_1),\ \int_{\mathcal{D}} \sigma\varepsilon(\vec{v}-\vec{u})\,d\Omega$$

$$+ \int_{\partial\mathcal{D}\cap\Gamma} \{\vec{R}(-(\vec{v}-\vec{u})) + (\gamma-\beta)F\}\,d\Gamma \leq \Psi(d,T) - \Psi(d_1,T), \tag{1}$$

Let us give some useful technical results.

**7.1. Technical results**

Since the functions $\mathcal{A}_1$ and $\mathcal{A}_2$ are quadratic, the relation (1) gives

$$\left.\begin{aligned}
&\forall d \in \mathcal{W}(d_1) \cap K,\ \forall \mathcal{D} \subset \Omega,\\
&\frac{1}{2}\mathcal{A}_1(\mathcal{D},\vec{v}-\vec{u},\vec{v}-\vec{u}) + \frac{1}{2}\mathcal{A}_2(\mathcal{D},\vec{v}-\vec{u},\vec{v}-\vec{u}) \geq \int_{\mathcal{D}}(\sigma-\tau')\varepsilon(\vec{v}-\vec{u})\,d\Omega\\
&- \int_{\mathcal{D}\cap\Gamma}(\vec{R}+k\vec{u})(\vec{v}-\vec{u})\,d\Gamma + \int_{\mathcal{D}\cap F}(F+w)(\gamma-\beta)\,d\Gamma,
\end{aligned}\right\} \quad (2)$$

where $\tau' = \lambda(T)\operatorname{tr}\varepsilon(\vec{u})\mathbb{1} + 2\mu\varepsilon(\vec{u}) - (2\lambda_0+\mu_0)\alpha(T-T_0)\mathbb{1}$.

Let $\mathcal{U}_\beta$ be the subspace of the $(\vec{u},\gamma)$ such that the support of $\gamma$ is contain in that of $\beta$, and $\mathcal{U}_u$ be the subspace of the $(\vec{v},\beta)$ such that the support of $\vec{v}$ is contained in that of $\vec{u}$. It is worth noting that the function $d \longrightarrow \Psi(d)$ is convex on these subspaces. It can therefore be shown that the relation (1) or (2) is satisfied not only on a neighbourhood of $d_1$ but on $\mathcal{U}_\beta \cap K$ and $\mathcal{U}_u \cap K$ as well.

Let us now examine the properties of the internal forces.

**7.2. Constitutive law for** $\sigma$.

Let $\mathcal{D}$ be a domain such that $\mathcal{D} \cap \Gamma = \emptyset$. We have

$$\forall \vec{v} \in \mathcal{U}_0^1,\quad \frac{1}{2}\mathcal{A}_1(\mathcal{D},\vec{v}-\vec{u},\vec{v}-\vec{u}) \geq \int_{\mathcal{D}}(\sigma-\tau')\varepsilon(\vec{v}-\vec{u})\,d\Omega, \quad (3)$$

where $\mathcal{U}_0^1$ is the first space component of $\mathcal{U}_0$.

As $\mathcal{A}_1$ is quadratic relation (3) is equivalent to

$$\forall \vec{q} \in \mathcal{U}_0^1,\ \int_{\mathcal{D}}(\sigma-\tau')\varepsilon(\vec{q})\,d\Omega = 0.$$

This relation gives : $\operatorname{div}(\sigma-\tau') = 0$ in $\mathcal{D}$ and $(\sigma-\tau')\vec{n} = 0$

on $\partial\mathcal{D}$ ($\vec{n}$ is the outward normal to $\mathcal{D}$). The directionn $\vec{n}$ being arbitrary because $\mathcal{D}$ is (provided $\mathcal{D} \cap \Gamma = \emptyset$), we then have

$$\sigma = \tau' = \lambda(T)\mathrm{tr}\,\varepsilon(\vec{u})\mathbb{1} + 2\mu\varepsilon(\vec{u}) - \alpha(T-T_0)(3\lambda_0+2\mu_0)\mathbb{1}, \quad (4)$$

which is the classical elastic constitutive law.

### 7.3. Constitutive laws for $\vec{R}$ and F.

#### 7.3.1. Contact ($\vec{u} = 0$)

We shall show that where contact exists

$$\begin{aligned} F &= -w, \text{ if, } 0 < \beta < 1, \\ F &\geq -w, \text{ if, } \beta = 1, \\ F &\leq -w, \text{ if, } \beta = 0. \end{aligned} \quad (5)$$

We will also show that the positive part of the normal traction $R_N^+$ and the tangential traction $\vec{R}_T = \vec{R} - R_N\vec{n}$ are not too large ($R_N = \vec{R}\vec{n}$, $\vec{n}$ is the ouward normal to $\Omega$ on $\Gamma$ ; $R_N^+ = R_N$ if $R_N \geq 0$, $R_N^+ = 0$ if $R_N \leq 0$).

In order to prove these relations, one should first perform variations of $(\vec{v},\gamma)$ that will not change the nature of the contact and use the classical tests of the variational methods. In this way the properties of F are obtained. Variations of $(\vec{v},\gamma)$ that do change the nature of the contact (contact replaced by separation) give the properties of $\vec{R}$.

Let $\mathcal{D}$ be a domain such that $\vec{u} = 0$ on $\mathcal{D} \cap \Gamma$. Let us first choose $(\vec{u},\gamma) \in \mathcal{U}_\beta \cap K$ in relation (1). We have

$$0 \geq \int_{\mathcal{D}\cap\Gamma}\{F+w\}(\gamma-\beta)\,d\Gamma.$$

This relation is satisfied by any $\gamma$ such that $0 \leq \gamma \leq 1$. It gives immediately the relations (5).

Let $\mathcal{O}$ be a domain of $\mathcal{D} \cup \Gamma$. We choose $\gamma = 0$ on $\mathcal{O}$ and $\vec{v} = \vec{u}$ outside (figure 4).

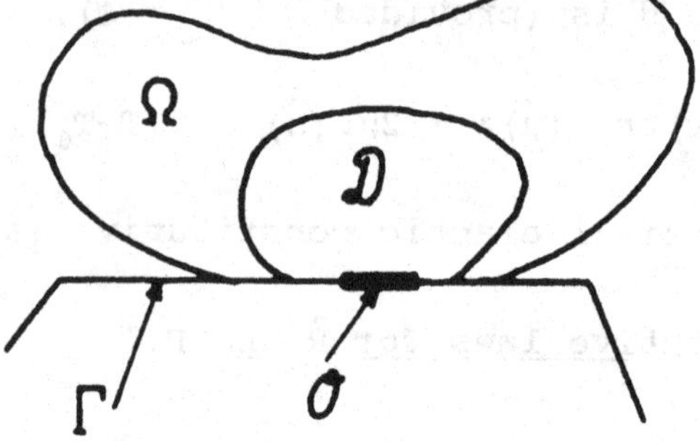


Figure 4
A domain $\mathcal{D}$ and a neighbourhood $\mathcal{O}$ of a point x of $\Gamma$

We assume that $\mathcal{W}(d_1)$ $(d_1 = (\vec{u},\beta))$ is large enough for $(\vec{v},\gamma) \in \mathcal{W}(d_1)$. Relation (2) gives

$$\frac{1}{2}\mathcal{A}_1(\mathcal{D}, \vec{v}-\vec{u}, \vec{v}-\vec{u}) + \frac{1}{2}\int_{\mathcal{O}} k_2(T)\vec{v}^2\, d\Gamma + \int_{\mathcal{O}} \vec{R}\vec{v}\, d\Gamma \geq \int_{\mathcal{D}\cap\Gamma} (w+F)(\gamma-\beta)\, d\Gamma. \tag{6}$$

We let

$$c^2(\mathcal{O}) = \mathrm{Inf}\left\{-\int_{\mathcal{D}\cap\Gamma}(w+F)(\gamma+\beta)\,d\Gamma \mid \gamma = 0 \text{ on } \mathcal{O} \;;\; \gamma\vec{v} = \beta\vec{u} \text{ outside } \mathcal{O})\right\}.$$

By relations (5) and $\vec{v} = \vec{u}$ outside $\mathcal{O}$, we obtain

$$c^2(\mathcal{O}) = \int_{\mathcal{O}} (w+F)\beta\; d\Gamma \geq 0.$$

Let us now note that

$$\mathrm{Inf}\left\{\frac{1}{2}\mathcal{A}_1(\mathcal{D},\vec{w}-\vec{u},\vec{w}-\vec{u}) + \frac{1}{2}\int_{\mathcal{O}} k_2(T)\vec{v}^2\, d\Gamma \mid \vec{w} = \vec{v} \text{ on } \mathcal{D}\cap\Gamma\right\}$$

$$= \frac{1}{2}\|\vec{v}-\vec{u}\|^2_{\mathcal{D}\cap\Gamma} = \frac{1}{2}\|\vec{v}\|^2_{\mathcal{O}},$$

is the square of a norm of the difference $\vec{u}-\vec{v}$ on $\mathcal{D} \cap \Gamma$ which differs from $\vec{u}$ only on $\mathcal{O}$ (it is a $H^{1/2}$ Sobolev norm).

Let us return to relation (6) which we can now write

$$c^2(\mathcal{O}) + \frac{1}{2}\|\vec{v}\|^2_{\mathcal{O}} + \int_{\mathcal{O}} \vec{R}\vec{v}\, d\Gamma \geq 0, \tag{7}$$

Let us first suppose that $R_N$ is a tension ($R_N \geq 0$). Then there exist a $\vec{\varphi}$ with $\vec{\varphi}\vec{n} \leq 0$ on $\mathcal{O}$ and $\vec{\varphi} = 0$ outside $\mathcal{O}$ such that

$$\|\vec{\varphi}\|_{\mathcal{O}}\|\vec{R}\|_{\mathcal{O}} = -\int_{\mathcal{O}} \vec{R}\vec{\varphi}\, d\Gamma,$$

where the norm of $\vec{R}$ is the dual norm of the norm of $\vec{v}$. Let us take $\vec{v} = \alpha\vec{\varphi}(\alpha > 0)$ and thus (7) gives

$$\|\vec{R}\|_{\mathcal{O}} \leq \frac{c^2(\mathcal{O})}{\alpha\|\vec{\varphi}\|_{\mathcal{O}}} + \frac{\alpha}{2}\|\vec{\varphi}\|_{\mathcal{O}}.$$

Taking the minimum of the right hand side with respect to $\alpha$ ($\alpha$ can be bounded since $d = (\alpha\vec{\varphi},\gamma)$ must be an element of $\mathcal{W}(d_1)$, we see that $\|\vec{R}\|_{\mathcal{O}}$ is bounded. Thus the traction $\vec{R}$ is not too large if $R_N$ is a tension.

Let us now suppose that $R_N$ is a compression ($R_N \leq 0$). Relation (7) can be written

$$-\int_{\mathcal{O}} \vec{R}_T\vec{v}_T\, d\Gamma \leq c^2(\mathcal{O}) + \frac{1}{2}\|\vec{v}\|^2_{\mathcal{O}} + \int_{\mathcal{O}} v_N R_N\, d\Gamma,$$

where $v_N = \vec{v}\vec{n}$ is the normal displacement ($v_N \leq 0$ since $(\vec{v},\gamma) \in K$) and $\vec{v}_T = \vec{v}-v_n\vec{n}$ is the tangential displacement. It follows that

$$-\int_{\mathcal{O}} \vec{R}_T\vec{v}_T\, d\Gamma \leq c^2(\mathcal{O}) + \frac{1}{2}\|\vec{v}_T\|^2_{\mathcal{O}}.$$

Defining

$$\||\vec{R}_T|\| = \sup\left\{\int_{\mathcal{O}} \frac{\vec{R}_T}{\|\vec{v}\|_{\mathcal{O}}} \;\middle|\; v_n = 0 \text{ on } \mathcal{O}\right\}, \tag{8}$$

the preceding relation implies that

$$\||R_T|\| \leq \frac{c^2(\mathcal{O})}{\alpha\|\vec{\varphi}\|_{\mathcal{O}}} + \frac{\alpha}{2}\|\vec{\varphi}\|_{\mathcal{O}}, \tag{9}$$

where $\vec{\varphi}$ is the function which achieves the supremum in (8). Taking the infimum of the right hand side of (9) with respect to $\alpha > 0$ ($\alpha$ can be bounded since $d = (\alpha\vec{\varphi},\gamma)$ must be an element of $\mathcal{W}(d_1)$), we prove that $\vec{R}_T$ is bounded.

**Note.** If $c^2(\mathcal{O}) = 0$, relations (7) and (9) show that $\vec{R}_T = 0$ and $R_N \leq 0$ is arbitrary. This is the classical unilateral constitutive law. This occurs for example if $\beta$ and $\vec{u}$ are simultanously zero on $\mathcal{O}$ or if $\beta < 1$ (since $F+w = 0$, relation (5)). The quantity $c^2(\mathcal{O})$ is strictly positive only if $\beta = 1$.

The properties of $\vec{R}$ comply well with what is expected for contact behaviour : if is possible to press down without any limit but it is not possible to pull too much.

If the external action $A = 0$, the equilibrium equation (2.3) $F = A = 0$ and the constitutive law (5) showthat the only possible value of $\beta$ is 1. This corresponds to the following experimental result : if there is contact, the adhesion is total. This is the case for glass-polyurethane contact [8].

<u>7.3.2. No adhesion</u> ($\beta = 0$)

We will show that where there is no adhesion the classical unilateral conditions are satisfied

$$R_N + k_2 u_N \le 0, \quad (R_N + k_2 u_N) u_N = 0, \quad \vec{R}_T + k_2 \vec{u}_T = 0, \tag{10}$$

We will further show that

$$w + F - H \le 0$$

where H is the energy release rate which is defined below.

In order to prove these properties, we follow the same lines as in the previous paragraph ; first introduce variations of $(\vec{u},\beta)$ that do not alter the state of contact, then introduce variations that affect the nature of the contact state.

Let $\mathcal{D}$ be a domain of $\Omega$ and $\mathcal{O}$ a subdomain of $\mathcal{D} \cap \Gamma$ where $\beta = 0$. Let us first take $(\vec{v},\beta) \in K$ with $\vec{v} = \vec{u}$ ouside $\mathcal{O}$. By virtue of (2), we have

$$\frac{1}{2} \mathcal{A}_1 (\mathcal{D}, \vec{v}-\vec{u}, \vec{v}-\vec{u}) + \frac{1}{2} \mathcal{A}_2 (\mathcal{D}, \vec{v}-\vec{u}, \vec{v}-\vec{u}) \ge -\int_{\mathcal{O}} (\vec{R} + k_2 \vec{u}) (\vec{v}-\vec{u}) \, d\Gamma .$$

It results easily from this relation that

$$\int_{\mathcal{O}} (\vec{R} + k_2 \vec{u}) \vec{u} \, d\Gamma = 0,$$

$$\forall \vec{\varphi} \in \mathcal{U}_0^1 \text{ with } \varphi_N \le 0 \text{ on } \Gamma, \quad 0 \ge -\int_{\mathcal{O}} (\vec{R} + k_2 \vec{u}) \vec{\varphi} \, d\Gamma ,$$

which gives the relations (10).

Let us choose $(\vec{v}, \gamma)$ such that $\vec{v} = 0$ on $\mathcal{O}$ and $\gamma = \beta$ outside $\mathcal{O}$ (figure 5).

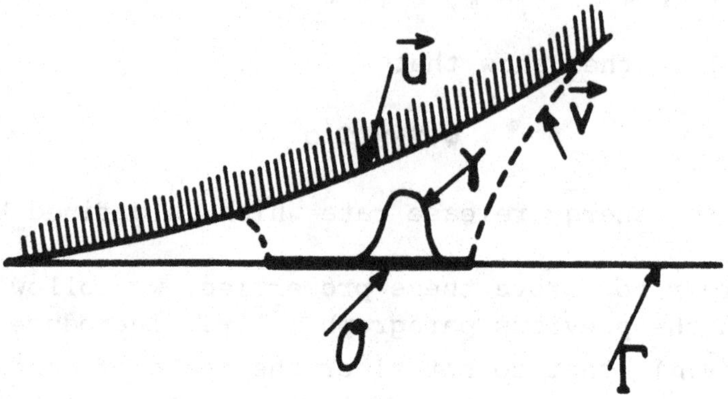


Figure 5
A neighbourhood $\mathcal{O}$ of a point x of $\Gamma$ where there is not contact and functions $(\vec{v},\gamma)$

We suppose that $(\vec{v},\gamma) \in \mathcal{W}(d_1) \cap K$. It results from (2) that

$$\frac{1}{2}\mathcal{A}_1(\mathcal{D},\vec{v}-\vec{u},\vec{v}-\vec{u}) + \frac{1}{2}\mathcal{A}_2(\mathcal{D},\vec{v}-\vec{u},\vec{v}-\vec{u}) + \int_{\mathcal{D}\cap\Gamma}(\vec{R}+k_2\vec{u})(\vec{v}-\vec{u})\,d\Gamma$$
$$- \int_{\mathcal{O}}(w+F)\gamma\,d\Gamma \geq 0.$$

Let us define

$$c^2(\mathcal{O}) =$$
$$\mathrm{Inf}\left\{\frac{1}{2}\mathcal{A}_1(\mathcal{D},\vec{v}-\vec{u},\vec{v}-\vec{u}) + \frac{1}{2}\mathcal{A}_2(\mathcal{D},\vec{v}-\vec{u},\vec{v}-\vec{u}) + \int_{\mathcal{D}\cap\Gamma}(\vec{R}+k\vec{u})(\vec{v}-\vec{u})\,d\Gamma \right. \tag{11}$$
$$\left. \middle| \ \vec{v} = 0, \text{ on } \mathcal{O} \ ; \ \vec{v}\beta = 0 \ ; \ v_N \leq 0 \text{ outside } \mathcal{O}\right\},$$

By (10), this expression is positive. We therefore have

$$\int_{\mathcal{O}}(w+F)\gamma\,d\Gamma \leq c^2(\mathcal{O}),$$

Let us suppose that $\mathcal{O}$ is a neighbourhood of some point x of $\Gamma$. Dividing this relation by the measure of $\mathcal{O}$, meas$(\mathcal{O})$, and defining

$$H(x) = \lim_{\mathcal{O} \to x} \frac{c^2(\mathcal{O})}{\text{meas}(\mathcal{O})}, \tag{12}$$

we obtain

$$w+F \leq H.$$

**Note.** We have in fact, $(w+F)^+ \leq H$, because H is positive.

Let us call H the energy release rate. This term, the energy release rate, is already used in crack theory : it is the derivative of the potential energy with respect to the length of the crack [2]. We prove in the next section that our definition is almost the same : H is the derivative of the potential energy with respect to the contact surface.

7.3.3. A property of the energy release rate.

Let us consider a domain $\mathcal{D}$ of the structure such that $\mathcal{D} \cap \Gamma$ contains a part where the structure is fixed to $\Gamma$, i.e. points where $\beta > 0$. Due to the constitutive equations and the equilibrium equations, the displacement $\vec{u}$ is the actual displacement of $\mathcal{D}$ loaded by body forces div $\sigma = (\sigma_{ij,j})$ and surface forces $\sigma_{ij} n_j$ on $\partial\mathcal{D}-\Gamma$, and submitted to unilateral boundary conditions on the part of $\mathcal{D} \cap \Gamma$ where it is not fixed. The stress field $\sigma = \sigma(\vec{u},T)$, given by relation (4), is the actual stress field. This property results also from (1) with $\gamma = \beta$ and $\vec{v}$ being fixed to $\Gamma$ where $\vec{u}$ is, i.e., where $\beta > 0$.

Let us define

$$\mathcal{F}[\mathcal{D},\vec{v},C] = \frac{1}{2} \mathcal{A}_1(\mathcal{D},\vec{v},\vec{v}) + \frac{1}{2} \mathcal{A}_2(\mathcal{D},\vec{v},\vec{v})$$

$$- \int_{\mathcal{D}} (\sigma_{ij,j}) v_i \, d\Omega - \int_{\partial\mathcal{D}-\Gamma} \sigma_{ij} n_j v_i \, d\Gamma,$$

with $\vec{v} = 0$ on the part C of $\Gamma$.

Let us go back to $c_2(\mathcal{O})$ (relation (11)). We can write

$$B = \frac{1}{2}\mathcal{A}_1(\mathcal{D},\vec{v}-\vec{u},\vec{v}-\vec{u}) + \frac{1}{2}\mathcal{A}_2(\mathcal{D},\vec{v}-\vec{u},\vec{v}-\vec{u}) + \int_{\mathcal{D}\cup\Gamma}(\vec{R}+k_2\vec{u})(\vec{v}-\vec{u})\,d\Gamma$$

$$= \frac{1}{2}\mathcal{A}_1(\mathcal{D},\vec{v},\vec{v}) + \frac{1}{2}\mathcal{A}_2(\mathcal{D},\vec{v},\vec{v}) - \left\{\frac{1}{2}\mathcal{A}_1(\mathcal{D},\vec{u},\vec{u}) + \frac{1}{2}\mathcal{A}_2(\mathcal{D},\vec{u},\vec{u})\right\}$$

$$- \int_{\mathcal{D}}\sigma(\varepsilon(\vec{v}) - \varepsilon(\vec{u}))\,d\Omega + \int_{\mathcal{D}\cap\Gamma}\vec{R}(\vec{v}-\vec{u})\,d\Gamma =$$

$$= \left\{\frac{1}{2}\mathcal{A}_1(\mathcal{D},\vec{v},\vec{v}) + \frac{1}{2}\mathcal{A}_2(\mathcal{D},\vec{v},\vec{v}) - \int_{\mathcal{D}}\sigma_{ij,j}v_i\,d\Omega - \int_{\partial\mathcal{D}-\Gamma}\sigma_{ij}n_jv_i\,d\Gamma\right\}$$

$$- \left\{\frac{1}{2}\mathcal{A}_1(\mathcal{D},\vec{u},\vec{u}) + \frac{1}{2}\mathcal{A}_2(\mathcal{D},\vec{u},\vec{u}) - \int_{\mathcal{D}}\sigma_{ij,j}u_i\,d\Omega - \int_{\partial\mathcal{D}-\Gamma}\sigma_{ij}n_ju_i\,d\Gamma\right\}$$

$$= \mathcal{F}[\mathcal{D},\vec{v},Z(\vec{u}) \cup \mathcal{O})] - \mathcal{F}[\mathcal{D},\vec{u},Z(\vec{u})]$$

because $\sigma\vec{n} = \vec{R}$ on $\mathcal{D} \cap \Gamma$, relation (2.4). The function $\vec{u}$ is zero on the part $\mathcal{Z}(\vec{u}) = \{x \mid x \in \Gamma\ ;\ \beta(x) > 0\}$ and the function $\vec{v}$ is zero on the part $\mathcal{Z}(\vec{u}) \cup \mathcal{O}$ of $\Gamma$.

We then have

$$c^2(\mathcal{O}) = \mathrm{Inf}\{B \mid \vec{v} = 0 \text{ on } \mathcal{O}\ ;\ \vec{v}\beta = \vec{u}\beta = 0,\ v_N \leq 0 \text{ outside } \mathcal{O}\}$$

$$= \mathrm{Inf}\{\mathcal{F}[\mathcal{D},\vec{v},\mathcal{Z}(\vec{u}) \cup \mathcal{O} \mid \vec{v} = 0 \text{ on } \mathcal{Z}(\vec{u}) \cup \mathcal{O}\ ;\ v_N \leq 0 \text{ on } \mathcal{D} \cap \Gamma\}$$

$$- \mathcal{F}[\mathcal{D},\vec{u},\mathcal{Z}(\vec{u})].$$

The quantity $\mathcal{F}[\mathcal{D},\vec{v},\mathcal{Z}(\vec{u})\cup \mathcal{O}]$ is the potential energy for the displacement $\vec{v}$ of the part $\mathcal{D}$ of the structure being fixed on the part $\mathcal{Z}(\vec{u}) \cup \mathcal{O}$ of $\Gamma$ and being in unilateral contact with the support on the remaining part of $\Gamma$. The infimum with respect to $\vec{v}$ is the actual potential energy of the part $\mathcal{D}$ of the structure being loaded by the volume forces $\sigma_{ij,j}$, the surface forces $\sigma_{ij}n_j$ on $\partial\mathcal{D}-\Gamma$, being fixed on $\mathcal{Z}(\vec{u}) \cup \mathcal{O}$ and being in unilateral contact on the remaining part of $\Gamma$. The infimum is also $\mathcal{F}[\mathfrak{D},\hat{u},\mathfrak{Z}(\vec{u}) \cup \mathcal{O}]$ where $\hat{u}$ is the actual equilibrium displacement of the structure with these boundary conditions.

It is obvious that $\mathcal{F}[\mathfrak{D},\vec{u},\mathfrak{Z}(\vec{u})]$ is the actual potential energy of the part $\mathfrak{D}$ of the structure being loaded by the same forces but fixed only on the part $\mathfrak{Z}(\vec{u})$ and in unilateral contact on the remaining part of $\partial\mathfrak{D} \cap \Gamma$.

The energy release rate

$$H(x) = \lim_{\mathcal{O}\to x} \frac{\mathcal{F}[\mathfrak{D},\hat{u},\mathfrak{Z}(\vec{u}) \cup \mathcal{O}] - \mathcal{F}[\mathfrak{D},\vec{u},\mathfrak{Z}(\vec{u})]}{\operatorname{meas}(\mathcal{O})}$$

is the derivative of the potential energy with respect to the part on which the structure is fixed. It is the amount of energy by unit area necessary to modify the state of contact at the point x. This property justifies H to be called energy release rate. The usual definition is the derivative of the potential energy with respect to some crack length [2].

It seems that the energy release rate depends on $\mathfrak{D}$. Let us prove that this is not the case.

**THEOREM 7.1.** The energy release rate H does not depend on the part $\mathfrak{D}$ under consideration.

**Proof.** Let us consider two domains $\mathfrak{D}$ and $\mathfrak{D}'$ such that $\mathfrak{D} \subset \mathfrak{D}'$ (figure 6). Let there be $\tilde{\Gamma}$ the interior boundary between $\mathfrak{D}$ and $\mathfrak{D}'-\mathfrak{D}$.

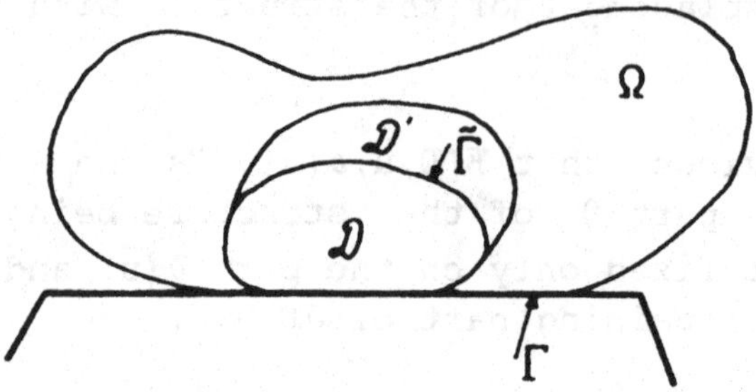


Figure 6
The domains $\mathfrak{D}$ and $\mathfrak{D}'$ ($\mathfrak{D} \subset \mathfrak{D}'$).
The interior boundary $\tilde{\Gamma}$ between $\mathfrak{D}$ and $\mathfrak{D}'-\mathfrak{D}$.

Let us define

$$\mathfrak{B}(\mathfrak{D},\vec{v}-\vec{u}) = \frac{1}{2}\mathcal{A}_1(\mathfrak{D},\vec{v}-\vec{u},\vec{v}-\vec{u}) + \frac{1}{2}\mathcal{A}_2(\mathfrak{D},\vec{v}-\vec{u},\vec{v}-\vec{u}) + \int_{\mathfrak{D}\cup\Gamma}(\vec{R}+k\vec{u})(\vec{v}-\vec{u})\,d\Gamma,$$

and

$$c^2(\mathcal{O},\mathfrak{D}) = \mathrm{Inf}\{\mathfrak{B}(\mathfrak{D},\vec{v}-\vec{u}) \mid \vec{v} = 0 \text{ on } \mathfrak{Z}(\vec{u}) \cup \mathcal{O}\ ; \ v_N \le 0 \text{ on } \mathfrak{D}\cap\Gamma\} = \mathfrak{B}(\mathfrak{D},\vec{v}(\mathfrak{D})-\vec{u}).$$

It is the quantity $c^2(\mathcal{O})$ for the domain $\mathfrak{D}$. We have

$$\begin{aligned} c^2(\mathcal{O},\mathfrak{D}') &= \mathfrak{B}(\mathfrak{D},\vec{v}(\mathfrak{D}')-\vec{u}) + \mathfrak{B}(\mathfrak{D}'-\mathfrak{D},\ \vec{v}(\mathfrak{D}')-\vec{u}) \\ &\ge c^2(\mathcal{O},\mathfrak{D}) + \mathfrak{B}(\mathfrak{D}'-\mathfrak{D},\ \vec{v}(\mathfrak{D}')-\vec{u}). \end{aligned} \tag{13}$$

We have also

$$\begin{aligned} c^2(\mathcal{O},\mathfrak{D}) &+ \mathrm{Inf}\{\mathfrak{B}(\mathfrak{D}'-\mathfrak{D},\vec{v}-\vec{u}) \mid \vec{v} = \vec{v}(\mathfrak{D}) \text{ on } \tilde{\Gamma}\} \\ &= \mathfrak{B}(\mathfrak{D}',\vec{w}-\vec{u}) \ge c^2(\mathcal{O},\mathfrak{D}'), \end{aligned} \tag{14}$$

because the fonction $\vec{w}$ is a candidate for the minimum of $\mathcal{B}(\mathcal{D}')$.

Let us define

$$L(\mathcal{D}'-\mathcal{D},\vec{v}(\mathcal{D})) = \mathrm{Inf}\{\mathcal{B}(\mathcal{D}'-\mathcal{D}, \vec{v}-\vec{u}) \mid \vec{v} = \vec{v}(\mathcal{D}) \text{ on } \tilde{\Gamma}\}.$$

It is obvious that

$$L(\mathcal{D}'-\mathcal{D},\vec{v}(\mathcal{D}')) = \mathcal{B}(\mathcal{D}'-\mathcal{D},\vec{v}(\mathcal{D}')-\vec{u}).$$

From (13) and (14), we get

$$\begin{aligned} c^2(\mathcal{O},\mathcal{D}') &\geq c^2(\mathcal{O},\mathcal{D}) + L(\mathcal{D}'-\mathcal{D},\vec{v}(\mathcal{D}')), \\ c^2(\mathcal{O},\mathcal{D}) + L(\mathcal{D}'-\mathcal{D},\vec{v}(\mathcal{D})) &\geq c^2(\mathcal{O},\mathcal{D}'). \end{aligned} \tag{15}$$

We have

$$L(\mathcal{D}'-\mathcal{D},\vec{v}) = \frac{1}{2}\,\|\vec{v}-u\|^2_{H^{1/2}(\tilde{\Gamma})}$$

where the norm is the Sobolev $H^{1/2}$ norm of $\vec{v}-\vec{u}$ on $\tilde{\Gamma}$.

We assume that

$$(\vec{v}(\mathcal{D})-\vec{u})(x) = \int_{\mathcal{O}} G(x-y)(-\vec{u}(y))dy$$

where G is a Green function. The function $(\vec{v}(\mathcal{D})-\vec{u})$ and its derivatives then are proportionnal to meas($\mathcal{O}$). Therefore it can be proved that

$$\lim_{\mathcal{O}\to x} \frac{L(\mathcal{D}'-\mathcal{D},\vec{v}(\mathcal{D}))}{\text{meas}\mathcal{O}} = \lim_{\mathcal{O}\to x} \frac{L(\mathcal{D}'-\mathcal{D},\vec{v}(\mathcal{D}'))}{\text{meas}\mathcal{O}} = 0.$$

We then conclude from (15) that

$$\lim_{\mathcal{O}\to x} \frac{c^2(\mathcal{O},\mathcal{D})}{\text{meas}(\mathcal{O})} = \lim_{\mathcal{O}\to x} \frac{c^2(\mathcal{O},\mathcal{D}')}{\text{meas}(\mathcal{O})} = H.$$

We have proved (at least for the situation of figure 6) that H does not depend on $\mathcal{D}$. We assume that this is true in the other situations.

The energy release rate can be also defined when there is contact. Let us consider a subdomain $\mathcal{O}$ of $\mathcal{D} \cap \Gamma$ where there is contact ($\vec{u} = 0$ on $\mathcal{O}$). Let us define

$$- c^{2\,\prime}(\mathcal{O}) = \operatorname{Inf}\left\{\frac{1}{2}\, \mathcal{A}_1(\mathcal{D}, \vec{v}-\vec{u}, \vec{v}-\vec{u}) + \frac{1}{2}\, \mathcal{A}_2(\mathcal{D}, \vec{v}-\vec{u}, \vec{v}-\vec{u}) + \int_{\mathcal{O}} \vec{R}\vec{v}\, d\Gamma \mid v_N \leq 0 \;;\; \vec{v}-\vec{u} = 0 \text{ outside } \mathcal{O}\right\}.$$

This quantity is negative since $\vec{v} = \vec{u}$ is a possible choice. By relation (6) or (7), we have

$$c^2(\mathcal{O}) = \int_{\mathcal{O}} (w+F)\beta \, d\Gamma \geq c'^{\,2}(\mathcal{O}).$$

Dividing by the measure of $\mathcal{O}$, meas($\mathcal{O}$), and making $\mathcal{O} \longrightarrow x$ we obtain

$$(w+F)\beta \geq H$$

where

$$H(x) = \lim_{\mathcal{O} \to x} \frac{c'^{\,2}(\mathcal{O})}{\text{meas}(\mathcal{O})} \geq 0,$$

is the energy release rate. When $w+F > 0$ (it is the case when $0 = A = F$, relation (2.3)), we have $\beta = 1$ and

$$w+F \geq H.$$

When $w+F = 0$, we have

$$H = 0.$$

In this situation we have $0 \leq \beta \leq 1$ by relation (5).

We have therefore obtained the following relations

$$\begin{aligned} &w+F \leq H, && \text{where } \beta = 0, \\ &w+F = H = 0, && \text{where } 0 < \beta < 1, \\ &w+F \geq H, && \text{where } \beta = 1. \end{aligned} \tag{16}$$

Let us remark that, if $A = 0$, by the equilibrium equation $F = A = 0$, we have

$$w \leq H, \text{ where } \beta = 0,$$

$$w \geq H, \text{ where } \beta = 1,$$

which is the classical Griffith criterium of fracture mechanics. Let us also again remark that $\beta$ cannot have values between 0 and 1 since $w \neq 0$ (relation (16), for $0 < \beta < 1$).

## 8. ISOTHERMAL EQUILIBRIUM OF A NON DISSIPATIVE STRUCTURE. VARIATIONAL FORMULATION.

Let us consider a structure $\Omega$ on which act volume forces $\vec{f}$, surface forces $\vec{g}$ on the part $\Gamma_1$ and possibly the work A on the part $\Gamma$ where the structure is in contact with the support $\Gamma$. The structure is fixed on the part $\Gamma_0$. We assume that the behaviour of the structure and the contact on $\Gamma$ are non dissipative or elastic ($\Phi \equiv 0$). We assume that the temperature is constant and known everywhere ($T = T_0$) (figure 7).

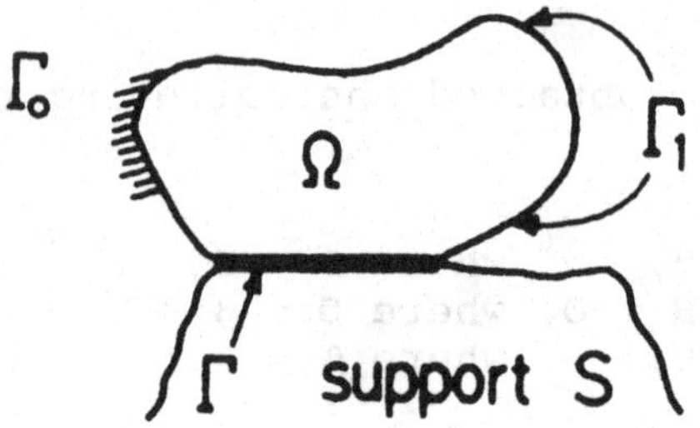


Figure 7
The structure loaded by surface forces on $\Gamma_1$.

The equations that have to be satisfied by an equilibrium position $d_1 = (\vec{u},\beta)$ are the equilibrium equations (2.1), (2.2), (2.3), (2.4) and the constitutive law

$$f_1 = (\sigma,\vec{R},F) \in \bar{\partial}\Psi(d_1).$$

The equilibrium equations are equivalent to the virtual work theorem

**THEOREM 8.1.** The equilibrium equations (2.1), (2.2), (2.3), (2.4), are equivalent to

$$\forall \mathcal{D} \subset \Omega,\ \forall d \in \mathcal{U}_0,\ \mathcal{T}_e(\mathcal{D},d) + \mathcal{T}_i(\mathcal{D},d) = 0,$$

where the work of the internal (resp. external) forces $\mathcal{T}_i(\mathcal{D},d)$ (resp. $\mathcal{T}_e(\mathcal{D},d)$) is equal to $\hat{\mathcal{P}}_i(\mathcal{D},d)$ (resp. $\hat{\mathcal{P}}_e(\mathcal{D},d)$).

**Proof.** It suffices to follow backwards the calculations of paragraph 2.4.

The data of the problem are $\Psi$, $T_0$, $\vec{f}$, $\vec{g}$, A. The unknown is $d_1 = (\vec{u},\beta)$. Let us give a variational formulation of the equations. We have

$$\forall d \in \mathcal{W}(d_1) \cap K, \quad \Psi(\Omega,d) \geq \Psi(\Omega,d_1) + \langle f_1, d-d_1 \rangle_\Omega .$$

Moreover

$$<f_1, d>_\Omega = \int_\Omega \sigma\varepsilon(\vec{v})d\Omega + \int_\Gamma \{F\gamma - \vec{R}\vec{v}\}d\Omega = -\mathcal{T}_i(\Omega, d)).$$

By means of the virtual work theorem, we have

$$\forall d \in \mathcal{W}(d_1) \cap K, \quad \Psi(\Omega,d) - \mathcal{T}_e(\Omega,d) \geq \Psi(\Omega,d_1) - \mathcal{T}_e(\Omega,d_1).(1).$$

We then have :

**THEOREM 8.2.** The equilibrium positions of the structure are the local minimum on $\mathcal{U}_0 \cap K$ of the potential energy

$$\mathcal{F}[d] = \Psi(\Omega,d) - \mathcal{T}_e(\Omega,d) = \frac{1}{2}\mathcal{A}_1(\Omega,\vec{v},\vec{v})$$

$$+ \frac{1}{2}\mathcal{A}_2(\Omega,\vec{v},\vec{v}) + \int_\Gamma w(1-\gamma)d\Gamma - \int_\Omega \vec{f}\vec{v}\, d\Omega - \int_{\Gamma_1} g\vec{v}\, d\Gamma - \int_\Gamma A\gamma\, d\Gamma.$$

By appropriately defining the respective functional spaces we can prove an existence result :

**THEOREM 8.3.** Assuming the set $\Omega$ to be smooth, meas $\Gamma_0 > 0$, meas $\Gamma > 0$, $\mu > 0$, $3\lambda+2\mu > 0$, $k_2 \geq 0$,

$$\mathcal{U}_0^1 = \{\vec{v} \mid \vec{v} \in H^1(\Omega) \ ; \ \vec{v}|\Gamma_0 = 0\}, \ \mathcal{U}_0^2 = L^2(\Gamma) \ ;$$

$$\vec{g} \in L^2(\Gamma_1) \ ; \ \vec{f} \in L^2(\Omega) \ ; \ A \in L^2(\Gamma),$$

problem (1) has at least one solution. If $w+A > 0$ (this is the case if $A = 0$), $\beta = 1$, where $\vec{u} = 0$. Il $w+A < 0$, $\beta = 0$ everywhere on $\Gamma$.

If $A = 0$, we find that as long as there is contact, the adhesion is total ($\beta = 1$). This point which conforms with experiment, has already been mentionned. This remark leads to the classical formulation of adherence problems.

## 9. CLASSICAL ADHERENCE PROBLEMS

We have just shown that if A = 0 the adhesion intensity $\beta$ assumes only the values 0 and 1. Taking advantage of this result we restrict ourselves to function $\gamma$ with values 0 or 1. Let us define

$$\tilde{K} = \{\vec{v} \mid \vec{v} \in \mathcal{U}_0^1 \ ; \ v_N \leq 0 \text{ on } \Gamma \text{ on } \Gamma\},$$

$$\mathcal{K}(\vec{v}) = \text{meas}\{x \mid x \in \Gamma \ ; \ \vec{v}(x) = 0\},$$

where $\mathcal{K}(\vec{v})$ is the contact area corresponding to the displacement $\vec{v}$ ;

$$\tilde{\mathcal{F}}[\vec{v}] = \frac{1}{2} \mathcal{A}_1 (\Omega,\vec{v}) + \frac{1}{2} \mathcal{A}_2 (\Omega,\vec{v})$$

$$- \int_\Omega \vec{f}\vec{v} \, d\Omega - \int_{\Gamma_1} \vec{g}\vec{v} \, d\Gamma + w(\text{meas}(\Gamma) - \mathcal{K}(\vec{v})).$$

Problem (8.1) is therefore equivalent to finding the local minima of $\tilde{\mathcal{F}}$ on $\tilde{K}$. That is to say, finding a $\vec{u} \in \tilde{K}$ and a neighboorhood $\tilde{W}$ of $\vec{u}$ (in the sense of the topology induced by that of $\mathcal{U}_0$) such that

$$\vec{u} \in \tilde{K}, \ \forall v \in \tilde{W}(\vec{u}), \ \tilde{\mathcal{F}}[\vec{u}] \leq \tilde{\mathcal{F}}[\vec{v}].$$

This relation is the basic relation introducing the adhesion theory in physics [8], [9].

## 10. DISSIPATIVE BEHAVIOUR

Let us examine the structure of the pseudo potential of dissipation (6.5). The dissipation effects that are of interest to us are those related to $\dot{\beta}$. We therefore limit ourselves to the case where $\varphi = 0$ and $\bar{\varphi}(\dot{\vec{u}},\dot{\beta}) = \bar{\varphi}(\dot{\beta})$.

The constitutive equations are

$$(\sigma,\vec{R},G_2) \in \bar{\partial}\Psi(d_1), \quad (F-G_2) \in \partial\bar{\varphi}(\dot{\beta}). \tag{1}$$

They imply that

$$\left.\begin{array}{l} w+G_2-H \le 0, \text{ where } \beta = 0 \\ w+G_2 = H = 0, \text{ where } 0 < \beta < 1 \\ w+G_2-H \ge 0, \text{ where } \beta = 1 \end{array}\right\}, \tag{2}$$

(in relation (7.1) and (7.16) the non dissipative force $G_2$ is equal to the internal force F. Here F is equal to the sum of $G_2$ and of the dissipative force element of $\partial\bar{\varphi}(\dot{\beta})$).

The velocity field $\dot{\beta}$ should take into account eventual jumps from $\beta = 0$ to $\beta = 1$. When e.g. tearing wall paper off, $\beta$ passes from 1 to zero along the debonding line. The velocity $\dot{\beta}$ is therefore a Dirac function. We consequently assume $\dot{\beta}$ to be a measure on $\Gamma$. The irreversible force $F-G_2$, the product of which with $\dot{\beta}$ is a power, is then an element of a linear space in duality with the measures on $\Gamma$ (for example, $F-G_2$ is a continuous function on $\Gamma$). Thus the subgradients of $\partial\bar{\varphi}(\dot{\beta})$must be element of this linear space in dualy with the measures on $\Gamma$ (for example continuous functions). Let us give three examples of such pseudo-potentials.

## 10.1. Linear viscous behaviour

Let us choose

$$\bar{\varphi}_1(\dot{\beta}) = \frac{1}{2}(g * \dot{\beta})\dot{\beta},$$

where $g * \dot{\beta}$ is the convolution of the continuous function g with $\dot{\beta}$ on the surface $\Gamma$

$$g * \dot{\beta}(x) = \int_\Gamma g(x-y)\dot{\beta}(y)dy.$$

The pseudo-potential

$$\Phi_1(\dot\beta) = \int_\Gamma (g * \dot\beta)(x) \times \dot\beta(x)dx$$

is convex, positive and zero at the origin, if g is the Fourier transform of a positive, bounded measure (e.g. $g = e^{-|x|}$, $1/(1+x^2)$, $1,\ldots$). We therefore have

$$F-G_2 = g * \dot\beta.$$

The behaviour is linearly viscous. The convolution implies a kinematical interaction between neighbouring points. The merit of this choice, guided by the need of mathematical consistency is to be in accordance with well established experimental results, as we shall see further. Let us further remark that the unusual from of the dissipation is related to the absence of reversible interaction between neighbouring points due to the zero gradient theory for $\beta$. The first gradient theory allows fort classical quadratic dissipation with respect to $\dot\beta$.

<u>**10.2. Norton-Hoff viscous behaviour**</u>

Let $p > 1$ be a number and define the pseudo-potential

$$\Phi_2^p(\dot\beta) = \frac{1}{p}\left\{\int_\Gamma (g * \dot\beta)\dot\beta \; d\Gamma\right\}^{p/2},$$

we then have

$$F = G_2 = \left\{\int_\Gamma (g * \dot\beta)\dot\beta\right\}^{\frac{p}{2}-1} g * \dot\beta,$$

which gives a nonlinear viscous behaviour. If $p = 2$, we have again the linear viscous behaviour.

<u>**10.3. Viscous behaviour with impossible bond restitution**</u>

In this situation only debonding is possible. From the moment it takes place, it is not possible to restore the bonds no matter how large the applied forces are. Such is the case with glued wall paper.

Let $\mathcal{M}^-$ be the set of negative measures on $\Gamma$. We choose

$$\Phi_3(\dot{\beta}) = \frac{1}{2}\int_\Gamma (d * \dot{\beta})\dot{\beta}\, d\Gamma + I_{\mathcal{M}^-}(\dot{\beta}),$$

where $I_{\mathcal{M}^-}$ is the indicator function of the set $\mathcal{M}^-$. We therefore have, letting $\mathcal{O}$ to be the interior of the support of $\dot{\beta}$,

$$F(x) - G_2(x) = (g * \dot{\beta})(x), \text{ if } x \in \mathcal{O},$$

$$F(x) - G_2(x) \geq (g * \dot{\beta})(x), \text{ if } x \notin \mathcal{O}.$$

## 10.4. Evolution equation

Let us consider the same structure $\Omega$ (figure 7) which we now suppose to be made of a dissipative material, loaded by the same external actions, possibly time-dependant. The unknowns are $d_1(t)$ and $T(t)$ which must satisfy the equilibrium equations (2.1), (2.2), (2.3), (2.4), the constitutive laws

$$(\sigma, \vec{R}, F) \in \bar{\partial}\Psi(d_1), \qquad F - G_2 \in \partial\Phi(\dot{\beta}),$$

the energy balance equation, the Fourier's law, as well as boundary and initial conditions.

## 10.5. Consequences of the dissipation

As we have already pointed out that we can consider two processes for the bonding and the debonding :

a - Bonding and debonding along a line

The phenomenon takes place along a moving line which separates the bonded from the debonded areas. Just like when tearing wall paper off, the debonding happens along a line.

b - Surface or block bonding and debonding

The phenomenon takes place over the whole of a surface. When e.g. we glue a postage stamp by pressing it down, the gluing takes place in block. On the other hand the debonding of the stamp takes place along a line. This behaviour seems to be general : bonding is mostly block bonding and debonding is always line debonding.

Let us show that the dissipative constitutive laws account for these experimental facts. For the sake of simplicity, we assume it is a two dimensions problem and that the temperature remains constant.

**THEOREMS 10.1.** Let us consider a structure with contact being viscous with impossible bond restitution. If $w+A > 0$ (in particular if $A = 0$) only line debonding is possible.

**Proof.** Let us assume the structure to debond in block on a subdomain $\mathcal{O}$ of its support. Due to the equilibrium equations and the constitutive law (1), (2) we have

$$u = 0, \quad 0 < \beta < 1, \quad F-G_2 = g * \dot{\beta}, \quad G_2 = -w, \quad F = A \text{ on } \mathcal{O}.$$

This gives

$$w+A = g * \dot{\beta}. \tag{3}$$

Since $\dot{\beta} \le 0$ everywhere on $\Gamma$, $g * \dot{\beta} \le 0$ which is impossible because $w+A > 0$. Thus block debonding is impossible.

Let us examine line debonding (figure 8) and denote by $s(t)$ the abscissa of the point of debonding.

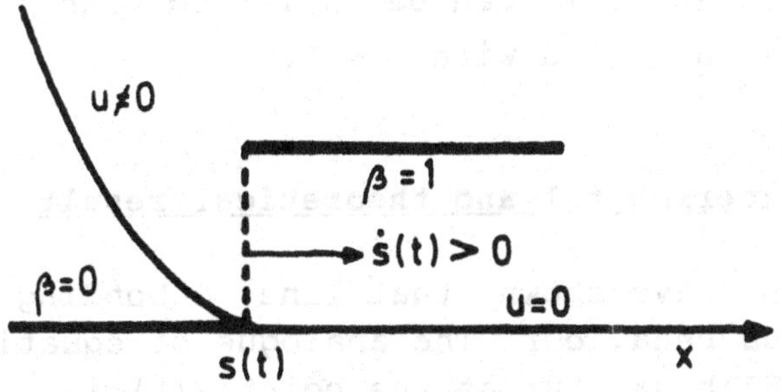


Figure 8
Line debonding ($\dot{s}(t) > 0$).

We have $\beta(x,t) = H(x-s(t))$ where $H$ is the Heaviside's function and

$$\dot{\beta}(x,t) = \delta(x-s(t))(-\dot{s}(t)),$$

where $\delta$ is the Dirac's function and $\dot{s}(t) > 0$ if debonding occurs.

The equations (2), the constitutive law $F-G_2 = g * \dot{\beta}$ and the equilibrium equations imply

$$\left.\begin{array}{lll} w+A-H(x,t) \le -g(x-s(t))\dot{s}(t), & \text{if } x < s(t), & (\beta = 0), \\ w+A-H(s(t),t) = -g(0)\dot{s}(t), & \text{if } x = s(t), & \\ w+A-H(s(t),t) \ge -g(x-s(t))\dot{s}(t), & \text{if } x > s(t), & (\beta = 1) \end{array}\right\} ,(4)$$

It is possible for these equations to be satisfied if H is large where debonding has already taken place and H is small where the bonds are still intact. Thus line debonding is possible.

It is also possible to prove the following theorem.

**THEOREM 10.2.** If $w+A > 0$ (in particular if $A = 0$) both line and surface bonding are possible.

**Proof.** Equation (3) can be satisfied with $\dot{\beta} \geq 0$. Equations (4) can be satisfied with $\dot{s} < 0$.

## 10.6. An experimental and theoretical result

Experiments have shown that line debonding can have a nonlinear viscous behaviour. The analogue of equation (3) for a Norton-Hoff [5] behaviour at the point s(t) is

$$w-H(s(t),t) = -\ (g(0))^{p/2}\ (\dot{s}(t))^{p-1},\ (s > 0).$$

This relation was experimentally established with p = 1.6 for glass-polyurethane contact [1].

### References

[1] M. BARQUINS.- Sur le pelage spontané des élastomères, C.R. Acad. Sci., Paris, Série II, 298, 1984, p. 725-730.

[2] M. D. BUI.- Mécanique de la rupture fragile. Masson, Paris, 1978.

[3] M. FREMOND.- Contact unilateral avec adhérence. Proceedings on the Second Meeting on Unilateral Problems in Structural Analysis, Ravello (Italie), September 22/24/1983, Edited by G. del Piero, F. Maceri) Springer-Verlag, New York, 1985.

[4] M. FREMOND. -Adhérence des solides. Journal de Mécanique Théorique et Appliquée, vol. 6, N°3; 1987, p. 383-407.

[5] A. FRIAA.- La loi de Norton-Hoff généralisée en plasticité et viscoplasticité. Thèse d'Etat, Université Pierre et Marie Curie, 1979.

[6] P. GERMAIN.- Mécanique des milieux continus, Masson, Paris, 1973.

[7] B. HALPHEN, C.S. NGUYEN.- Sur les lois de comportement élastoviscoplastique à potentiel généralisé. C.R. Acad. Sci., Paris, Série A, 277, (1973), p. 319-321.

[8] D. MAUGIS, M. BARQUINS.- Fracture Mechanics and Adherence of Viscoelastic solids, in Adhesion and Adsorption of Polymers, Part A, L.H. Lee, ed., Plenum, New York, 198O.

[9] D. MAUGIS.- Adherence of Solids, in Macroscopic Aspects of Adhesion and Lubrification, J.M. Georges, ed. Elsevier, Amsterdam, 1982.

[10] J.J. MOREAU.-Fonctionnelle Convexes. Séminaire sur les équations aux dérivées partielles. Collège de France, Paris, 1966.

[11] J.J. MOREAU.- Fonctions de résistance et fonctions de dissipation. Séminaire d'Analyse Convexe, Montpellier, 1971.

# APPROXIMATION OF CONTACT PROBLEMS.
# SHAPE OPTIMIZATION IN CONTACT PROBLEMS

**J. Haslinger**
**KFK MFFUK, Praha, Czechoslovakia**

ABSTRACT

The main part of this contribution deals with the approximation of variational inequalities, with special emphasize to contact problems. Finite element technique is used, starting from different variational formulations (primal, mixed). More details, concerning the approximation and numerical realization of contact problems can be found in [**9**]. Second part of this contribution is devoted to the optimization of the shape of contact zone of an elastic body, unilaterally supported by a rigid foundation, in order to obtain an even distribution of normal forces along contact part (for more details see [**10**]).

## 1. PRIMAL, DUAL AND MIXED VARIATIONAL FORMULATION OF ELLIPTIC INEQUALITIES

Let $V$ be a real Hilbert space equipped with the norm $\| \ \|$ and a scalar product. denoted by $( \ , \ )$. Let $V'$ be the dual space over $V$ with duality pairing denoted by $\langle \ , \ \rangle$. Let $a : V \times V \to R_1$ be a bilinear form, which is

- *continuous on* $V$:

$$\exists M = \text{const.} > 0 \quad \forall u, v \in V \quad |a(u,v)| \leq M\|u\|\|v\| \ ;$$

- $V$*-elliptic on* $V$:

$$\exists \alpha = \text{const.} > 0 \quad \forall v \in V \quad a(v,v) \geq \alpha\|v\|^2 \ ;$$

- *symmetric on* $V$:

$$a(u,v) = a(v,u) \quad \forall u, v \in V \ .$$

Finally let $j : V \to R_1$ be a convex, continuous and nonnegative functional and set

$$L(v) = J(v) + j(v) \quad \forall v \in V \ ,$$

where

$$J(v) = \frac{1}{2}a(v,v) - \langle f, v\rangle \ , \quad f \in V'$$

is a quadratic functional.

Now, let $K \neq \emptyset$ be a closed, convex subset of $V$. We shall assume the following minimization problem

$$\begin{cases} \text{find } u \in K \text{ such that} \\ L(u) \leq L(v) \quad \forall v \in K \ . \end{cases} \tag{P}$$

Using classical arguments of calculus of variations and assumptions, concerning $L$, one has

**Theorem 1.1.** *There exists a unique solution $u$ of* (**P**). *Moreover, this solution can be equivalently characterized through the relation*

$$u \in K \ : \quad a(u, v-u) + j(v) - j(u) \geq \langle f, v-u\rangle \quad \forall v \in K \ . \tag{1.1}$$

**Remark 1.1.** Let $\mathcal{A} : V \mapsto V'$ be the linear, continuous mapping defined by

$$a(u,v) = \langle \mathcal{A}u, v\rangle \quad \forall u, v \in V \ .$$

Then $u \in K$ solves (1.1) iff $f - \mathcal{A}u \in \partial j(u)$, where $\partial j(u)$ is the subgradient of $j$ at $u$.

REMARK 1.2. In our next considerations, $L$ will play the role of the *total potential energy* of the system, $V$ the space of *virtual displacements* and $K$ the set of *kinematically admissible displacements*.

Sometimes it becomes that formulation (**P**) is not suitable, sometimes from theoretical, sometimes from practical point of view. In these cases, reformulation of (**P**) is useful. Below we describe one way how to do it, making use of the saddle-point approach. Using the form of $L$, one can rewrite (**P**) in the following form:

$$\inf_{v \in K} L(v) = \inf_{v \in A} \sup_{\mu \in B} \mathcal{L}(v, \mu) , \tag{1.2}$$

$A \subseteq V$, $B \subseteq Y$ with a suitable choice of $A$, $B$ ($Y$ being another Hilbert space) and $\mathcal{L} : A \times B \to R_1$. Here is a typical example of such a reformulation.

EXAMPLE: (releasing of constraints)
Let $\Phi : V \times \Lambda \to R_1$, where $\Lambda$ is a convex cone with vertex at $\theta$ (zero element of $Y$), $\Lambda \subseteq Y$ and

$$\Phi(v, \rho\mu) = \rho\Phi(v, \mu) \quad \forall \rho \geq 0 , \quad \forall [v, \mu] \in V \times \Lambda$$

(i.e. $\Phi$ is positively 1-homogeneous in the $2^{\text{nd}}$ component).

Now, we shall suppose that the following characterization of $K$ holds:

$$v \in K \iff v \in V \;\; \& \;\; \Phi(v, \mu) \leq 0 \quad \forall \mu \in \Lambda .$$

LEMMA 1.1. *It holds*

$$I(v) \equiv \sup_{\mu \in \Lambda} \Phi(v, \mu) = \begin{cases} 0 & \text{if } v \in K \\ +\infty & \text{if } v \notin K, \end{cases}$$

*i.e. $I$ is the indicator function of $K$.*

PROOF:
1° If $v \in K$, then $\Phi(v, \mu) \leq 0 \; \forall \mu \in \Lambda$ and $\Phi(v, \theta) = 0$ so that $I(v) = 0$.
2° Let $v \notin K$. Then there exists $\bar{\mu} \in \Lambda$ such that

$$\Phi(v, \bar{\mu}) = c > 0 .$$

But

$$I(v) \geq \Phi(v, \rho\bar{\mu}) = \rho\Phi(v, \bar{\mu}) = \rho c \to +\infty , \quad \rho \to \infty .$$

□

It is readily seen that one can formally write

$$\inf_{v \in K} L(v) = \inf_{v \in V} \sup_{\mu \in \Lambda} \{L(v) + \Phi(v, \mu)\} , \tag{1.3}$$

i.e. $\mathcal{L}(v,\mu) \overset{\text{def}}{\equiv} L(v) + \Phi(v,\mu)$, $A = V$, $B = \Lambda$.

REMARK 1.3. The functional $L$ is possibly non-differentiable, due to the presence of $j$. Similar approach, namely dualization of $j$ can be used in order to achieve smooth formulation of (**P**). This will be used in the case of contact problems with friction.

According to the definition of (**P**) and (1.2), the problem

$$\inf_{v\in A} \sup_{\mu\in B} \mathcal{L}(v,\mu) = \inf_{v\in K} L(v) \tag{1.4}$$

will be called *primal*, while

$$\sup_{\mu\in B} \inf_{v\in A} \mathcal{L}(v,\mu) \tag{1.5}$$

will be *dual* to (1.4).

Let

$$\mathcal{S}(\mu) = \inf_{v\in A} \mathcal{L}(v,\mu)\ .$$

Then $\mathcal{S} : B \to R_1$ is called *dual* functional to $L$ and $\mu$ the *dual variable* to $v$.

An interesting question arises, namely: what relation holds between the *primal* and the *dual* formulation. This relation can be established on the basis of the *saddle-point theory*, some results of which we present below (for details see [**1**]).

Let $\mathcal{H} : A \times B \to R_1$ be a general functional.

DEFINITION 1.1. *A point $\{u,\lambda\} \in A\times B$ is said to be a saddle point of $\mathcal{H}$ on $A \times B$ if and only if*

$$\mathcal{H}(u,\mu) \le \mathcal{H}(u,\lambda) \le \mathcal{H}(v,\lambda) \tag{1.6}$$

*holds for any $v \in A$, $\mu \in B$.*

An equivalent characterization of a saddle point is given by

THEOREM 1.2. *$\{u,\lambda\}$ is a saddle point of $\mathcal{H}$ on $A\times B$ if and only if*

$$\mathcal{H}(u,\lambda) = \min_{v\in A} \sup_{\mu\in B} \mathcal{H}(v,\mu) = \max_{\mu\in B} \inf_{v\in A} \mathcal{H}(v,\mu)\ . \tag{1.7}$$

PROOF:

1° Let $v \in A$, $\mu \in B$ be arbitrary. Then

$$\inf_{v\in A} \mathcal{H}(v,\mu) \le \mathcal{H}(v,\mu) \implies$$

$$\sup_{\mu\in B} \inf_{v\in A} \mathcal{H}(v,\mu) \le \sup_{\mu\in B} \mathcal{H}(v,\mu) \quad \forall v \in A\ ,$$

from which

$$\sup_{\mu\in B} \inf_{v\in A} \mathcal{H}(v,\mu) \le \inf_{v\in A} \sup_{\mu\in B} \mathcal{H}(v,\mu)\ . \tag{1.8}$$

Let $\{u, \lambda\} \in A \times B$ be a saddle point of $\mathcal{H}$ on $A \times B$. Then

$$\inf_{v\in A} \sup_{\mu\in B} \mathcal{H}(v,\mu) \le \sup_{\mu\in B} \mathcal{H}(u,\mu) = \mathcal{H}(u,\lambda)$$
$$= \inf_{v\in A} \mathcal{H}(v,\lambda) \le \sup_{\mu\in B} \inf_{v\in A} \mathcal{H}(v,\mu) .$$

Comparing this with (1.8) we see that

$$\mathcal{H}(u,\lambda) = \sup_{\mu\in B} \inf_{v\in A} \mathcal{H}(v,\mu) = \inf_{v\in A} \sup_{\mu\in B} \mathcal{H}(v,\mu) . \tag{1.9}$$

But

$$\mathcal{H}(u,\lambda) = \inf_{v\in A} \mathcal{H}(v,\lambda) = \mathcal{S}(\lambda) = \sup_{\mu\in B}[\inf_{v\in A} \mathcal{H}(v,\mu)] = \sup_{\mu\in B} \mathcal{S}(\mu) . \tag{1.10}$$

Analogously

$$\mathcal{H}(u,\lambda) = \sup_{\mu\in B} \mathcal{H}(u,\mu) = \inf_{v\in A} \sup_{\mu\in B} \mathcal{H}(v,\mu) \le \sup_{\mu\in B} \mathcal{H}(v,\mu) .$$

This, together with (1.9) and (1.10), leads to (1.7).
2° Contrary, let $\{u, \lambda\}$ be a point from $A \times B$, satisfying (1.7). Then

$$\max_{\mu\in B} \inf_{v\in A} \mathcal{H}(v,\mu) = \inf_{v\in A} \mathcal{H}(v,\lambda) \le \mathcal{H}(u,\lambda)$$

and

$$\min_{v\in A} \sup_{\mu\in B} \mathcal{H}(v,\mu) = \sup_{\mu\in B} \mathcal{H}(u,\mu) \ge \mathcal{H}(u,\lambda)$$

so that

$$\mathcal{H}(u,\mu) \le \sup_{\mu\in B} \mathcal{H}(u,\mu) = \mathcal{H}(u,\lambda) = \inf_{v\in A} \mathcal{H}(v,\lambda) \le \mathcal{H}(v,\lambda)$$

holds for any $\mu \in B$, $v \in A$. □

CONSEQUENCE 1.1. *Let us assume the situation given by Example. Then the first component of a saddle point of $\mathcal{L}$ on $V \times \Lambda$ belongs to $K$ and it is a solution of the constraint minimization problem:*

$$u \in K : \quad L(u) \le L(v) \quad \forall v \in K ,$$

*while $\lambda \in \Lambda$ is a solution of the dual problem:*

$$\mathcal{S}(\lambda) = \inf_{v\in V} \mathcal{L}(v,\lambda) = \sup_{\mu\in\Lambda} \inf_{v\in V} \mathcal{L}(v,\mu) = \sup_{\mu\in\Lambda} \mathcal{S}(\mu) .$$

Next we prove some useful results concerning the properties of saddle points, their existence, eventually uniqueness.

To this end we shall impose the following supplementary conditions, concerning $A$, $B$ and $\mathcal{H}$: namely let $A$, $B$ be *closed, convex* and

(j) $\forall \mu \in B$, $v \mapsto \mathcal{H}(v,\mu)$ is convex and weakly lower semicontinuous;
(jj) $\forall v \in A$, $\mu \mapsto \mathcal{H}(v,\mu)$ is concave and weakly upper semicontinuous.

As far as the existence of a saddle point is concerned, we have

THEOREM 1.3. *Let assumptions (j), (jj) concerning $\mathcal{H}$ be satisfied and let $A$, $B$ be moreover **bounded**. Then there exists at least one saddle point of $\mathcal{H}$ on $A \times B$.*

If $A$ and $B$ are not bounded, one can use

THEOREM 1.4. *Let (j), (jj) be satisfied and moreover let*

$$\tilde{\alpha}) \qquad \exists \mu_0 \in B \text{ such that } \lim_{\substack{\|v\| \to \infty \\ v \in A}} \mathcal{H}(v, \mu_0) = +\infty$$

*(coerciveness with respect to the first component)*

$$\tilde{\beta}) \qquad \exists v_0 \in A \text{ such that } \lim_{\substack{\|\mu\| \to \infty \\ \mu \in B}} \mathcal{H}(v_0, \mu) = -\infty \, .$$

*Then there exists at least one saddle point of $\mathcal{H}$ on $A \times B$.*

REMARK 1.4. If only one of the sets $A$, $B$ is bounded, then the coercivness of $\mathcal{H}$ with respect to unbounded set has to be required.

The most general result, which we shall need, is given in

THEOREM 1.5. *Let (j), (jj) be satisfied. Moreover, let*

- *$A$ be bounded or there exists $\mu_0 \in B$ such that*

$$\lim_{\substack{\|v\| \to \infty \\ v \in A}} \mathcal{H}(v, \mu_0) = \infty$$

- *$B$ be bounded or*

$$\lim_{\substack{\|\mu\| \to \infty \\ \mu \in B}} \inf_{v \in A} \mathcal{H}(v, \mu) \to -\infty \, .$$

*Then there exists at least one saddle point of $\mathcal{H}$ on $A \times B$.*

In what follows, we present one very important application of the previous results. Let $J : V \to R_1$ be a quadratic functional introduced at the beginning of this chapter. We define the closed, convex set $K \subseteq V$ as the kernel of linear and continuous mapping $\mathcal{B} : V \mapsto Y$ ($Y$ another Hilbert space), i.e.

$$K = \{v \in V \mid \mathcal{B}v = 0\} \, .$$

Finally, let $\mathcal{H} : V \times Y' \to R_1$ be a functional given by

$$\mathcal{H}(v, \mu) = \frac{1}{2} a(v, v) - \langle f, v \rangle + [\mu, \mathcal{B}v] \, ,$$

where $\mu \in Y'$ and $[\ ,\ ]$ denotes the duality pairing between $Y'$ and $Y$.

THEOREM 1.6. *Let there exist a constant $\beta > 0$ such that*

$$\sup_{v \in V} \frac{[\mu, \mathcal{B}v]}{\|v\|} \geq \beta \|\mu\|_{Y'} \quad \forall \mu \in Y' . \tag{1.11}$$

*Then there exists a unique saddle point of $\mathcal{H}$ on $V \times Y'$.*

PROOF:
1° *Uniqueness.* Uniqueness of the first component follows from strict convexity of the mapping $v \mapsto \mathcal{H}(v, \mu)$, $v \in V$. Let $\{u, \lambda\}$, $\{u, \bar{\lambda}\} \in V \times Y'$ be two saddle points of $\mathcal{H}$ on $V \times Y'$ or equivalently

$$\begin{aligned} &a(u,v) + [\lambda, \mathcal{B}v] = \langle f, v \rangle \\ &a(u,v) + [\bar{\lambda}, \mathcal{B}v] = \langle f, v \rangle \quad \forall v \in V . \end{aligned}$$

Subtracting the second equation from the first one, we get

$$\begin{aligned} &[\lambda - \bar{\lambda}, \mathcal{B}v] = 0 \quad \forall v \in V \\ \iff &\sup_{v \in V} \frac{[\lambda - \bar{\lambda}, \mathcal{B}v]}{\|v\|} = 0 \\ \implies &\lambda = \bar{\lambda} \end{aligned}$$

as follows from (1.11).
2° *Existence.* We shall apply the results of Theorem 1.5. First of all

$$\lim_{\substack{\|v\| \to \infty \\ v \in V}} \mathcal{H}(v, \mu) = +\infty \tag{1.12}$$

holds for any $\mu \in Y'$. We have to prove that

$$\lim_{\substack{\|\mu\| \to \infty \\ \mu \in Y'}} \inf_{v \in V} \mathcal{H}(v, \mu) = -\infty .$$

Let $\mu \in Y'$ be fixed and let us solve the problem

$$\begin{cases} \text{find } u = u(\mu) \in V \text{ such that} \\ \mathcal{H}(u, \mu) = \inf\limits_{v \in V} \mathcal{H}(v, \mu) \end{cases} \tag{1.13}$$

or

$$\begin{cases} \text{find } u \in V \text{ such that} \\ a(u,v) + [\mu, \mathcal{B}v] = \langle f, v \rangle \quad \forall v \in V . \end{cases} \tag{1.13'}$$

Let $\mathcal{A}$ be the mapping $V \mapsto V'$ defined by means of

$$a(u,v) = \langle \mathcal{A}u, v \rangle .$$

From Lax-Milgram theorem it follows that $\mathcal{A}$ maps $V$ onto $V'$ and

$$\|\mathcal{A}^{-1}f\| \le \frac{1}{\alpha}\|f\|_{V'} \quad \forall f \in V' .$$

If $\mathcal{B}^*$ denotes the adjoint mapping to $\mathcal{B}$, the solution $u$ of (1.13) can be written:

$$u = \mathcal{A}^{-1}(f - \mathcal{B}^*\mu) .$$

Let $\tilde{\mathcal{H}} : V \mapsto R_1$ be given by

$$\tilde{\mathcal{H}}(v) = \langle f, v\rangle - [\mu, \mathcal{B}v] .$$

As $a(u,u) = \tilde{\mathcal{H}}(u)$, one has

$$\begin{aligned}
\mathcal{H}(u,\mu) &= -\frac{1}{2}a(u,u) = -\frac{1}{2}\langle \mathcal{A}u, u\rangle \\
&= -\frac{1}{2}\langle f, \mathcal{A}^{-1}f\rangle + \frac{1}{2}\langle f, \mathcal{A}^{-1}\mathcal{B}^*\mu\rangle + \frac{1}{2}\langle \mathcal{B}^*\mu, \mathcal{A}^{-1}f\rangle - \frac{1}{2}\langle \mathcal{B}^*\mu, \mathcal{A}^{-1}\mathcal{B}^*\mu\rangle \qquad (1.14)\\
&\le -c_1\|\mathcal{B}^*\mu\|^2_{V'} + c_2\|\mathcal{B}^*\mu\|_{V'} + c_3
\end{aligned}$$

with $c_1,\ c_2,\ c_3 > 0$.

From (1.11) it follows that

$$\sup_{v\in V}\frac{[\mu, \mathcal{B}v]}{\|v\|_V} = \sup_{v\in V}\frac{[\mathcal{B}^*\mu, v]}{\|v\|_V} = \|\mathcal{B}^*\mu\|_{V'} \ge \beta\|\mu\|_{Y'} .$$

Finally, this and (1.14) yield

$$\mathcal{H}(u,\mu) \le -\tilde{c}_1\|\mu\|^2_{Y'} + \tilde{c}_2\|\mu\|_{Y'} + c_3 \to -\infty$$

if $\|\mu\|_{Y'} \to \infty$. All assumptions of Theorem 1.5 are satisfied. □

Let $\mathcal{L} : A \times B \mapsto R_1$ be a given functional, which is related to the problem (**P**) by means of (1.2).

DEFINITION 1.2. *The problem of finding a saddle point of $\mathcal{L}$ on $A \times B$ will be called a mixed formulation of* (**P**).

## 2. APPROXIMATION OF VARIATIONAL INEQUALITIES

### 2.1. Approximation of primal formulation

Let $\{K_h\}$, $h \to 0+$ be a family of convex, closed subsets of $V$, not necessarily imbedded into $K$. Instead of $(\mathbf{P})$ we shall assume a family of minimization problems:

$$\text{find } u_h \in K_h \ : \quad L(u_h) \le L(v_h) \quad \forall v_h \in K_h \,. \tag{$\mathbf{P}_h$}$$

Recall that under our assumptions $(\mathbf{P}_h)$ has a unique solution $u_h$ for any $h > 0$.

REMARK 2.1. As in continuous case, $u_h \in K_h$ solving $(\mathbf{P}_h)$ can be equivalently characterized through the relation:

$$u_h \in K_h \ : \quad a(u_h, v_h - u_h) + j(v_h) - j(u_h) \ge \langle f, v_h - u_h \rangle \quad \forall v_h \in K_h \,. \tag{2.1}$$

A natural question arises: what is the relation between $u$ and $u_h$. First we prove

LEMMA 2.1. *Let $u$, $u_h$ be solutions of* $(\mathbf{P})$, $(\mathbf{P}_h)$, *respectively. Then*

$$\begin{aligned} \alpha\|u - u_h\|^2 \le & \langle f, u - v_h\rangle + \langle f, u_h - v\rangle \\ & + a(u_h - u, v_h - u) + a(u, v - u_h) + a(u, v_h - u) \\ & + j(v) - j(u) + j(v_h) - j(u_h) \end{aligned} \tag{2.2}$$

*holds for any $v \in K$, $v_h \in K_h$.*

PROOF: Let $v \in K$, $v_h \in K_h$. Using $V$-ellipticity of $a$ we can write:

$$\begin{aligned} \alpha\|u - u_h\|^2 &\le a(u - u_h, u - u_h) \\ &= a(u,u) + a(u_h,u_h) - a(u,u_h) - a(u_h,u) \\ &\le a(u,v) + \langle f, u - v\rangle + j(v) - j(u) + a(u_h, v_h) \\ &\quad + \langle f, u_h - v_h\rangle + j(v_h) - j(u_h) - a(u,u_h) - a(u_h,u) \\ &= \langle f, u - v_h\rangle + \langle f, u_h - v\rangle + a(u, v - u_h) \\ &\quad + a(u_h - u, v_h - u) + a(u, v_h - u) \\ &\quad + j(v) - j(u) + j(v_h) - j(u_h) \,. \end{aligned}$$

□

REMARK 2.2. If $K_h \subset K$ $\forall h$, then we can set $v = u_h$ in (2.2) and we obtain:

$$\begin{aligned} \alpha\|u - u_h\|^2 \le & \langle f, u - v_h\rangle + a(u_h - u, v_h - u) + a(u, v_h - u) + j(v_h) - j(u) \\ & \forall v_h \in K_h \,. \end{aligned} \tag{2.3}$$

REMARK 2.3. (2.2) and (2.3) will be used for estimating the discretization error $\|u - u_h\|$.

Concerning the convergence of $u_h$ to $u$, the following result will be needed:

THEOREM 2.1. *Let the following conditions will be satisfied:*

$$\forall v \in K \quad \exists v_h \in K_h \text{ such that } v_h \to v,\ h \to 0+ ; \tag{2.4}$$

$$v_h \in K_h, \quad v_h \rightharpoonup v \text{ (weakly) implies } v \in K . \tag{2.5}$$

*Then*

$$\|u - u_h\| \to 0, \quad h \to 0+ .$$

PROOF: From the definition of $L$ and (2.4), the boundness of $\{u_h\}$ follows, i.e.

$$\exists c > 0 \text{ such that } \|u_h\| \le c \quad \forall h > 0 .$$

As $V$ is a Hilbert space, one can find a subsequence $\{u_{h'}\} \subset \{u_h\}$ such that

$$u_{h'} \rightharpoonup u^* \in K$$

because of (2.5). Let us prove that $u^*$ is the solution of (**P**). Let $v \in K$ be given. Then there exists a sequence $\{\bar{v}_h\}$, $\bar{v}_h \in K_h$ such that

$$\bar{v}_h \to v \tag{2.6}$$

(see (2.4)). From the definition of $(\mathbf{P}_{h'})$:

$$L(u_{h'}) \le L(\bar{v}_{h'}) . \tag{2.7}$$

Passing to the limit in (2.7) and using weak lower semicontinuity of $L$, we arrive at

$$L(u^*) \le L(v) \quad \forall v \in K , \tag{2.8}$$

i.e. $u^*$ solves (**P**). As $u^*$ is unique, the whole sequence $\{u_h\}$ tends weakly to $u = u^*$. In order to prove that $u_h \to u$, $h \to 0+$ in $V$, we use (2.2). Let $\bar{\bar{v}}_h \in K_h$ be such that

$$\bar{\bar{v}}_h \to u, \quad h \to 0+ \tag{2.9}$$

and substitute $v := u$. Using the fact that $u_h \rightharpoonup u$, $h \to 0+$ as well as the weak lower semicontinuity of $j$ and (2.9), we arrive at the assertion. □

## 2.2. Approximation of mixed formulation of elliptic inequalities

Here we shall assume a special case of functional $\mathcal{L} : V \times Y \mapsto R_1$,

$$\mathcal{L}(v, \mu) = J(v) + b(v, \mu) - [g, \mu] , \tag{2.10}$$

where $J$ is a quadratic functional, given by a symmetric, $V$-elliptic and continuous bilinear form $a$, $b : V \times Y \to R_1$ is a continuous bilinear form, $g \in Y'$ and $[g, \mu]$ denotes the value of $g$ at $\mu$. The norm in $Y$ will be denoted by $|\ |$. Finally, let $K \subseteq V$, $\Lambda \subseteq Y$ be two non-empty, closed convex subsets (for details see [**2**]). We make the following assumptions, concerning $K$ and $\Lambda$: $\Lambda$ is either

(CC) *a convex cone with vertex at $\theta$ and $K = V$*

or

(BC) *a bounded convex subset of* $Y$.

Let $\{u, \lambda\}$ be a saddle point of $\mathcal{L}$ on $K \times \Lambda$, i.e.

$$\mathcal{L}(u, \mu) \leq \mathcal{L}(u, \lambda) \leq \mathcal{L}(v, \lambda) \quad \forall v \in K, \ \forall \mu \in \Lambda . \tag{$\tilde{\mathcal{P}}$}$$

In accordance with the theory presented in Chapter 1, the first component $u \in K$ solves the problem

$$J(u) + j(u) = \inf_{v \in K} \{J(v) + j(v)\} ,$$

where $j(v) = \sup_{\Lambda} \{b(v, \mu) - [g, \mu]\}$.

Now, let $h, H \to 0+$ be two parameters, tending to zero. With any couple $h, H$ we associate finite dimensional subspaces $V_h \subset V$ and $Y_H \subset Y$, respectively. Let $K_h$ and $\Lambda_H$ be closed, convex subsets of $V_h$ and $Y_H$, respectively. Similarly as in the continuous case we make the following assumptions: $\Lambda_H$ is either

($\mathrm{CC}_H$) *a convex cone with vertex at* $\theta$ *and* $K_h = V_h$

or

($\mathrm{BC}_H$) *a convex subset of* $Y_H$*, bounded uniformly in* $Y$, i.e. there exists a positive constant $c$ such that

$$|\mu_H| \leq c \quad \forall \mu_H \in \Lambda_H, \ \forall H \in (0, 1) .$$

By the *approximation of* $(\tilde{\mathcal{P}})$ we mean the problem of finding a saddle-point $\{u_h, \lambda_H\}$ of $\mathcal{L}$ on $K_h \times \Lambda_H$:

$$\begin{cases} \text{find } \{u_h, \lambda_H\} \in K_h \times \Lambda_H \text{ such that} \\ \mathcal{L}(u_h, \mu_H) \leq \mathcal{L}(u_h, \lambda_H) \leq \mathcal{L}(v_h, \lambda_H) \quad \forall v_h \in K_h, \ \forall \mu_H \in \Lambda_H , \end{cases} \tag{$\tilde{\mathcal{P}}_{hH}$}$$

or equivalently

$$\begin{cases} \text{find } \{u_h, \lambda_H\} \in K_h \times \Lambda_H \text{ such that} \\ a(u_h, v_h - u_h) + b(v_h - u_h, \lambda_H) \geq \langle f, v_h - u_h \rangle \quad \forall v_h \in K_h \\ b(u_h, \mu_H - \lambda_H) \leq [g, \mu_H - \lambda_H] \quad \forall \mu_H \in \Lambda_H . \end{cases} \tag{$\tilde{\mathcal{P}}'_{hH}$}$$

Analogously to the continuous case, the first component $u_h \in K_h$ minimizes the functional $J(v_h) + j_H(v_h)$ over $K_h$, where $j_H(v_h) = \sup_{\Lambda_H} \{b(v_h, \mu_H) - [g, \mu_H]\}$.

As far as the existence, eventually the uniqueness of $\{u_h, \lambda_H\}$ is concerned, we can directly apply results of Chapter 1. We shall distinguish both cases ($\mathrm{CC}_H$) and ($\mathrm{BC}_H$) separately.

THEOREM 2.2. *If* $(CC_H)$ *and* $b(v_h, \mu_H) = 0 \ \forall v_h \in V_h \implies \mu_H = 0$, *then there exists a unique solution of* $(\tilde{\mathcal{P}}_{hH})$.

If $\{\Lambda_H\}$ are uniformly bounded, situation is simpler. One has

THEOREM 2.3. *Let* $(BC_H)$ *be satisfied. Then there exists a solution of* $(\tilde{\mathcal{P}}_{hH})$, *the first component of which is uniquely determined.*

Our aim will be to establish relations between $u_h$, $u$ and $\lambda_H$, $\lambda$. To this end we give another, equivalent formulation of $(\tilde{\mathcal{P}})$.

Let $\mathcal{H} = V \times Y$ be a Hilbert space, equipped with the norm

$$\|\mathcal{V}\|_{\mathcal{H}} = \{\|v\|^2 + |\mu|^2\}^{\frac{1}{2}} , \quad \mathcal{V} = (v, \mu) \in \mathcal{H} ,$$

$\mathcal{A} : \mathcal{H} \times \mathcal{H} \mapsto R_1$ a bilinear form defined by

$$\mathcal{A}(\mathcal{U}, \mathcal{V}) = a(u, v) + b(v, \lambda) - b(u, \mu) , \quad \mathcal{U} = (u, \lambda) \in \mathcal{H} , \quad \mathcal{V} = (v, \mu) \in \mathcal{H} ,$$

and $\mathcal{F} : \mathcal{H} \to R_1$ a linear functional

$$\langle \mathcal{F}, \mathcal{V} \rangle = \langle f, v \rangle - [g, \mu] , \quad \mathcal{V} = (v, \mu) \in \mathcal{H} .$$

The definition of $\mathcal{A}$ immediately implies

$$\mathcal{A}(\mathcal{V}, \mathcal{V}) = a(v, v) \quad \forall \mathcal{V} = (v, \mu) \in \mathcal{H} ; \tag{2.11}$$

$$\exists M = \text{const.} > 0 : \quad |\mathcal{A}(\mathcal{U}, \mathcal{V})| \le M \|\mathcal{U}\|_{\mathcal{H}} \|\mathcal{V}\|_{\mathcal{H}} \quad \forall \mathcal{U}, \mathcal{V} \in \mathcal{H} . \tag{2.12}$$

It is readily seen that $(\tilde{\mathcal{P}})$ is equivalent to

$$\begin{cases} \text{find } \mathcal{U} = (u, \lambda) \in \mathcal{K} = K \times \Lambda \text{ such that} \\ \mathcal{A}(\mathcal{U}, \mathcal{V} - \mathcal{U}) \ge \langle \mathcal{F}, \mathcal{V} - \mathcal{U} \rangle \quad \forall \mathcal{V} \in \mathcal{K} . \end{cases} \tag{$\tilde{\mathbf{P}}$}$$

Next, let $\mathcal{K}_{hH} = K_h \times \Lambda_H$ be a closed, convex subset of $\mathcal{H}$; $\mathcal{K}_{hH} \not\subset \mathcal{K}$, in general. The problem

$$\begin{aligned} &\text{find } \mathcal{U}_{hH} = \{u_h, \lambda_H\} \in \mathcal{K}_{hH} \text{ such that} \\ &\mathcal{A}(\mathcal{U}_{hH}, \mathcal{V}_{hH} - \mathcal{U}_{hH}) \ge \langle \mathcal{F}, \mathcal{V}_{hH} - \mathcal{U}_{hH} \rangle \quad \forall \mathcal{V}_{hH} \in \mathcal{K}_{hH} \end{aligned} \tag{$\tilde{\mathbf{P}}_{hH}$}$$

represents the approximation of $(\tilde{\mathbf{P}})$, equivalent to $(\tilde{\mathcal{P}}'_{hH})$ (or $(\tilde{\mathcal{P}}_{hH})$).

First we prove an auxiliary lemma.

LEMMA 2.2. *Let $\{u,\lambda\}$, $\{u_h,\lambda_H\}$ be solutions of $(\tilde{\mathcal{P}})$ and $(\tilde{\mathcal{P}}_{hH})$, respectively. Then*

$$\begin{aligned}\|u-u_h\|^2 \le c\Big[&\{\|u-v_h\|^2+|\lambda-\mu_H|^2\}+A_1(v_h)\\ &+A_2(v)+\{b(u,\lambda_H-\mu)-[g,\lambda_H-\mu]\}\\ &+\{b(u,\lambda-\mu_H)-[g,\lambda-\mu_H]\}+|\lambda-\lambda_H|^2\Big]\end{aligned} \tag{2.13}$$

*holds for any $v_h\in K_h$, $v\in K$, $\mu_H\in\Lambda_H$, $\mu\in\Lambda$, where*

$$\begin{aligned}A_1(v_h)&=a(u,v_h-u)+b(v_h-u,\lambda)+\langle f,u-v_h\rangle\\ A_2(v)&=a(u,v-u_h)+b(v-u_h,\lambda)+\langle f,u_h-v\rangle\end{aligned}$$

*and $c$ is positive constant independent of $h$, $H$.*

PROOF: By virtue of (2.11) and the definition of $(\mathbf{P})$ and $(\mathbf{P}_{hH})$ we get:

$$\begin{aligned}\alpha\|u-u_h\|^2 &\le \mathcal{A}(\mathcal{U}-\mathcal{U}_{hH},\mathcal{U}-\mathcal{U}_{hH})\\ &=\mathcal{A}(\mathcal{U},\mathcal{U})-\mathcal{A}(\mathcal{U}_{hH},\mathcal{U})-\mathcal{A}(\mathcal{U},\mathcal{U}_{hH})+\mathcal{A}(\mathcal{U}_{hH},\mathcal{U}_{hH})\\ &\le\langle\mathcal{F},\mathcal{U}-\mathcal{V}\rangle+\mathcal{A}(\mathcal{U},\mathcal{V})+\langle\mathcal{F},\mathcal{U}_{hH}-\mathcal{V}_{hH}\rangle+\mathcal{A}(\mathcal{U}_{hH},\mathcal{V}_{hH})\\ &\quad-\mathcal{A}(\mathcal{U}_{hH},\mathcal{U})-\mathcal{A}(\mathcal{U},\mathcal{U}_{hH})\\ &=\langle\mathcal{F},\mathcal{U}-\mathcal{V}_{hH}\rangle+\langle\mathcal{F},\mathcal{U}_{hH}-\mathcal{V}\rangle+\mathcal{A}(\mathcal{U},\mathcal{V}-\mathcal{U}_{hH})\\ &\quad+\mathcal{A}(\mathcal{U}_{hH}-\mathcal{U},\mathcal{V}_{hH}-\mathcal{U})+\mathcal{A}(\mathcal{U},\mathcal{V}_{hH}-\mathcal{U})\\ &=A_1(v_h)+A_2(v)+\{b(u,\lambda_H-\mu)-[g,\lambda_H-\mu]\}\\ &\quad+\{b(u,\lambda-\mu_H)-[g,\lambda-\mu_H]\}\\ &\quad+a(u_h-u,v_h-u)+b(v_h-u,\lambda_H-\lambda)-b(u_h-u,\mu_H-\lambda)\,.\end{aligned} \tag{2.13'}$$

The boundedness of $a$, $b$ together with the inequality $2hf\le\frac{1}{\varepsilon}h^2+\varepsilon f^2$ yield ($\varepsilon>0$)

$$\begin{aligned}&\alpha\|u-u_h\|^2\\ &\le A_1(v_h)+A_2(v)+\{b(u,\lambda_H-\mu)-[g,\lambda_H-\mu]\}\\ &+\{b(u,\lambda-\mu_H)-[g,\lambda-\mu_H]+M_1\varepsilon\|u-v_h\|^2+\frac{M_1}{\varepsilon}\|u-v_h\|^2\\ &+\frac{M_2}{\varepsilon}\|v_h-u\|^2+M_2\varepsilon|\lambda_H-\lambda|^2+M_2\varepsilon\|u-u_h\|^2+\frac{M_2}{\varepsilon}|\lambda-\mu_H|^2\,.\end{aligned} \tag{2.14}$$

For $\varepsilon>0$ sufficiently small, we arrive at (2.13). □

In order to prove the relation between $\lambda$ and $\lambda_H$ we shall need the following assumption:

$$\begin{cases}\exists\hat\beta>0 \text{ independent on } h,H>0 \text{ such that}\\ \sup\limits_{V_h}\dfrac{b(v_h,\mu_H)}{\|v_h\|}\ge\hat\beta|\mu_H| \text{ holds for any } \mu_H\in Y_H\,.\end{cases} \tag{S}$$

A direct consequence of Lemma 2.2 is

THEOREM 3.2. *There exists a unique solution $u$ of $(\mathcal{P})$. Moreover, the solution can be equivalently characterized through the relation*

$$\begin{aligned} u \in K \ : \quad & (\tau(u), \varepsilon(v-u)) + \int_{\Gamma_K} g(|v_t| - |u_t|)\, ds \\ & \geq (F_i, v_i - u_i)_0 + (P_i, v_i - u_i)_0 \quad \forall v \in K \, . \end{aligned} \tag{3.19}$$

REMARK 3.1. If $g \equiv 0$ on $\Gamma_K$, (3.18) reduces to

$$T_t(x) = 0 \quad \text{on } \Gamma_K \, , \tag{3.18'}$$

which corresponds to frictionless case.

### 3.4. Formulation of contact problems in terms of stresses

Let $S$ be a space of $2 \times 2$ symmetric matrices:

$$S = \{\mathcal{N} = (\mathcal{N}_{ij})_{i,j=1}^2, \ \mathcal{N}_{ij} \in L^2(\Omega), \ \mathcal{N}_{ij} = \mathcal{N}_{ji} \text{ a.e. in } \Omega\}$$

and set

$$\mathcal{N}_{ij} = \varepsilon_{ij}(v) \, . \tag{3.20}$$

Then, using (3.20), we can formally write

$$\mathcal{J}(v) = \frac{1}{2} \int_\Omega c_{ijkl} \mathcal{N}_{ij} \mathcal{N}_{kl}\, dx + j(v) - \langle f, v \rangle \equiv \mathcal{H}_1(\mathcal{N}, v)$$

with notation of Section 3.3. We may look at (3.20) as constraints, which will be removed by introducing the following lagrangian:

$$\mathcal{H}([\mathcal{N}, v], \lambda) = \mathcal{H}_1(\mathcal{N}, v) + \int_\Omega \lambda_{ij}(\varepsilon_{ij}(v) - \mathcal{N}_{ij})\, dx \, ,$$

$\{[\mathcal{N}, v], \lambda\} \in S \times K \times S$.

It is easy to see that

$$\sup_{\lambda \in S} \int_\Omega \lambda_{ij}(\varepsilon_{ij}(v) - \mathcal{N}_{ij}) dx = \begin{cases} 0 & \text{if } \mathcal{N} = \varepsilon(v) \\ +\infty & \text{otherwise.} \end{cases}$$

Thus

$$\inf_{v \in K} \mathcal{J}(v) = \inf_{\substack{v \in K \\ \mathcal{N} \in S}} \sup_{\lambda \in S} \mathcal{H}([\mathcal{N}, v], \lambda) \, . \tag{3.21}$$

Dual problem to (3.21) now reads:

$$\sup_{\lambda \in S} \inf_{\substack{v \in K \\ \mathcal{N} \in S}} \mathcal{H}([\mathcal{N}, v], \lambda) \, . \tag{3.22}$$

Using a direct calculation, we obtain

LEMMA 3.1. *There exists a unique saddle-point* $([\tilde{\mathcal{N}},\tilde{v}],\tilde{\lambda})$ *of* $\mathcal{H}$ *on* $S\times K\times S$. *Moreover*

$$\tilde{\mathcal{N}}=\varepsilon(u)\ ,\quad \tilde{v}=u\ ,\quad \tilde{\lambda}=\tau(u)\ ,$$

*where* $u\in K$ *is a solution of primal formulation* $(\mathcal{P})$.

Denote by

$$\tilde{\mathcal{S}}(\lambda)=\inf_{\substack{v\in K\\ \mathcal{N}\in S}}\mathcal{H}([\mathcal{N},v],\lambda)$$

the dual formulation to $\mathcal{J}$. One can easy prove that

$$\tilde{\mathcal{S}}(\lambda)=-\frac{1}{2}\int_{\Omega}a_{ijkl}\lambda_{ij}\lambda_{kl}\,dx\ ,$$

where $a_{ijkl}$ are coefficients of inverse Hooke's law. Moreover

$$\sup_{\lambda\in S}\tilde{\mathcal{S}}(\lambda)=\sup_{\lambda\in K^{+}_{F,P,g}(\Omega)}\tilde{\mathcal{S}}(\lambda)\ ,$$

with $K^{+}_{F,P,g}(\Omega)$ given by

$$K^{+}_{F,P,g}(\Omega)=\{\lambda\in S\mid\int_{\Omega}\lambda_{ij}\varepsilon_{ij}(v)\,dx+j(v)\geq\langle f,v\rangle\ \forall v\in K\}\ .$$

Instead of $\tilde{\mathcal{S}}$ we introduce $\mathcal{S}=-\tilde{\mathcal{S}}$.

The explicit form of the dual problem to $(\mathcal{P})$ now reads

$$\text{find }\lambda^{*}\in K^{+}_{F,P,g}(\Omega)\ :\quad \mathcal{S}(\lambda^{*})\leq\mathcal{S}(\lambda)\quad\forall\lambda\in K^{+}_{F,P,g}(\Omega)\ . \qquad (\mathcal{P}^{*})$$

As $\mathcal{S}$ is a quadratic functional, one immediately has:

THEOREM 3.3. *There exists a unique solution* $\lambda^{*}$ *of* $(\mathcal{P}^{*})$. *Moreover*

$$\lambda^{*}=\tau(u)\ ,$$

*where* $u\in K$ *is the solution of the primal formulation* $(\mathcal{P})$.

REMARK 3.2. Using integration by parts, we can show that $\lambda\in K^{+}_{F,P,g}(\Omega)$ if and only if

$$\begin{aligned}
&\frac{\partial\lambda_{ij}}{\partial x_j}+F_i=0 &&\text{in }\Omega\ , && i=1,2\ ;\\
&\lambda_{ij}n_j=P_i &&\text{on }\Gamma_P\ , && i=1,2\ ;\\
&\lambda_{ij}n_jn_i\leq 0 &&\text{on }\Gamma_K\ ;\\
&|\lambda_{ij}n_jt_i|\leq g &&\text{on }\Gamma_K\ .
\end{aligned}\qquad(3.23)$$

By virtue of (2.23), (2.24), $u^* \in K$, $\lambda^* \in \Lambda$. Let us show that $\{u^*, \lambda^*\}$ is a solution of $(\tilde{\mathcal{P}})$. Let $\{v, \mu\} \in K \times \Lambda$ be an arbitrary element. From (2.21), (2.22) the existence $\{v_h\}$, $\{\mu_H\}$, $v_h \in K_h$, $\mu_h \in \Lambda_H$ such that

$$v_h \to v \quad \text{in } V \,, \quad \mu_H \to \mu \quad \text{in } Y \tag{2.26}$$

follows. $\{u_{h'}, \lambda_{H'}\}$ being solution of $(\tilde{\mathcal{P}}_{h'H'})$ (or $(\tilde{\mathcal{P}}'_{h'H'})$) satisfies

$$a(u_{h'}, u_{h'} - v_{h'}) + b(u_{h'} - v_{h'}, \lambda_{H'}) \le \langle f, u_{h'} - v_{h'} \rangle \quad \forall v_{h'} \in K_{h'} \tag{2.27}$$

$$b(u_{h'}, \mu_{H'} - \lambda_{H'}) \le [g, \mu_{H'} - \lambda_{H'}] \quad \forall \mu_{H'} \in \Lambda_{H'} \,. \tag{2.28}$$

Passing to the limit with $h'$, $H' \to 0+$ in (2.27) and using (2.25), (2.26) we obtain

$$a(u^*, u^* - v) + \liminf_{h',H' \to 0+} b(u_{h'}, \lambda_{H'}) - b(v, \lambda^*) \le \langle f, u^* - v \rangle \quad \forall v \in K \,. \tag{2.29}$$

The same procedure can be applied to (2.28):

$$b(u^*, \mu) - [g, \mu - \lambda^*] \le \liminf_{h',H' \to 0+} b(u_{h'}, \lambda_{H'}) \quad \forall \mu \in \Lambda \,. \tag{2.30}$$

Setting $\mu = \lambda^*$ in (2.30), we obtain

$$b(u^*, \lambda^*) \le \liminf_{h',H' \to 0+} b(u_{h'}, \lambda_{H'}) \,. \tag{2.31}$$

Substitution of (2.31) into (2.29) yields:

$$a(u^*, u^* - v) + b(u^* - v, \lambda^*) \le \langle f, u^* - v \rangle \quad \forall v \in K \,.$$

The choice $v = u^*$ in (2.29) implies

$$\liminf_{h',H' \to 0+} b(u_{h'}, \lambda_{H'}) \le b(u^*, \lambda^*) \,.$$

From this and (2.30), we have

$$b(u^*, \mu - \lambda^*) \le [g, \mu - \lambda^*] \quad \forall \mu \in \Lambda \,.$$

Thus $\{u^*, \lambda^*\}$ is a solution of $(\tilde{\mathcal{P}})$. By virtue of its uniqueness, the whole sequences $\{u_h\}$, $\{\lambda_H\}$ tend weakly to $u$, $\lambda$, respectively. Let us show that $u_h \to u$ strongly in $V$. Let $\{\bar{v}_h\}$, $\bar{v}_h \in K_h$, $\{\bar{\mu}_H\}$, $\bar{\mu}_H \in \Lambda_H$ be such that

$$\bar{v}_h \to u \,, \quad \bar{\mu}_H \to \lambda \,.$$

Applying (2.20) with $v = u$, $\mu = \lambda$, $v_h = \bar{v}_h$, $\mu_H = \bar{\mu}_H$ and using the weak convergence $u_h \rightharpoonup u$, $\lambda_H \rightharpoonup \lambda$, we obtain $u_h \to u$ in $V$. □

REMARK 2.6. If $K_h \subset K$, $\Lambda_H \subset \Lambda$, the conditions (2.23), (2.24), respectively, are satisfied.

THEOREM 2.7. *Let (CC), (CC$_H$) and (S) be satisfied. Let $\{u,\lambda\}$ be the unique solution of $(\tilde{\mathcal{P}})$. Moreover, let us suppose that*

$$\forall v \in V \quad \exists v_h \in V_h : \quad v_h \to v \text{ in } V ; \tag{2.32}$$

$$\forall \mu \in \Lambda \quad \exists \mu_H \in \Lambda_H : \quad \mu_H \to \mu \text{ in } L ; \tag{2.33}$$

$$\mu_H \in \Lambda_H , \quad \mu_H \rightharpoonup \mu \quad \text{in } L \text{ implies } \mu \in \Lambda ; \tag{2.34}$$

*there exist a real number $d$, a positive number $c$ and a bounded sequence $\{\bar{v}_h\}$, $\bar{v}_h \in V_h$ such that $j_H(v_h) \geq d$ $\forall v_h \in V_h$, $\forall h, H \in (0,1)$, $j_H(\bar{v}_h) \leq c \quad \forall h, H \in (0,1)$ .* (2.35)

*Then $u_h \to u$, $\lambda_H \to \lambda$.*

PROOF: We shall prove the boundedness of $\{u_h\}$ and $\{\lambda_H\}$ only. The rest of the proof is analogous to that of Theorem 2.6. The convergence of $\lambda_H$ to $\lambda$ follows from (2.16).

According to the interpretation of $(\tilde{\mathcal{P}}'_{hH})$, $u_h \in V_h$ satisfies

$$a(u_h, v_h - u_h) + j_H(v_h) - j_H(u_h) \geq \langle f, v_h - u_h \rangle \quad \forall v_h \in V_h .$$

Hence

$$a(u_h, u_h) + j_H(u_h) \leq a(u_h, v_h) + j_H(\bar{v}_h) - \langle f, \bar{v}_h - u_h \rangle .$$

This and (2.35) imply the boundness of $\{u_h\}$ and by virtue of (2.16) we deduce the boundedness of $\{\lambda_H\}$. □

REMARK 2.7. If $\Lambda_H \subset \Lambda$ $\forall H \in (0,1)$, (2.34) is automatically satisfied.

Condition (S), guaranteeing the convergence of $\lambda_H$ to $\lambda$, is very restrictive. If we are interested in the convergence of $u_h$ to $u$, condition (S) may be avoided. Indeed let us suppose that the functions

$$j(v) = \sup_{\Lambda}\{b(v,\mu) - [g,\mu]\}$$

$$j_H(v_h) = \sup_{\Lambda_H}\{b(v_h,\mu_H) - [g,\mu_H]\}$$

take their values from the set $\{0, +\infty\}$. We shall denote by

$$\mathcal{K} = \{v \in V \mid j(v) = 0\}$$

$$\mathcal{K}_{hH} = \{v_h \in V_h \mid j_H(v_h) = 0\} ,$$

i.e. $j$ and $j_H$ are the indicator functions of the closed convex sets $\mathcal{K}$ and $\mathcal{K}_{hH}$, respectively. Let $\{u,\lambda\} \in V \times \Lambda$ and $\{u_h, \lambda_H\} \in V_h \times \Lambda_H$ be solutions of $(\tilde{\mathcal{P}})$ and $(\tilde{\mathcal{P}}_{hH})$, respectively. Following the interpretation of these problems we already know that $u \in K$ and $u_h \in \mathcal{K}_{hH}$ are solutions of the minimizing problems:

$$J(u) \leq J(v) \qquad \forall v \in \mathcal{K}$$

and

$$J(u_h) \leq J(v_h) \qquad \forall v_h \in \mathcal{K}_{hH} ,$$

respectively. As far as the convergence of $u_h$ to $u$ is concerned, we have

THEOREM 2.8. *Let (CC), ($CC_H$) be satisfied and there exist solutions $\{u, \lambda\}$ and $\{u_h, \lambda_H\}$ of ($\tilde{\mathcal{P}}$) and ($\tilde{\mathcal{P}}_{hH}$), respectively, the first components of which are uniquely determined. Let*

$$\forall v \in \mathcal{K} \quad \exists v_h \in \mathcal{K}_{hH} \quad v_h \to v \quad \text{in } V\,; \tag{2.36}$$

$$v_h \in \mathcal{K}_{hH}\,, \quad v_h \rightharpoonup v \quad \text{in } V \text{ implies } v \in \mathcal{K}\,. \tag{2.37}$$

*Then $u_h \to u$ in $V$.*

PROOF is a direct consequence of Theorem 2.1.

## 3. PRIMAL, DUAL AND MIXED VARIATIONAL FORMULATION OF CONTACT PROBLEMS

### 3.1. Basic notations

We begin with some basic concepts of the theory of linear elacticity in two dimensions (see [3]). Practical situations of three dimensions can often be reduced into two-dimensional cases by symmetry, etc.

Let us consider a two-dimensional elastic body represented by a bounded domain $\Omega \subset R_2$ subjected to internal and external forces. These forces cause the body to deformate so that the point $x$, $x = (x_1, x_2)$ of the undeformed body becomes the point $x'$, $x' = (x_1', x_2')$ of the deformed body; $x'$ can be written as $x' = x + u(x)$, where $u = (u_1, u_2)$ denotes the vector of deformation.

The strain caused by displacements $u$ is characterized by the so called *strain tensor*[1]

$$\varepsilon = \{\varepsilon_{ij}(u)\}_{i,j=1}^2, \quad \varepsilon_{ij}(u) = \frac{1}{2}\left(\frac{\partial u_i}{\partial x_j} + \frac{\partial u_j}{\partial x_i}\right). \tag{3.1}$$

Similarly, the stresses can be characterized by the so called *stress tensor* (symmetric)

$$\tau = \{\tau_{ij}\}_{i,j=1}^2. \tag{3.2}$$

In the equilibrium state the sresses $\tau$ are related to the volumic forces $F = (F_1, F_2)$ by the system of equilibrium equations

$$\left.\begin{aligned} \frac{\partial \tau_{11}}{\partial x_1} + \frac{\partial \tau_{12}}{\partial x_2} + F_1 = 0 \\ \frac{\partial \tau_{21}}{\partial x_1} + \frac{\partial \tau_{22}}{\partial x_2} + F_2 = 0 \end{aligned}\right\} \quad \text{in } \Omega\,. \tag{3.3}$$

[1] Here we assume the linearized strain tensor.

In the case of *linear elasticity* the stress and strain tensors are connected by the linearized Hooke's law [1]

$$\tau_{ij}(u) = c_{ijkl}\varepsilon_{kl}(u) \equiv \sum_{k,l=1}^{2} c_{ijkl}\varepsilon_{kl}(u), \tag{3.4}$$

where $c_{ijkl}$ are the elasticity coefficients satisfying symmetry conditions

$$c_{ijkl} = c_{jikl} = c_{klij} \qquad \forall i,j,k,l = 1,2 \quad \text{a.e. in } \Omega \,. \tag{3.5}$$

In the case of *homogenous and isotropic* material Hooke's law reads

$$\begin{pmatrix} \tau_{11} \\ \tau_{22} \\ \tau_{12} \end{pmatrix} = \begin{pmatrix} 2\mu+\lambda & \lambda & 0 \\ \lambda & 2\mu+\lambda & 0 \\ 0 & 0 & \mu \end{pmatrix} \begin{pmatrix} \varepsilon_{11} \\ \varepsilon_{22} \\ \varepsilon_{12} \end{pmatrix}, \tag{3.6}$$

where $\lambda$ and $\mu$ are Lame's coefficients.

By multiplying (3.3) with $v$ and integrating both sides over $\Omega$ and applying Green's formula we obtain for $u, v \in (H^1(\Omega))^2$

$$-\int_\Omega \frac{\partial \tau_{ij}(u)}{\partial x_j} v_i \, dx = \int_\Omega \tau_{ij}(u)\varepsilon_{ij}(v)\, dx - \int_{\partial\Omega} \tau_{ij}(u) n_j v_i \, ds \tag{3.7}$$

so that

$$\int_\Omega \tau_{ij}(u)\varepsilon_{ij}(v)\, dx - \int_{\partial\Omega} \tau_{ij}(u) n_j v_i \, ds = \int_\Omega F_i v_i \, dx \,, \tag{3.8}$$

making use of (3.3). Here $n = (n_1, n_2)$ denotes the outward normal unit vector to $\partial\Omega$.

To guarantee the uniqueness of the displacement field $u$, we impose the condition $u = 0$ on a given non-empty part $\Gamma_u$ open in $\partial\Omega$. On the remaining part $\Gamma_P$ of the boundary the surface tractions $P = (P_1, P_2)$ are given, i.e. $T_i \equiv \tau_{ij} n_j = P_i$, $i = 1, 2$.

We denote by $V$ the space of admissible displacements:

$$V = \{v = (v_1, v_2) \in (H^1(\Omega))^2 \mid v_i = 0 \text{ on } \Gamma_u,\ i = 1,2\} \,. \tag{3.9}$$

For any $v \in V$ the equation (3.8) yields:

$$\int_\Omega \tau_{ij}(u)\varepsilon_{ij}(v)\, dx = \int_\Omega F_i v_i \, dx + \int_{\Gamma_P} P_i v_i \, ds \equiv \langle f, v\rangle \,. \tag{3.10}$$

[1] In the sequel this summation convention will be used.

In order to obtain the existence and uniqueness of the solution of (3.10), it is necessary to prove the $V$-ellipticity of the bilinear form $a(\cdot,\cdot)$, given by $a(u,v) = \int_\Omega \tau_{ij}(u)\varepsilon_{ij}(v)\,dx \equiv (\tau(u),\varepsilon(v))$, i.e.

$$\exists\gamma = \text{const.} > 0 : \quad a(v,v) \geq \gamma\|v\|_{1,\Omega}^2 \quad \forall v \in V, \tag{3.11}$$

where $\|v\|_{1,\Omega} := \sqrt{\|v_1\|_{1,\Omega}^2 + \|v_2\|_{1,\Omega}^2}$.

Assuming that the elasticity coefficients satisfy the algebraic ellipticity condition

$$\exists q_0 > 0 : \ c_{ijkl}(x)\xi_{ij}\xi_{kl} \geq q_0\xi_{ij}\xi_{ij} \quad \forall \xi_{ij} = \xi_{ji} \in R_n \tag{3.12}$$

a.e. in $\Omega$, then (3.11) follows from Korn's inequality

$$\exists C > 0 : \quad \int_\Omega \varepsilon_{ij}(v)\varepsilon_{ij}(v)dx \geq C\|v\|_{1,\Omega}^2 \quad \forall v \in V. \tag{3.13}$$

### 3.2. Classical formulation of contact problems

Let us assume a structure, composed of two deformable bodies in mutual contact and submitted to a body force $F$ and surface tractions $P$ on a part of the boundary. The aim is to determine the behaviour of this structure. The problem is non-classical in the sense that the partition of the boundary into the contact and non-contact zones is not known a-priori as well as reactive forces between bodies. The deformation of each body depends not only on given applied forces but also on contact tractions along contact surfaces. There are many applications of this problem in several branches of engineering, e.g. in soil and rock mechanics, in mechanical engineering etc. To simplify the explanation, we restrict ourselves to the study of a contact problem between a *plane elastic* body and a rigid foundation.

Let a body be represented in its non-deformable state by a bounded domain $\Omega$, the Lipschitz boundary of which is decomposed as follows: $\partial\Omega = \bar{\Gamma}_u \cup \bar{\Gamma}_P \cup \bar{\Gamma}_K$, where $\Gamma_u$, $\Gamma_P$ and $\Gamma_K$ are non-empty parts of $\partial\Omega$, open in $\partial\Omega$. By a *classical solution* of a contact problem we mean a displacement field $u = (u_1, u_2)$, which is in the equilibrium state with $F$ and $P$, i.e.:

$$\text{(equilibrium equations):} \quad \frac{\partial\tau_{ij}}{\partial x_j}(u) + F_i = 0 \quad \text{in } \Omega \quad i = 1,2\ ; \tag{3.14}$$

and it satisfies the following set of boundary conditions:

$$\tau_{ij}(u)n_j = P_i \quad \text{on } \Gamma_P\ ; \tag{3.15}$$

$$u_i = 0 \quad \text{on } \Gamma_u \qquad i = 1,2\ ; \tag{3.16}$$

$$T_n(u) \equiv \tau_{ij}(u)n_in_j \leq 0\,, \quad u_n \equiv u\cdot n \leq 0\,, \quad T_n(u)u_n = 0 \quad \text{on } \Gamma_K\ ; \tag{3.17}$$

$$\left.\begin{array}{l} |T_t(x)| \leq g(x) \\ |T_t(x)| < g(x) \implies u_t(x) \equiv u(x)\cdot t = 0 \\ |T_t(x)| = g(x) \implies \exists\lambda(x) \geq 0 \text{ such that } u_t(x) = -\lambda(x)T_t(x) \end{array}\right\} \quad \text{on } \Gamma_K\,. \tag{3.18}$$

Here $T_n(x)$, $T_t(x)$ denotes the normal, tangential component of the stress vector $T(u)$ at a point $x$, $t$ is the tangential vector to $\partial\Omega$, $g(x)$ is bounded, measurable and non-negative function defined on $\partial\Omega$. (3.17) is a condition of unilateral contact of $\Omega$ with a rigid support.[1] Influence of friction is modeled by (3.18). We consider a very simple model, namely the so called "model with a given friction". Later, we shall briefly discuss the case of coulombien friction.

### 3.3. Primal variational formulation of contact problems

In order to give a variational form of (3.14)–(3.18), we introduce the following notations:

$$
\begin{aligned}
&V = \{v \in (H^1(\Omega))^2 \mid v_i = 0 \text{ on } \Gamma_u,\ i = 1,2\} \\
&K = \{v \in V \mid v_n = v \cdot n \leq 0 \text{ on } \Gamma_K\} \\
&\mathcal{J}(v) = \frac{1}{2}\int_\Omega \tau_{ij}(v)\varepsilon_{ij}(v)\,dx - \int_\Omega F_i v_i\,dx - \int_{\Gamma_P} P_i v_i\,ds + \int_{\Gamma_K} g|v_t|\,ds\,.
\end{aligned}
$$

In accordance with notations of Chapter 1,

$$
\begin{aligned}
a(u,v) &\equiv \int_\Omega \tau_{ij}(u)\varepsilon_{ij}(v)\,dx \equiv (\tau(u),\varepsilon(v))\,, \\
\langle f, v\rangle &\equiv \int_\Omega F_i v_i\,dx + \int_{\Gamma_P} P_i v_i\,ds \equiv (F_i, v_i)_0 + (P_i, v_i)_0\,, \\
j(v) &\equiv \int_{\Gamma_K} g|v_t|\,ds\,, \\
L(v) &\equiv \mathcal{J}(v)\,.
\end{aligned}
$$

$K$ is the closed, convex subset of $V$ of all admissible displacements, $\mathcal{J}$ is a total potential energy functional. Next, we shall assume that $F = (F_1, F_2) \in (L^2(\Omega))^2$, $P = (P_1, P_2) \in (L^2(\Gamma_P))^2$.

By a *variational solution* we mean a function $u \in K$, satisfying

$$
\mathcal{J}(u) \leq \mathcal{J}(v) \qquad \forall v \in K\,. \tag{$\mathcal{P}$}
$$

The relation between a classical and a variational formulation is given by

THEOREM 3.1. *($\mathcal{P}$) and (3.14)–(3.18) are formally equivalent.*

Applying Theorem 1.1, we immediately have

[1] Here we assume that $\Omega$ is unilaterally supported by a rigid foundation along the whole $\Gamma_K$, i.e. dist($\Gamma_K$,rigid foundation) $= 0$ and $\Gamma_K$ cannot enlarge during deformation.

THEOREM 2.4. *Let* (CC), ($CC_H$) *and* (S) *be satisfied. Let there exist a solution* $\{u,\lambda\}$ *of* $(\tilde{\mathcal{P}})$. *Then*

$$\|u-u_h\|^2 \le c\Big[\{\|u-v_h\|^2+|\lambda-\mu_H|^2\} \tag{2.15}$$
$$+\{b(u,\lambda_H-\mu)-[g,\lambda_H-\mu]\}+\{b(u,\lambda-\mu_H)-[g,\lambda-\mu_H]\}\Big]\,,$$

$$|\lambda-\lambda_H|\le c\{\|u-u_h\|+|\lambda-\mu_H|\} \tag{2.16}$$

*hold for any* $v_h\in V_h$, $\mu\in\Lambda$, $\mu_H\in\Lambda_H$ *with a positive constant c.*

PROOF: Since (CC) and ($CC_H$) are satisfied, $K=V$, $K_h=V_h$, i.e. $K$ and $K_h$ are linear sets. Therefore

$$A_1(v_h)=0 \quad \forall v_h\in V_h\,. \tag{2.17}$$

As $K=V$ and $V_h\subset V$ $\forall h\in(0,1)$, we can choose $v=u_h$ in (2.13). Hence

$$A_2(v)=0\,. \tag{2.17'}$$

Let $\mu_H\in\Lambda_H$ be arbitrary. Applying (S) we obtain

$$\hat{\beta}|\lambda_H-\mu_H|\le\sup_{V_h}\frac{b(v_h,\mu_H-\lambda_H)}{\|v_h\|}\,. \tag{2.18}$$

Now we may write

$$\begin{aligned}
b(v_h,\mu_H-\lambda_H)&=b(v_h,\mu_H)-b(v_h,\lambda_H)\\
&=b(v_h,\mu_H)+a(u_h,v_h)-\langle f,v_h\rangle\\
&=b(v_h,\mu_H)+a(u_h,v_h)-a(u,v_h)-b(v_h,\lambda)\\
&=b(v_h,\mu_H-\lambda)+a(u_h-u,v_h)\\
&\le c\{|\mu_H-\lambda|+\|u_h-u\|\}\|v_h\|\,.
\end{aligned}$$

This, together with (2.18) lead to

$$|\mu_H-\lambda_H|\le c\{\|u-u_h\|+|\lambda-\mu_H|\}\quad\forall\mu_H\in\Lambda_H\,.$$

Using the triangle inequality

$$|\lambda-\lambda_H|\le|\lambda-\mu_H|+|\mu_H-\lambda_H|\quad\forall\mu_H\in\Lambda_H\,,$$

we obtain (2.16). Finally, replacing the term $M_2\varepsilon|\lambda_H-\lambda|$ on the right hand side of (2.14) by (2.16) and making use of (2.17), (2.17′), we obtain (2.15) for $\varepsilon>0$ sufficiently small. □

REMARK 2.4. If $\Lambda_H\subset\Lambda$ for $\forall H\in(0,1)$, we can insert $\mu=\lambda_H$ into (2.15). Therefore, (2.15) takes the following simpler form:

$$\|u-u_h\|^2\le c\Big[\{\|u-v_h\|^2+|\lambda-\mu_H|^2\}$$
$$+\{b(u,\lambda-\mu_H)-[g,\lambda-\mu_H]\}\Big]\quad\forall v_h\in V_h,\ \mu_H\in\Lambda_H\,. \tag{2.19}$$

THEOREM 2.5. *Let* (BC) *and* ($BC_H$) *be satisfied. Then*

$$\begin{aligned}\|u - u_h\|^2 \le c\Big[&A_1(v_h) + A_2(v) + \{\|u - v_h\|^2 + |\lambda - \mu_H|^2\}\\ &+ \|u - v_h\| + \{b(u, \lambda_H - \mu) - [g, \lambda_H - \mu]\}\\ &+ \{b(u, \lambda - \mu_H) - [g, \lambda - \mu_H]\}\Big]\end{aligned} \tag{2.20}$$

*holds for any* $v_h \in K_h$, $v \in K$, $\mu \in \Lambda$, $\mu_H \in \Lambda_H$. *Moreover, if* $K = V$, $K_h = V_h$ *and* (S) *is satisfied, then (2.15) and (2.16) hold.*

PROOF: We have to prove (2.20) only. As $\Lambda$, $\Lambda_H$ are bounded in $Y$,

$$|b(v_h - u, \lambda_H - \lambda)| \le c\|v_h - u\| \quad \forall v_h \in K_h .$$

Hence (2.20) follows due to (2.13′). □

REMARK 2.5. If $K_h \subset K$, $\Lambda_H \subset \Lambda$, $\forall h, H \in (0,1)$, then setting $v = u_h$, $\mu = \lambda_H$ we obtain $A_2(v) = 0$, $b(u, \lambda_H - \mu) - [g, \lambda_H - \mu] = 0$.

Next, let us suppose that the pair of real parameters $h$, $H$ satisfies

$$h \to 0+ \iff H \to 0+ .$$

Relations (2.15), (2.16) and (2.20) can be used to estimate the rate of convergence of $u_h$ to $u$ and $\lambda_H$ to $\lambda$, provided the exact solution is smooth enough. Other applications are given by the following convergence theorems.

THEOREM 2.6. *Let* (BC), ($BC_H$) *be satisfied and, moreover, let*

$$\forall v \in K \quad \exists v_h \in K_h : \quad v_h \to v \quad \text{in } V ; \tag{2.21}$$

$$\forall \mu \in \Lambda \quad \exists \mu_H \in \Lambda_H : \quad \mu_H \to \mu \quad \text{in } Y ; \tag{2.22}$$

$$v_h \in K_h , \quad v_h \rightharpoonup v \quad \text{(weakly) in } V \text{ implies } v \in K ; \tag{2.23}$$

$$\mu_H \in \Lambda_H , \quad \mu_H \rightharpoonup \mu \quad \text{(weakly) in } Y \text{ implies } \mu \in \Lambda . \tag{2.24}$$

*Let the solution* $\{u, \lambda\} \in K \times \Lambda$ *of* $(\tilde{\mathcal{P}})$ *be unique. Then*

$$u_h \to u \quad \text{in } V , \quad \lambda_H \rightharpoonup \lambda \quad \text{(weakly) in } Y .$$

PROOF: First, $\{u_h\}$, $\{\lambda_H\}$ are bounded. For $\{\lambda_H\}$ this follows from ($BC_H$), for $\{u_h\}$ from (2.21) and $(\tilde{\mathcal{P}}'_{hH})_2$. Hence, there exist a subsequence $\{u_{h'}, \lambda_{H'}\} \subset \{u_h, \lambda_H\}$ and an element $\{u^*, \lambda^*\} \in V \times Y$ such that

$$u_{h'} \rightharpoonup u^* , \quad \lambda_{H'} \rightharpoonup \lambda^* \quad \text{in } Y . \tag{2.25}$$

### 3.5. Mixed formulation of contact problems

From numerical point of view there are two difficulties in treatment of $(\mathcal{P})$, namely

- $(\mathcal{P})$ is a *constrained* problem;
- $(\mathcal{P})$ is *non-smooth.*

To overcome these difficulties, duality approach can be applied. First we present some auxiliary results, useful in what follows. For the sake of simplicity we assume that $\Gamma_P = \emptyset$ and Hooke's law (3.4) will be written as

$$\tau = \Lambda \varepsilon, \tag{3.4'}$$

where $\Lambda$ is the symmetric and positive definite mapping from the space $S$ (see Section 3.4) into itself. By $\Lambda^{\frac{1}{2}}$ we denote the square root of $\Lambda$. $\Lambda^{-1}$, $\Lambda^{-\frac{1}{2}}$ are the inverse mappings to $\Lambda$, $\Lambda^{\frac{1}{2}}$, respectively.

Let $V$ be endowed with the equivalent norm

$$|||v||| = (\Lambda \varepsilon(v), \varepsilon(v)). \tag{3.24}$$

The set $H^{\frac{1}{2}}(\Gamma_K)$ of all traces of functions, belonging to $V$, will be equipped with the norm

$$\|\varphi\|_{\frac{1}{2},\Gamma_K} = \inf_{\substack{v\in V\\ \gamma v=\varphi}} |||v|||, \quad \varphi = (\varphi_1, \varphi_2) \in H^{\frac{1}{2}}(\Gamma_K).$$

The symbol $\gamma v$ denotes the trace of $v \in V$. From this definition, it follows immediately that

$$\|\varphi\|_{\frac{1}{2},\Gamma_K} \equiv |||u|||,$$

where $u \in H^1(\Omega)$ is the unique solution of

$$\begin{cases} (\Lambda\varepsilon(u), \varepsilon(v)) = 0 & \forall v \in V, \\ \gamma v = \varphi & \text{on } \Gamma_u. \end{cases} \tag{3.25}$$

Let $\delta : V \to H^{\frac{1}{2}}(\Gamma_K)$ be a mapping defined by the relation $\delta v = (v_n, v_t)$. Finally, by $H(\mathrm{div}, \Omega)$, we denote the space of symmetric tensors, the divergence of which is square integrable in $\Omega$:

$$H(\mathrm{div}, \Omega) = \left\{ \sigma = (\sigma_{ij})_{i,j=1}^2 \in S, \mathrm{div}\ \sigma = \left( \frac{\partial \sigma_{1j}}{\partial x_j}, \frac{\partial \sigma_{2k}}{\partial x_k} \right) \in \left[L^2(\Omega)\right]^2 \right\}.$$

We equip $H(\mathrm{div}, \Omega)$ with the norm

$$\|\sigma\|_{H(\mathrm{div},\Omega)} = \left\{ (\Lambda^{-1}\sigma, \sigma) + (\mathrm{div}\ \sigma, \mathrm{div}\ \sigma)_0 \right\}^{\frac{1}{2}}. \tag{3.26}$$

The closed subspace of $H(\text{div}, \Omega)$ consisting of functions with zero divergence will be denoted by $H_0(\text{div}, \Omega)$. Clearly, $H_0(\text{div}, \Omega)$ is a Hilbert space with norm

$$\|\sigma\|_{H_0(\text{div},\Omega)} = (\Lambda^{-1}\sigma, \sigma)^{\frac{1}{2}}. \tag{3.27}$$

The following Green's formulas hold for any $\tau \in H(\text{div}, \Omega)$ and $v \in V$:

(i) there exists a unique mapping $T = (T_1, T_2) \in \mathcal{L}(H(\text{div}, \Omega), H^{-\frac{1}{2}}(\Gamma_K))$ such that

$$(\tau, \varepsilon(v)) + (\text{div } \tau, v)_0 = \langle T(\tau), \gamma v\rangle = \langle T_1(\tau), \gamma v_1\rangle + \langle T_2(\tau), \gamma v_2\rangle.$$

(ii) there exists a unique mapping $\tilde{T} = (T_n, T_t) \in \mathcal{L}(H(\text{div}, \Omega), H^{-\frac{1}{2}}(\Gamma_K))$ such that

$$(\tau, \varepsilon(v)) + (\text{div } \tau, v)_0 = \langle \tilde{T}(\tau), \delta v\rangle = \langle T_n(\tau), v_n\rangle + \langle T_t(\tau), v_t\rangle.$$

Now we describe certain properties of $T$ and $\tilde{T}$.

LEMMA 3.2. *$T$ maps $H_0(\text{div}, \Omega)$ onto $H^{-\frac{1}{2}}(\Gamma_K)$.*

PROOF. Let $\varphi^* \in H^{-\frac{1}{2}}(\Gamma_K)$ be given. Then the problem

$$\begin{cases} \text{find } u \in V \text{ such that} \\ (\Lambda\varepsilon(u), \varepsilon(v)) = \langle \varphi^*, \gamma v\rangle, \quad \forall v \in V, \end{cases} \tag{3.28}$$

has a unique solution $u = u(\varphi^*)$. On setting $\sigma = \Lambda\varepsilon(u)$ and using (i), we get immediately that $\sigma \in H_0(\text{div}, \Omega)$ and $T(\sigma) = \varphi^*$.

The symbol $\|\varphi^*\|_{-\frac{1}{2},\Gamma_K}$ denotes the usual dual norm, namely

$$\|\varphi^*\|_{-\frac{1}{2},\Gamma_K} = \sup_{\substack{H^{\frac{1}{2}}(\Gamma_K) \\ \varphi\neq 0}} \frac{\langle \varphi^*, \varphi\rangle}{\|\varphi\|_{\frac{1}{2},\Gamma_K}}.$$

Next we derive some more useful expressions for $\|\varphi^*\|_{-\frac{1}{2},\Gamma_K}$.

LEMMA 3.3. *The following equalities,*

$$\|\varphi^*\|_{-\frac{1}{2},\Gamma_K} = \|\sigma\|_{H_0(\text{div},\Omega)} = |||u(\varphi^*)|||, \quad \forall \varphi^* \in H^{-\frac{1}{2}}(\Gamma_K),$$

*hold, where $u = u(\varphi^*)$ is the solution of (3.28) and $\sigma = \Lambda\varepsilon(u)$.*

PROOF. Let $\sigma \in H_0(\text{div}, \Omega)$ satisfy $T(\tau) = \varphi^*$. Then, for any $\varphi \in H^{-\frac{1}{2}}(\Gamma_K)$, (i) yields

$$\langle \varphi^*, \varphi\rangle = \langle T(\tau), \varphi\rangle = \langle \tau, \varepsilon(v)\rangle = (\Lambda^{-\frac{1}{2}}\tau, \Lambda^{\frac{1}{2}}\varepsilon(v)) \quad \forall v \in V, \quad \gamma v = \varphi.$$

Hence

$$\langle\varphi^*,\varphi\rangle \le \|\tau\|_{H_0(\mathrm{div},\Omega)}|||v||| \quad \forall v\in V,\ \gamma v=\varphi,$$

so that

$$\|\varphi^*\|_{-\frac{1}{2},\Gamma_K} \le \|\tau\|_{H_0(\mathrm{div},\Omega)} \quad \forall\tau\in H_0(\mathrm{div},\Omega),\ T(\tau)=\varphi^*. \tag{3.29}$$

Let $u=u(\varphi^*)$ be the solution of (3.28) and let $\sigma=\Lambda\varepsilon(u)\in H_0(\mathrm{div},\Omega)$. Then

$$\|\sigma\|^2_{H_0(\mathrm{div},\Omega)} = (\Lambda^{-1}\sigma,\sigma) = (\varepsilon(u),\ \Lambda\varepsilon(u)) = |||u(\varphi^*)|||^2. \tag{3.30}$$

By inserting $v=u(\varphi^*)$ into (3.28), using (3.30) and the definition of $\|\ \|_{\frac{1}{2},\Gamma_K}$, we obtain

$$\begin{aligned}\langle\varphi^*,\gamma u\rangle &= (\Lambda\varepsilon(u),\varepsilon(u)) = |||u(\varphi^*)|||^2 = \|\sigma\|_{H_0(\mathrm{div},\Omega)}|||u|||\\ &\ge \|\sigma\|_{H_0(\mathrm{div},\Omega)}\|\gamma u\|_{\frac{1}{2},\Gamma_K}.\end{aligned}$$

Hence

$$\|\varphi^*\|_{-\frac{1}{2},\Gamma_K} \ge \|\sigma\|_{H_0(\mathrm{div},\Omega)}.$$

This inequality and (3.29) yield the following lemma.

LEMMA 3.4. *The following expression holds:*

$$\|\varphi^*\|_{-\frac{1}{2},\Gamma_K} = \sup_V \frac{\langle\varphi^*,\gamma v\rangle}{|||v|||}, \quad \forall\varphi^*\in H^{-\frac{1}{2}}(\Gamma_K).$$

PROOF. From the definition of $\|\varphi^*\|_{-\frac{1}{2},\Gamma_K}$ as the dual norm, it follows that

$$\langle\varphi^*,\gamma v\rangle \le \|\varphi^*\|_{-\frac{1}{2},\Gamma_K}\|\gamma v\|_{\frac{1}{2},\Gamma_K} \le \|\varphi^*\|_{-\frac{1}{2},\Gamma_K}|||v|||,$$

and so

$$\sup_V \frac{\langle\varphi^*,\gamma v\rangle}{|||v|||} \le \|\varphi^*\|_{-\frac{1}{2},\Gamma_K}. \tag{3.31}$$

On the other hand, let $u=u(\varphi^*)$ be the solution of (3.28). Then, for $\sigma=\Lambda\varepsilon(u)$, we get

$$\begin{aligned}\langle\varphi^*,\gamma u\rangle = |||u(\varphi^*)|||^2 &= |||u(\varphi^*)|||\cdot\|\sigma\|_{H_0(\mathrm{div},\Omega)}\\ &= |||u(\varphi^*)|||\cdot\|\varphi^*\|_{-\frac{1}{2},\Gamma_K},\end{aligned}$$

by virtue of Lemma 3.3. From this, and from (3.31), the lemma follows.

In a similar way to the proof of Lemma 3.2, one can prove

LEMMA 3.5. *$\tilde{T}$ maps $H_0(div,\Omega)$ onto $H^{-\frac{1}{2}}(\Gamma_K)$.*

Let $\varphi^*$ be an arbitrary element of $H^{-\frac{1}{2}}(\Gamma_K)$. By virtue of Lemma 3.2, there exists $\tau\in H_0(\mathrm{div},\Omega)$ such that $T(\tau)=\varphi^*$. Let us set $\mu^*=T(\tau)\in H^{-\frac{1}{2}}(\Gamma_K)$ and write $\mu^*=\beta^*\varphi^*$ with $\beta^*: H^{-\frac{1}{2}}(\Gamma_K)\to H^{-\frac{1}{2}}(\Gamma_K)$. It is readily seen that $\mu^*$ does not depend on the choice of $\tau\in H_0(\mathrm{div},\Omega)$ where $T(\tau)=\varphi^*$. By comparing Green's formulas (i), (ii), we obtain

LEMMA 3.6. *For any* $\varphi^* \in H^{-\frac{1}{2}}(\Gamma_K)$,

$$\|\varphi^*\|_{-\frac{1}{2},\Gamma_K} = \|\beta^*\varphi^*\|_{-\frac{1}{2},\Gamma_K}. \tag{3.32}$$

In order to transform problem $(\mathcal{P})$ into an unconstrained and smooth one, we introduce the following sets of Lagrange multipliers $\Lambda_1,\ \Lambda_2$:

$$\Lambda_1 = H_-^{-\frac{1}{2}}(\Gamma_k) = \left\{\mu_1 \in H^{-\frac{1}{2}}(\Gamma_k) | \langle \mu_1, v_n\rangle \geq 0 \quad \forall v \in K\right\},$$
$$\Lambda_2 = \left\{\mu_2 \in L^2(\Gamma_K) \mid |\mu_2| \leq g \quad \text{a.e. on } \Gamma_K\right\},$$

be two closed convex subsets of $H^{-\frac{1}{2}}(\Gamma_K)$ of Lagrange multipliers and let $\mathcal{L} : V \times \Lambda_1 \times \Lambda_2 \to R_1$ be a Lagrange function defined by means of the relation

$$\mathcal{L}(v, \mu_1, \mu_2) = \frac{1}{2}(\Lambda\varepsilon(v), \varepsilon(v)) - \langle \mu_1, v_n\rangle - \langle \mu_2, v_t\rangle - (F_i, v_i)_0,$$

where $\langle , \rangle$ is the duality pairing between $H^{-\frac{1}{2}}(\Gamma_K)$ and $H^{\frac{1}{2}}(\Gamma_K)$.

The mixed variational formulation of the Signorini problem with given friction is defined as the problem of finding a saddle–point $\{w, \lambda_1, \lambda_2\} \in V \times \Lambda_1 \times \Lambda_2$ of $\mathcal{L}$ on $V \times \Lambda_1 \times \Lambda_2$ satisfying

$$\mathcal{L}(w, \mu_1, \mu_2) \leq \mathcal{L}(w, \lambda_1, \lambda_2) \leq \mathcal{L}(v, \lambda_1, \lambda_2) \quad \forall v \in V, (\mu_1, \mu_2) \in \Lambda_1 \times \Lambda_2. \tag{$\tilde{\mathcal{P}}$}$$

The relation between $(\mathcal{P})$ and $(\tilde{\mathcal{P}})$ is given by

THEOREM 3.4. *There exists a unique solution* $(w, \lambda_1, \lambda_2)$ *of* $(\tilde{\mathcal{P}})$. *Moreover,*

$$w = u, \ \lambda_1 = T_n(u), \lambda_2 = T_t(u),$$

*where* $u \in K$ *is the solution of* $(\mathcal{P})$ *and* $T_n(u),\ T_t(u)$ *are the corresponding normal and tangential components of the stresses on* $\Gamma_K$.

REMARK 3.3. If we wish to remove the non–differentiability of $(\mathcal{P})$ only, it is sufficient to introduce the set $\Lambda = \Lambda_2$ and to define

$$\mathcal{L}(v, \mu) = \frac{1}{2}(\Lambda\varepsilon(v), \varepsilon(v)) - (F_i, v_i)_0 + \int_{\Gamma_K} \mu v_t \, ds, \quad (v, \mu) \in K \times \Lambda.$$

It is very easy to prove that $\mathcal{L}$ posses a unique saddle–point $(\tilde{v}, \tilde{\lambda})$ on $K \times \Lambda$. Moreover,

$$\tilde{v} = u, \ \tilde{\lambda} = T_t(u),$$

where $u \in K$ solves $(\mathcal{P})$.

### 3.6. Reciprocal formulation of contact problems. Case I.

In the mixed variational formulation $(\tilde{\mathcal{P}})$, presented in the last section, normal and tangential tractions along $\Gamma_K$ play the role of Lagrange multipliers associated with the constraint and the nondifferentiability of $\mathcal{J}$. Elimination of the displacement field $v$ leads to a new variational problem (the so-called reciprocal) for $T_n(u)$, $T_t(u)$ only. The derivation of this formulation is the aim of the present section.

Let $\{w, \lambda_1, \lambda_2\} \in V \times \Lambda_1 \times \Lambda_2$ be the solution of $(\tilde{\mathcal{P}})$. Then:

$$\mathcal{L}(w, \lambda_1, \lambda_2) = \inf_V \sup_{\Lambda_1 \times \Lambda_2} \mathcal{L}(v, \mu_1, \mu_2) = \sup_{\Lambda_1 \times \Lambda_2} \inf_V \mathcal{L}(v, \mu_1, \mu_2). \tag{3.33}$$

Let us set

$$\tilde{\mathcal{S}}(\mu_1, \mu_2) = \inf_V \mathcal{L}(v, \mu_1, \mu_2). \tag{3.34}$$

We derive the explicit form of $\tilde{\mathcal{S}}$. For $(\mu_1, \mu_2) \in \Lambda_1 \times \Lambda_2$ fixed, (3.34) is connected with the solution of the following linear elliptic boundary value problem:

$$\left. \begin{aligned} &\text{find } z = z(\mu_1, \mu_2) \in V \text{ such that} \\ &(\Lambda\varepsilon(z), \varepsilon(v)) = \langle \mu_1, v_n \rangle + \langle \mu_2, v_t \rangle + (F_i, v_i)_0, \quad \forall v \in V. \end{aligned} \right. \tag{3.35}$$

In view of the linearity of (3.35), one can split $z$ into $\hat{z}$ and $\tilde{z}$ and write $z = \hat{z} + \tilde{z}$, where $\hat{z}, \tilde{z} \in V$ are the unique solutions of

$$(\Lambda\varepsilon(\hat{z}), \varepsilon(v)) = (F_i, v_i)_0 \quad \forall v \in V, \tag{3.36}$$

$$(\Lambda\varepsilon(\tilde{z}), \varepsilon(v)) = \langle \mu_1, v_n \rangle + \langle \mu_2, v_t \rangle \quad \forall v \in V. \tag{3.37}$$

In the sequel, we denote by $G : V' \to V$ the Green's operator, corresponding to $V$ and the bilinear form $(\Lambda\varepsilon(z), \varepsilon(v))$. For brevity, instead of (3.36), (3.37), we shall write

$$\hat{z} = G(F), \ \tilde{z} = G(\mu_1, \mu_2), \ (\mu_1, \mu_2) \in \Lambda_1 \times \Lambda_2.$$

Since $z \in V$ is a minimizer of the quadratic functional $\mathcal{L}(v, \mu_1, \mu_2)$, we have

$$\begin{aligned} \mathcal{L}(z, \mu_1, \mu_2) = &-\frac{1}{2}\langle \mu_1, z_n \rangle - \frac{1}{2}\langle \mu_2, z_t \rangle - \frac{1}{2}(F_i, z_i)_0 = -\frac{1}{2}\langle \mu_1, \tilde{z}_n \rangle \\ &-\frac{1}{2}\langle \mu_2, \tilde{z}_t \rangle - \frac{1}{2}\langle \mu_1, \hat{z}_n \rangle - \frac{1}{2}\langle \mu_2, \hat{z}_t \rangle - \frac{1}{2}(F_i, \tilde{z}_i)_0 - \frac{1}{2}(F_i, \hat{z}_i)_0. \end{aligned} \tag{3.38}$$

On inserting $v = \tilde{z}$ in (3.36), using the symmetry of $(\Lambda\varepsilon(z), \varepsilon(v))$ and (3.37), we obtain

$$\langle \mu_1, \hat{z}_n \rangle + \langle \mu_2, \hat{z}_t \rangle = (\Lambda\varepsilon(\tilde{z}), \varepsilon(\hat{z})) = (\Lambda\varepsilon(\hat{z}), \varepsilon(\tilde{z})) = (F_i, \tilde{z}_i)_0.$$

Thus (3.38) can be written as follows:

$$\begin{aligned}\mathcal{L}(z,\mu_1,\mu_2) = &-\frac{1}{2}\langle\mu_1,\tilde{z}_n\rangle - \frac{1}{2}\langle\mu_2,\tilde{z}_t\rangle - \langle\mu_1,\hat{z}_n\rangle - \langle\mu_2,\hat{z}_t\rangle - \frac{1}{2}(F_i,\hat{z}_i)_0 \\ \equiv &-\frac{1}{2}\langle\mu_1, G(\mu_1,\mu_2)n\rangle - \frac{1}{2}\langle\mu_2, G(\mu_1,\mu_2)t\rangle - \langle\mu_1, G(F)n\rangle \\ &- \langle\mu_2, G(F)t\rangle - \frac{1}{2}(F_i,\hat{z}_i)_0.\end{aligned} \tag{3.39}$$

Denote by $\beta : [H^{-\frac{1}{2}}(\Gamma_K)]^2 \times [H^{-\frac{1}{2}}(\Gamma_K)]^2 \to R_1$ a bilinear form

$$\beta(\mu,\nu) = \langle\mu_1, G(\nu_1,\nu_2)n\rangle + \langle\mu_2, G(\nu_1,\nu_2)t\rangle, \mu = (\mu_1,\mu_2), \nu = (\nu_1,\nu_2) \tag{3.40}$$

and let $\xi : [H^{-\frac{1}{2}}(\Gamma_K)]^2 \to R_1$ be given by

$$\xi(\mu) = -\langle\mu_1, G(F)n\rangle - \langle\mu_2, G(F)t\rangle.$$

By making use of this notation, we can rewrite $\tilde{\mathcal{S}}$ as follows:

$$\tilde{\mathcal{S}}(\mu) = \tilde{\mathcal{S}}(\mu_1,\mu_2) = -\frac{1}{2}\beta(\mu,\mu) + \xi(\mu) - \frac{1}{2}(F_i,\hat{z}_i)_0.$$

Then the term $(F_i,\hat{z}_i)$ does not depend on $(\mu_1,\mu_2)$ and, therefore, it can be omitted. Finally, let us set

$$\mathcal{S}(\mu) = -\tilde{\mathcal{S}}(\mu) - \frac{1}{2}(F_i,\hat{z}_i)_0 = \frac{1}{2}\beta(\mu,\mu) - \xi(\mu).$$

Now we are in a position to define the *reciprocal variational formulation.* It is the problem of finding $\lambda = (\lambda_1,\lambda_2) \in \Lambda_1 \times \Lambda_2$ such that

$$\mathcal{S}(\lambda) \le \mathcal{S}(\mu) \quad \forall\mu = (\mu_1,\mu_2) \in \Lambda_1 \times \Lambda_2. \tag{$\mathcal{P}_r$}$$

The relation between $(\mathcal{P})$ and $(\mathcal{P}_r)$ is given by

THEOREM 3.5. *There exists a unique solution $(\lambda_1,\lambda_2)$ of $(\mathcal{P}_r)$. Moreover,*

$$\lambda_1 = T_n(u), \ \lambda_2 = T_t(u), \tag{3.41}$$

*where $u \in K$ is the solution of $(\mathcal{P})$.*

By using result of Section 3.5, one can give an equivalent form of $\mathcal{S}$:

$$\begin{aligned}\mathcal{S}(\mu) =&\frac{1}{2}\beta(\mu,\mu) - \xi(\mu) = \frac{1}{2}\langle\mu_1, G(\mu_1,\mu_2)n\rangle + \frac{1}{2}\langle\mu_2, G(\mu_1,\mu_2)t\rangle - \xi(\mu) \\ =&\frac{1}{2}\langle\mu_1, z_n\rangle + \frac{1}{2}\langle\mu_2, z_t\rangle - \xi(\mu) = \frac{1}{2}\langle\mu,\delta z\rangle - \xi(\mu) \\ =&\frac{1}{2}|||z(\mu)|||^2 - \xi(\mu), \quad \mu = (\mu_1,\mu_2),\end{aligned}$$

where $z \in V$ is a solution of

$$(\Lambda\varepsilon(z), \varepsilon(v)) = \langle \mu, \delta v\rangle = \langle \beta^{*-1}\mu, \gamma v\rangle \quad \forall v \in V,$$

and $\beta^{*-1} : H^{-\frac{1}{2}}(\Gamma_K) \to H^{-\frac{1}{2}}(\Gamma_K)$ denotes the inverse of $\beta^*$. The results of Lemmas 3.3 and 3.6 make it possible to write

$$|||z(\mu)|||^2 = \|\beta^{*-1}\mu\|^2_{-\frac{1}{2},\Gamma_K} = \|\mu\|^2_{-\frac{1}{2},\Gamma_K}.$$

Hence $\mathcal{S}$ can be written in the form

$$\mathcal{S}(\mu) = \frac{1}{2}\|\mu\|^2_{-\frac{1}{2},\Gamma_K} - \xi(\mu). \tag{3.42}$$

### 3.7. Reciprocal formulation of contact problems. Case II.

Starting point for this case will be the formulation presented in Section 3.4. For the sake of simplicity we assume that $g \equiv 0$ on $\Gamma_K$, i.e. frictionless case is considered. Let us repeat notations:

$$\mathcal{S}(\tau) = (\Lambda^{-1}\tau, \tau)\,, \quad \tau \in S\,,$$
$$K^+_{F,P}(\Omega) \equiv K^+_{F,P,0}(\Omega) = \{\tau \in S \mid (\tau, \varepsilon(v)) \geq \langle f, v\rangle \ \forall v \in K\}\,.$$

Let

$$V_0 = \{v \in (H^1(\Omega))^2 \mid v_i = 0,\ i = 1,2 \text{ on } \Gamma_u,\ v_n = 0 \text{ on } \Gamma_K\}$$

and

$$K_{F,P}(\Omega) = \{\tau \in S \mid (\tau, \varepsilon(v)) = \langle f, v\rangle \ \forall v \in V_0\}\,.$$

It is easy to see that $\tau \in K_{F,P}(\Omega)$ if and only if

$$\frac{\partial \tau_{ij}}{\partial x_j} + F_i = 0 \qquad \text{in } \Omega\,, \quad i = 1,2\,;$$
$$\tau_{ij}n_j = P_i \qquad \text{on } \Gamma_P\,, \quad i = 1,2\,;$$
$$T_t(\tau) \equiv \tau_{ij}n_jt_i = 0 \qquad \text{on } \Gamma_K\,.$$

Moreover

$$\tau \in K^+_{F,P}(\Omega) \iff \tau \in K_{F,P}(\Omega) \ \& \ \langle T_n(\tau), v_n\rangle \geq 0 \quad \forall v \in K$$

and

$$\sup_{v\in K}[-\langle T_n(\tau), v_n\rangle] \qquad (\tau \in K_{F,P}(\Omega))$$

is the indicator function of $K^+_{F,P}(\Omega)$. Thus

$$\inf_{\tau\in K^+_{F,P}(\Omega)} \mathcal{S}(\tau) = \inf_{\tau\in K_{F,P}(\Omega)} \sup_{v\in K} \left[\frac{1}{2}(\Lambda^{-1}\tau, \tau) - \langle T_n(\tau), v_n\rangle\right]\,.$$

Denote by

$$\mathcal{L}(\tau, v) = \frac{1}{2}(\Lambda^{-1}\tau, \tau) - \langle T_n(\tau), v_n\rangle$$

the lagrangian, defined on $K_{F,P}(\Omega) \times K$. One has

THEOREM 3.6. *There exists a unique saddle-point of $\mathcal{L}$ on $K_{F,P}(\Omega) \times K$. Moreover*

$$\sigma^* = \tau(u) = \Lambda\varepsilon(u) ,$$
$$u^* = u \quad \text{on } \Gamma_K ,$$

*where $u \in K$ solves $(\mathcal{P})$ (with $g \equiv 0$).*

Using the fact that $(\sigma^*, u^*)$ is a saddle-point of $\mathcal{L}$ on $K_{F,P}(\Omega) \times K$, we may write:

$$\mathcal{L}(\sigma^*, u^*) = \inf_{\tau \in K_{F,P}(\Omega)} \sup_{v \in K} \mathcal{L}(\tau, v) = \sup_{v \in K} \inf_{\tau \in K_{F,P}(\Omega)} \mathcal{L}(\tau, v) .$$

For $v \in K$, we denote

$$\tilde{\Phi}(v) = \inf_{\tau \in K_{F,P}(\Omega)} \mathcal{L}(\tau, v) \tag{3.43}$$

the dual functional to $S$. Next we derive its explicit form.

Let $\sigma = \sigma(v)$, $v \in K$, be the solution of (3.43). Using well-known results, we immediately obtain that

$$\sigma = \Lambda\varepsilon(w) ,$$

where the deplacement field $w$ solves the following elasticity problem:

$$\begin{aligned}
&\frac{\partial \sigma_{ij}}{\partial x_j} + F_i = 0 && \text{in } \Omega , && i = 1,2 ;\\
&w_i = 0 && \text{on } \Gamma_u , && i = 1,2 ;\\
&\sigma_{ij} n_j = P_i && \text{on } \Gamma_P , && i = 1,2 ;\\
&T_t(\sigma) \equiv \sigma_{ij} n_j t_i = 0 && \text{on } \Gamma_K ; &&\\
&w_n = v_n && \text{on } \Gamma_K . &&
\end{aligned}$$

Let $\tau_0 \in S$ be such that

$$(\tau_0, \varepsilon(v)) = \langle f, v \rangle \qquad \forall v \in V_0 .$$

Splitting $K_{F,P}(\Omega)$ we may write

$$K_{F,P}(\Omega) = \tau_0 + K_{00}(\Omega) ,$$

where

$$K_{00}(\Omega) = \{\tau \in S \mid (\tau, \varepsilon(v)) = 0 \ \forall v \in V_0\} .$$

Then

$$\inf_{\tau \in K_{F,P}(\Omega)} \mathcal{L}(\tau, v) = \inf_{\tau \in K_{00}(\Omega)} \mathcal{L}(\tau_0 + \tau, v) .$$

A direct calculation yields:

$$\mathcal{L}(\tau_0 + \tau, v) = \mathcal{H}(\tau) + \text{term depending on } \tau_0 \text{ only},$$

with

$$\mathcal{H}(\tau) = \frac{1}{2}(\Lambda^{-1}\tau, \tau) + (\Lambda^{-1}\tau, \tau_0) - \langle T_n(\tau), v_n \rangle \ ,$$

$\tau \in K_{00}(\Omega)$. Next we shall analyze problem

$$\inf_{\tau \in K_{00}(\Omega)} \mathcal{H}(\tau) \ . \tag{3.44}$$

Relation between (3.43) and (3.44) is clear: if $\bar{\tau}$ solves (3.44), $\bar{\tau} + \tau_0$ solves (3.43). $\bar{\tau}$ satisfies the following relation

$$(\Lambda^{-1}\bar{\tau}, \tau) + (\Lambda^{-1}\tau_0, \tau) - \langle T_n(\tau), v_n \rangle = 0 \qquad \forall \tau \in K_{00}(\Omega) \ .$$

Let $u_0$ be a displacement field, related to $\tau_0$:

$$\tau_0 = \Lambda\varepsilon(u_0) \ .$$

Then

$$(\Lambda^{-1}\tau_0, \tau) = \langle u_0, T(\tau) \rangle_{\Gamma_u} + \langle u_{0n}, T_n(\tau) \rangle_{\Gamma_K} \ ,$$

because of $\tau \in K_{00}(\Omega)$. Next we shall assume that

$$\begin{aligned} u_0 &= 0 \quad \text{on } \Gamma_u \\ u_{0n} &= 0 \quad \text{on } \Gamma_K \ . \end{aligned}$$

This choice leads to:

$$(\Lambda^{-1}\tau_0, \tau) = 0 \qquad \forall \tau \in K_{00}(\Omega)$$

and

$$\mathcal{H}(\bar{\tau}) = \inf_{\tau \in K_{00}(\Omega)} \mathcal{H}(\tau) = -\frac{1}{2}\langle T_n(\bar{\tau}), v_n \rangle \ .$$

Again, classical result says that

$$\bar{\tau} = \Lambda\varepsilon(\bar{u}) \ ,$$

where $\bar{u}$ is a displacement field solving the linear elasticity problem:

$$\begin{aligned} &\frac{\partial \bar{\tau}_{ij}}{\partial x_j} = 0 && i = 1,2 \quad \text{in } \Omega \ ; \\ &\bar{\tau}_{ij} n_j = 0 && i = 1,2 \quad \text{on } \Gamma_P \ ; \\ &\bar{u}_i = 0 && i = 1,2 \quad \text{on } \Gamma_u \ ; \\ &\left.\begin{aligned} \bar{u}_n &= v_n \\ T_t(\bar{\tau}) &= 0 \end{aligned}\right\} && \quad \text{on } \Gamma_K \ . \end{aligned}$$

Let $H : H^{\frac{1}{2}}(\Gamma_K) \mapsto K_{00}(\Omega)$ be a mapping defined by

$$H(v_n) = \bar{\tau} \ ,$$

where $\bar{\tau}$ is a solution of (3.44). Then

$$\mathcal{H}(\bar{\tau}) = -\frac{1}{2}\langle H(v_n) \cdot n, v_n \rangle \ .$$

Going back to the problem

$$\inf_{\tau \in K_{F,P}(\Omega)} \mathcal{L}(\tau, v) = \inf_{\tau \in K_{00}(\Omega)} \mathcal{L}(\tau + \tau_0, v) = \mathcal{L}(\bar{\tau} + \tau_0, v) \ ,$$

we see that

$$\mathcal{L}(\bar{\tau} + \tau_0, v_n) = -\frac{1}{2}\langle H(v_n) \cdot n, v_n \rangle - \langle T_n(\tau_0), v_n \rangle$$

(omitting the term $(\Lambda^{-1}\tau_0, \tau_0)$). Thus

$$\tilde{\Phi}(v) = -\frac{1}{2}\langle H(v_n) \cdot n, v_n \rangle - \langle T_n(\tau_0), v_n \rangle \ .$$

Denote by $\gamma(v_n, v_n) = \frac{1}{2}\langle H(v_n) \cdot n, v_n \rangle$ the quadratic part of $\tilde{\Phi}$, $\delta(v_n) = -\langle T_n(\tau_0), v_n \rangle$ its linear term, and

$$\Phi(v_n) = \gamma(v_n, v_n) - \delta(v_n) \ .$$

Now we show that the problem

$$\begin{cases} \text{find } v_n^* \in H_+^{\frac{1}{2}}(\Gamma_K) \text{ such that} \\ \Phi(v_n^*) \le \Phi(v_n) \quad \forall v_n \in H_+^{\frac{1}{2}}(\Gamma_K) \end{cases} \tag{3.45}$$

has a unique solution. Here

$$H_+^{\frac{1}{2}}(\Gamma_K) = \{\mu \in H^{\frac{1}{2}}(\Gamma_K) \mid \mu = v_n \text{ on } \Gamma_K, \ v \in K\} \ .$$

To this end we show that

$$\gamma(v_n, v_n) = \frac{1}{2}\langle H(v_n) \cdot n, v_n \rangle = \frac{1}{2}\|v_n\|^2_{\frac{1}{2},\Gamma_K} \ .$$ [1]

From this, the existence and the uniqueness of $v_n^*$ immediately follows.

[1] Definition of $\| \, \|_{\frac{1}{2},\Gamma_K}$ see below.

Let $V$ be equipped with the norm

$$|||v||| = (\Lambda\varepsilon(v), \varepsilon(v)) ,$$

which is equivalent to the usual norm in $(H^1(\Omega))^2$ due to Korn's inequality. Let $\varphi \in H^{\frac{1}{2}}(\Gamma_K)$ be an arbitrary element. Then

$$\|\varphi\|_{\frac{1}{2},\Gamma_K} \stackrel{\text{def}}{\equiv} \inf_{\substack{v\in V \\ v_n=\varphi \text{ on } \Gamma_K}} |||v||| = |||\bar{u}||| \tag{3.46}$$

defines the norm in $H^{\frac{1}{2}}(\Gamma_K)$.

It is easy to see that $\bar{u}$, realizing minimum in (3.46), is a solution of the linear elasticity problem

$$\begin{aligned}
&\frac{\partial\tau_{ij}(\bar{u})}{\partial x_j} = 0 && i = 1,2 \quad \text{in } \Omega ; \\
&\tau(\bar{u}) = \Lambda\varepsilon(\bar{u}) && \text{in } \Omega ; \\
&\bar{u}_i = 0 && i = 1,2 \quad \text{on } \Gamma_u ; \\
&\tau_{ij}(\bar{u})n_j = 0 && i = 1,2 \quad \text{on } \Gamma_P ; \\
&\left.\begin{aligned}\bar{u}_n = \varphi \\ T_t(\tau(\bar{u})) = 0\end{aligned}\right\} && \text{on } \Gamma_K .
\end{aligned} \tag{3.47}$$

If $\bar{\tau} \in K_{00}(\Omega)$ solves (3.44) then

$$\begin{aligned}
\mathcal{H}(\bar{\tau}) &= -\frac{1}{2}\langle H(v_n)\cdot n, v_n\rangle = -\frac{1}{2}(\Lambda^{-1}\bar{\tau}, \bar{\tau}) \\
&= -\frac{1}{2}(\varepsilon(\bar{u}), \tau(\bar{u})) ,
\end{aligned}$$

where $\bar{u}$ solves (3.47) with $\varphi$ replaced by $v_n$. Thus

$$\gamma(v_n, v_n) = \frac{1}{2}\langle H(v_n)\cdot n, v_n\rangle = \frac{1}{2}|||\bar{u}|||^2 = \frac{1}{2}\|v_n\|^2_{\frac{1}{2},\Gamma_K} .$$

From previous analysis the physical meaning of $v_n^*$ follows: $v_n^*$ is equal to the *normal component* of the *displacement field* $u$ on $\Gamma_K$.

## 3.8. Variational formulation of contact problems with coulombien friction

Up to now, a function $g$ appearing in (3.18) was given a priori. In classical Coulomb's law the known function $g$ is replaced by an *unknown* value $\mathcal{F}|T_n(u)|$, where $\mathcal{F}$ is the coefficient of Coulomb's friction. So, instead of (3.18) we have the following conditions:

$$\begin{aligned}
&|T_t(u)| \le \mathcal{F}|T_n(u)| \quad \text{a.e. on } \Gamma_K ; \\
&\text{if } |T_t(u)(x)| < \mathcal{F}|T_n(u)(x)| \implies u_t(x) = 0 ; \\
&\text{if } |T_t(u)(x)| = \mathcal{F}|T_n(u)(x)| \\
&\qquad \implies \exists\lambda = \lambda(x) : \quad u_t(x) = -\lambda(x)T_t(u)(x) .
\end{aligned} \tag{3.18'}$$

In order to give the variational formulation of such a problem, we use the model with "a given" friction as an auxiliary one. Nevertheless, some modification is necessary: instead of $g \in L^\infty(\Gamma_K)$, we shall write $\mathcal{F}g$, where $g \in H_+^{-\frac{1}{2}}(\Gamma_K) = -\Lambda_1$. Then term $\int_{\Gamma_K} g|v_t|\,ds$ will be replaced by $\langle \mathcal{F}g, |v_t|\rangle$. Let $u = u(g)$ be the solution of the following Signorini problem with given friction equal to $\mathcal{F}g$:

$$\begin{cases} \text{find } u = u(g) \in K \text{ such that} \\ (\Lambda\varepsilon(u), \varepsilon(v-u)) + \langle \mathcal{F}g, |v_t| - |u_t|\rangle \\ \qquad \geq (F_i, v_i - u_i)_0 + (P_i, v_i - u_i)_0 \quad \forall v \in K\,. \end{cases} \tag{$\mathcal{P}_{\mathcal{F}g}$}$$

It is easy to prove that $-T_n(u) \in H_+^{-\frac{1}{2}}(\Gamma_K)$. This last enables us to define the mapping $\phi : H_+^{-\frac{1}{2}}(\Gamma_K) \to H_+^{-\frac{1}{2}}(\Gamma_K)$ by means of the relation

$$\phi(g) = -T_n(u)\,.$$

*A weak solution of a Signorini problem with friction obeying Coulomb's law is defined as a fixed point of* $\phi$ in $H_+^{-\frac{1}{2}}(\Gamma_K)$. The existence of a fixed point of $\phi$ for small values of $\mathcal{F}$ and special geometry of $\Omega$ has been proved (see [4]).

Let $\mathcal{S}$ be a quadratic functional given by (3.42) and $K_{\mathcal{F}g}$ by

$$K_{\mathcal{F},g} = \{\mu = (\mu_1, \mu_2) \in (H^{-\frac{1}{2}}(\Gamma_K))^2 \mid \mu_1 \leq 0,$$
$$\langle \mu_2, v_t\rangle + \langle \mathcal{F}g, |v_t|\rangle \geq 0 \ \forall v \in V\}$$

with $g \in H_+^{-\frac{1}{2}}(\Gamma_K)$. Problem

$$\begin{cases} \text{find } (\lambda_1(g), \lambda_2(g)) \in K_{\mathcal{F}g} \text{ such that} \\ \mathcal{S}(\lambda_1(g), \lambda_2(g)) \leq \mathcal{S}(\mu_1, \mu_2) \quad \forall (\mu_1, \mu_2) \in K_{\mathcal{F}g} \end{cases} \tag{3.48}$$

is a reciprocal variational formulation to the contact problem with a given friction equal to $\mathcal{F}g$ (studied in Section 3.6). Again, one can prove that

$$\lambda_1(g) = T_n(u)\,, \quad \lambda_2(g) = T_t(u)\,.$$

Here $(\lambda_1, \lambda_2)$ depends on $g$, as $K_{\mathcal{F}g}$ depends on $g$.

Introducing the mapping $\tilde{\Phi} : H_+^{-\frac{1}{2}}(\Gamma_K) \mapsto H_+^{-\frac{1}{2}}(\Gamma_K)$ by means of

$$\tilde{\Phi}(g) = -\lambda_1(g)\,,$$

we see that coulombien friction corresponds to the case, when $g = -\lambda_1(g)$, i.e. when $g$ is a fixed point of $\tilde{\Phi}$. As a result, we are led to the so called *quasivariational inequality* for $-T_n(u)$.

# 4. APPROXIMATION OF CONTACT PROBLEMS BY FINITE ELEMENTS

In this chapter we shall analyse the approximation of contact problems without friction by finite element technique. Results of Chapter 2 will be employed. In the first part we shall discuss the case of a quite general domain. So, curved elements for the approximation will be used. In the second part we shall use mixed formulation for the approximation of contact problems. In both cases error estimates will be derived, provided the exact solution is smooth enough.

Notations are the same as in Section 3.3.

## 4.1. Approximation of contact problems – frictionless case. Primal formulation.

We start with the construction of finite-dimensional approximations of $K$. For the sake of simplicity we restrict ourselves to the case, when only $\Gamma_K$ is curved. Let $\Psi$ be a continuous concave (it is not necessary) function defined on $\langle a,b\rangle$, the graph of which is $\Gamma_K$. We choose $(m+1)$ points $A_1,\dots,A_{m+1}$ on $\Gamma_K$ in such a way that $A_1, A_{m+1}$ are boundary points of $\Gamma_K$. Let $A_i, A_{i+1} \in \Gamma_K$, $Q \in \Omega$. By a curved element $T$ we call a closed set bounded by the streight lines $QA_i$, $QA_{i+1}$ and the arc $\widehat{A_iA_{i+1}}$. The minimal interior angle of the curved element $T$ is called the minimal angle of the curved element $T$. A triangulation $\mathcal{T}_h$ of $\overline{\Omega}$ contains curved elements along $\Gamma_K$ and internal triangular ones. By the symbols $h$ and $\vartheta$ we denote the *maximal diameter* and the *minimal interior angle*, respectively, of all elements $T \in \mathcal{T}_h$. We shall assume only the so called *regular systems* of triangulations: a constant $\vartheta_0 > 0$ exists, independent of $h$ and such that

$$\vartheta \geq \vartheta_0 \quad \text{if } h \to 0+ \ .$$

A family of triangulations will be called *$\alpha$-$\beta$ regular*, if $\vartheta_0 = \alpha$ and

$$\frac{h}{h_{\min}} \leq \beta \ ,$$

where $h_{\min}$ is the minimal diameter of all $T \in \mathcal{T}_h$. Define

$$V_h = \{v \in (C(\overline{\Omega}))^2 \cap V \mid v|_T \in (P_1(T))^2 \ , \quad \forall T \in \mathcal{T}_h\}$$

and

$$K_h = \{v \in V_h \mid v \cdot n(A_i) \leq 0,\ i = 1,\dots,m+1\} \ ,$$

where $P_1(T)$ denotes the set of linear polynomials, defined on $T$. It is easy to see that $K_h$ represents a finite dimensional approximation of $K$ and $K_h \not\subset K$, in general.

An approximation of our problem is defined as the solution of the following problem:

$$\begin{cases} \text{find } u_h \in K_h \text{ such that} \\ \mathcal{J}(u_h) \leq \mathcal{J}(v) \quad \forall v \in K_h \end{cases} \qquad (\mathcal{P}_h)$$

with $\mathcal{J}$ defined in Section 3.3 (with $g \equiv 0$ on $\Gamma_K$).

Now we shall need some auxiliary results:

LEMMA 4.1. *Let $\Omega \subset R_2$ be a bounded convex domain, the boundary of which is twice continuously differentiable and let $\{\mathcal{T}_h\}$ be a $\alpha$-$\beta$ regular system of triangulations with $\alpha < \pi/8$, $\beta = 2$. Then*

$$\|u - u_I\|_{0,\partial\Omega} \le ch^{\frac{3}{2}} \|u\|_{2,\Omega} \qquad \forall u \in H^2(\Omega) ,$$

*where $u_I$ denotes the piecewise-linear Lagrange interpolate of $u$, $c > 0$ is independent of $h > 0$.*

PROOF: See [5].

LEMMA 4.2. *Let $v \in P_1(T)$, where $T$ is a closed triangular element. Let $T_h$ be the triangle generated by replacing the curved side by its chord. Then*

$$\|v\|^2_{1,\Delta(T,T_h)} \le ch\|v\|^2_{1,T_h} ,$$

*where $\Delta(T, T_h) = (T \setminus T_h) \cup (T_h \setminus T)$ and $c > 0$ is independent of $h$.*

PROOF: See [6].

The main result of this section is

THEOREM 4.1. *Let the solution $u$ of $(\mathcal{P})$ belong to $(H^2(\Omega))^2$, $\tau(u) \in H(div, \Omega)$ and $T_n(u) \in L^2(\Gamma_K)$. Let the system of triangulations $\{\mathcal{T}_h\}$ satisfy the assumptions of Lemma 4.1 and $\Psi$, describing $\Gamma_K$, be from $C^3(\langle a, b\rangle)$. Then*

$$\|u - u_h\| \le c(u)h^{\frac{3}{4}} .$$

PROOF: Using the definition of $(\mathcal{P})$, Green's formula and (2.2), we deduce

$$\frac{1}{2}a(u - u_h, u - u_h) \le \frac{1}{2}a(v_h - u, u_h - u) + \int_{\Gamma_K} T_n(v_n - u_{hn})\, ds$$
$$+ \int_{\Gamma_K} T_n(v_{hn} - u_n)\, ds \quad \forall v \in K,\ v_h \in K_h .$$

Let $v_h = u_I$, where $u_I$ is the piecewise linear Lagrange interpolate of $u$ on $\Omega$. It is easy to see that $u_I \in K_h$ and

$$\left\{\begin{aligned} &a(u_h - u, v_h - u) \le M\varepsilon\|u_h - u\|^2_{1,\Omega} + \frac{M}{\varepsilon}\|u - u_I\|^2_{1,\Omega} \\ &\qquad = M\varepsilon\|u_h - u\|^2_{1,\Omega} + \mathcal{O}(h^2) \quad \forall\varepsilon > 0 \\ &\int_{\Gamma_K} T_n(u_I - u)\cdot n\, ds \le c\|u_I - u\|_{0,\Gamma_K} \le ch^{\frac{3}{2}}\|u\|_{2,\Omega} , \end{aligned}\right. \tag{4.1}$$

where the assertion of Lemma 4.1 and the regularity assumption concerning $u$ have been used. The most difficult is to estimate the term

$$\int_{\Gamma_K} T_n(v_n - u_{hn})\, ds . \tag{4.2}$$

In what follows we shall construct a function $v \in K$ such that (4.2) is small. We identify the origin of coordinate system $(x_1, x_2)$ with the point $A_i$. Let $\Sigma_i$ be a closed set bounded with the arc $\widehat{A_iA_{i+1}} \equiv s_i \subset \Gamma_K$ and the chord $A_iA_{i+1}$. Let $x \in \Sigma_i$. By the symbol $P(x)$ and $Q(x)$, respectively, we denote the intersection of the perpendicular line through the point $x$ with $s_i$ and $A_iA_{i+1}$, respectively. Let us define functions $U_h, \tilde{U}_h$ on $\bigcup_{i=1}^{m} \Sigma_i$ by means of the following relations:

$$U_h(x) = u_h(x) \cdot n(P(x)) ,$$
$$\tilde{U}_h(x) = u_h(Q(x)) \cdot n(P(x)) = \tilde{u}_h(x) \cdot n(P(x)) ,$$

where we set $\tilde{u}_h(x) = u_h(Q(x))$. Clearly

$$U_h(x) = \tilde{U}_h(x) , \quad x \in A_iA_{i+1} .$$

Let $\Phi_i(x)$, $x \in A_iA_{i+1}$ be the linear Lagrange interpolate of $U_h$ on $A_iA_{i+1}$ and let us define $\tilde{\Phi}$ on $\bigcup_{i=1}^{m} \Sigma_i$ as follows:

$$\tilde{\Phi}(x) = \Phi_i(Q(x)) , \quad x \in \Sigma_i , \quad i = 1, \dots, m+1.$$

It is readily seen that $\tilde{\Phi} \le 0$ on $\Gamma_K$. We shall estimate

$$\|\tilde{\Phi} - U_h\|_{0,\Gamma_K} .$$

We may write:

$$\begin{aligned} \|\tilde{\Phi} - U_h\|_{0,\Gamma_K} &\le \|\tilde{\Phi} - \tilde{U}_h\|_{0,\Gamma_K} + \|\tilde{U}_h - U_h\|_{0,\Gamma_K} , \\ \|\tilde{U}_h - U_h\|^2_{0,\Gamma_K} &= \sum_{i=1}^{m} \|\tilde{U}_h - U_h\|^2_{0,s_i} \le 2 \sum_{i=1}^{m} \|u_h - \tilde{u}_h\|^2_{0,s_i} . \end{aligned} \tag{4.3}$$

Let $q$ be the arc's parameter of the point $P(x) = (P_1(x), P_2(x))$ and denote $Q_1(x) = x_1$. Then for $j = 1, 2$ we have

$$\begin{aligned} u_{hj}(q) - \tilde{u}_{hj}(q) &= \int_0^{P_1(x)} \frac{\partial}{\partial x_2}(u_{hj} - \tilde{u}_{hj})\, dx_2 \\ &= \int_0^{P_1(x)} \frac{\partial}{\partial x_2} u_{hj}(x_1, x_2)\, dx_2 . \end{aligned}$$

Integrating and using Fubini's theorem we obtain

$$\|u_{hj} - \tilde{u}_{hj}\|^2_{0,s_i} \le ch^2 |u_h|^2_{1,\Sigma_i} \qquad j = 1, 2.$$

From this and Lemma 4.2 we have

$$\|U_h - \tilde{U}_h\|_{0,\Gamma_K}^2 \le ch^2 \sum_{i=1}^{m} |u_h|_{1,\Sigma_i}^2 \le ch^3 |u_h|_{1,\Omega}^2 \,. \tag{4.4}$$

Let us estimate $\|\tilde{\Phi} - \tilde{U}_h\|_{0,\Gamma_K}^2$.

$$\|\tilde{\Phi} - \tilde{U}_h\|_{0,\Gamma_K}^2 = \sum_{i=1}^{m} \|\tilde{\Phi} - \tilde{U}_h\|_{0,s_i}^2 \,.$$

$$\begin{aligned}\tilde{\Phi}(q) - \tilde{U}_h(q) &= \int_0^{Q_1(x)} \frac{d}{dx_1}(\Phi_i(x_1,0) - \tilde{U}_h(x_1,0))\,dx_1 \\ &\quad + \int_0^{P_2(x)} \frac{d}{dx_2}(\Phi_i(Q_1(x),x_2) - \tilde{U}_h(Q_1(x),x_2)\,dx_2 \\ &= \int_0^{Q_1(x)} \frac{d}{dx_1}(\Phi_i(x_1,0) - \tilde{U}_h(x_1,0))\,dx_1 \,.\end{aligned}$$

Since $\Psi \in C^3(\langle a,b\rangle)$, we have $\tilde{U}_h \in H^2(A_iA_{i+1})$. Hence

$$\begin{aligned}|\tilde{\Phi}(q) - \tilde{U}_h(q)|^2 &\le ch|\Phi_i - \tilde{U}_h|_{1,A_iA_{i+1}}^2 \\ &\le ch^3 |\tilde{U}_h|_{2,A_iA_{i+1}}^2 \,.\end{aligned} \tag{4.5}$$

As $\tilde{U}_h(x) = \tilde{u}_h(x) \cdot n(P(x))$ and $\tilde{u}_h \in P_1(A_iA_{i+1})$, we may write

$$|\tilde{U}_h|_{2,A_iA_{i+1}}^2 \le c|u_h|_{1,A_iA_{i+1}}^2 \,.$$

Thus, (4.5) and the inverse inequality between $H^1(A_iA_{i+1})$ and $H^{\frac{1}{2}}(A_iA_{i+1})$ yield:

$$\begin{aligned}\|\tilde{\Phi} - \tilde{U}_h\|_{0,s_i}^2 &\le ch^4 \|u_h\|_{1,A_iA_{i+1}}^2 \\ &\le ch^3 \|u_h\|_{\frac{1}{2},A_iA_{i+1}}^2 \,.\end{aligned} \tag{4.6}$$

Adding (4.6) for $i = 1, \ldots, m$ we obtain:

$$\|\tilde{\Phi} - \tilde{U}_h\|_{0,\Gamma_K}^2 \le ch^3 \|u_h\|_{\frac{1}{2},\Gamma_h}^2 \,, \tag{4.7}$$

where $\Gamma_h = \bigcup_{i=1}^{m} A_iA_{i+1}$ is the polygonal approximation of $\Gamma_K$. Using the trace theorem we obtain:

$$\|u_h\|_{\frac{1}{2},\Gamma_h}^2 \le c\|u_h\|_{1,\Omega\setminus\bigcup_i\Sigma_i}^2 \le c\|u_h\|_{1,\Omega}^2 \,,$$

where $c > 0$ does not depend on $h$ for $h$ sufficiently small. Using these estimates (4.3), (4.4) and (4.7) we deduce

$$\|\tilde{\Phi} - U_h\|_{0,\Gamma_K} \leq ch^{\frac{3}{2}}\|u_h\|_{1,\Omega} . \tag{4.8}$$

Next, let $v \in V$ be such that

$$v \cdot n = \tilde{\Phi} \qquad \text{on } \Gamma_K .$$

Then $v \cdot n \leq 0$ on $\Gamma_K$, consequently $v \in K$. Finally we may write

$$\begin{aligned} \int_{\Gamma_K} T_n(v_n - u_{hn})\,ds &= \int_{\Gamma_K} T_n(\tilde{\Phi} - U_h)\,ds \\ &\leq ch^{\frac{3}{2}}\|u_h\|_{1,\Omega} . \end{aligned} \tag{4.9}$$

Since the norms $\|u_h\|_{1,\Omega}$ remain bounded, the assertion of Theorem now follows from (4.1), (4.9).

In the above error estimates we needed strong regularity assumptions, concerning the solution $u$. Unfortunately, there are no reasons to expect such a great smoothness. This is why we are going to prove the convergence of $u_h$ to $u$ without estimating the rate of convergence, using no regularity assumptions. According to Theorem 2.1, it is necessary to verify (2.4) and (2.5).

LEMMA 4.3. *(verification of (2.4))*
*Let us suppose that $\bar{\Gamma}_u \cap \bar{\Gamma}_K = \emptyset$ and there exists only a finite number of boundary points $\bar{\Gamma}_P \cap \bar{\Gamma}_K$, $\bar{\Gamma}_u \cap \bar{\Gamma}_P$. Then the set*

$$\mathcal{M} = K \cap (C^\infty(\overline{\Omega}))^2$$

*is dense in $K$ in $\mathcal{H}^1$-norm.*

PROOF: The proof for polygonal domains is given in [7], but its slight modification gives the same density result also in our case.

LEMMA 4.4. *The condition (2.5) holds.*

PROOF: Let $v_h \in K_h$ be such that

$$v_h \rightharpoonup v \quad \text{in } V , \quad h \to 0+ . \tag{4.10}$$

It is sufficient to show that $v \cdot n \leq 0$ on $\Gamma_K$ or equivalently

$$\int_{\Gamma_K} v \cdot n \cdot \varphi\,ds \leq 0$$

for any $\varphi \in C^1(\langle a,b\rangle)$, $\varphi \geq 0$ on $\langle a,b\rangle$.

Since the trace mapping is completely continuous from $V$ into $(L^2(\Gamma_K))^2$, we have

$$v_h \to v \quad \text{in } (L^2(\Gamma_K))^2 \,, \quad h \to 0+ \tag{4.11}$$

hence

$$v_{hn} \to v_n \quad \text{in } L^2(\Gamma_K)\,, \quad h \to 0+\,.$$

Let $\Psi_h$ be the piecewise linear function defined on $\langle a,b\rangle$, nodes of which are the points $A_1,\ldots,A_{m+1}$. Then

$$\Gamma_{Kh} = \{(x_1,x_2),\ x_1 \in \langle a,b\rangle,\ x_2 = \Psi_h(x_1)\}$$

is the linear approximation of $\Gamma_K$. Let us set

$$\begin{aligned}\vartheta(x_1) &= v(x_1,\Psi(x_1))\,,\\ \vartheta_h(x_1) &= v_h(x_1,\Psi(x_1))\,,\\ \vartheta_{hh}(x_1) &= v_h(x_1,\Psi_h(x_1))\,, \quad x_1 \in \langle a,b\rangle\,.\end{aligned}$$

By virtue of (4.11)

$$\vartheta_h \to \vartheta \quad \text{in } (L^2((a,b)))^2\,, \quad h \to 0+\,. \tag{4.12}$$

Let us prove also

$$\vartheta_{hh} \to \vartheta \quad \text{in } (L^2((a,b)))^2\,, \quad h \to 0+\,. \tag{4.13}$$

We may write

$$\|\vartheta_{hh}-\vartheta\|_{0,(a,b)} \leq \|\vartheta-\vartheta_h\|_{0,(a,b)} + \|\vartheta_h-\vartheta_{hh}\|_{0,(a,b)}\,. \tag{4.14}$$

From the definition of $\vartheta_{hh}$ it follows that these ones are piecewise linear Lagrange interpolates of $\vartheta_h$ on $\langle a,b\rangle$. Corresponding division of $\langle a,b\rangle$ will be denoted by $a = t_1^m < t_2^m < \cdots < t_{m+1}^m = b$. Using the approximative property of $\vartheta_{hh}$ we have

$$\begin{aligned}\|\vartheta_h-\vartheta_{hh}\|_{0,(a,b)} &\leq ch^{\frac12}\|\vartheta_h\|_{\frac12,(a,b)}\\ &\leq ch^{\frac12}\|v_h\|_{1,\Omega} \leq ch^{\frac12}\,,\end{aligned}$$

where $c>0$ is independent of $h$ for $h$ sufficiently small and (4.10) has been used. From this, (4.12) and (4.14), (4.13) follows. Now, let us prove that

$$\int_a^b \vartheta(x_1)\cdot n(x_1,\Psi(x_1))\varphi(x_1)\,dx_1 \leq 0$$

for any $\varphi \in C^1(\langle a,b\rangle)$, $\varphi \geq 0$ on $\langle a,b\rangle$. Using (4.13) we have

$$\int_a^b \vartheta_{hh} \cdot n\varphi \, dx_1 \to \int_a^b \vartheta \cdot n\varphi \, dx_1 \ , \quad h \to 0+ \ . \tag{4.15}$$

For the numerical computation of $\int_a^b \vartheta_{hh} \cdot n\varphi \, dx_1$ we use the trapezoid formula:

$$\begin{aligned}\int_a^b \vartheta_{hh} \cdot n\varphi \, dx_1 \approx & h((\vartheta_{hh} \cdot n)(t_1^m)\varphi(t_1^m) + 2(\vartheta_{hh} \cdot n)(t_2^m)\varphi(t_2^m) \\ & + \cdots + (\vartheta_{hh} \cdot n)(t_{m+1}^m)\varphi(t_{m+1}^m)) \equiv [\vartheta_{hh} \cdot n, \varphi] \ .\end{aligned}$$

Since $(\vartheta_{hh} \cdot n)(t_j^m)\varphi(t_j^m) = (v_h \cdot n)(A_j)\varphi(t_j^m) \leq 0 \ \forall j = 1, \ldots, m+1$ we have

$$[\vartheta_{hh} \cdot n, \varphi] \leq 0 \qquad \forall h > 0 \ .$$

The proof will be finished, if

$$[\vartheta_{hh} \cdot n, \varphi] \to \int_a^b \vartheta \cdot n\varphi \, dx_1 \ , \quad h \to 0+ \ . \tag{4.16}$$

We may write

$$\begin{aligned}&\left| \int_a^b \vartheta \cdot n\varphi \, dx_1 - [\vartheta_{hh} \cdot n, \varphi] \right| \\ &\leq \left| \int_a^b \vartheta \cdot n\varphi \, dx_1 - \int_a^b \vartheta_{hh} \cdot n\varphi \, dx_1 \right| + \left| \int_a^b \vartheta_{hh} \cdot n\varphi \, dx_1 - [\vartheta_{hh} \cdot n, \varphi] \right| \ .\end{aligned} \tag{4.17}$$

By virtue of the inverse inequality between $H^{\frac{1}{2}}(\langle a,b\rangle)$ and $H^1(\langle a,b\rangle)$:

$$\begin{aligned}&\left| \int_a^b \vartheta_{hh} \cdot n\varphi \, dx_1 - [\vartheta_{hh} \cdot n, \varphi] \right| \leq ch|\vartheta_{hh} \cdot n\varphi|_{1,(a,b)} \\ &\leq c(n,\varphi)h\|\vartheta_{hh}\|_{1,(a,b)} \leq c(n,\varphi)h^{\frac{1}{2}}\|\vartheta_{hh}\|_{\frac{1}{2},(a,b)} \\ &\leq c(n,\varphi)h^{\frac{1}{2}}\|v_h\|_{1,\Omega_h} \leq c(n,\varphi)h^{\frac{1}{2}} \ ,\end{aligned}$$

where $\Omega_h$ is the polygonal domain bounded with $\Gamma_u$, $\Gamma_P$ and $\Gamma_{Kh}$. From this, (4.10), (4.15) and (4.17) we obtain (4.16). Hence

$$(\vartheta \cdot n)(x_1) = (v \cdot n)(x_1, \Psi(x_1)) \leq 0 \ , \quad x_1 \in \langle a,b\rangle \ .$$

THEOREM 4.2. *Let the assumptions of Lemma 4.3 be satisfied. Then*

$$\|u - u_h\| \to 0 \ , \quad h \to 0+ \ .$$

PROOF: The assertion of the Theorem is an immediate consequence of Theorem 2.1.

## 4.2. Approximation of contact problems without friction by a mixed finite element method

In this section we shall analyse the approximation of mixed variational formulation of contact problems without friction. Let us start with the formulation of the continuous case.

By $\mathcal{L}$ we denote the Lagrangian, given by

$$\mathcal{L}(v,\mu) = \frac{1}{2}(\tau(v),\varepsilon(v)) - (F_i, v_i)_0 - (P_i, v_i)_0 + \langle \mu, v_n \rangle ,$$

defined for $(v,\mu) \in V \times \Lambda$, where

$$V \equiv \{v \in (H^1(\Omega))^2 \mid v = 0 \text{ on } \Gamma_u\} ,$$
$$\Lambda \equiv H_+^{-\frac{1}{2}}(\Gamma_K) = \{\mu \in H^{-\frac{1}{2}}(\Gamma_K) \mid \langle \mu, v_n \rangle \leq 0 \ \forall v \in V, \ v_n \leq 0 \text{ on } \Gamma_K\} .$$

*Mixed formulation* of the contact problem without friction is defined as the problem of finding a saddle-point $(u^*, \lambda) \in V \times \Lambda$ of $\mathcal{L}$ over $V \times \Lambda$, i.e.:

$$\mathcal{L}(u^*,\mu) \leq \mathcal{L}(u^*,\lambda) \leq \mathcal{L}(v,\lambda) \quad \forall v \in V, \ \forall \mu \in \Lambda \tag{$\tilde{\mathcal{P}}$}$$

(see also Section 3.5). It is easy to prove that there exists a unique saddle-point $(u^*, \lambda)$ of $\mathcal{L}$ over $V \times \Lambda$. Moreover, the following interpretation holds:

$$u^* = u \ , \quad \lambda = -T_n(u) , \tag{4.18}$$

where $u \in K$ solves $(\mathcal{P})$.

Approximation of $(\tilde{\mathcal{P}})$ will be based on the theory, presented in Section 2.2, case (CC). For the sake of simplicity we assume that $\Omega$ is a *polygonal domain.* Let $\{\mathcal{T}_h\}$, $h \to 0+$ be a *regular family* of triangulations of $\overline{\Omega}$ satisfying usual requirements, concerning the mutual position of triangles. With any $\mathcal{T}_h$, the following finite dimensional space will be associated:

$$V_h = \{v_h \in (C(\overline{\Omega}))^2 \mid v_h|_T \in (P_1(T))^2 \ \forall T \in \mathcal{T}_h, \ v_h = 0 \text{ on } \Gamma_u\} ,$$

i.e. $V_h$ contains all piecewise linear functions over a given triangulation, vanishing on $\Gamma_u$. Let $\{\mathcal{T}_H\}$ be a partition of the contact part $\Gamma_K$, nodes of which will be denoted by $b_1, \ldots, b_M$. Let us emphasize that $b_i$ don't coincide with nodes of $\mathcal{T}_h$ lying on $\Gamma_K$, in general. With any $\mathcal{T}_H$ the following approximation of $\Lambda$ will be associated:

$$\Lambda_H = \{\mu_H \in L^2(\Gamma_K) \mid \mu_H|_{\overline{b_i b_{i+1}}} \in P_0(\overline{b_i b_{i+1}}), \ \mu_H \geq 0 \text{ on } \Gamma_K\} ,$$

i.e. $\Lambda_H$ contains all non-negative and piecewise constant functions (over $\mathcal{T}_H$). The approximation of $(\tilde{\mathcal{P}})$ is now defined as follows:

$$\left\{\begin{array}{l}\text{find } (u_h^*, \lambda_H) \in V_h \times \Lambda_H \text{ such that}\\ \mathcal{L}(u_h^*, \mu_H) \leq \mathcal{L}(u_h^*, \lambda_H) \leq \mathcal{L}(v_h, \lambda_H) \quad \forall v_h \in V_h,\ \forall \mu_H \in \Lambda_H\end{array}\right. \qquad (\tilde{\mathcal{P}}_{hH})$$

or equivalently

$$\left\{\begin{array}{l}\text{find } (u_h^*, \lambda_H) \text{ such that}\\ a(u_h^*, v_h) + \langle \lambda_H, v_{hn}\rangle = (F_i, v_{ih})_0 + (P_i, v_{ih})_0 \quad \forall v_h \in V_h\\ \langle u_{hn}^*, \mu_H - \lambda_H\rangle \leq 0 \quad \forall \mu_H \in \Lambda_H\,.\end{array}\right.$$

*Interpretation of* $u_h^*$. Let

$$\mathcal{K}_{hH} = \{v_h \in V_h \mid \int_{\overline{b_i b_{i+1}}} v_{hn}\, ds \leq 0\ \forall i = 1,\ldots,M\}\,,$$

i.e. $\mathcal{K}_{hH}$ contains all piecewise linear functions $v_h$, the mean value of which in non-positive on $\overline{b_i b_{i+1}}$. Using the definition of $\mathcal{K}_{hH}$ and $u_h^*$, it is easy to see that $u_h^*$ belongs to $\mathcal{K}_{hH}$ and it minimizes the total potential energy functional $\mathcal{J}$ over $\mathcal{K}_{hH}$. Next we shall try to estimate $\|u - u_h^*\|$ and $\|T_n(u) - \lambda_H\|_{-\frac{1}{2},\Gamma_K}$, making use of the general results of Section 2.2.

Let us recall that in notations of Section 2.2:

$$b(v_h, \mu_H) = \langle v_{hn}, \mu_H\rangle = \int_{\Gamma_K} v_{hn}\mu_H\, ds\,,$$

$$g \equiv 0\,.$$

As already known, the crucial point is the verification of the stability conditions (S):

$$\sup_{\substack{v_h \neq 0\\ v_h \in V_h}} \frac{\langle v_{hn}, \mu_H\rangle}{\|v_h\|} \geq \hat{\beta}|\mu_H|_{-\frac{1}{2},\Gamma_K}\,, \qquad (4.19)$$

with $\hat{\beta} > 0$ independent of $h$, $H$. Let us repeat some results, guaranteeing (4.19), proofs of which can be found in [8].

We introduce an auxiliary elliptic problem:

$$\left\{\begin{array}{l}\text{find } w \in V \text{ such that}\\ (\varepsilon_{ij}(w), \varepsilon_{ij}(v)) = \langle \varphi^*, v_n\rangle \quad \forall v \in V\,,\end{array}\right. \qquad (4.20)$$

where $\varphi^* \in H^{-\frac{1}{2}}(\Gamma_K)$. Let (4.20) be *regular in the following sense*: if $\varphi^* \in H^{-\frac{1}{2}+\varepsilon}(\Gamma_K)$ ($\varepsilon > 0$ a positive number), then the solution $w$ of (4.20) belongs to $(H^{1+\varepsilon}(\Omega))^2$ and

$$\|w\|_{1+\varepsilon} \leq c(\varepsilon)\|\varphi^*\|_{-\frac{1}{2}+\varepsilon,\Gamma_K} \qquad (4.21)$$

holds with a constant $c(\varepsilon) > 0$, depending on $\varepsilon > 0$, in general. Now, the following lemma can be proved:

LEMMA 4.5. *If (4.20) is regular in above mentioned manner, then (4.19) is satisfied, provided the ratio $h/H$ is sufficiently small.*

The main result of this section is:

THEOREM 4.3. *Let the solution of $(\mathcal{P})$ belong to $K \cap (H^{1+q}(\Omega))^2$ for some $q > 0$ and let $T_n(u) \in L^2(\Gamma_K)$. Then*

$$\|u - u_h^*\| = \mathcal{O}(H^{\tilde{q}})$$
$$\|T_n(u) - \lambda_H\|_{-\frac{1}{2},\Gamma_K} = \mathcal{O}(H^{\tilde{q}}) \,, \quad H \to 0+ \,,$$

*where $\tilde{q} = \min\{q, \frac{1}{4}\}$.*

PROOF: Using (2.15) and (2.16) we see that

$$\|u - u_h^*\|^2 \le C\{\|u - v_h\|^2 + |T_n(u) - \mu_H|^2_{-\frac{1}{2},\Gamma_K} + \langle u_n, T_n(u) - \mu_H\rangle\} \tag{4.22}$$

$$\|T_n(u) - \lambda_H\|_{-\frac{1}{2},\Gamma_K} \le C\{\|u - u_h^*\| + |T_n(u) - \mu_H|_{-\frac{1}{2},\Gamma_K}\} \tag{4.23}$$

with a constant $C > 0$, which doesn't depend on $h$, $H$. Let $v_h = u_I \in V_h$ be the piecewise linear Lagrange interpolate of $n$ and $\mu_H$ the orthogonal $L^2$-projection of $-T_n(u)$ on $\Lambda_H$. Then

$$\|u - u_I\|^2 = \mathcal{O}(h^{2q}) \,,$$
$$|T_n(u) - \mu_H|_{-\frac{1}{2},\Gamma_K} = \mathcal{O}(H^{\frac{1}{2}}) \,,$$
$$|\langle u_n, T_n(u) - \mu_H\rangle| \le \|u_n\|_{\frac{1}{2},\Gamma_K} |T_n(u) - \mu_H|_{-\frac{1}{2},\Gamma_K} = \mathcal{O}(H^{\frac{1}{2}}) \,,$$

making use of classical approximation results. This, together with (4.22) and (4.23) leads to the assertion of Theorem 4.3.

Higher regularity assumptions make possible to improve the rate of convergence (see [8]).

## 5. SHAPE OPTIMIZATION IN CONTACT PROBLEMS

Let us suppose that a part of an elastic body $\Omega$ is close to a rigid foundation. Thus a part $\Gamma_K$ of $\Omega$'s boundary may get into contact with the foundation.

Let us assume that the boundary of $\Omega$ is decomposed as follows:

$$\partial\Omega = \overline{\Gamma}_u \cup \overline{\Gamma}_P \cup \overline{\Gamma}_K, \tag{5.1}$$

where different boundary conditions will be assumed, namely $u = 0$ on $\Gamma_u$, $\tau_{ij}(u)n_j = P_i$, $i = 1, 2$ on $\Gamma_P$, and $\Gamma_K$ includes the contact area. We suppose that $\Gamma_u$ and $\Gamma_K$ are non-empty and open in $\partial\Omega$.

To concretize the situation we assume that the shape of $\Omega$ will be given by Figure 5.1 and $\Gamma_K$ is described by a graph of a function $\alpha$, the choice of which will be specified later:

$$\Gamma_K(\alpha) = \{(x_1, x_2) \mid x_2 = \alpha(x_1),\ x_1 \in (a, b)\}. \tag{5.2}$$

Let the rigid foundation be given by the set $\{(x_1, x_2) \mid x_2 \le 0\}$. Then we have the following system of unilateral conditions on $\Gamma_K$:

$$\begin{cases} u_2(x_1, \alpha(x_1)) \ge -\alpha(x_1), & \text{(non-penetration of } \Omega \text{ into rigid foundation)} \\ T_2(x_1, \alpha(x_1)) \ge 0, & \\ (u_2 + \alpha)T_2 = 0, & \text{(no contact-no forces)} \\ T_1 \equiv 0, & \text{(no friction).} \end{cases} \tag{5.3}$$

In correspondence with unilateral conditions (5.3), we define the closed, convex set $K$ of kinematically admissible displacements,

$$K = \{v \in V \mid v_2(x_1, \alpha(x_1)) \ge -\alpha(x_1) \quad \forall x_1 \in (a, b)\}\ . \tag{5.4}$$

The variational solution of a contact problem is then defined in usual way:

$$\text{find } u \in K:\ \mathcal{J}(u) \le \mathcal{J}(v) \quad \forall v \in K. \tag{5.5}$$

**5.1. Setting of the optimal shape problem. Existence result.**

Up to now we considered the situation when the shape of an elastic body is given. Now we shall assume a family of contact problems. As a (control) parameter we consider the function $\alpha$, describing the shape of the contact part $\Gamma_K$ of the boundary $\partial\Omega$.

Let $C_0$, $C_1$, $C_2$ be given positive constants chosen in such a way that a set $U_{ad}$ defined by

$$U_{ad} = \{\alpha \in C^{0,1}([a, b]) \mid 0 \le \alpha \le C_0,\ |\alpha'(x_1)| \le C_1,\ \operatorname{meas}\Omega(\alpha) = C_2\}, \tag{5.6}$$

is non-empty and define

$$\Omega(\alpha) = \{(x_1, x_2) \in R_2 \mid x_1 \in (a, b),\ \alpha(x_1) \le x_2 \le \gamma\}\ .$$

Let $\hat{\Omega} = (a, b) \times (0, \gamma)$ (see Figure 5.1).

Let us note that $\hat{\Omega} \supset \Omega(\alpha)\ \forall \alpha \in U_{ad}$. Next we shall assume that all physical datas are defined in $\hat{\Omega}$.

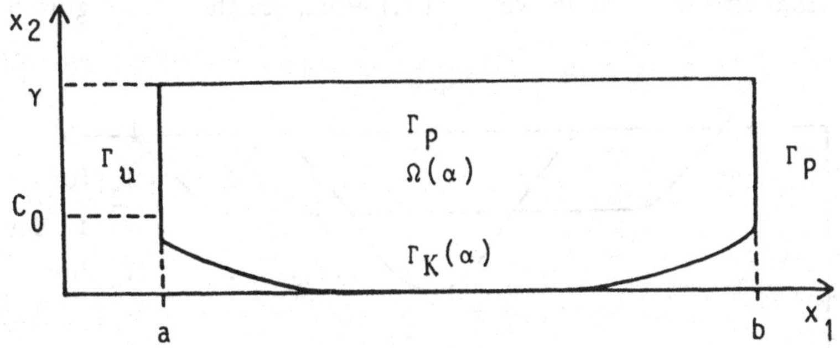


Figure 5.1.

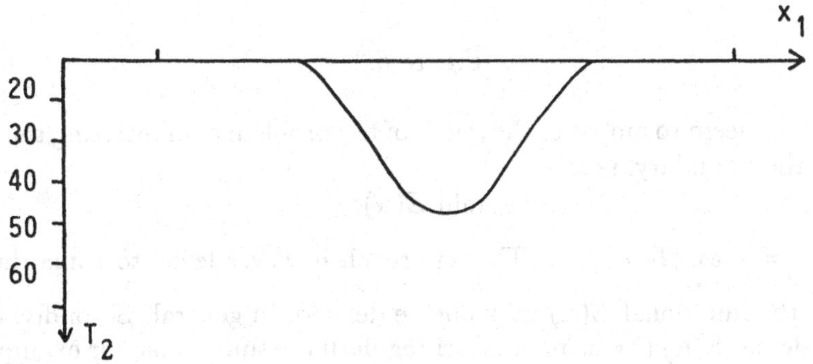


Figure 5.2.

By $V(\alpha)$, $K(\alpha)$ we denote corresponding sets of functions, defined over $\Omega(\alpha)$ and set

$$\begin{aligned}\mathcal{J}_\alpha(v) &\equiv \frac{1}{2}\big(\tau(v), \varepsilon(v)\big)_{\Omega(\alpha)} - (F_i, v_i)_{0,\Omega(\alpha)} - (P_i, v_i)_{0,\Gamma_P} \\ &= \frac{1}{2}\int_{\Omega(\alpha)} \tau_{ij}(v)\varepsilon_{ij}(v)\,dx - \int_{\Omega(\alpha)} F_i v_i\,dx - \int_{\Gamma_P} P_i v_i\,ds\,.\end{aligned}$$

For any $\alpha$ we consider a contact problem $(P(\alpha))$:

$$\begin{cases}\text{find } u(\alpha) \in K(\alpha) \text{ such that} \\ \mathcal{J}_\alpha(u(\alpha)) \le \mathcal{J}_\alpha(v) \quad \forall v \in K(\alpha).\end{cases} \tag{$P(\alpha)$}$$

A physical phenomenon closely connected with the contact problems is the behaviour of the contact stress $T_2(\alpha)$. A typical configuration of the contact stresses along the contact boundary $\Gamma_K(\alpha)$ is given in Figure 5.2.

As the stress peaks are undesirable from the practical point of view it is natural to ask whether it is possible, and how to find such a shape for the contact surface that the normal stresses become evenly distributed, see the joint Figure 5.3.

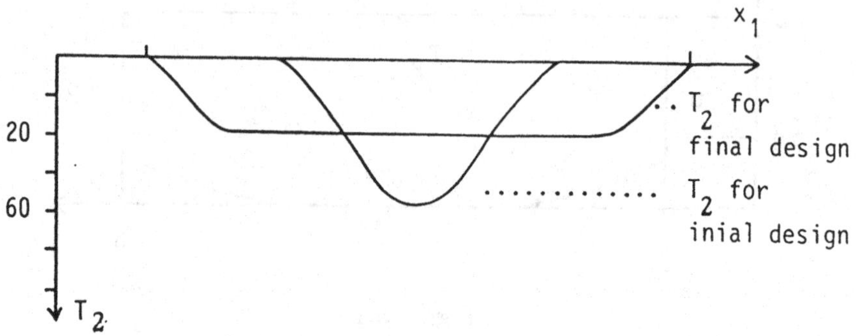


Figure 5.3

This would seem to motivate the study of the problem of minimizing the maximal stresses on the boundary, i.e.

$$\min E(\alpha),$$

where $E(\alpha) = \max_{\Gamma_K(\alpha)} |T_2(\alpha)(x)|$. This approach however leads to some difficulties. First of all, the functional $E(\alpha)$ may not be defined, in general. Secondly, if it were possible to define $E(\alpha)$ (by using a-priori regularity assumptions, for example), then $E(\alpha)$ would not be differentiable, in general.

So another choice of the cost functional is necessary. A good choice (frequently used by engineers) is the total potential energy functional, evaluated in the equilibrium state $u(\alpha)$

$$E(\alpha) = \mathcal{J}_\alpha(u(\alpha)) \ . \tag{5.7}$$

For a suitable $U_{ad}$ it is possible to show that the minimizer of $E(\alpha)$ in $U_{ad}$ yields an "even" distribution of contact forces. Moreover, $E(\alpha)$ is differentiable with respect to $\alpha$ despite the fact that the mapping $\alpha \mapsto u(\alpha)$ is non-differentiable, in general.

So our problem reads:

$$\begin{cases} \text{find } \alpha^* \in U_{ad} \text{ such that} \\ E(\alpha^*) \leq E(\alpha) \quad \forall \alpha \in U_{ad}, \end{cases} \tag{P}$$

with $E(\alpha)$ given by (5.7) and $U_{ad}$ given by (5.6).

For the existence of the solution $\alpha^*$ of (**P**) we have

THEOREM 5.1. *There exists at least one solution of* (**P**).

Before we prove the result, we shall need some auxiliary results.

LEMMA 5.1. *Let $\alpha_n \overrightarrow{\rightarrow} \alpha$ in $[a,b]$ and let $\varphi \in K(\alpha)$, $\varphi = (\varphi_1, \varphi_2)$ be given. Then there exist $\varphi_j \in (H^1(\hat{\Omega}))^2$ and a subsequence $\{\alpha_{n(j)}\} \subset \{\alpha_n\}$ such that $\varphi_j|_{\Omega(\alpha_{n(j)})} \in K(\alpha_{n(j)})$ and $\varphi_j \to \tilde{\varphi}$ in $(H^1(\hat{\Omega}))^2$, where $\tilde{\varphi} = (\tilde{\varphi}_1, \tilde{\varphi}_2)$ denotes the Calderon extension of $\varphi$ from $\Omega(\alpha)$ on $\hat{\Omega}$.*

PROOF: Define a function

$$\psi_2(x_1, x_2) = \max\{\tilde{\varphi}_2(x_1, x_2), -x_2\}, \quad (x_1, x_2) \in \hat{\Omega}\,.$$

We find that $\psi_2 \in H^1(\hat{\Omega})$, $\psi_2|_{\Gamma_u} = 0$ and

$$\psi_2(x_1, \alpha(x_1)) = \varphi_2(x_1, \alpha(x_1)) \geq -\alpha(x_1),$$

i.e. $\psi|_{\Omega(\alpha)} \in K(\alpha)$, where $\psi = (\psi_1, \psi_2)$ with $\psi_1 = \tilde{\varphi}_1$. Let us split $\tilde{\varphi}$ as follows: $\tilde{\varphi} = \psi + \Phi$. From the construction of $\psi$ we see that

$$\Phi_1|_{\Gamma_u} = \Phi_2|_{\Gamma_u} = \Phi_2|_{\Gamma_K(\alpha)} = 0, \quad \Phi = (\Phi_1, \Phi_2).$$

Applying the classical density result, one can find a sequence $\{\Phi_j\}$, $\Phi_j = (\Phi_{1j}, \Phi_{2j}) \in (C^\infty(\overline{\hat{\Omega}}))^2$ with $\Phi_{1j}, \Phi_{2j}$ vanishing in a neighbourhood of $\overline{\Gamma}_u, \overline{\Gamma_u \cup \Gamma_K(\alpha)}$, respectively, and such that

$$\Phi_j \to \Phi \ \text{ in } (H^1(\hat{\Omega}))^2.$$

Let us define now:

$$\varphi_j = \psi + \Phi_j.$$

Then

$$\|\tilde{\varphi} - \varphi_j\|_{1,\hat{\Omega}} = \|\Phi - \Phi_j\|_{1,\hat{\Omega}} \to 0, \quad j \to \infty.$$

Let us denote $d_j \equiv \operatorname{dist}\{\operatorname{supp}\Phi_{2j}, \overline{\Gamma_u \cup \Gamma_K(\alpha)}\}$ and let $j_0$ be fixed. As $\alpha_j \overrightarrow{\rightarrow} \alpha$ in $[a,b]$, then there exists $n_0 = n(j_0)$ such that the graph of the function $\alpha_{n_0}$ does not meet $\operatorname{supp}\Phi_{2j_0}$. As

$$\varphi_{2j_0} = \psi_2 \quad \text{on} \quad \hat{\Omega} \setminus \operatorname{supp}\Phi_{2j_0}$$

and

$$\psi_2(x_1, x_2) \geq -x_2 \quad \forall (x_1, x_2) \in \hat{\Omega}\,,$$

(as follows from the definition of $\psi_2$) we immediately get

$$\psi_2(x_1, \alpha_{n_0}(x_1)) \geq -\alpha_{n_0}(x_1).$$

Hence

$$\varphi_{j_0}|_{\Omega(\alpha_{n_0})} \in K(\alpha_{n_0})\,.$$

LEMMA 5.2. *Let $\alpha_n \rightrightarrows \alpha$ in $[a,b]$ and let $u_n \in K(\alpha_n)$ be the solution of $P(\alpha_n)$, i.e.*[1]

$$\begin{aligned}&(\tau(u_n),\varepsilon(z-u_n))_{\Omega(\alpha_n)}\\ &\geq (F,z-u_n)_{0,\Omega(\alpha_n)} + (P,z-u_n)_{0,\Gamma_P} \qquad \forall z \in K(\alpha_n)\,.\end{aligned} \tag{5.8}$$

*Then there exists a subsequence of $\{u_n\}$, denoted again by $\{u_n\}$ such that*

$$u_n \rightharpoonup u \qquad \text{in } (H^1(G_m))^2$$

*for any integer $m$, with $G_m = \{(x_1,x_2) \in \mathbf{R}^2 \mid x_1 \in (a,b),\ \alpha(x_1) + \frac{1}{m} < x_2 < \gamma\}$, where $u \in K(\alpha)$ is a solution of $(P(\alpha))$.*

PROOF: As $F \in (L^2(\hat{\Omega}))^2$, $P \in (L^2(\Gamma_P))^2$, from (5.8) and Korn's inequality it follows that[2]

$$\|u_n\|_{1,\Omega(\alpha_n)} \leq C \qquad \forall n\,. \tag{5.9}$$

Let $m$ be fixed and let $G_m$ be given as above. As $\alpha_n \rightrightarrows \alpha$ in $[a,b]$, there exists $n_0(m)$ such that

$$\overline{G_m} \subset \Omega(\alpha_n) \qquad \forall n \geq n_0(m)\,.$$

From this and (5.9) it follows

$$\|u_n\|_{1,G_m} \leq \|u_n\|_{1,\Omega(\alpha_n)} \leq C\,. \tag{5.10}$$

Hence, one may extract a subsequence $\{u_{n_1}\} \subset \{u_n\}$ such that

$$u_{n_1} \rightharpoonup u^{(m)} \qquad \text{in } (H^1(G_m))^2\,.$$

A function $u^{(m)} \in (H^1(G_m))^2$ is such that $u^{(m)} = 0$ on $\Gamma_u \cap \partial G_m$. Similarly, as $G_{m+1} \supset \overline{G_m}$, there exists $n_0(m+1) \geq n_0(m)$ such that

$$\overline{G_{m+1}} \subset \Omega(\alpha_n) \qquad \forall n \geq n_0(m+1)\,.$$

Hence a subsequence $\{u_{n_2}\} \subset \{u_{n_1}\}$ can be chosen in such a way that

$$u_{n_2} \rightharpoonup u^{(m+1)} \qquad \text{in } (H^1(G_{m+1}))^2\,,$$

where $u^{(m+1)} \in (H^1(G_{m+1}))^2$ and $u^{(m+1)} = 0$ on $\Gamma_u \cap \partial G_{m+1}$. Clearly

$$u^{(m+1)} = u^{(m)} \qquad \text{in } G_m\,.$$

---

[1] We assume that $F \in [L^2(\hat{\Omega})]^2$, $P \in [L^2(\partial\hat{\Omega})]^2$ and $(F,v)_{0,\Omega(\alpha)} \equiv (F_i,v_i)_{0,\Omega(\alpha)}$, $(P,v)_{0,\Gamma_P} \equiv (P_i,v_i)_{0,\Gamma_P}$.

[2] Here we use the fact that the constant in Korn's inequality can be chosen independently of $\alpha_n \in U_{ad}$ (see [11]).

Proceeding in this way for any integer $k$, one can construct a subsequence $\{u_{n_k}\} \subset \{u_{n_{k-1}}\}$ such that

$$u_{n_k} \rightharpoonup u^{(m+k-1)} \qquad \text{in } (H^1(G_{m+k-1}))^2 \ ,$$

with $u^{(m+k-1)} \in (H^1(G_{m+k-1}))^2$, $u^{(m+k-1)} = 0$ on $\Gamma_u \cap \partial G_{m+k-1}$ and

$$u^{(m+k-1)} = u^{(j)} \qquad \text{in } G_j$$

for any $j < k+m-1$. Let $\{u_n^D\}$ be a diagonal sequence determined by $\{u_{n_k}\}$. From the above analysis it follows that

$$u_n^D \rightharpoonup u|_{G_m} \qquad \text{in } (H^1(G_m))^2 \tag{5.11}$$

for any integer $m$, where

$$u|_{G_m} \equiv u^{(m)} \qquad \text{in } G_m \ .$$

Clearly, $u \in (H^1(\Omega(\alpha)))^2$, $u = 0$ on $\partial\Omega(\alpha) \cap \Gamma_u$. Instead of $\{u_n^D\}$, $\{\alpha_n^D\}$, we shall next write simply $\{u_n\}$, $\{\alpha_n\}$. As $u_n \in K(\alpha_n)$, $\alpha_n \overrightarrow{\rightarrow} \alpha$ in $[a, b]$, (5.9) and (5.11) hold, a function $u \in K(\alpha)$. It remains to verify that $u = u(\alpha)$ is a solution of $(P(\alpha))$.

Let $\xi \in K(\alpha)$ be an arbitrary function. Let $\xi_j \in K(\alpha_{n_j})$ be such that

$$\xi_j \to \tilde{\xi} \qquad \text{in } (H^1(\hat{\Omega}))^2 \tag{5.12}$$

($\tilde{\xi}$ is the Calderon extension of $\xi$ on $\hat{\Omega}$). The existence of such a sequence is ensured by Lemma 5.1. Let $u_{n_j} \equiv u(\alpha_{n_j})$ be the corresponding solution of $(P(\alpha_{n_j}))$:

$$\begin{aligned} &\big(\tau(u_{n_j}), \varepsilon(\xi_j - u_{n_j})\big)_{\Omega(\alpha_{n_j})} \\ &\geq \big(F, \xi_j - u_{n_j}\big)_{0,\Omega(\alpha_{n_j})} + \big(P, \xi_j - u_{n_j}\big)_{0,\Gamma_P} \ . \end{aligned} \tag{5.13}$$

In (5.13), a function $\xi_j \in K(\alpha_{n_j})$ satisfying (5.12) is assumed. Now

$$\begin{aligned} &\big(\tau(u_{n_j}), \varepsilon(\xi_j - u_{n_j})\big)_{\Omega(\alpha_{n_j})} \\ &= \big(\tau(u_{n_j}), \varepsilon(\xi_j - u_{n_j})\big)_{G_m} + \big(\tau(u_{n_j}), \varepsilon(\xi_j - u_{n_j})\big)_{\Omega(\alpha_{n_j}) \setminus \Omega(\alpha)} \\ &\quad + \big(\tau(u_{n_j}), \varepsilon(\xi_j - u_{n_j})\big)_{(\Omega(\alpha) \setminus G_m) \cap \Omega(\alpha_{n_j})} \\ &\leq \big(\tau(u_{n_j}), \varepsilon(\xi_j - u_{n_j})\big)_{G_m} + \big(\tau(u_{n_j}), \varepsilon(\xi_j)\big)_{\Omega(\alpha_{n_j}) \setminus \Omega(\alpha)} \\ &\quad + \big(\tau(\tilde{u}_{n_j}), \varepsilon(\xi_j)\big)_{\Omega(\alpha) \setminus G_m} \ . \end{aligned}$$

From this, (5.10), (5.11) and (5.12) we easily get[1]

$$\begin{aligned}&\limsup_{j\to\infty}\ \big(\tau(u_{n_j}),\varepsilon(\xi_j-u_{n_j})\big)_{\Omega(\alpha_{n_j})}\\ &\le (\tau(u),\varepsilon(\xi-u))_{G_m}+C\,\|\xi\|_{1,\Omega(\alpha)\setminus G_m}\,.\end{aligned}\tag{5.14}$$

Similarly,

$$\begin{aligned}&\big(F,\xi_j-u_{n_j}\big)_{0,\Omega(\alpha_{n_j})}\\ &=\big(F,\xi_j-u_{n_j}\big)_{0,G_m}+\big(F,\xi_j-u_{n_j}\big)_{0,\Omega(\alpha_{n_j})\setminus\Omega(\alpha)}\\ &\quad+\big(F,\xi_j-u_{n_j}\big)_{0,(\Omega(\alpha)\setminus G_m)\cap\Omega(\alpha_{n_j})}\,.\end{aligned}$$

From this, (5.10), (5.11) and (5.12) we get

$$\begin{aligned}&\liminf_{j\to\infty}\ \big(F,\xi_j-u_{n_j}\big)_{0,\Omega(\alpha_{n_j})}\\ &\ge (F,\xi-u)_{0,G_m}-C(\|F\|_{0,\Omega(\alpha)\setminus G_m}+\|\xi\|_{0,\Omega(\alpha)\setminus G_m})\,.\end{aligned}\tag{5.15}$$

Finally,

$$\big(P,\xi_j-u_{n_j}\big)_{0,\Gamma_P}=\big(P,\xi_j-u_{n_j}\big)_{0,\Gamma_P\setminus M_m}+\big(P,\xi_j-u_{n_j}\big)_{0,M_m}\,,$$

where the one-dimensional Lebesgue measure of

$$M_m=\{(x_1,x_2)\in R_2\mid x_2\in(\alpha(a),\alpha(a)+\frac{1}{m}),\ x_2\in(\alpha(b),\alpha(b)+\frac{1}{m})\}$$

is $1/m$ (see Figure 5.1) (this consideration can be omitted if $\operatorname{dist}(\overline{\Gamma_P},\overline{\Gamma_K(\alpha)})>0$). Then

$$\begin{aligned}&\liminf_{j\to\infty}\ \big(P,\xi_j-u_{n_j}\big)_{0,\Gamma_P}\\ &\ge (P,\xi-u)_{0,\Gamma_P\setminus M_m}-C(\|P\|_{0,M_m}+\|\xi\|_{0,M_m})\,.\end{aligned}\tag{5.16}$$

Combining (5.14), (5.15) and (5.16) we have

$$\begin{aligned}&(\tau(u),\varepsilon(\xi-u))_{G_m}+C\,\|\xi\|_{1,\Omega(\alpha)\setminus G_m}\\ &\ge (F,\xi-u)_{0,G_m}+(P,\xi-u)_{0,\Gamma_P\setminus M_m}\\ &\quad-C(\|F\|_{0,\Omega(\alpha)\setminus G_m}+\|\xi\|_{0,\Omega(\alpha)\setminus G_m})-C(\|P\|_{0,M_m}+\|\xi\|_{0,M_m})\,.\end{aligned}$$

Letting $m\to\infty$ we obtain

$$\begin{aligned}&(\tau(u),\varepsilon(\xi-u))_{\Omega(\alpha)}\\ &\ge (F,\xi-u)_{0,\Omega(\alpha)}+(P,\xi-u)_{0,\Gamma_P}\qquad\forall\xi\in K(\alpha)\,,\end{aligned}\tag{5.17}$$

[1] Here we use the uniform extension property of $\{\Omega(\alpha)\}$, $\alpha\in U_{ad}$.

i.e. $u$ is a solution of $(P(\alpha))$.

PROOF OF THEOREM 5.1: Let us denote

$$q = \inf_{\alpha \in U_{ad}} E(\alpha)$$

and let $\alpha_n \in U_{ad}$ be a minimizing sequence, i.e.

$$q = \lim_{n\to\infty} E(\alpha_n) \ .$$

Using the compactness of $U_{ad}$, we can extract a subsequence $\{\alpha_{n_j}\} \subset \{\alpha_n\}$ such that

$$\alpha_{n_j} \rightrightarrows \alpha^* \qquad \text{in } [a,b] \ . \tag{5.18}$$

Clearly $\alpha^* \in U_{ad}$. Let $u_{n_j}(\alpha_{n_j}) \in K(\alpha_{n_j})$ be solutions of $(P(\alpha_{n_j}))$. From Lemma 5.2 it follows immediately that there exists a subsequence of $\{u_{n_j}(\alpha_{n_j})\}$ (denoted again by $\{u_{n_j}(\alpha_{n_j})\}$) such that

$$u_{n_j}(\alpha_{n_j}) \rightharpoonup u(\alpha^*) \qquad \text{in } (H^1(G_m(\alpha^*)))^2 \tag{5.19}$$

for any $m$, where

$$G_m(\alpha^*) = \{(x_1,x_2) \in \Omega(\alpha^*) \mid x_2 \geq \alpha^*(x_1) + \frac{1}{m},\ x_1 \in (a,b)\}$$

and $u(\alpha^*) \in K(\alpha^*)$ is the solution of $(P(\alpha^*))$. From this it follows that

$$E(\alpha^*) \geq q \ . \tag{5.20}$$

On the other hand, for $m$ fixed and $n_j$ sufficiently large,

$$E(\alpha_{n_j}) = E_{G_m(\alpha^*)}(\alpha_{n_j}) + E_{\Omega(\alpha_{n_j})\setminus G_m(\alpha^*)}(\alpha_{n_j}) \ ,$$

where

$$\begin{aligned} E_{G_m(\alpha^*)}(\alpha_{n_j}) =& \frac{1}{2}\left(\tau(u_{n_j}),\varepsilon(u_{n_j})\right)_{G_m(\alpha^*)} \\ & - \left(F,u_{n_j}\right)_{0,G_m(\alpha^*)} - \left(P,u_{n_j}\right)_{0,\Gamma_P\setminus M_m} \end{aligned}$$

and

$$\begin{aligned} E_{\Omega(\alpha_{n_j})\setminus G_m(\alpha^*)}(\alpha_{n_j}) &= \frac{1}{2}\left(\tau(u_{n_j}),\varepsilon(u_{n_j})\right)_{\Omega(\alpha_{n_j})\setminus G_m(\alpha^*)} \\ &\quad - \left(F,u_{n_j}\right)_{0,\Omega(\alpha_{n_j})\setminus G_m(\alpha^*)} - \left(P,u_{n_j}\right)_{0,M_m} \\ &\geq - \left(F,u_{n_j}\right)_{0,\Omega(\alpha_{n_j})\setminus G_m(\alpha^*)} - \left(P,u_{n_j}\right)_{0,M_m} \ , \end{aligned}$$

where the one-dimensional Lebesgue measure of $M_m$ is $1/m$ (see Figure 5.1) (with eventual modification if $\operatorname{dist}(\overline{\Gamma_P}, \overline{\Gamma_K(\alpha)}) > 0$). Then

$$\begin{aligned}
&\liminf_{j\to\infty} E(\alpha_{n_j}) \\
&\geq \liminf_{j\to\infty} E_{G_m(\alpha^*)}(\alpha_{n_j}) + \liminf_{j\to\infty} E_{\Omega(\alpha_{n_j})\setminus G_m(\alpha^*)}(\alpha_{n_j}) \\
&\geq E_{G_m(\alpha^*)}(\alpha^*) + \liminf_{j\to\infty} \left( -\left(F, u_{n_j}\right)_{0,\Omega(\alpha_{n_j})\setminus G_m(\alpha^*)} - \left(P, u_{n_j}\right)_{0,M_m} \right) \qquad (5.21)\\
&\geq E_{G_m(\alpha^*)}(\alpha^*) - C\left( \|F\|_{0,\Omega(\alpha^*)\setminus G_m(\alpha^*)} + \|P\|_{0,M_m} \right)
\end{aligned}$$

holds for any $m$. Here

$$E_{G_m(\alpha^*)}(\alpha^*) = \frac{1}{2}\left(\tau(u), \varepsilon(u)\right)_{G_m(\alpha^*)} - (F, u)_{0,G_m(\alpha^*)} - (P, u)_{0,\Gamma_P\setminus M_m} .$$

The lower weak semicontinuity of $E_{G_m(\alpha^*)}$, (5.19), and (5.10) have been used. Letting $m \to \infty$, we have from (5.21):

$$q = \liminf_{j\to\infty} E(\alpha_{n_j}) \geq E(\alpha^*) .$$

This, together with (5.20), gives the assertion of Theorem 5.1.

REFERENCES:


1. Ekeland, I. and R. Temam, "Analyse convexe et problèmes variationnels," Dunod, Gauthier-Villars, Paris-Bruxelles-Montrèal, 1974.
2. Haslinger, J., *Mixed formulation of variational inequalities and its approximation*, Apl. Mat. **26**, 462–475.
3. Nečas, J. and I. Hlaváček, "Mathematical theory of elastic and elasto-plastic bodies: an introduction," Elsevier Scientific Publishing Company, Amsterdam-Oxford-New York, 1981.
4. Nečas, J., J. Jarusèk and J. Haslinger, *On the solution of the variational equality to the Signorini problem with small friction*, Boll. Unione Mat. Ital. (5) **17-B**, 796–811.
5. Nitzsche, J.A., *Über ein Variationsprinzip zur Lösung von Dirichlet-Problemen bei Verwendung von Teilräumen, die keinen Randbedingungen unterworfen sind*, Abh. Math. Sem. Univ. Hamburg **36**, 9–15.
6. Fix G. and G. Strang, "An analysis of the finite element method," Prentice-Hall, Englewood Cliffs.
7. Hlaváček, I. and J. Lovíšek, *A finite element analysis for the Signorini problem in plane elastostatics*, Apl. Mat. **22** (1977), 215–228.
8. Haslinger, J. and I. Hlaváček, *Approximation of the Signorini problem with friction by a mixed finite element method*, JMAA, No 1 **86** (1982), 99–121.

9. Hlaváček, I., J. Haslinger, J. Nečas and J. Lovíšek, "Solution of variational inequalities in mechanics," to appear in Springer-Verlag, 1988.
10. Haslinger, J. and P. Neittaanmäki, "Optimal shape design in systems governed by elliptic variational inequalities," to appear in John Wiley & Sons.
11. Haslinger, J., P. Neittaanmäki and T. Tiihonen, *Shape optimization of an elastic body in contact based on penalization of the state inequality*, Apl. Mat. **31** (1986), 54–77.

# DISCONTINUITIES AND PLASTICITY

**P.M. Suquet**
**Université des Sciences et Techniques du Languedoc, Montepellier, France**

# 1. INTRODUCTION

The guide line of these lectures is that Plasticity and spatial discontinuities are two companion phenomena :

a) on the one hand displacements (or displacement rates) solutions of Plasticity problems are naturally discontinuous. This conclusion is easily drawn from the asymptotically linear growth of the functionals arising in the variational formulation of Plasticity problems :

$$\operatorname{Inf}_u J(u) = \int_\Omega j(\varepsilon(u))dx - L(u) \qquad (1.1)$$

where $\varepsilon$ stands for the usual deformation operator (infinitesimal strains), $L$ is a linear form and $j$ satisfies :

$$j(e) \geq k_0|e| - k_1 \qquad (1.2)$$

Accounting for (1.2) the natural space of definition for $J$ contains discontinuous vector fields. But this is only *one* of the difficulties arising from the variational problems (1.1) . Another one is the non lower semi continuity of functionals, which has for main consequence a *relaxation procedure* necessary to obtain well posed variational problems. Our presentation of this subject follows and extends TEMAM & STRANG's [1] original work.

b) on the other hand it is known from experimental observation that Plasticity takes its source in microscopic slips along specific cristallographic planes. To give a more mathematical basis to this assertion we examine a model of rigid blocks slipping ones over others. The blocks size is assumed to be a small parameter $1/n$ , and the problem can be described by minimizing a functional $J^n$ . In order to derive the *homogenized problem* (when $1/n$ tends to 0), we use the theory of *epi-convergence* which allows to compute the limits of variational problems. We shall point out a few singular aspects of epi-convergence in Plasticity mainly arising from (roughly speaking) the *non commutativity of relaxation and epi convergence.*

c) located midway between these two aspects of "Plasticity and discontinuities" the problems of *thin plastic layers* will be studied in these notes.

Our aim is to bring *attention of readers* on recent mathematical developments which could open significantly wide fields of investigation for researchers interested in mathematical aspects of Mechanics. Therefore detailled proofs are often omitted, but conjectures and theorems are, as far as possible, distinguished. The author's personal contribution in theoretical developments is most of the time rather limited, but the goal is to attract readers to a *mechanical approach* of TEMAM's or ATTOUCH's books [2] [3] , to BOUCHITTE's thesis [4] , or to recent works of the Italian School. The background necessary to follow the text includes a basic knowledge of convex analysis, of measure theory and of classical functional analysis, and can be found, together with numerous other subjects, in PANAGIOTOPOULOS [9] .

## ACKNOWLEDGMENTS

Several discussions with Guy BOUCHITTE and numerous uses of his results are gratefully acknowledged.

# 2. LIMIT ANALYSIS

## 2.1 STATICAL AND KINEMATICAL FORMULATIONS

In all further developments we consider that the material under consideration occupies a domain $\Omega$ of $\mathbb{R}^3$ and is submitted to body forces $\lambda f$ and boundary forces $\lambda g$ on a part $\Gamma_1$ of its boundary $\partial\Omega$. $\Gamma_o$, complementary of $\Gamma_1$ in $\partial\Omega$, will be submitted to imposed displacements.

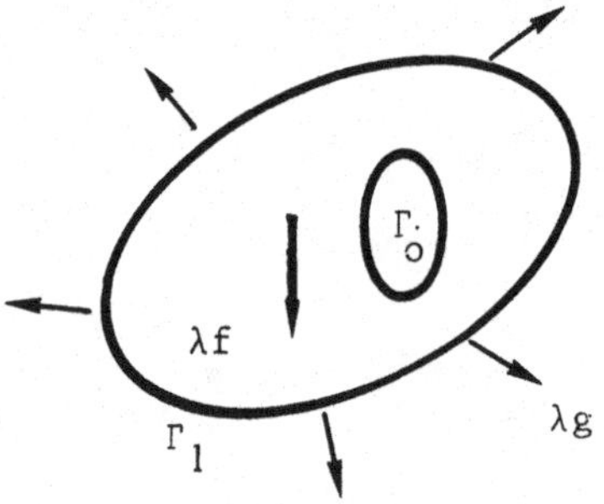


- Figure 2.1 -

Throughout the text we shall assume that the following assumption is met

$$\left.\begin{array}{ll}(1) & \partial\Omega \text{ is } C^1 .\ \Gamma_o \text{ and } \Gamma_1 \text{ are closed in } \partial\Omega ,\\ & \Gamma_o \cap \Gamma_1 = \emptyset ,\ \Gamma_o \cup \Gamma_1 = \partial\Omega \\ (2) & f \in L^\infty(\Omega)^3 ,\quad g \in L^\infty(\Gamma_1)^3 \end{array}\right\} \qquad (2.1)$$

Assumption (1) can be weakened but we do not seak here for minimal hypothesis.

In this section the only information used on the constitutive law of $\Omega$ is its strength capacity defined, at every particle $x$ of $\Omega$, through a set $P(x)$ in $\mathbb{R}^9_s$ (+) delimiting allowable stress tensors $\sigma(x)$

(+) $\mathbb{R}^9_s$ : 3 × 3 symmetric tensors

$$\sigma \text{ admissible} \iff \sigma(x) \in P(x) \quad \text{a.e.} \quad x \text{ in } \Omega \tag{2.2}$$

Throughout the following we shall assume that :

$$\begin{array}{l} P(x) \text{ is a closed convex set of } \mathbb{R}^9_s, \text{ containing} \\ 0 \text{ in its interior, and symmetric with respect to } 0 \end{array} \tag{2.3}$$

Although physically relevant examples of $P(x)$ are given by Tresca or Von Mises criterions we shall tacitly assume (unless explicitely specified) that $P(x)$ is a ball of radius $k(x)$ :

$$P(x) = \{\sigma \in \mathbb{R}^9_s \quad |\sigma| \leq k(x)\} \quad (0 < k_o \leq k(x) \leq k_1) \tag{2.4}$$

We emphasize that the knowledge of $P(x)$ does not presume of the remaining material properties (elasticity, viscosity...) since several different materials, exhibiting different behaviors, might have the same strength capacity (cf Figure 2.2) and therefore should be considered as identical in this section.

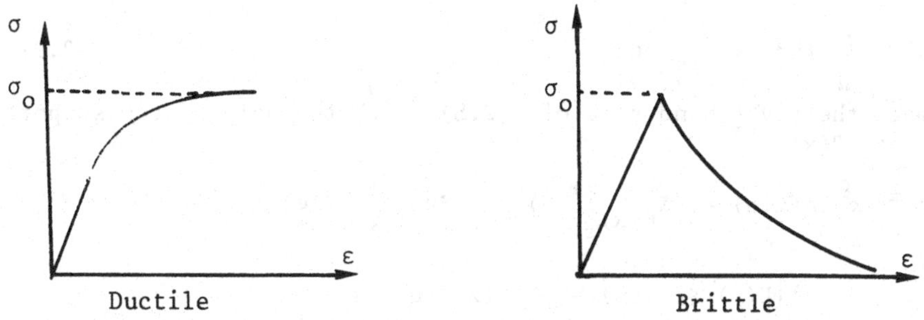


*Two different materials (ductile and brittle) with the same strength capacity illustrated by the ultimate stress* $\sigma_o$ .

- Figure 2.2 -

The equilibrium equations, independent of the material properties, read as

$$\operatorname{div} \sigma + \lambda f = 0 \text{ in } \Omega \quad , \quad \sigma.n = \lambda g \text{ on } \Gamma_1 \tag{2.5}$$

and the space of statically admissible stress fields is defined as :

$$S(\lambda f , \lambda g) = \{\sigma \in \mathbb{L}^2(\Omega) , \ \sigma \text{ satisfies } (2.5)\} , \text{ where } \mathbb{L}^p(\Omega) = L^p(\Omega)^9_s .$$

The central problem of *limit analysis* (+) is to find the maximum value of the loading parameter $\lambda$ for which a stress field $\sigma$ satisfying both material constraints (2.2) and equilibrium equations (2.5) can be found [5]

$$\lambda^s = \text{Sup}\{\lambda \in \mathbb{R} \,|\, \exists \sigma \in S(\lambda f, \lambda g) \;,\; \sigma(x) \in P(x) \text{ a.e. } x \text{ in } \Omega\}. \quad (2.6)$$

$\lambda^s$ is the limit load, and (2.6) is its statical definition, since it relies on stress fields $\sigma$ only. A dual definition of the limit load, involving only kinematical fields is classically given [5] . Let $\lambda < \lambda^s$ , $\sigma$ a statically and plastically admissible stress field associated with $\lambda$ by (2.6) , and let $u$ be any kinematically admissible displacement rate field

$$u \in V = \{v \in H^1(\Omega)^3 \;,\; v = 0 \text{ on } \Gamma_0\} \;. \quad (2.7)$$

The principle of virtual work, elsewhere called Green's formula, yields

$$\lambda\, L(u) = \int_\Omega \sigma\, \varepsilon(u) dx \quad (2.8)$$

where $L(u)$ is related (multiplied by $\lambda$) to the potential of given forces

$$L(u) = \int_\Omega f u dx + \int_{\Gamma_1} g u ds \quad (2.9)$$

We estimate the right hand side of (2.8) with the help of the support function of $P(x)$ (++)

$$\forall\, e \in \mathbb{R}^9_s \,, \pi(x,e) = (I_{P(x)})^*(e) = \sup_{\tau \in P(x)} (\tau . e) \;,$$

$$\lambda |L(u)| \leq \Pi(u) = \int_\Omega \pi(x, \varepsilon(u)) dx$$

i.e.

$$\lambda^s \leq \lambda^k = \inf_{u \in V} \frac{\Pi(u)}{|L(u)|} \;. \quad (2.10)$$

Noting that $\Pi$ is positively homogeneous of degree 1 (consequence of the same property valid for $\pi$) and that $\Pi$ is even (consequence of assumption (2.3)) we obtain an alternative definition of the *kinematical limit load* $\lambda^k$

$$\lambda^k = \text{Inf}\{\Pi(u) \;,\; u \in V \;,\; L(u) = 1\}. \quad (2.11)$$

The inequality $\lambda^s \leq \lambda^k$ is a simple consequence of the derivation of $\lambda^k$

---

(+) also called *yield design* when no reference to Plasticity is made (cf [5]).

(++) $I_X$ is the indicator function of the set $X$ . $f^*$ is the conjugate of $f$ in the appropriate duality.

(see (2.10)) . The equality $\lambda^s = \lambda^k$ is a more difficult result, extensively discussed in the mechanical literature (see for instance NAYROLES [6]) and to which we devote the next section.

To close these preliminaries we sum up basic properties of the functional $\Pi$

$$\Pi(\alpha u) = |\alpha| \Pi(u) , \tag{2.12}$$

$$\Pi(u_1 + u_2) \leq \Pi(u_1) + \Pi(u_2) , \tag{2.13}$$

and under assumption (2.4)

$$k_o |\varepsilon(u)|_{\mathbb{L}^1(\Omega)} \leq \Pi(u) \leq k_1 |\varepsilon(u)|_{\mathbb{L}^1(\Omega)} . \tag{2.14}$$

(2.12)(2.14) are immediate, (2.13) is a consequence of (2.11) and of the convexity of $\Pi$ .

## 2.2 HOMOGENEOUS MATERIAL. DUALITY RESULT

The following theorem is due to TEMAM-STRANG [1] , in a more general setting since assumption (2.4) simplifies several technical points of [1] .

*Theorem 2.1 (TEMAM-STRANG). We assume that assumptions* (2.1)(2.3)(2.4) *are met. Then*

$$\lambda^s(\mathrm{Sup}(2.6)) = \lambda^k(\mathrm{Inf}(2.11)) \tag{2.15}$$

*Proof.* Theorem 2.1 results from a general argument of Convex Analysis [7] [8] [9] .

<u>Abstract result</u>. Let V and Y be two Banach spaces et $\Lambda$ be a linear mapping from V into Y . We consider the following variational problem (P) and its dual $(P^*)$

$$(P) \quad \operatorname*{Inf}_{v \in V} [F(v) + G(\Lambda v)]$$

$$(P^*) \quad \operatorname*{Sup}_{p^* \in Y^*} [- F^*(\Lambda^* p^*) - G^*(-p^*)]$$

where F(resp. G) are convex, lower semi continuous (l.s.c.) , proper functions from V (resp. Y) into $\mathbb{R} \cup \{+\infty\}$ . Assume that

i) $\mathrm{Inf}(P) \in \mathbb{R}$

ii) $(\exists u_o \in V)\ F(u_o) < +\infty$ , G continuous at $\Lambda u_o$ .

Then

iii) $\mathrm{Inf}(P) = \mathrm{Sup}(P^*)$ and iv) $(P^*)$ possesses at least one solution.

This abstract result is applied to limit analysis, specifying V , Y , $\Lambda$ , F , G as follows :

V given by (2.7) , $Y = \mathbb{L}^2(\Omega)$ , $\Lambda = \varepsilon$

$F(v) = 0$ if $L(v) = 1$ , $+\infty$ otherwise,

$G(p) = \int_\Omega \pi(p)dx$ .

With these notations (P) is exactly (2.11) . It can be checked that[(+)]

$$F^*(\Lambda^* p^*) = \begin{cases} -\lambda \text{ if } \exists \ \lambda \in \mathbb{R} \text{ such that } -p^* \in S(\lambda f, \lambda g), \\ +\infty \text{ otherwise}, \end{cases}$$

$$G^*(p^*) = \begin{cases} 0 \text{ if } p^*(x) \in P \quad \text{a.e. } x \text{ in } \Omega, \\ +\infty \text{ otherwise}. \end{cases}$$

Therefore $(P^*)$ reads as

$$-\operatorname{Sup}\{-\lambda \mid \exists \sigma(=-p^*) \in S(\lambda f, \lambda g), \ \sigma(x) \in P \text{ a.e. } x \text{ in } \Omega\}$$

which is exactly (2.6) . Since G is everywhere continuous and finite on Y , since Inf P is obviously finite, general conclusions iii) and iv) of the abstract result apply and complete the proof of theorem 2.1 .

The kinematical approach (2.11) is more convenient for a numerical computation of the limit load $\lambda^s = \lambda^k$ , since most of finite elements approximations are based on displacements. For this purpose any descent method will provide a minimizing sequence and one has to check that this procedure eventually converges to a minimizer of (2.11) , the existence of which is *not* ensured by theorem 2.1 . A more thorough study of (2.11) is therefore required, and we recall the most classical arguments used to obtain solutions of minimization problems in the form

$$\operatorname*{Inf}_{x \in K} \Phi(x) \tag{2.16}$$

where K is a convex subset of a topological vector space X . Usual assumptions are

i) X is the dual of a Banach space Y (X is often assumed to be reflexive).

ii) K is closed for the weak * topology of X .

iii) $\Phi$ is convex, proper, l.s.c. for the weak * topology of X .

iv) K is bounded or

$$\lim_{x \in K, \ |x| \to +\infty} \Phi(x) = +\infty \tag{2.17}$$

By experience we know that the choice of X is often governed by the growth condition (2.17) . The program of study for the variational problem (2.16) in a specific situation is therefore the following :

(+) a similar reasoning will be detailed in section 2.4 .

a) Find the optimal space $X$ in order to ensure iv) : in the case of limit analysis the answer is provided by $BD(\Omega)$ to which section 2.3 is devoted. Check i).
b) Check ii) and iii) : in the case of limit analysis $K$ *is not closed for the weak ★ topology of* $X$, and we need to introduce the l.s.c. hull of $\Phi + I_K$ through a so-called *relaxed problem.*

## 2.3 VECTOR FIELDS WITH BOUNDED DEFORMATION : $BD(\Omega)$

A first guess for $X$, suggested by the growth condition (2.14) would be that it consists of vector fields $v$ with a deformation $\varepsilon(v)$ in $\mathbb{L}^1(\Omega)$. However, since $L^1(\Omega)$ is not the dual of a Banach space (and of course not the dual of $L^\infty(\Omega)$), such a first guess fails.

The correct functional framework, introduced independently by G. STRANG [10] and the author [11] [12], is

$$BD(\Omega) = \{u \in L^1(\Omega)^N, \quad \varepsilon_{ij}(u) \in M^1(\Omega), \quad 1 \leqslant i, j \leqslant N\}^{(+)} \tag{2.18}$$

where $M^1(\Omega)$ stands for the space of bounded measures on $\Omega$, dual of $C_c^o(\Omega)$ space of continuous functions on $\Omega$ vanishing on $\partial\Omega$ [13]. One could naturally ask whether $BD(\Omega)$ is equal to $BV(\Omega)^N$ or not : the answer is negative as proved by MATTHIES & al [14] who established that Korn's inequality fails in $W^{1,1}(\Omega)^N$. This result motivated further studies of $BD(\Omega)$ and we shall briefly(++) sum up a few results on $BD(\Omega)$ which help to understand the (non)regularity of possible solutions of Limit Analysis and Plasticity equations.

1) <u>Structure</u> $BD(\Omega)$ endowed with the following norms is a Banach space

$$|u|_{BD(\Omega)} = \sum_{i=1}^N |u_i|_{L^1(\Omega)} + \sum_{i,j=1}^N |\varepsilon_{ij}(u)|_{M^1(\Omega)},$$

or equivalently [2]

$$|u|_{BD(\Omega)} = r(u) + \sum_{i,j=1}^N |\varepsilon_{ij}(u)|_{M^1(\Omega)}, \tag{2.19}$$

where $r(u)$ is any continuous semi-norm on $BD(\Omega)$ strictly positive on rigid displacements : $r(u) > 0$ if $u = a + b \wedge x$ $a,b \in \mathbb{R}^3$, $u \neq 0$.

$BD(\Omega)$ *is not a reflexive Banach space* (similarly to $M^1(\Omega)$) however an important result due to MATTHIES & al [14] asserts that :

$$BD(\Omega) \text{ is the dual space of } C(\bar{\Omega})_s^9/(S(0,0) \cap C(\bar{\Omega})_s^9) \tag{2.20}$$

Therefore it can be endowed with a weak ★ topology for which bounded sets are relatively compact.

---

(+) For practical applications $N = 3$.

(++) For further details see [2] [9] [12].

2) Regularity Following the classical derivation of Sobolev's inequalities one can prove [2] [12] that $BD(\Omega)$ is continuously embedded into $L^p(\Omega)^N$ for $1 \leq p \leq N/(N-1)$ ($N=3$ in $\mathbb{R}^3$), with a compact embedding for $1 \leq p < N/(N-1)$. Moreover, under the assumption (2.1) ($\partial\Omega$ is $C^1$) one can state [2] the following result (of Deny-Lions type)

$$[u \in \mathcal{D}'(\Omega)^N \ , \ \varepsilon_{ij}(u) \in M^1(\Omega)] \Longrightarrow u \in BD(\Omega) \ .$$

Vector fields in $BD(\Omega)$ admit a trace on $\partial\Omega$ as specified by the following theorem [2] [12]

*Theorem 2.2 Assume that (2.1) is met. Then*
*i) There exists a linear and continuous operator $\gamma$ mapping $BD(\Omega)$ onto $L^1(\partial\Omega)^N$ such that for every $u$ in $BD(\Omega)$ and every $\varphi$ in $C^1(\bar{\Omega})^{N^2}_S$ the following Green's formula holds true :*

$$\int_\Omega u_i \varphi_{ij,j} dx + \int_\Omega \varphi_{ij} \, d\varepsilon_{ij}(u) = \int_{\partial\Omega} \gamma(u)_i \varphi_{ij} n_j ds \ . \tag{2.21}$$

*When $u$ is a regular field (e.g. $C^0(\bar{\Omega})$) $\gamma(u)$ coincides with $u|_{\partial\Omega}$.*
*ii) Let $S$ be an orientable $C^1$ variety included in $\Omega$, separating $\Omega$ into $\Omega_-$ and $\Omega_+$. There exists two linear and continuous trace operators $\bar{\gamma}_S$ and $\overset{+}{\gamma}_S$ mapping $BD(\Omega)$ onto$_2$ $L^1(S)^N$, such that for every $u$ in $BD(\Omega)$ and for every $\varphi$ in $\mathcal{D}(\bar{\Omega})^N_S$ the following Green's formula holds true*

$$\int_{\Omega_-} u_i \varphi_{ij,j} dx + \int_{\Omega_-} \varphi_{ij} d\varepsilon_{ij}(u) = \int_S \overset{+}{\gamma}_S(u)_i \varphi_{ij} n_j ds^{(+)}$$

$$\int_{\overline{\Omega}_-} u_i \varphi_{ij,j} dx + \int_{\overline{\Omega}_-} \varphi_{ij} d\varepsilon_{ij}(u) = \int_S \overset{+}{\gamma}_S(u)_i \varphi_{ij} n_j ds \tag{2.22}$$

*$\bar{\gamma}_S(u)$ and $\overset{+}{\gamma}_S(u)$ are the internal and external traces of $u$ on $S$. For a $C^0(\bar{\Omega})^N$ vector field, $\bar{\gamma}_S(u)$ and $\overset{+}{\gamma}_S(u)$ coincide with $u_{|S}$.*

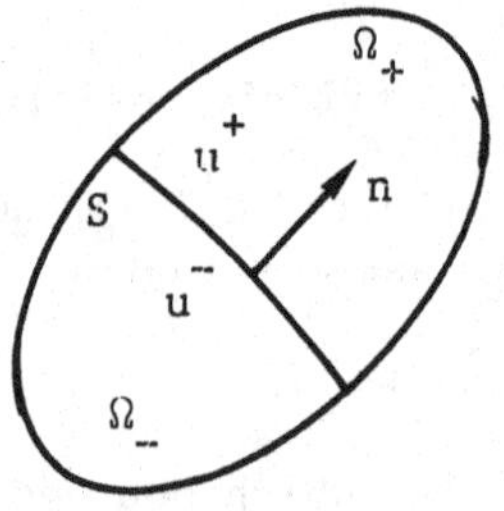


- Figure 2.3 -

For the sake of simplicity we shall denote

$$\bar{u} = \bar{\gamma}_S(u) \qquad \overset{+}{u} = \overset{+}{\gamma}_S(u) \quad \text{and}$$

$$u|_{\partial\Omega} = \gamma(u)$$

---

(+) $n$ is oriented from $\Omega_-$ to $\Omega_+$

Let us note that Green's formula (2.21) plays an essential role in the definition of the trace operator $\gamma$ . Indeed we shall point out that $\mathcal{D}(\bar{\Omega})^N$ *is not dense* in $BD(\Omega)$ for the topology defined by (2.19) . Therefore the usual linear trace operator defined on $\mathcal{D}(\bar{\Omega})^N$ can be extended to $BD(\Omega)$ in infinitely many operators. However only one of these possible continuations satisfies (2.21) .

Another worth noting result is that elements of $BD(\Omega)$ might admit discontinuities on arbitrary surfaces included in $\bar{\Omega}$ . A simple formula relates discontinuities of $u$ and a mass of $\varepsilon(u)$ on a surface $S$ . By difference between the two members of (2.22) we obtain

$$\int_S \varphi_{ij} d\varepsilon_{ij}(u) = \int_S (u_i^+ - u_i^-)\varphi_{ij} n_j ds = \int_S \varphi_{ij}([\![u]\!] \otimes_s n)_{ij} ds \tag{2.23}$$

where the symmetry of $\varphi$ has been used in the last equality, and where

$$[\![u]\!] = u^+ - u^- \ , \ (q \otimes_s n)_{ij} = 1/2(q_i n_j + q_j n_i) \tag{2.24}$$

As a simple consequence, if $u \in C(\bar{\Omega}_-)^3 \cup C(\bar{\Omega}_+)^3$ (with different traces on $S$), then $u \in BD(\Omega)$ with

$$\varepsilon(u) = \varepsilon(u)\big|_{\Omega_-} \chi_{\Omega_-}(x) + \varepsilon(u)\big|_{\Omega_+} \chi_{\Omega_+}(x) + [\![u]\!] \otimes_s n \, \delta_S \ , \tag{2.25}$$

$\chi_X(x)$ is the characteristic function of $X$ , and $\delta_S$ is the Dirac measure on $S$ .

3) Approximation We already mentionned the fact that $\mathcal{D}(\bar{\Omega})^N$ *is not dense in* $BD(\Omega)$ , since its closure consists of vector fields $u$ in $L^1(\Omega)^N$ with a deformation $\varepsilon(u)$ in $\mathbb{L}^1(\Omega)$ . However $\mathcal{D}(\bar{\Omega})^N$ is dense in $BD(\Omega)$ for topologies which are weaker than the strong one (the weak ⋆ topology for instance).

One of these convenient topologies is defined by the "total deformation" (term used by analogy with the total variation of an element of BV). More specifically

$$(\forall\, u \in BD(\Omega)) (\exists\, u_n \in \mathcal{D}(\bar{\Omega})^N) \ \lim_{n \to +\infty} (u_n - u) = 0 \ \text{ in } L^1(\Omega)^N$$
$$\lim_{n \to +\infty} (u_n - u)\big|_{\partial\Omega} = 0 \ \text{ in } L^1(\partial\Omega)^N, \ \lim_{n \to +\infty} \left| \int_\Omega |\varepsilon(u_n)| - \int_\Omega |\varepsilon(u)| \right| = 0 \tag{2.26}$$

4) Extension Let $\omega \supset\supset \Omega$ , $u_1 \in BD(\Omega)$ and $u_2 \in BD(\omega - \bar{\Omega})$ . Then $\tilde{u}$ defined below belongs to $BD(\omega)$

$$\tilde{u}(x) = u_1(x) \ \text{ if } \ x \in \Omega \ , \ \tilde{u}(x) = u_2(x) \ \text{ if } \ x \in \omega - \bar{\Omega} \ .$$

## 2.4 HOMOGENEOUS MATERIAL. RELAXED PROBLEM I

Coming back to the variational problem (2.11) in which we want to recognize the structure of (2.16) with $X = BD(\Omega)$ , we check that the growth condition (2.17) is likely to be satisfied by virtue of (2.14) . $\Phi$ is a proper, l.s.c., convex functional on $BD(\Omega)$ . However another problem arises when examing the properties of $K$ :

$$K = K_1 \cap K_2 \quad , \quad K_1 = \{u \in BD(\Omega) \ , \ u = 0 \ \text{on} \ \Gamma_o\}$$

$$K_2 = \{u \in BD(\Omega) \ , \ L(u) = 1\} \ .$$

Indeed neither $K_1$ nor $K_2$ are closed sets for the weak ★ topology since the trace map $\gamma$ , although continuous for the strong topologies of $BD(\Omega)$ and $L^1(\partial\Omega)$ , is not continuous for the weak ★ topologies of these spaces. This is readily seen in dimension 1 . $\Omega = ]0,1[$ , consider the sequence :

$$u_n(x) = 1 \ \text{if} \ x \geq 1/n \ , \ u_n(x) = nx \ \text{if} \ 0 \leq x \leq \frac{1}{n}$$

Then $u_n(0) = 0$ and $|u_n|_{BD} \leq 2$ . From this last estimate we deduce that $u_n$ converges in $BD(\Omega)$ weak ★ to $u = 1$ the trace of which at $x = 0$ is 1 and not 0 . This example proves that $K_1$ is not closed but it also proves that the linear form

$$u \in BD(\Omega) \longrightarrow \int_{T_1} guds$$

is not continuous on $BD(\Omega)$ endowed with the weak ★ topology. Therefore $K_2$ is not closed either. Consequently $\Phi + I_K$ is not a l.s.c. functional on $X$ , and we have to compute its l.s.c. hull, also called *its closure* defined as the largest l.s.c. function minorizing $\Phi + I_K$ ,

$$\overline{\Phi + I_K}(u) = \lim_{u_n \xrightarrow{\tau} u} \inf \ (\Phi + I_K)(u_n) \ , \tag{2.27}$$

where the topology $\tau$ is understood to be the weak ★ topology on $BD(\Omega)$ . Computing the above closure is a difficult mathematical problem, and heuristic mechanical considerations greatly help to guess the result. For this purpose we interpret more thoroughly the terms entering (2.11) :

1) $\Pi(u)$ is the energy dissipated in a rigid plastic body by the mechanism $u$ . We classically know that $u$ can exhibit discontinuities across the body ($BD(\Omega)$ accounts for this fact). However discontinuities might also appear on the boundary as a result of its possible plastification. Therefore we must abandon the commonly admitted idea that the material body occupies *an open set,* and consider that *the body occupies the closed set* $\overline{\Omega}$ , since the boundary may contribute to the dissipation. In other terms the mechanism should be defined on $\overline{\Omega} = \Omega \cup \partial\Omega$ by a couple $(u|_\Omega \ , u^+|_{\partial\Omega})$ and the total dissipation should read as

$$\hat{\Pi}(u,u^+) = \int_\Omega \pi(\varepsilon(u))dx + \int_{\partial\Omega} \pi^b(u^+ - u)ds \tag{2.28}$$

where $\pi^b(x,\lambda) = \pi(\lambda \otimes_s n(x))$ (+) (2.29)

(+) It should be noted that, even if $\pi$ does not depend on the space variable $x$ , $\pi^b$ may depend on $x$ through the normal unit vector $n(x)$ .

$\pi^b$ is the dissipation due to a strain localization on the boundary $(u^+ \neq u^-)$. Expressions (2.28) and (2.29) are consistent with (2.25).

2) Boundary conditions are to be satisfied by $u^+|_{\partial\Omega}$ since they are imposed by a "tool" external to the body $\Omega$. Therefore $u^+|_{\Gamma_o} = 0$ and a displacement rate field is defined by $(u|_\Omega, u^+|_{\Gamma_1})$.

3) L(u) is (up to the factor $\lambda$) the power of external forces. Therefore the power of body forces f should be evaluated in the displacement rate $u|_\Omega$, while the power of surface forces g should be evaluated in the displacement rate $u^+|_{\Gamma_1}$

$$\hat{L}(u, u^+) = \int_\Omega fu dx + \int_{\Gamma_1} gu^+ ds$$

Finally, heuristic mechanical arguments lead us to the following kinematical definition of the limit load

$$\hat{\lambda}^k = \text{Inf}\,\{\hat{\Pi}(u,u^+),\ \hat{L}(u,u^+) = 1\ ,\ (u,u^+) \in H^1(\Omega)^3 \times H^{1/2}(\Gamma_1)^3\} \quad (2.30)$$

$$\hat{\Pi}(u,u^+) = \int_\Omega \pi(\varepsilon(u))dx + \int_{\Gamma_o} \pi^b(-u)ds + \int_{\Gamma_1} \pi^b(u^+ - u)ds\ .$$

This definition may also be recovered basing on mathematical developments. Let $\lambda < \lambda^s$, $\sigma$ associated with $\lambda$ by (2.6), u and $u^+$ be elements of $H^1(\Omega)^3$ and $H^{1/2}(\Gamma_1)^3$.

$$\int_{\Gamma_1} \sigma.n\, u^+ ds = \lambda \int_{\Gamma_1} gu^+ ds\ ,$$

$$\int_\Omega \sigma\varepsilon(u)dx - \int_{\Gamma_o} \sigma.n\, u\, ds = \lambda \int_\Omega fudx + \int_{\Gamma_1} \sigma.n\, u^+ ds\ .$$

Taking the difference between the above equalities, noting that by virtue of the symmetry of $\sigma$ we have $\sigma.n\, u = \sigma(u \otimes_s n)$, taking into account the definition of the support function $\pi$, we recover (2.30) where $(u,u^+)$ belong to $H^1(\Omega)^3 \times H^{1/2}(\Gamma_1)^3$. But our initial goal is not yet reached, and at least two questions need an answer :

i) Does the equality $\hat{\lambda}^k = \lambda^k$ hold ?

ii) Do we obtain a well posed variational problem (lower semi continuity of the "relaxed" functional) ?

The answer to the first question is provided by theorem 2.3 .

*Theorem 2.3 Assume that assumptions* (2.1)(2.3)(2.4) *are met. Then*

$$\hat{\lambda}^k(\text{Inf}(2.30)) = \lambda^k(\text{Inf}(2.11)) \quad (2.31)$$

*Proof.* In order to prove (2.31) it is sufficient to prove that (2.30) and (2.11) have the same dual problem. Computation of the dual problem of (2.30) relies on the abstract result used in theorem 2.1 . We set :

$$V = H^1(\Omega)^3 \times H^{1/2}(\Gamma_1)^3,\ v = (u,u^+)\ ,\ Y = \mathbb{L}^2(\Omega) \times H^{1/2}(\Gamma_o)^3 \times H^{1/2}(\Gamma_1)^3$$

$$\Lambda = (\Lambda_1 \times \Lambda_2 \times \Lambda_3)(u,u^+) \longrightarrow (\varepsilon(u), u|_{\Gamma_o}, u^+ - u|_{\Gamma_1})$$

$$F(v) = 0 \text{ if } \hat{L}(u,u^+) = 1 , \quad +\infty \text{ otherwise}$$

$$G(p) = G_1(p_1) + G_2(p_2) + G_3(p_3)$$

$$G_1(p_1) = \int_\Omega \pi(p_1)dx , \quad G_2(p_2) = \int_{\Gamma_0} \pi^b(p_2)ds , \quad G_3(p_3) = \int_{\Gamma_1} \pi^b(p_3)ds$$

Problem (2.30) is now under the standard form :

$$\operatorname*{Inf}_{v \in V} \left[F(v) + G(\Lambda v)\right]$$

*Lemma 2.1 Under assumptions* (2.1)

a) $F^*(\Lambda^* p^*) = -\lambda$ *if there exists* $\lambda \in \mathbb{R}$ *such that*

$$\left.\begin{aligned} & -\operatorname{div} p_1^* + \lambda f = 0 \\ & p_3^* = p_1^* n = -\lambda g \quad \textit{on } \Gamma_1 \\ & p_1^* n = -p_2^* \quad \textit{on } \Gamma_0 \end{aligned}\right\} \tag{2.32}$$

b) $F^*(\Lambda^* p^*) = +\infty$ *otherwise*

*Proof.* By definition

$$F^*(\Lambda^* p^*) = \operatorname*{Sup}_{v \in V} \left[(\Lambda^* p^*, v) - F(v)\right] = \operatorname*{Sup}_{\hat{L}(v) = 1} (p^*, \Lambda v)$$

$$= \operatorname*{Sup}_{\substack{(u,u^+) \in V \\ \hat{L}(u,u^+) = 1}} \int_\Omega p_1^* \varepsilon(u)dx + \int_{\Gamma_0} p_2^* u \, ds + \int_{\Gamma_1} p_3^*(u^+ - u)ds \tag{2.33}$$

Let $(u_0,u_0^+)$ be an element of $V$ such that $\hat{L}(u_0,u_0^+) = 1$ . For every $(u,u^+)$ in $V$ we define

$$(\overline{u},\overline{u}^+) = (u,u^+) + (1 - \hat{L}(u,u^+))(u_0,u_0^+)$$

and we check that :

$$\hat{L}(\overline{u},\overline{u}^+) = 1 , \quad (\overline{u},\overline{u}^+) \in V$$

Thus $(\overline{u},\overline{u}^+)$ can be used in (2.33) :

$$F^*(\Lambda^* p^*) = \operatorname*{Sup}_{(u,u^+)} \left[\int_\Omega p_1^* \varepsilon(u)dx + \int_{\Gamma_0} p_2^* u \, ds + \int_{\Gamma_1} p_3^*(u^+ - u)ds - \lambda(1 - \hat{L}(u,u^+))\right] \tag{2.34}$$

where $\lambda = -\int_\Omega p_1^* \varepsilon(u_0)dx - \int_{\Gamma_0} p_2^* u_0 ds - \int_{\Gamma_1} p_3^*(u_0^+ - u_0)ds$ .

In a first step the sup in (2.34) is limited to $u$ in $\mathcal{D}(\Omega)^3$ and $u^+ = 0$ , and we obtain

$$F^*(\Lambda^*p^*) \geq -\lambda + \sup_{u \in \mathcal{D}(\Omega)^3} \int_\Omega (p_1^* \,\varepsilon(u) + \lambda f u)\,dx$$

The last supremum is $+\infty$ except if

$$-\operatorname{div} p_1^* + \lambda f = 0 \tag{2.35}$$

Therefore $F^*(\Lambda^*p^*) = +\infty$ except if (2.35) holds, and we evaluate $F^*(\Lambda^*p^*)$ for those $p^*$ only. Coming back to (2.33) we obtain with the help of Green's formula

$$F^*(\Lambda^*p^*) = -\lambda + \sup_{(u,u^+) \in V} \Big[\int_{\Gamma_o} (p_1^*.n + p_2^*)u\,ds + \int_{\Gamma_1} (p_3^* + \lambda g)u^+ ds$$
$$+ \int_{\Gamma_1} (p_1^*.n - p_3^*)u\,ds\Big]$$

$u|_{\Gamma_o}$, $u|_{\Gamma_1}$, $u^+|_{\Gamma_1}$ can be choosen independently and the above supremum is $+\infty$ except if all the conditions (2.32) are satisfied, in which case $F^*(\Lambda^*p^*) = -\lambda$. This completes the proof of lemma 1. To end the proof of theorem 3 we have to compute $G^*(p^*)$, but this computation is standard (and rather technical for $G_2^*$ and $G_3^*$) and we obtain :

$$G_1^*(p_1^*) = \begin{cases} 0 & \text{if } p_1^*(x) \in P \text{ a.e. } x \text{ in } \Omega \\ +\infty & \text{otherwise,} \end{cases} \tag{2.36}$$

and we use lemma 6.2 of TEMAM [2] to show that

$$G_2^*(-p_1^*.n|_{\Gamma_o}) = 0 \text{ if } p_1^*(x) \in P,\ p_1^* \in \mathbb{L}^2(\Omega),\ \operatorname{div} p_1^* \in L^2(\Omega)^3 \tag{2.37}$$

$$G_3^*(p_1^*.n|_{\Gamma_1}) = 0 \text{ under the same conditions} \tag{2.38}$$

The abstract result (§ 2.2) together with lemma 2.1, (2.36),(2.37), (2.38), completes the proof of theorem 2.3 .■

A simplified form of the relaxed problem can be given. Acheving first the infimum with respect to $u^+$ we compute part of the relaxation term :

$$\inf_{u^+}\{\int_{\Gamma_1} \pi^b(u^+ - u)ds\ ,\ \hat{L}(u, u^+) = 1\ ,\ u^+ \in H^{1/2}(\Gamma_1)^3\}$$
$$= \mu|1 - L(u)| \tag{2.39}$$

where

$$\mu = \sup\{\alpha \in \mathbb{R} \mid \alpha g(x) \in C(x) \text{ a.e. } x \in \Gamma_1\},$$

$$C(x) = \operatorname{dom}(\pi^b(x,.)^*) \tag{2.40}$$

$\pi^b(x,.)$ is a positively homogeneous function of degree one and thus is the support function of a closed convex set in $\mathbb{R}^3$ denoted by $C(x)$. Lemma 6.3 of TEMAM allows to characterize $C(x)$ as the set of all stress vectors $\sigma.n$, when $\sigma$ is plastically admissible and has a smooth divergence. In other terms

$$\mu = \operatorname{Sup}\{\alpha \in \mathbb{R} \mid \exists \sigma\ ,\ \sigma(x) \in P \text{ a.e. } x \text{ in } \Omega\ ,$$
$$\operatorname{div} \sigma \in L^3(\Omega)^3\ ,\ \sigma.n(x) = \alpha\, g(x) \text{ a.e. } x \text{ in } \Gamma_1\} \tag{2.41}$$

The proof of (2.39) (left as an exercise to the reader) relies on the abstract result recalled in section 2.2 .

We are now in a position to simplify (2.30) with the help of (2.39)

$$\left.\begin{aligned} \lambda^k &= \operatorname{Inf}\{\Pi_R(u) + \mu |1 - L(u)|\ ,\ u \in H^1(\Omega)^3\} \\ \Pi_R(u) &= \Pi(u) + \int_{\Gamma_0} \pi^b(-u)\, ds \end{aligned}\right\} \tag{2.42}$$

(2.42) makes contact with TEMAM & STRANG's work who first introduced the following "relaxed" problem

$$\lambda^k = \operatorname{Inf}\{\Pi_R(u)\ ,\ u \in H^1(\Omega)^3\ ,\ L(u) = 1\} \tag{2.43}$$

Equality between the infimum in (2.42) and (2.43) results from equality of the dual problems. However it should be noted that the constraint $L(u) = 1$ in (2.43) is not closed.

*Remark.* (2.30), introduced in BOUCHITTE & SUQUET [20] seems to be original (to the author's knowledge), although the heuristic argument leading to (2.30) was already exposed in SUQUET [21] [22] . However the simplified form (2.41) bears a strong resemblance with the relaxed problem introduced by DEMENGEL [23] .

## 2.5 RELAXED PROBLEM II . FUNCTION OF A MEASURE

We argued in section 2.3 that displacement rates naturally live in $BD(\Omega)$ , and it is readily seen that $u^+$ should be taken element of $M^1(\Gamma_1)^3$ . Consequently we need to extend the definition of $\Pi$ to elements of these spaces. The case of an homogeneous material satisfying (2.4) offers no difficulty since

$$\pi(e) = k|e| \qquad , \qquad \pi^b(q) = k|q \otimes_s n| \tag{2.44}$$

Every bounded measure on $\Omega$ admits an absolute value (BOURBAKI [13]) and this allows to define $\Pi(u)$ for $u$ in $BD(\Omega)$ . Assuming that $\Omega$ has a $C^1$ boundary (assumption (2.1)) we can define $(q \otimes_s n)_{ij}$ as a bounded measure on $\Gamma_1$ as soon as $q$ belongs to $M^1(\Gamma_1)$ . Therefore this particular case offers no difficulty. The general case ($P(x)$ in a general form and depending on $x$) is more difficult and the appropriate tool to extend $\hat{\Pi}$ to irregular displacement rates fields is the so called theory of *functions of measures.*

This theory takes its roots in GOFFMAN & SERRIN's work (no dependence on the space variable). It has been improved by DEMENGEL-TEMAM [16] , GIAQUINTA & al [17] , VALADIER [18] , HADHRI [19] , BOUCHITTE [4] and applies now to non homogeneous materials. The following result is *not optimal* but contains main desirable informations for mechanical purposes in the present context.

Function of a measure. An abstract result

Consider a functional $J$ defined on $M^1(\Omega)^m$ by

$$J(\mu) = \begin{cases} \int_\Omega j(x,h)dx & \text{if } \mu = h\,dx,\ h \in L^1(\Omega)^m \\ +\infty & \text{otherwise}, \end{cases} \tag{2.45}$$

where $j$ obeys a "linear-growth condition"

$$(\exists \lambda_o > 0)(\exists b \in L^1(\Omega)) \quad j(x,z) \geq \lambda_o |z| - b(x) . \tag{2.46}$$

We wish to extend the definition of $J$ to all $\mu$ in $M^1(\Omega)^m$, through the l.s.c. closure of $J$ for the weak $\star$ topology of $M^1(\Omega)^m$. For this purpose we need additional notations :

i) $j_\infty$ is the *principal part* of $j$

$$\pi(x,z) = \lim_{t \to +\infty} \frac{1}{t} j(x,tz) \tag{2.47}$$

$\pi(x,.)$ is the support function of $I_{P(x)}$ where

$$P(x) = \text{dom}(j^\star(x,.)) \tag{2.48}$$

ii) On $C_c^o(\Omega)^m$ we define

$$J^\star(\varphi) = \int_\Omega j^\star(x,\varphi(x))dx ,$$

$$E = \text{dom } J^\star = \{\varphi \in C_c^o(\Omega)^m ,\ \int_\Omega j^\star(x,\varphi(x))dx < +\infty\} \tag{2.49}$$

and $h(x,z) = \sup_{\varphi \in E} (\varphi(x).z)$ . (2.50)

Note that $h$ can be different from $\pi$ .

iii) Given a bounded measure $\mu$ on $\Omega$, Radon-Nikodym's theorem allows to define its integrable and its singular parts with respect to the Lebesgue measure dx

$$\mu = \mu_a\, dx + \mu_s ,\quad \mu_a \in L^1(\Omega) ,\quad \mu_s \perp dx$$

Then we can state the following (not optimal but sufficient here) theorem.

*Theorem 2.4 Assume that* $j$ *obeys* (2.46) *and moreover that*

*i)* $j$ *is l.s.c. on* $\Omega \times \mathbb{R}^m$

*ii)* $\exists\ C \in L^1(\Omega)$ $|j(x,0)| \leq C(x)$ a.e. $x \in \Omega$

*Then the* J's *closure in* $M^1(\Omega)^m$ *weak* $\star$ *is*

$$\bar{J}(\mu) = \int_\Omega j(x,\mu_a(x))dx + \int_\Omega h\left(x, \frac{d\mu_s}{d|\mu_s|}\right) d|\mu_s| \tag{2.51}$$

*If moreover*

$(\forall\, z \in \mathbb{R}^m)$ $\pi(.,z)$ *is upper semi continuous (u.s.c.) then* $h = \pi$ (2.52)

*Comments.* Proof of theorem 2.4 is beyond the scope of these notes. However we can interpret a few of its conclusions :
a) If "the material is homogeneous", i.e. if $j$ does not depend on $x$ we require only that $j$ is l.s.c. with respect to $z$. $\bar{J}$ is given by (2.51)(2.52). Following examples illuminates the role played by the principal part of $j$ in the definition of $\bar{J}$ on singular measures with respect to the Lebesgue measure. Let $m = 1$, $\Omega = ]-1, +1[$, and consider the usual approximation of the Dirac measure $\delta_o$ by stepwise functions

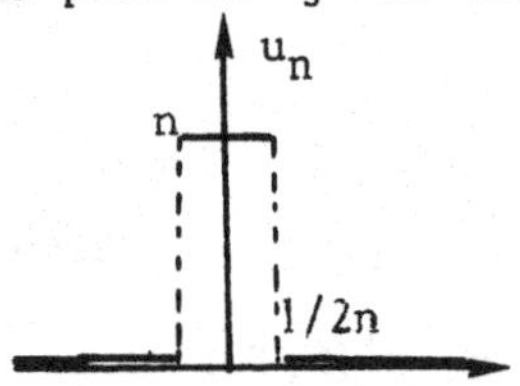


$$u_n(x) = \begin{cases} 0 & \text{if } |x| \geq \frac{1}{2n} \\ n & \text{if } |x| \leq \frac{1}{2n} \end{cases}$$

Then $J(u_n)$ is finite and

$$J(u_n) = \frac{1}{n}\, j(n) \quad \text{which tends to} \quad \pi(1)$$

Therefore

$$J(\delta_o) = \pi(1)$$

b) $\bar{J}$ is obtained as the bi-conjugate function of $J$ $(\bar{J} = J^{**})$ in the duality $(M^1(\Omega)^m, C_c^o(\Omega)^m)$

$$\bar{J}(\mu) = \sup_{\varphi \in C_c^o(\Omega)^m} \left[ \langle \mu, \varphi \rangle - \int_\Omega j^*(x, \varphi(x))\,dx \right] \tag{2.53}$$

This duality relation explains the role played by *continuous* functions in the definition of $h$. We give an example (taken from BOUCHITTE [4]) which illustrates this point :

$$j(x,z) = k(x)|z| = \pi(x,z) \quad \text{where} \quad k(x) \in L^\infty(\Omega), \ k \geq 0$$

$$J(u) = \int_\Omega k(x)|u|\,dx \qquad u \in L^1(\Omega)$$

Then [4]

$$\bar{J}(\mu) = \int_\Omega \hat{k}(x)\,d|\mu| \tag{2.54}$$

where $\hat{k}$ is the closure of $\tilde{k} : \Omega \to \mathbb{R}^+ \cup \{+\infty\}$

$$\tilde{k}(x) = \lim_{\rho \to 0} \frac{1}{|B(x,\rho)|} \int_{B(x,\rho)} k(y)\,dy$$

$(B(x,\rho)$ is the ball centered in $x$ with radius $\rho)$. Indeed

$$E = \{\varphi \in C_c^o(\Omega), \quad |\varphi(x)| \leq k(x) \ \text{a.e. in } \Omega\} \tag{2.55}$$

Averaging the inequality contained in (2.55) on balls yields easily

$$|\varphi(x)| \leq \tilde{k}(x) \quad \text{for } \textit{every} \ x \text{ in } \Omega$$

and by lower semi continuity

$$|\varphi(x)| \leq \hat{k}(x) \quad \text{for } \textit{every} \quad x \text{ in } \Omega$$

Therefore an alternate (and more precise) definition of $E$ (2.55) is

$$E = \{\varphi \in C_c^o(\Omega) \ , \ |\varphi(x)| \leq \hat{k}(x) \ \text{ for every } x \text{ in } \Omega\} \qquad (2.56)$$

(2.54) follows easily from (2.56).

c) From its true definition, $h$ is a l.s.c. function with respect to $z$. Note that it is also a l.s.c. function with respect to $x$. Indeed, taking $\mu_n = z\,\delta_{x_n}$ in (2.51) with $x_n$ converging to $x_\infty$ in $\Omega$, yields (by l.s.c. of $\bar{J}$)

$$\bar{J}(z\delta_{x_\infty}) = h(x_\infty, z) \leq \liminf_{n\to+\infty} \bar{J}(z\,\delta_{x_n}) = h(x_n, z) .$$

The following consequence is worth noting : we come back to the preceeding example and assume that $k(x)$ is piecewise constant : $k(x) = k_1$ if $x \in \Omega_1$, $k(x) = k_2$ if $x \in \Omega_2$ where $\Omega_1, \Omega_2$ are open subsets of $\Omega$, $\Omega_1 \cap \Omega_2 = \emptyset$ $\bar{\Omega}_1 \cup \bar{\Omega}_2 = \bar{\Omega}$. $k$ is not precisely defined on $S = \bar{\Omega}_1 \cap \bar{\Omega}_2$ (which a set of zero measure) but it results from the l.s.c. of $h$ (or directly in this case from (2.54)) that if we want to define $k$ everywhere its correct value on $S$ is $\inf(k_1, k_2)$. Extending this remark to materials with a limited strength, contact is made with the following HADHRI's result [24] : consider a nonhomogeneous material, the strength domain of which takes two values, $P_1$ in $\Omega_1$, $P_2$ in $\Omega_2$, then the strength domain for $x$ in $S = \bar{\Omega}_1 \cap \bar{\Omega}_2$ is $P_1 \cap P_2$.

d) Theorem 2.4 is devoted to closures of functionals defined on $M^1(\Omega)^m$, but we are mainly interested in functionals defined on $BD(\Omega)$. Although these notions are obviously related, there is a gap between them : let $\mathcal{J}$ be defined on $M^1(\Omega)^9_S$ by (2.45) and set

$$J(u) = \mathcal{J}(\varepsilon(u)) \ , \ \bar{J}\ ,(\bar{\mathcal{J}}) \text{ closures of } J, (\mathcal{J}) \text{ in } BD(\Omega)\,, \,(M^1(\Omega)^9_s)$$

It is clear that $\bar{J} \geq \bar{\mathcal{J}}$ but the equality is not obvious and requires additional assumptions on the dependence of $j$ with respect to the space variable $x$ (see for instance DAL MASO [25]). However in case of an homogeneous material the following theorem due to BOUCHITTE [26] provides the desired closure.

*Theorem 2.5 (BOUCHITTE [26]) Let* $j : \mathbb{R}^9_s \to \mathbb{R} \cup \{+\infty\}$ *be a convex, l.s.c., proper function satisfying*

$(\exists\, \alpha, \beta, \gamma \in \mathbb{R}^3_+)$ $\alpha|z| - \gamma \leq j(z) \leq \beta(|z| + 1)$ $\quad \forall z \in \mathbb{R}^9_s$

*and* $J$ *defined on* $L^1(\Omega)^3$ *by*

$$J(u) = \begin{cases} \int_\Omega j(\varepsilon(u))\,dx & \textit{if } u \in H^1(\Omega)^3,\ u = 0 \textit{ on } \partial\Omega \\ +\infty & \textit{otherwise} \end{cases}$$

*Then the closure of* $J$ *in* $L^1(\Omega)^3$ *is*

$$\overline{J}(u) = \begin{cases} \overline{J}(\varepsilon(u)) + \int_{\partial\Omega} \pi(-u \otimes_s n)\,ds & \textit{if} \quad u \in BD(\Omega) \\ +\infty & \textit{otherwise} \end{cases}$$

Theorems 2.4 and 2.5 have two consequences, in case of an homogeneous material submitted to body forces only

a) $\Pi_R(u)$ can be extended to elements of $BD(\Omega)$

b) the obtained extension defines through (2.30) (where $u$ lies in $BD(\Omega)$, and $u^+$ does not appear since $\Gamma_1 = \emptyset$) the relaxed problem associated with (2.11) (uniquely defined as the variational problem formulated on the closure of the initial functional).

## 2.6 BIBLIOGRAPHICAL COMMENTS. OPEN PROBLEMS.

1) We have often assumed a very particular form (2.4) for $P$. Examples of more appropriate strength domains are provided by Von Mises or Tresca criterions in which only the deviatoric part of $\sigma$ is limited. By a duality argument, kinematically admissible vector fields for the limit analysis problem are *incompressible*. This constraint complicates significantly the analysis, mainly in approximations by smooth functions of kinematical fields. The reader is referred to TEMAM [2] for additional details and references.

2) To the author's knowledge (2.30) is a new form of the relaxed problem of limit analysis. However, in case of a *nonhomogeneous material* the author is not aware of minimal assumption to be imposed on material variations in order to extend (2.30) with the help of functions of measures, into a variational problem formulated on a l.s.c. functional.

3) Numerical resolution of limit analysis problems often require *regularization*, on which we shall comment in the context of elasto-plasticity (see next sections). The reader is referred to MERCIER [27], FREMOND-FRIAA [28], JOHNSON [29], and their references which are related to the approach developped here, although mostly based on continuous approximations. Other relevant references on limit analysis are CHRISTIANSEN [30], and NGUYEN DANG HUNG [31].

4) Regularization is mostly used to obtain a *differentiable* functional to minimize. However advantage can be taken of the linear growth of the original functional, by using linear programming techniques. For additional information on this point of view, mostly developped by an Italian school, the reader is referred for instance to GAVARINI [32].

# 3. ELASTO-PLASTICITY

The previous section discussed the maximal value of an applied load schematized by the loading parameter $\lambda$ . The present section deals with a question which naturally follows that of the limit load : can a load $\lambda < \lambda^k = \lambda^s$ be supported by the material ?

The answer to this question is *no* in general, as, proved by the following example, but *yes* in the context of elasto-plastic materials as described below. Let us show however that limit analysis applied to brittle materials overestimates the load capacity of a structure.

Example

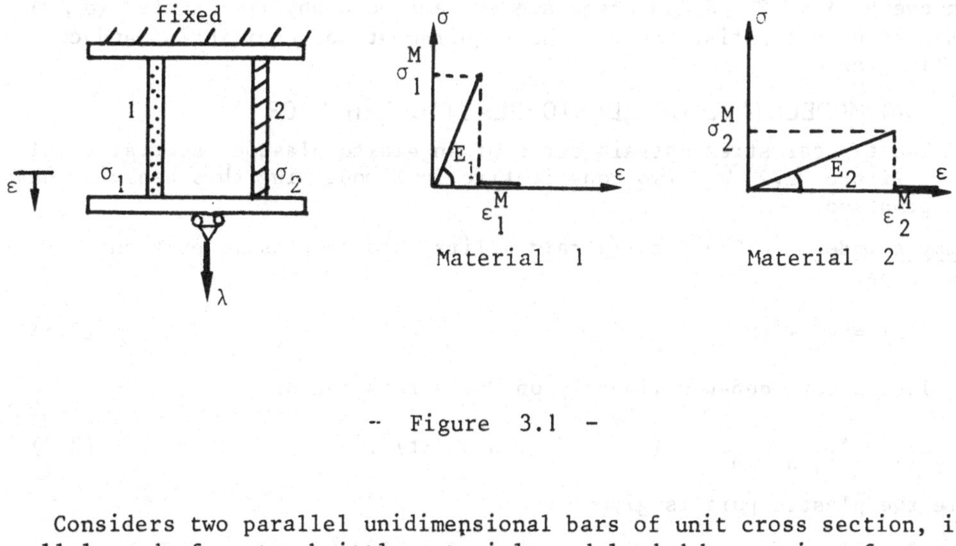


– Figure 3.1 –

Considers two parallel unidimensional bars of unit cross section, in parallel, made from two brittle materials and loaded by a unique force . Ultimate stresses and strains of each phase are denoted by $\sigma_i^M$ and $\varepsilon_i^M$ . Equilibrium reads as

$$\sigma = \text{constant} \quad (\sigma_i) \quad \text{in each phase ,}$$

$$\sigma_1 + \sigma_2 = \lambda$$

Therefore the limit load defined (statically) by (2.6) is

$$\lambda^s = \sigma_1^M + \sigma_2^M$$

But (2.6) does not account for the limitation of strains in both phases

$$\varepsilon_1 = \varepsilon_2 \leqslant \mathrm{Inf}(\varepsilon_1^M , \varepsilon_2^M) = \varepsilon_1^M \quad .$$

Therefore the maximal stress which can be reached by phase 2 , (as long as phase 1 is unbroken), is

$$\sigma_2 = E_2 \, \varepsilon_1^M \ ,$$

and the resulting maximal load (two phases unbroken) is

$$\lambda = \sigma_1^M + E_2 \varepsilon_1^M < \sigma_1^M + \sigma_2^M \quad .$$

However a higher load could be reached if phase 1 breaks :

$$\lambda = \sigma_2^M \text{ (phase 1 broken : } \sigma_1 = 0) \quad .$$

In conclusion the maximal value of $\lambda$ which can be associated with a physical state of the material (both $\sigma$ and $\varepsilon$ satisfying equilibrium equations *and* constitutive law) is

$$\lambda^{\mathrm{Max}} = \mathrm{Sup}(\sigma_2^M , \sigma_1^M + E_2 \varepsilon_1^M) < \lambda^s \quad .$$

This example shows that the limit load determined in section 2 *overestimates* the strength capacity of a brittle structure. However in case of elasto-plastic constituents the limit load is *exact*, in the sense that with every $\lambda < \lambda^s$ we can associate at least one physical state $(\sigma , \varepsilon)$ of the structure satisfying all the requirements of equilibrium and constitutive laws.

## 3.1 TWO MODELLINGS OF ELASTO-PLASTIC BEHAVIOR

The typical stress-strain curve of an elasto-plastic material is plotted on Figure 2.2 and two constitutive laws modelling this behavior have been proposed.

Hencky's model : The total strain splits into an elastic part and a plastic part ,

$$\varepsilon(u) = \varepsilon^e + \varepsilon^p \quad . \tag{3.1}$$

The elastic part depends linearly on the stress tensor

$$\varepsilon_{ij}^e = A_{ijkh} \, \sigma_{kh} \ , \quad (\varepsilon^e = A\sigma \quad \text{in short}) \ , \tag{3.2}$$

while the plastic part is given by

$$\varepsilon^p \in \partial I_P(\sigma) \tag{3.3}$$

$P$ is the strength domain, its interior is the elasticity domain ($\varepsilon^p = 0$ if $\sigma \in \mathrm{Int}\ P$) , its boundary is the yield surface. Note that (3.3) forces $\sigma$ to stay in $P$, otherwise $\partial I_P$ would be empty. $A = (A_{ijkh})$ and $P$ may depend on $x$ when the material is nonhomogeneous.

Prandtl-Reuss' model : The total *strain-rate* splits into an elastic part and a plastic part

$$\varepsilon(\dot{u}) = \dot{\varepsilon}^e + \dot{\varepsilon}^p \tag{3.4}$$

(an over dot stands for time derivative). The elastic part depends linearly on the stress rate

$$\dot{\varepsilon}^e = A\,\dot{\sigma} \ , \tag{3.5}$$

while the plastic part is given by the flow law

$$\dot{\varepsilon}^p \in \partial I_P(\sigma) \ . \tag{3.6}$$

The main difference between these two models is that Hencky's law is a model of (physically) nonlinear elasticity rather than a model of plasticity, and therefore does not account for the irreversible character of yielding, while Prandtl-Reuss' law really accounts for the difference of behavior between loading or unloading paths.

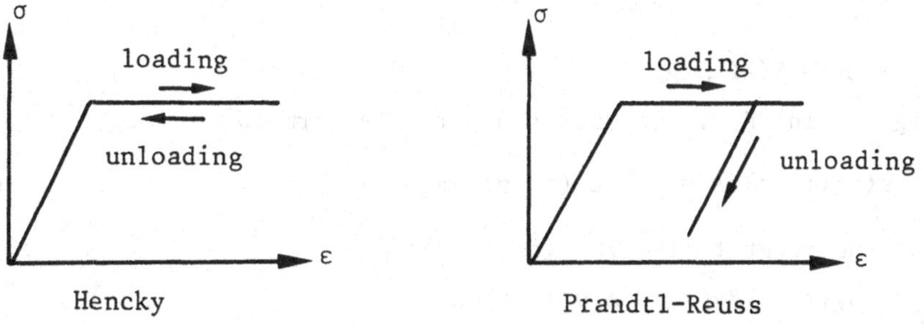


- Figure 3.2 -

Mathematically these two models are again different. Hencky's law leads to minimization problems for convex functionals, while Prandtl-Reuss' equations lead to evolution problems. However they have several common points : discontinuities of displacements (or displacement-rates) arise in both models, and time-discretized Prandtl-Reuss equations are close to Hencky's formulation. Therefore we study in a first step Hencky's model.

*Remark.* Alternate writings of (3.3) or (3.6) can be proposed. The more significant one takes the form of HILL's *principle of maximum work*

$$(3.3) \iff \sigma \in P \ , \ (\forall\, \tau \in P) \quad (\varepsilon^p, \tau - \sigma) \leq 0 \tag{3.7}$$

$$(3.6) \iff \sigma \in P \ , \ (\forall\, \tau \in P) \quad (\dot{\varepsilon}^p, \tau - \sigma) \leq 0 \tag{3.8}$$

## 3.2 HENCKY'S MODEL

Governing equations of the problem are summarized below

$$\left.\begin{aligned} &\operatorname{div}\sigma + \lambda f = 0 \quad \text{in } \Omega \\ &\sigma.n = \lambda g \quad \text{on } \Gamma_1 \quad , \quad u = u_o \quad \text{on } \Gamma_o \\ &\varepsilon(u) = A\sigma + \varepsilon^p \qquad \varepsilon^p \in \partial I_P(\sigma) \quad . \end{aligned}\right\} \qquad (3.9)$$

Variational formulation of (3.9) in terms of stresses

The stress analysis of (3.9) is classical (cf DUVAUT-LIONS [30]). We assume that $P(x)$ is constant on borelian subsets of $\Omega$ and we set :

$$\mathbb{P} = \{\tau \in \mathbb{L}^2(\Omega) \ , \ \tau(x) \in P(x) \quad \text{a.e. } x \in \Omega\}$$

where throughout the text we denote as previously

$$\mathbb{L}^p(\Omega) = L^p(\Omega)^9_s \ ,$$

and recalling the space of statically admissible stress fields

$$S(\lambda f \ , \lambda g) = \{\tau \in \mathbb{L}^2(\Omega) \ , \operatorname{div}\tau + \lambda f = 0 \text{ a.e. in } \Omega, \tau.n = \lambda g \text{ a.e } \Gamma_1\}$$

we set

$$K_\lambda = \mathbb{P} \cap S(\lambda f \ , \lambda g)$$

Choosing $\tau$ in $K_\lambda$ , we obtain by Green's formula

$$\int_\Omega \varepsilon(u)(\tau - \sigma)dx = \int_{\Gamma_o} u_o(\tau - \sigma)n \, ds \quad , \qquad (3.10)$$

and from the constitutive law (cf (3.7))

$$\int_\Omega \varepsilon(u)(\tau - \sigma)dx \leq \int_\Omega A\,\sigma(\tau - \sigma)dx \ .$$

Therefore the variational formulation of (3.9) in terms of stresses is

$$\left.\begin{aligned} &\textit{Find } \sigma \in K_\lambda \textit{ such that} \\ &(\forall\, \tau \in K_\lambda) \quad A(\sigma \ , \tau - \sigma) \geq \ell(\tau - \sigma) \end{aligned}\right\} \qquad (3.11)$$

where

$$A(\sigma \ , \tau) = \int_\Omega A\,\sigma\,\tau dx \ , \ \ell(\tau) = \int_{\Gamma_o} u_o \ \tau.n \, ds \qquad (3.12)$$

In order to formulate a very classical result of existence and uniqueness for $\sigma$ we need additional assumptions

$$\left.\begin{aligned} &u_o \in H^{1/2}(\Gamma_o)^3 \\ &A \text{ is coercive, symmetric, bounded :} \\ &(\exists\, \alpha > 0)(\forall\, \xi \in \mathbb{R}^9_s)\ (\text{a.e. } x \in \Omega) \quad A_{ijkh}(x)\xi_{ij}\xi_{kh} \geq \alpha|\xi|^2_{\mathbb{R}^9_s} \\ &A_{ijkh} = A_{jikh} = A_{khij} \\ &A_{ijkh} \text{ is measurable and} \\ &(\exists\, \beta > 0)(\text{a.e. } x \in \Omega)(\forall\, ijkh)\,|(A_{ijkh}(x)| \leq \beta \end{aligned}\right\} \qquad (3.13)$$

*Proposition 3.1 Assume that (2.1)(2.3)(3.13) are met, and that $\lambda < \lambda^s$ defined by (2.6). Then there exists a unique $\sigma$ solution of (3.11). Moreover $\sigma$ is solution of the minimization problem*

$$\operatorname*{Inf}_{\tau \in K_\lambda} \frac{1}{2} A(\tau,\tau) - \ell(\tau) \tag{3.14}$$

The assumption $\lambda < \lambda^s$ ensures the non vacuity of $K_\lambda$. The proof of proposition (3.1) is classical (DUVAUT & LIONS [33], TEMAM [2], PANAGIOTOPOULOS [9], NECAS- HLAVACEK [34]). Denoting by $(\sigma^e, u^e)$ the elastic response of the material to the same loading $(\lambda f, \lambda g, u_o)$, it is readily seen that $\sigma$ is the projection of $\sigma^e$ on $K_\lambda$ for the scalar product on $\mathbb{L}^2(\Omega)$ associated with the bilinear form $A(.,.)$ (3.12).

Variational formulation in terms of displacements (TEMAM [2])

Let us set for $\xi$ in $\mathbb{R}^9_s$

$$j^*(x,\xi) = \frac{1}{2} A(x)\xi\, \xi + I_{P(x)}(\xi)$$

and for $e$ in $\mathbb{R}^9_s$

$$j(x,e) = (j^*(x.))^*(e) \quad (+)$$

$$J(u) = \begin{cases} \int_\Omega j(x,\varepsilon(u))dx & \text{if } u \in H^1(\Omega)^3 \\ +\infty & \text{otherwise} \end{cases} \tag{3.15}$$

It should be noted that $j$ has a linear growth at infinity, since $P = \operatorname{dom} j^*$ is bounded (cf (2.4))

$$(\exists k_o, k_1, C \in \mathbb{R}^3_+)\ (\text{a.e. } x \in \Omega) \quad k_o|e| - C \leqslant j(x,e) \leqslant k_1(|e|+1)$$

Therefore the natural domain of definition of $J$ is $BD(\Omega)$, provided that we take $J$'s closure for the weak $\star$ topology of $BD(\Omega)$. This is technical in case of a nonhomogeneous material and we shall therefore assume that

$A$ *and* $P$ *do not depend on* $x$,
and again to avoid technical difficulties, (3.16)
$A$ *is isotropic*

The minimization problem in terms of displacements (the dual of which is (3.14)) reads as

$$\operatorname*{Inf}_{u \in BD(\Omega)} \left(\bar{J}(u) + \int_{\Gamma_o} \pi^b(u_o - u)ds - \lambda L(u)\right) \tag{3.17}$$

(+) $j$ is the (nonquadratic) density of elastic energy for Hencky's model.

where $L(u) = \int_\Omega fu + \int_{\Gamma_1} gu\, ds$ .

The following theorem is a compendium of TEMAM's results [2] , slightly adapted to the present context :

*Theorem 3.1 Assume that* (2.1)(2.3)(2.4)(3.13)(3.16) *are met, and that* $\lambda < \lambda^s$ . *Then*

*i)* (3.14) *is the dual of* (3.17)

*ii) The functional* (3.17) *is l.s.c. on* $BD(\Omega)$ *weak* ★ .

*iii) The minimization problem* (3.17) *admits at least one solution.*

*Comments.* Proof of theorem 3.1 is beyond the scope of these lectures, however we can comment its results. The main point is that, although L is not l.s.c. or u.s.c. on $BD(\Omega)$ weak ★ (3.17) is l.s.c. under the assumption $\lambda < \lambda^s$ . Existence of a solution follows easily from this result, which is highly technical and follows mainly KOHN-TEMAM [35] . Extension of these results to nonhomogeneous materials, basing on recent results on functions of measures, is completed in HADHRI [36] .

Basing on the variational properties (3.14) or (3.17) , numerical approximations of solutions for elastic plastic problems can be proposed. The kinematical formulation is (as usual) the most convenient one, although is should be performed with discontinuous shape functions. The reader is referred to BEN DHIA [37] , HADHRI -BEN DHIA [38] and the bibliography of [2] for additional information on this subject. But the (relative) flexibility of the kinematical approach should not hide that the interesting unknown from a mechanical point of view is the *stress tensor* $\sigma$ , which yields the whole information on plastified zones. However the stress formulation (3.11) is not appropriate to (internal) finite element approximations since it contains a constraint in divergence form (equilibrium equations). Probably motivated by this argument, JOHNSON [39] introduced for the Prandtl-Reuss' material a mixed formulation (both $\sigma$ and u are unknown) which we describe now in the present context.

Mixed variational formulation

Define a Banach space S by :

$$S = \{\tau \in \mathbb{L}^2(\Omega) \ , \ \operatorname{div} \tau \in L^3(\Omega)^3\}$$

and choose $\tau$ in $S \cap \mathbb{P}$ . With the help of Green's formula and of the constitutive law (we use (3.7)) we obtain, on the one hand

$$\int_\Omega \varepsilon(u)(\tau - \sigma)dx = -\int_\Omega u(\operatorname{div} \tau - \operatorname{div} \sigma)dx + \int_{\Gamma_o} u_o(\tau - \sigma)n ds + \int_{\Gamma_1} u(\tau-\sigma).n ds \ ,$$

and on the other hand

$$\int_\Omega A\, \sigma(\tau - \sigma)dx \geq \int_\Omega \varepsilon(u)(\tau - \sigma)dx \qquad (\forall\, \tau \in \mathbb{P}) \ .$$

Since we expect solutions u to be discontinuous, even on the boundary $\partial\Omega$ , we introduce a separate displacement $u^+$ on $\Gamma_1$ , which is expected to be in $M^1(\Gamma_1)^3$ . Therefore we have to require additional regularity for g and $\tau.n$ ,

$$g \in C^o(\Gamma_1)^3 \ , \tag{3.18}$$

and we search for $\sigma$ in the Banach space

$$S^c = \{\tau \in \mathbb{L}^\infty(\Omega) \ , \ \operatorname{div} \tau \in L^3(\Omega)^3 \ , \ \tau.n \in C^o(\Gamma_1)^3\} \ .$$

The proposed mixed variational formulation is

$$\left.\begin{array}{l} \textit{Find} \ (\sigma , u , u^+) \in S^c \cap \mathbb{P} \times BD(\Omega) \times M^1(\Gamma_1)^3 \ \textit{such that} \\ A(\sigma , \tau - \sigma) + b_1(u , \tau - \sigma) + b_2(u^+ , \tau - \sigma) \geq \ell(\tau - \sigma) \ \forall \ \tau \in S^c \cap \mathbb{P} \\ b_1(v , \sigma) = \lambda L_1(v) \quad \forall \ v \in BD(\Omega) \\ b_2(\mu , \sigma) = \lambda L_2(\mu) \quad \forall \ \mu \in M^1(\Gamma_1)^3 \end{array}\right\} \quad (3.19)$$

where $b_1(u , \tau) = \int_\Omega u \operatorname{div} \tau \, dx$ , $b_2(\mu , \tau) = -\int_{\Gamma_1} \tau.n \, d\mu$

$$L_1(v) = -\int_\Omega f \, v \, dx \ , \quad L_2(\mu) = -\int_{\Gamma_1} g \, d\mu$$

*Theorem 3.1* [40] *Assume that* (2.1)(2.3)(2.4)(3.13)(3.18) *are met and that* $\lambda < \lambda^s$ . *Then* (3.19) *admits at least one solution* $(\sigma , u , u^+)$. $\sigma$ *is unique.*

*Proof.* The proof relies on a regularization procedure.

Regularization Define

$$\varphi_\mu(x,\tau) = \frac{1}{2\mu} \, |\tau - \Pi_{P(x)} \tau|^2_{\mathbb{R}^9_s} \ , \quad \Phi_\mu(.) = \int_\Omega \varphi_\mu(x,.)dx$$

$$V_{ad} = \{u \in H^1(\Omega)^3 \ , \ u = u_0 \ \text{on} \ \Gamma_o\}$$

and consider the following problem

$$\left.\begin{array}{l} \textit{Find} \ (\sigma^\mu , u^\mu) \in S(\lambda f , \lambda g) \times V_{ad} \ \textit{such that} \\ A\sigma^\mu + \partial\varphi_\mu(\sigma^\mu) = \varepsilon(u^\mu) \end{array}\right\} \quad (3.20)$$

(3.20) admits a unique solution $(\sigma^\mu , u^\mu)$ which has the following variational property

$$\sigma^\mu = \text{Arg Min} \ \{\frac{1}{2} A(\tau,\tau) + \Phi_\mu(\tau) - \ell(\tau) \ , \ \tau \in S(\lambda f,\lambda g))\}$$

$$u^\mu = \text{Arg Min}\{[\frac{1}{2} A(.,.) + \Phi_\mu(.)]^*(\varepsilon(v)) - \lambda L(v) \ , \ v \in V_{ad}\}$$

A priori estimates Note that inequality $\lambda < \lambda^s$ implies that the following *safety assumption* is met

$$\left.\begin{array}{l} (\exists \ \chi \in S(\lambda f,\lambda g) \cap \mathbb{P})(\exists \ \delta > 0) \ \textit{such that} \\ |\tau|_{\mathbb{L}^\infty(\Omega)} < \delta \Rightarrow \chi + \tau \in \mathbb{P} \end{array}\right. \quad (3.21)$$

To check (3.21) choose $\eta$ such that $\lambda < \eta < \lambda^s$ , and let $\sigma_o$ be an element of $S(\eta f,\eta g) \cap \mathbb{P}$ .
Then $\chi = \frac{\lambda}{\eta} \sigma_o$ satisfies all the requirements of (3.21) , with

$\delta = \frac{\eta-\lambda}{2\eta}|\sigma_o|_\infty$ (Recall that $\mathbb{P}$ contains 0 as an interior point).

Then taking $\sigma^\mu - \chi$ as trial function in (3.20), we obtain

$$A(\sigma^\mu,\sigma^\mu) + \int_\Omega \partial\varphi_\mu(\sigma^\mu)(\sigma^\mu - \chi)dx = \int_\Omega \varepsilon(v^\mu)(\sigma^\mu - \chi)dx + A(\sigma^\mu,\chi) \tag{3.22}$$

But monotonicity of $\partial\varphi_\mu$, and noting that $\partial\varphi_\mu(\chi) = 0$, we have

$$\int_\Omega \partial\varphi_\mu(\sigma^\mu)(\sigma^\mu - \chi)dx \geq 0 \quad ,$$

and by Green's formula

$$\int_\Omega \varepsilon(u^\mu)(\sigma^\mu - \chi)dx = \int_{\Gamma_o} u_0(\sigma^\mu - \chi)n\, ds = \int_\Omega \varepsilon(u^e)(\sigma^\mu - \chi)dx \, , \tag{3.23}$$

where $u^e$ denotes an element of $H^1(\Omega)^3$ with trace $u_0$ on $\Gamma_o$. Therefore we deduce from (3.22) that

$$\alpha|\sigma^\mu|^2_{\mathbb{L}^2(\Omega)} \leq C|\sigma^\mu|_{\mathbb{L}^2(\Omega)}{}^{(+)} \, ,$$

i.e.

$$\sigma^\mu \text{ is bounded in } \mathbb{L}^2(\Omega) \, . \tag{3.24}$$

Coming back to (3.22) with the help of (3.24)(3.23) yields

$$0 \leq \Phi_\mu(\sigma^\mu) = \Phi_\mu(\sigma^\mu) - \Phi_\mu(\chi) \leq \int_\Omega \partial\varphi_\mu(\sigma^\mu)(\sigma^\mu - \chi)dx \leq C \quad . \tag{3.25}$$

*Lemma 3.1 Assume that*

$$\int_\Omega \partial\varphi_\mu(\sigma^\mu)(\sigma^\mu - \chi)dx \leq C$$

*Then*

$$|\partial\varphi_\mu(\sigma^\mu)|_{\mathbb{L}^1(\Omega)} \leq C$$

*Proof.*

$$|\partial\varphi_\mu(\sigma^\mu)|_{\mathbb{L}^1(\Omega)} = \frac{1}{\delta} \sup_{|\tau|_{\mathbb{L}^\infty(\Omega)} \leq \delta} \int_\Omega \partial\varphi_\mu(\sigma^\mu).\tau\, du \tag{3.26}$$

But

$$\int_\Omega \partial\varphi_\mu(\sigma^\mu).\tau dx = \int_\Omega \partial\varphi_\mu(\sigma^\mu)(\chi + \tau - \sigma^\mu)dx + \int_\Omega \partial\varphi_\mu(\sigma^\mu)(\sigma^\mu - \chi)dx \tag{3.27}$$

The first integral of the right hand side of (3.27) is non positive : by the safety assumption $\chi + \tau \in \mathbb{P}$, by $\varphi_\mu$'s definition $\partial\varphi_\mu(\chi + \tau) = 0$ and the result follows from the monotonicity of $\partial\varphi_\mu$. The conjunction of (3.26)(3.27) yields

(+) Throughout the following C denotes a (varying !) constant which does not depend on $\mu$.

$$|\partial\varphi_\mu(\sigma^\mu)|_{\mathbb{L}^1(\Omega)} \leq \frac{1}{\delta} \int_\Omega \partial\varphi_\mu(\sigma^\mu)(\sigma^\mu - \chi)\,dx$$

and completes the proof of lemma 3.1 . ■

We deduce from (3.24) and lemma 3.1 that

$$\varepsilon(u^\mu) \text{ is bounded in } M^1(\Omega)^9_s .$$

From the boundary condition $u = u_o$ on $\Gamma_o$ and (2.19) (with $r(u) = \int_{\Gamma_o} |u|ds$) we obtain that

$$u^\mu \text{ is bounded in } BD(\Omega) \quad . \tag{3.28}$$

<u>Passing to the limit $\mu \to 0$</u> .

From (3.24) and (3.28) we deduce that there exists $(\sigma , u , u^+) \in \mathbb{L}^2(\Omega) \times BD(\Omega) \times M^1(\Gamma_1)^3$ such that (for a subsequence)

$$\sigma^\mu \rightharpoonup \sigma \text{ in } \mathbb{L}^2(\Omega) \text{ weak ,}$$

$$u^\mu \rightharpoonup u \text{ in } BD(\Omega) \text{ weak } \star ,$$

$$u^\mu|_{\Gamma_1} \rightharpoonup u^+ \text{ in } M^1(\Gamma_1)^3 \text{ weak } \star .$$

Moreover $\sigma$ belongs to $S(\lambda f , \lambda g)$ which is a closed convex subset of $\mathbb{L}^2(\Omega)$ . Therefore the two last assertions in (3.19) are satisfied. Then we note from (3.25) that :

$$\int_\Omega |\sigma^\mu - \Pi_p\sigma^\mu|^2 dx = 2\mu\ \Phi_\mu(\sigma^\mu) \leq 2\,\mu\,C \quad . \tag{3.29}$$

The lower semi continuity on $\mathbb{L}^2(\Omega)$ of the functional $F$ below, together with (3.29) , proves that $\sigma$ belongs to $\mathbb{P}$ ,

$$F(\tau) = \int_\Omega |\tau - \Pi_p\tau|^2\, dx .$$

It remains to establish the first assertion of (3.19) . For this purpose we note that for every $\tau$ in $S^c \cap \mathbb{P}$ we have

$$A(\sigma^\mu , \tau - \sigma^\mu) + \int_\Omega \partial\varphi_\mu(\sigma^\mu)(\tau - \sigma^\mu)dx + b_1(u^\mu , \tau - \sigma^\mu)$$

$$+ b_2(u^\mu|_{\Gamma_1} , \tau - \sigma^\mu) = \ell(\tau - \sigma^\mu) \quad . \tag{3.30}$$

By l.s.c. we have

$$- A(\sigma,\sigma) \geq \limsup_{\mu\to 0} \left[- A(\sigma^\mu , \sigma^\mu)\right] , \tag{3.31}$$

and by monotonicity

$$\int_\Omega \partial\varphi_\mu(\sigma^\mu)(\tau - \sigma^\mu)dx \leq 0 \quad . \tag{3.32}$$

Moreover

$$\lim_{\mu\to 0} b_1(u^\mu , \sigma^\mu) = \lim_{\mu\to 0} \lambda L_1(u^\mu) = \lambda L_1(u) = b_1(u,\sigma)$$

$$\lim_{\mu\to 0} b_2(u^\mu|_{\Gamma_1} , \sigma^\mu) = \lim_{\mu\to 0} \lambda L_2(u^\mu|_{\Gamma_1}) = \lambda L_2(u^+) = b_2(u^+,\sigma)$$

Passing to the limit in (3.30) with the help of (3.31)(3.32) establishes the first assertion of (3.19) . The proof of theorem 3.2 is complete.■

## 3.3 PRANDTL-REUSS MODEL

We summarize equations of the problem

$$\left.\begin{aligned} &\operatorname{div} \sigma + \lambda f = 0 \quad \text{in} \quad \Omega \\ &\sigma . n = \lambda g \quad \text{on} \quad \Gamma_1 \,, \quad u = u_o \quad \text{on} \quad \Gamma_o \\ &\varepsilon(\dot{u}) = A\dot{\sigma} + \dot{\varepsilon}^p \,, \quad \dot{\varepsilon}^p \in \partial I_P(\sigma) \end{aligned}\right\} \quad (3.33)$$

Note that (3.33) is now *an evolution problem* for $\sigma$ and therefore requires data of $\lambda f$ , $\lambda g$ , $u_o$ on a time interval $[0,T]$ , and initial conditions for $\sigma$ , while $u$ interfers in (3.33) by its time-derivative $\dot{u}$ only, which is the kinematical unknown. For simplicity only, we assume that $\lambda$ is a function of $t$ , while $f$ and $g$ are function of $x$ only

$$\lambda(.) \in W^{1,\infty}(0,T)$$

The initial stress state $\sigma_o$ is assumed to obey

$$\sigma_o \in S(\lambda(0)f\,,\,\lambda(0)g) \cap \mathbb{P} \qquad (3.34)$$

Moreover

$$u_o \in W^{1,\infty}(0,T\,;\,H^{1/2}(\Gamma_o)^3) \,, \quad \operatorname{mes}(\Gamma_o) > 0 \,.$$

Stress formulation

Since $\lambda$ is now a function of time, the set of plastically and statically admissible stress fields is a moving closed convex set in $\mathbb{L}^2(\Omega)$

$$K(t) = S(\lambda(t)f\,,\,\lambda(t)g) \cap \mathbb{P} \,.$$

Following a parallel path to that which led to (3.11) we can formulate (3.33) in terms of stresses only,

$$\left.\begin{aligned} &\textit{Find } \sigma(t) \in K(t) \textit{ such that for a.e. } t \textit{ in } ]0,T[ \\ &(\forall\, \tau \in K(t)) \quad A(\dot{\sigma}\,, \tau - \sigma) \geq \dot{\ell}(\tau - \sigma) \\ &\sigma(0) = \sigma_o \end{aligned}\right\} \quad (3.35)$$

We assume that the following safe load assumption is met.

Safe load assumption

$$(\exists \chi \in W^{1,\infty}(0,T;\mathbb{L}^\infty(\Omega))) \quad \chi(t) \in S(\lambda(t)f,\lambda(t)g) \quad \forall\, t \in ]0,T[ \qquad (3.36)$$

$$(\exists \delta > 0) \quad |\tau|_{\mathbb{L}^\infty(\Omega)} < \delta \Rightarrow \chi(t) + \delta \in \mathbb{P} \qquad \forall\, t \in ]0,T[$$

*Theorem 3.3 Assume that* (2.1)(2.3)(2.4)(3.13)(3.36) *are met. Then* (3.35) *admits a unique solution*

$$\sigma \in W^{1,2}(0,T\,;\,\mathbb{L}^2(\Omega)) \,.$$

Let us show how (3.35) can be given the form of a "sweeping problem". Define on $\mathbb{L}^2(\Omega)$ on new scalar product

$$((\sigma , \tau)) = A(\sigma,\tau) . \tag{3.37}$$

Then (3.35) amounts to a differential *inclusion* in $\mathbb{L}^2(\Omega)$

$$\left.\begin{array}{l}\dot{\sigma} + \partial I_{K(t)}(\sigma) \ni 0 \\ \sigma(0) = \sigma_o\end{array}\right\}$$

which has been studied in the seventies under the name of "sweeping problem". The reader is referred to BREZIS [41] and MOREAU [42] for further details. Theorem 3.3 is a consequence of these works.

Displacement rates formulation : A stress-rate formulation of (3.33) can also be established : it is the HODGE & PRAGER's variational principle (see for instance ANZELOTTI [43]). Its dual problem (termed GREENBERG's principle) formulates (3.19) in terms of displacement rates only. However rates principles, which can be useful in theoretical discussions such as stability of elastic plastic bodies, must be abandonned as for the numerical resolution of the problem.

Mixed variational formulation

JOHNSON [39] proposed a mixed variational formulation of (3.19), further extended by the author [40]. By similar arguments to those developped in section 3.2 we obtain

*Find* $(\sigma,\dot{u},\dot{u}^+)\ [0,T] \rightarrow S^c \cap \mathbb{P} \times BD(\Omega) \times M^1(\Gamma_1)^3$ *such that for a.e.* $t \in ]0,T[$

$$\left.\begin{array}{l}A(\dot{\sigma},\tau-\sigma) + b_1(\dot{u},\tau-\sigma) + b_2(\dot{u}^+,\tau-\sigma) \geq \dot{\ell}(\tau - \sigma)\ (\forall\tau \in S^c \cap\mathbb{P}) \\ b_1(v,\sigma) = \lambda L_1(v)\ (\forall\ v \in BD(\Omega)) \\ b_2(\mu,\sigma) = \lambda L_2(\mu)\ (\forall\ \mu\in M^1(\Gamma_1)^3) \\ \sigma(0) = \sigma_o\end{array}\right\} \tag{3.38}$$

where $\dot{\ell}(\tau) = \int_{\Gamma_o} \dot{u}_o\ \tau.n\ ds$

*Theorem* 3.4 [37] *Assume that* (2.1)(2.3)(2.4)(3.13)(3.36) *are met. Then* (3.38) *admits at least one solution* $(\sigma , \dot{u} , \dot{u}^+)$ . $\sigma$ *is unique.*

We briefly sketch the proof of theorem 3.4 .

Regularization Following DUVAUT-LIONS [33] and JOHNSON [39] we regularize (3.19) with the help of a viscoplastic law due to PERZYNA [44] .

$$\left.\begin{array}{l}A\dot{\sigma}^\mu + \partial\varphi_\mu(\sigma^\mu) = \varepsilon(\dot{u}^\mu) \quad , \quad \sigma^\mu(0) = \sigma_o \\ \sigma^\mu \in S(\lambda f,\lambda g) \ , \ \dot{u}^\mu = \dot{u}_o \quad \text{on} \quad \Gamma_o \quad .\end{array}\right\} \tag{3.39}$$

Existence and uniqueness of a solution to (3.39) are easy consequences of the CAUCHY-LIPSCHITZ theorem. Indeed let $(\sigma^e , u^e)$ be the elastic response of $\Omega$ to the loading $(\lambda f, \lambda g, u_o)$. Then $\bar{\sigma}^\mu = \sigma^\mu - \sigma^e$ is solution of a differential equation in the Hilbert space $S(0,0)$ endowed with the elastic scalar product $((.))$ (cf (3.37))

$$\left.\begin{aligned} &\dot{\bar{\sigma}}^\mu + B(t, \sigma^\mu) = h(t) \quad \text{in} \quad S(0,0) \\ &\bar{\sigma}^\mu(0) = \sigma_o - \sigma^e(0), \end{aligned}\right\} \tag{3.40}$$

where the Lipschitz operator $B(t,.)$ on $S(0,0)$, and the second member $h$ in $S(0,0)$ are defined, with the help of Riescz theorem, by

$$((B(t,\bar{\sigma}),\tau)) = \int_\Omega \partial\varphi_\mu(\bar{\sigma} + \sigma^e(t))\tau \, dx$$

$$((h(t),\tau)) = -\int_\Omega A\dot{\sigma}^e \tau \, dx + \int_{\Gamma_o} \dot{u}_o \, \tau.n \, ds .$$

(3.40) is a usual differential equation, with a Lipschitz operator, and admits a unique solution. We define an element of $L^2(0,T;\mathbb{L}^2(\Omega))$ by

$$e(t) = A\,\dot{\sigma}^\mu + \partial\varphi_\mu(\sigma^\mu) - \varepsilon(\dot{u}^e)$$

which satisfies for almost every $t$ in $]0,T[$

$$\int_\Omega e(t)\,\tau \, dx = 0 \qquad \forall\, \tau \in S(0,0)$$

Lemma 3.2, the proof of which is left to the reader, implies the existence of $\bar{u}^\mu$ in $L^2(0,T;H^1(\Omega)^3)$ such that

$$e = \varepsilon(\dot{\bar{u}}^\mu), \quad \dot{\bar{u}}^\mu = 0 \text{ on } \Gamma_o ,$$

and we set $\dot{u}^\mu = \dot{\bar{u}}^\mu + \dot{u}^e$. $(\sigma^\mu, \dot{u}^\mu)$ satisfy (3.39).

*Lemma 3.2 Let* $e$ *in* $\mathbb{L}^2(\Omega)$ *be such that* $\int_\Omega e\,\tau\,dv = 0$ *for every* $\tau$ *in* $S(0,0)$. *Then there exists a unique* $v$ *in* $H^1(\Omega)^3$ *such that*

$$e = \varepsilon(v), \quad v = 0 \text{ on } \Gamma_o$$

A priori estimates I Multiplying (3.39) by $\sigma^\mu - \chi$ (where $\chi$ satisfies the safety assumption (3.36)), yields with the help of Gromwall lemma

$$\sigma^\mu \text{ is bounded in } L^\infty(0,T;\mathbb{L}^2(\Omega)), \tag{3.41}$$

$\int_\Omega \partial\varphi_\mu(\sigma^\mu)(\sigma^\mu - \chi)dx$ is bounded in $L^1(0,T)$, and by lemma 3.1,

$$\partial\varphi_\mu(\sigma^\mu) \text{ is bounded in } L^1(0,T;\mathbb{L}^1(\Omega)) . \tag{3.42}$$

A priori estimates II We multiply (3.39) by $\dot{\sigma}^\mu - \dot{\chi}$, and note that (use the safe-load assumption)

$$\left|\int_0^T \int_\Omega \partial\varphi(\sigma^\mu)(\dot{\sigma}^\mu - \dot{\chi})dx\right| \leq \Phi_\mu(\sigma^\mu(T)) + C\int_0^T |\partial\varphi_\mu(\sigma^\mu)|_{\mathbb{L}^1(\Omega)} dt ,$$

$$\left|\int_\Omega \varepsilon(\dot{u}^\mu)(\dot{\sigma}^\mu - \dot{\chi})dx\right| = \left|\int_\Omega \varepsilon(\dot{u}^e)(\dot{\sigma}^\mu - \dot{\chi})dx\right| \leq c\,|\dot{\sigma}^\mu - \dot{\chi}|_{\mathbb{L}^2(\Omega)} ,$$

and it follows readily from (3.41)(3.42) that

$$\dot{\sigma}^\mu \text{ is bounded in } L^2(0,T;\mathbb{L}^2(\Omega)) \quad . \tag{3.43}$$

Multiplying (3.39) by $\sigma^\mu - \chi$ again, and using (3.41)(3.43) we obtain

$$\int_\Omega \partial\varphi_\mu(\sigma^\mu)(\sigma^\mu - \chi)dx \text{ is bounded in } L^2(0,T) \, ,$$

and by lemma 3.1 ,

$$\partial\varphi_\mu(\sigma^\mu) \text{ is bounded in } L^2(0,T;\mathbb{L}^1(\Omega)) \, ,$$

i.e. by (3.39) and (3.43) ,

$$\varepsilon(\dot{u}^\mu) \text{ is bounded in } L^2(0,T;\mathbb{L}^1(\Omega)) \, .$$

From the boundary condition $\dot{u}^\mu = \dot{u}_o$ we deduce

$$\dot{u}^\mu \text{ is bounded in } L^2(0,T;BD(\Omega)) \tag{3.44}$$

$$\dot{u}^\mu|_{\Gamma_1} \text{ is bounded in } L^2(0,T;M^1(\Gamma_1)^3) \, . \tag{3.45}$$

<u>Passing to the limit $\mu \to 0$</u> .

By (3.41)(3.43)(3.44)(3.45) we extract from $(\sigma^\mu , \dot{u}^\mu , \dot{u}^\mu|\Gamma_1)$ a subsequence converging in $W^{1,2}(0,T;\mathbb{L}^2(\Omega)) \times L^2_w(0,T;BD(\Omega)) \times L^2_w(0,T;M^1(\Gamma_1)^3)$ weak ⋆ to $(\sigma , \dot{u} , \dot{u})$ (+) . Moreover $\sigma(t)$ belongs to $S(\lambda f^w, \lambda g)$ and satisfies $\sigma(0) = \sigma_o$ . In addition we note that

$$\int_0^T \int_\Omega |\sigma - \Pi_p \sigma^\mu| dxdt \leqslant 2\mu \int_0^T \int_\Omega (\partial\varphi_\mu(\sigma^\mu), \sigma^\mu - \chi) dxdt \leqslant 2C\mu$$

By l.s.c. $\sigma(t)$ belongs to $\mathbb{P}$ for a.e. t in $]0,T[$ . It remains to prove the first inequality of (3.38) . The proof is similar to that of section 3.2 , with the help of the following remark :

$$\sigma^\mu \textit{ converges to } \sigma \textit{ in } L^\infty(0,T;\mathbb{L}^2(\Omega)) \textit{ strong} \tag{3.46}$$

Since we already know that *weak convergence* holds in (3.46) it is sufficient to prove that

$$\limsup_{\mu \to 0} A(\sigma^\mu, \sigma^\mu) \leqslant A(\sigma,\sigma) \quad \text{for every } t \text{ in } ]0,T[$$

Multiplying (3.39) by $\sigma^\mu - \sigma$ , integrating on $\Omega \times ]0,t[$ , and using the monotonicity of $\partial\varphi_\mu$ (note that $\partial\varphi_\mu(\sigma) = 0$) , yields

$$\int_0^t \int_\Omega A\dot{\sigma}^\mu \sigma^\mu dxds \leqslant \int_0^t \int_\Omega A\dot{\sigma}^\mu \sigma dxds + \int_0^t \int_\Omega \varepsilon(\dot{u}^e)(\sigma^\mu - \sigma)dxds \, .$$

We can pass to the limit in the above inequality and we obtain

$$\limsup_{\mu \to 0} \int_0^t A(\dot{\sigma}^\mu, \sigma^\mu) ds \leqslant \int_0^t A(\dot{\sigma}, \sigma) ds \, ,$$

and since $\sigma^\mu(0) = \sigma(0) = \sigma_o$

$$\limsup_{\mu \to 0} A(\sigma^\mu(t), \sigma^\mu(t)) \leqslant A(\sigma(t), \sigma(t)) \, ,$$

---

(+) if X' is the topological dual of a Banach space X , then the dual of $L^2(0,T;X)$ is $L^2_w(0,T;X')$ which is equal to $L^2(0,T;X')$ only if X is reflexive. Applying this result to $X' = BD(\Omega)$ and $X' = M^1(\Gamma_1)^3$ successively explains the previous functional spaces.

which completes the proof of (3.46). The remaining of the proof of theorem 3.4 is easy. More details can be found in PANAGIOTOPOULOS [9] or SUQUET [40].

## 3.4 EXAMPLES OF DISCONTINUOUS SOLUTIONS IN ELASTO-PLASTICITY

Theorem 3.4 lets open the question of *regularity and uniqueness* for $\dot{u}$. The two following simple examples show that $\dot{u}$ is *not unique*, and may be discontinuous even when the safe load assumption (3.36) is met.

a) Discontinuous solutions (+)

$$N = 1 \ , \quad \Omega = ]-1,+1[ \ , \quad A = 1 \ , \quad P = [-1,+1]$$

$$f(x) = (1+x) \ \text{if} \ x < 0 \ , \quad f(x) = -(1-x) \ \text{if} \ x > 0 \ , \quad \lambda(t) = t$$

The initial condition for the stress is $\sigma_o = 0$. The statically admissible stress fields have the form

$$\tau(x,t) = C_\tau(t) - t(x + \frac{x^2}{2}) \ \text{if} \ x \leq 0 \ , \quad C_\tau(t) + t(x - \frac{x^2}{2}) \ \text{if} \ x \geq 0 \qquad (3.47)$$

where $C_\tau$ is a function of $t$ only.

*Elastic solution.*

The elastic solution $\sigma^e$ is a minimizer for the elastic energy among statically admissible fields

$$\frac{1}{2}\int_{-1}^{+1} |\tau(x)|^2 \, dx \ .$$

We find $C = -t/3$. Therefore the body remains elastic up to $t = 3$ where the point $x = 0$ plastifies.

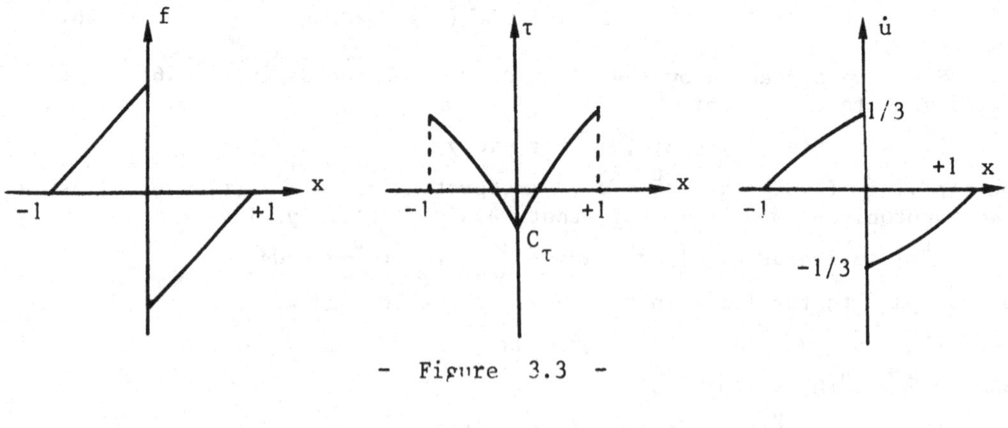


- Figure 3.3 -

(+) This example is an academic one and is not relevant from a physical point of view. However it is relevant in the discussion of the mathematical equations of plasticity.

*Elastic-plastic solution*

It can be easily checked that the safe-load condition is satisfied up to $t = 4$. For $t \in ]3,4[$ we know by the general theorem that the stress solution is unique. It reads as

$$\sigma(x,t) = -1 - t(x + \frac{x^2}{2}) \text{ if } x \leq 0 \text{ , } = -1 + t(x - \frac{x^2}{2}) \text{ if } x \geq 0 \quad (3.48)$$

To prove (3.48) it is sufficient to check that

$$\int_{-1}^{+1} \dot{\sigma}(\tau - \sigma)dx \geq 0$$

for all $\tau$'s in the form (3.47), which is an easy task. The only plastified point is $x = 0$ and outside this point the body remains elastic :

$$\varepsilon(\dot{u}) = \frac{d\dot{u}}{dx} = \dot{\sigma} \quad \text{for} \quad x \neq 0$$

i.e. by integration over $]-1,0[$ and $]0,1[$ :

$$\dot{u}(x,t) = \alpha_+(t) - \frac{x^2}{2} + \frac{x^3}{6} \text{ if } x < 0 \text{ , } = \alpha_-(t) + \frac{x^2}{2} - \frac{x^3}{6} \ x > 0 \quad (3.49)$$

$\alpha_+$ and $\alpha_-$ are functions of $t$ only and are determined by requiring that $\dot{u} = 0$ for $x = \pm 1$ : we find

$$\alpha_+ = -1/3 \qquad \alpha_- = 1/3 \ .$$

However we see from (A.3) that

$$\dot{u}^-(0,t) = \lim_{\substack{x<0 \\ x\to 0}} \dot{u}(x,t) = 1/3 \text{ , } \quad \dot{u}^+(0,t) = \lim_{\substack{x>0 \\ x\to 0}} \dot{u}(x,t) = -1/3 \ .$$

The displacement rate $\dot{u}$, *which is unique in this example* since the plastic zone reduces to one point, is discontinuous in $x = 0$.

b) Non uniqueness of the displacement rate

$N = 1$ , $\Omega = ]0,1[$ , $P = [-1,+1]$ , $A = 1$

The body is loaded by an imposed displacement

$u(0) = 0$ , $u(1) = t$ , $f = 0$

The initial condition for the stress is $\sigma_o = 0$.

*Elastic solution.* $u = tx$ , $\sigma = t$. The body remains elastic up to $t = 1$.

*Elastic plastic solution.* It can be easily checked that the safe load condition is satisfied *for every* $t$ by $\chi = 0$. For $t \geq 1$ the stress remains constant $\sigma(x) = 1$ a.e. $x$ in $\Omega$, and the body is entirely plastified. The mixed formulation (3.38) can be shown to reduce to

$$\int_0^1 \dot{u}(\mathrm{div}\,\tau - \mathrm{div}\,\sigma)du \geq \tau(1) - \sigma(1)$$

and since $\sigma = 1$

$$\int_0^1 \dot{u}\frac{d\tau}{dx}dx \geq \tau(1) - 1 \quad (3.50)$$

(3.50) admits an infinite number of solutions. For instance $\dot{u} = 0$, $\dot{u} = 1$, $\dot{u} = x$ are continuous solutions but discontinuous solutions can also be proposed (any stepwise increasing function. The solution $\dot{u} = 1$

gives an example of loss of boundary condition at $x = 0$ .

## 3.5 BIBLIOGRAPHICAL COMMENTS. OPEN PROBLEMS

1. It results from theorem 3.1 that the closure of the strain energy in a Hencky's model does not require to consider a displacement $u^+|_{\Gamma_1}$ different from $u|_{\Gamma_1}$ . This should also be the case for the mixed variational formulation, and one should be able to prove that the solution $(u,u^+)$ given by theorem 3.2 can be chosen such that $u^+ = u|_{\Gamma_1}$ . The author was unable to prove this point.

2. Lemma 3.1 is a fundamental contribution of JOHNSON [39] .

3. Regularization of elasto-plasticity can be achieved by introduction of viscous or hardening effects. A successful viscous law has been proposed by FRIAA [45] : let $j$ be the jauge function of $P$ and the associate potential $\psi_p$

$$j(\tau) = \text{Inf}\,\{k \geq 0 \quad \tau \in kP\}$$

$$\psi_p(\tau) = \frac{1}{p}\,(j(\tau))^p .$$

Instead of (3.39) consider the elastic visco-plastic material

$$A\dot{\sigma}^p + \partial\psi_p(\sigma^p) = \varepsilon(\dot{u}^p) .$$

Evolution problems for such materials are studied in DJAOUA & SUQUET [67] , and similar constitutive equations are considered in BLANCHARD & LE TALLEC [68] .

Plasticity with hardening is dealt with in JOHNSON [46] and NGUYEN QUOC SON [47] .

4. Numerical approximations of elasto-plastic problems by mixed finite elements do not seem so popular. Let us mention JOHNSON [48] , NGUYEN DANG HUNG [31] , and DJAOUA [49] (viscoplasticity). They should be more developped since they provide a good approximation of $\sigma$ .

# 4. THIN LAYERS

In this section we examin the asymptotic behavior of a thin layer of small width $\delta$ , between two elastic bodies, when $\delta$ goes to 0 . Such a problem can be considered as a preliminary to the study of bonds between two materials, the behavior of bonds being rather different (and often more complex) than the behavior of the elastic bodies stuck together. In the model problem considered below the layer is assumed to obey a non-linear elastic power law of Norton's type. Considering a viscoplastic layer (obeying FRIAA's constitutive equations for instance) would be more realistic but would lead to additional technical difficulties (since both $u$ and $\dot{u}$ interfer in constitutive equations) which are not our main point of interest.

## 4.1 THE THREE BODIES PROBLEM : $\delta > 0$

Main notations are displayed on figure 4.1 below

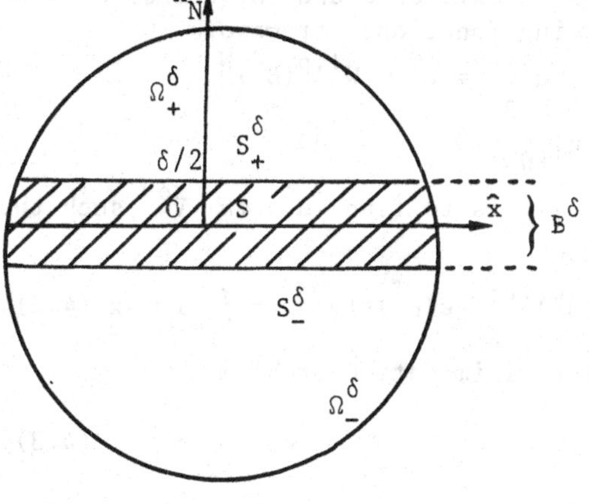


- Figure 4.1 -

$$\Omega_{\pm} = \{x \in \Omega \ ; \ \pm x_N > 0\}$$

$$\Omega^{\delta}_{\pm} = \{x \in \Omega \ ; \ \pm x_N > \frac{\delta}{2}\}$$

$$\Omega^{\delta} = \Omega^{\delta}_{+} \cup \Omega^{\delta}_{-}$$

$$B^{\delta} = \{x \in \Omega \ ; \ |x_N| < \frac{\delta}{2}\}$$

$$B^{\delta}_{\pm} = B^{\delta} \cap \Omega_{\pm}$$

$$S^{\delta}_{\pm} = \{x \in \Omega \ , \ x_N = \pm\frac{\delta}{2}\}$$

$$S = \{x \in \Omega \ , \ x_N = 0\}$$

For the sake of simplicity we have assumed that the layer develops in the vicinity of the plane $x_N = 0$ . Extension to curved interfaces seems possible but technical arguments have not yet been completed by the author.

In the following we consider the body fixed on its boundary, and submitted to body forces $f \in L^\infty(\Omega)^N$

$$u^\delta = 0 \text{ on } \partial\Omega \ , \quad \operatorname{div} \sigma^\delta + f = 0 \text{ in } \Omega \ .$$

The part $\Omega^\delta = \Omega_+^\delta \cup \Omega_-^\delta$ of the material is elastic

$$\sigma^\delta = a\, e(u^\delta) \text{ in } \Omega^\delta \ , \quad a = A^{-1}$$

while the layer $B^\delta$ is occupied by a nonlinear elastic material the behavior of which is given by the model Norton's law

$$\sigma^\delta = k^\delta |\varepsilon(u^\delta)|^{p-2}\, \varepsilon(u^\delta) = k^\delta \varphi_p'(\varepsilon(u^\delta)) \text{ in } B^\delta,$$

where $\varphi_p(e) = |e|^p/p$ . Exponent $p$ takes its values in the range $]1,2]$ . $p = 2$ describes an elastic material, while $p = 1$ gives a rigid plastic behavior ($p = 1.5$ for a superplastic material, $p = 1.2$ for a hot-worked metal, in the more realistic model where $u$ stands for the displacement rate).

The coefficient $k^\delta$ depends on $\delta$ in an essential manner

$$k^\delta = K\, \delta^\alpha \qquad 0 < K < +\infty \ .$$

The equations of the problem are summarized below

$$\left.\begin{array}{l} \sigma^\delta = a\, \varepsilon(u^\delta) \text{ in } \Omega^\delta \ , \quad \sigma^\delta = k^\delta |\varepsilon(u^\delta)|^{p-2}\, \varepsilon(u^\delta) \text{ in } B^\delta \\ \operatorname{div} \sigma^\delta + f = 0 \text{ in } \Omega \ , \quad u^\delta = 0 \text{ on } \partial\Omega \end{array}\right\} \tag{4.1}$$

Classical variational arguments ensure existence and uniqueness of a solution $(\sigma^\delta, u^\delta)$ , within the following functional framework

$$V^\delta = \{u \mid u|_{\Omega^\delta} = u^1 \in H^1(\Omega^\delta)^N \ , \ u|_{B^\delta} = u^2 \in W^{1,p}(B^\delta)^N$$
$$u^1 = u^2 \text{ on } S_\pm^\delta \ , \ u|_{\partial\Omega} = 0\}$$

The variational formulation of (4.1) is to find $u^\delta$ in $V^\delta$ such that for every $v$ in $V^\delta$

$$\int_{\Omega^\delta} a\, \varepsilon(u^\delta)\varepsilon(v)dx + k^\delta \int_{B^\delta} |\varepsilon(u^\delta)|^{p-2}\, \varepsilon(u^\delta)\varepsilon(v)dx = \int_\Omega f\, v\, dx \tag{4.2}$$

which is equivalent to the following minimization problem

$$\operatorname{Inf}_{v \in V^\delta} J^\delta(v) \tag{4.3}$$

$$J^\delta(v) = \frac{1}{2} \int_{\Omega^\delta} a\varepsilon(v)\varepsilon(v)dx + \frac{k^\delta}{p} \int_{B^\delta} |\varepsilon(v)|^p dx - \int_\Omega f\, v\, dx \tag{4.4}$$

$J^\delta$ is a strictly convex, continuous functional on $V^\delta$ and existence and uniqueness of a minimizer is ensured by classical theorems.

## 4.2 THE ASYMPTOTIC PROBLEM $\delta = 0$

We now investigate the behavior of the solution $(\sigma^\delta, u^\delta)$ when $\delta$ goes to $0$ , i.e. when the layer $B^\delta$ shrinks to the surface $S$ . In the limit a jump of u across S is to be expected for suitable values of exponent $\alpha$ , and we recall notation for such discontinuous fields

$$[\![u]\!] = u^+ - u^- \qquad u^{\pm}(\hat{x}) = \lim_{x_N \to 0^{\pm}} u(\hat{x}, x_N) \ .$$

On S the normal vector n is pointing from $\Omega_-$ to $\Omega_+$ (in the particular case under consideration it coincides with $e_N$) .

*Theorem 4.1* *Assume that* $1 < p \leq 2$ , *and that assumptions* (2.1) *are met. Then* $(\sigma^\delta, u^\delta)$ *converges, in a sense which will be specified in the text, towards* $(\sigma^o, u^o)$ *unique solution in* $\mathbb{L}^2(\Omega) \times H^1(\Omega_+ \cup \Omega_-)^N$ *of*

$$\left.\begin{array}{ll} \sigma^o = a\, \varepsilon(u^o) \ \ in\ \Omega_+ \cup \Omega_- & \\ \operatorname{div} \sigma^o + f = 0 \ \ in\ \Omega , & u^o = 0 \ \ on\ \partial\Omega \\ \sigma^o.n = 0 & on\ S\ \ if\ \ \alpha > p-1 \\ \sigma^o.n = K\, j_p'([\![u^o]\!]) & on\ S\ \ if\ \ \alpha = p-1 \end{array}\right\} \qquad (4.5)$$

*The potential* $j_p$ *is defined on* S *as*

$$j_p(\lambda) = \varphi_p(\lambda \otimes_s n)$$

*and*

$$j_p'(\lambda)_i = \frac{\partial j_p}{\partial \lambda_i}(\lambda) = \frac{1}{2} |\lambda \otimes_s n|^{p-2} (\lambda_i + \lambda_n n_i)^{(+)} \quad where \quad \lambda_n = \lambda_j n_j$$

*Proof.* We give the proof for $\alpha = p-1$ (if $\alpha < p-1$ take $K = 0$) .

Functional preliminaries The following functional space will be useful in the sequel

$$V = \{u \in H^1(\Omega_+ \cup \Omega_-)^N , \ u = 0 \ \text{ on } \partial\Omega\}$$

V is a Hilbert space for the norm

$$|u|_V = |u|_{H^1(\Omega_+)^N} + |u|_{H^1(\Omega_-)^N}$$

Let $\sigma$ be a stress field in S(f) , then the following Green's formula holds true for every v in V

$$\int_{\Omega_+ \cup \Omega_-} \sigma\, \varepsilon(v)\, dx + \int_\Omega f\, v\, dx = - \int_S \sigma.n[\![v]\!]\, ds \qquad (4.6)$$

We turn back to the study of $(\sigma^\delta, u^\delta)$.

A priori estimates Multiply (4.1) by $u^\delta$ and integrate over $\Omega$ to obtain

---

(+) Note that $\dfrac{\partial j_p}{\partial \lambda_i}(\lambda) = |\lambda \otimes_s n|^{p-2} \left(\dfrac{\lambda_i n_j + \lambda_j n_i}{2}\right) n_j$ since $\lambda_i$ appears in $(\lambda \otimes_s n)_{ij}$ and in $(\lambda \otimes_s n)_{ji}$ .

$$\int_{\Omega^\delta} a\varepsilon(u^\delta)\varepsilon(u^\delta)dx + K\,\delta^{p-1}\int_{B^\delta}|\varepsilon(u^\delta)|^p dx = \int_\Omega f\,u^\delta\ . \tag{4.7}$$

Then, taking advantage of the fact that $u^\delta$ vanishes on $\partial\Omega$ :

$$\left|\int_\Omega f\,u^\delta dx\right| \leq |f|_{L^\infty(\Omega)^N}\,|u^\delta|_{L^1(\Omega)^N} \leq C|\varepsilon(u^\delta)|_{\mathbb{L}^1(\Omega)} \tag{4.8}$$

By Hölder's inequality we obtain

$$\begin{aligned}|\varepsilon(u^\delta)|_{\mathbb{L}^1(\Omega)} &\leq |\Omega^\delta|^{1/2}|\varepsilon(u^\delta)|_{\mathbb{L}^2(\Omega^\delta)} + |B^\delta|^{1/p'}|\varepsilon(u^\delta)|_{\mathbb{L}^p(B^\delta)}\\ &\leq C|\varepsilon(u^\delta)|_{\mathbb{L}^2(\Omega^\delta)} + \delta^{p-1/p}|\varepsilon(u^\delta)|_{\mathbb{L}^p(B^\delta)}\end{aligned} \tag{4.9}$$

Coercivity of the elasticity tensor, together with (4.7)(4.8)(4.9) yields

$$x_1^2 + x_2^p \leq C(x_1 + x_2) \tag{4.10}$$

where $x_1 = |\varepsilon(u^\delta)|_{\mathbb{L}^2(\Omega^\delta)}$ , $x_2 = \delta^{p-1/p}|\varepsilon(u^\delta)|_{\mathbb{L}^p(B^\delta)}$ .
A classical argument of growth at infinity shows that (4.10) implies :

$$|\varepsilon(u^\delta)|_{\mathbb{L}^2(\Omega^\delta)} \leq C \tag{4.11}$$

$$\delta^{p-1/p}|\varepsilon(u^\delta)|_{\mathbb{L}^p(B^\delta)} \leq C \tag{4.12}$$

We deduce from (4.9)(4.11)(4.12) and the boundary conditions that

$$u^\delta \text{ is bounded in } BD(\Omega) \tag{4.13}$$

Convergence of $(\sigma^\delta, u^\delta)$ On the one hand $u^\delta$ contains a subsequence which converges in $BD(\Omega)$ weak ★ to an element $u^o$ of this space

$$u^\delta \longrightarrow u^o \text{ in } BD(\Omega) \text{ weak ★} \tag{4.14}$$

On the other hand for every $\eta > 0$ we deduce from (4.11) and the boundary conditions that $u^\delta$ is bounded in $H^1(\Omega^\eta)^N$ , and therefore that $u^o$ belongs to $H^1(\Omega^\eta)^N$ for every $\eta > 0$ with

$$|u^o|_{H^1(\Omega^\eta)^N} \leq C \tag{4.15}$$

Letting $\eta$ tend to 0 in (4.15) shows that $u^o$ belongs to $V$ . Moreover, denoting by $\chi_{\Omega^\delta}$ the characteristic function of $\Omega^\delta$ we obtain

$$\chi_{\Omega^\delta}\,\varepsilon(u^\delta) \longrightarrow \varepsilon(u^o) \text{ in } \mathbb{L}^2(\Omega_+ \cup \Omega_-) \text{ weak} \tag{4.16}$$

The compact embedding of $BD(\Omega)$ into $L^1(\Omega)^N$ shows that

$$u^\delta \longrightarrow u^o \text{ in } L^1(\Omega)^N \text{ strong} \tag{4.17}$$

We deduce from (4.11) that $\sigma^\delta$ is a bounded sequence in $\mathbb{L}^2(\Omega^\delta)$ , and from (4.12) that

$$\int_{B^\delta} |\sigma^\delta|^{p'} dx = \int_{B^\delta} K^{p'} \delta^p |\varepsilon(u^\delta)|^p dx \leq C\,\delta$$

Since $p' = \frac{p}{p-1} \geq 2$ we obtain that $\sigma^\delta$ is bounded in $\mathbb{L}^2(\Omega)$, with a weakly converging subsequence (without loss of generality we can assume that it converges for the same subsequence of $\delta$ than the subsequence extracted from $u^\delta$), and we denote by $\sigma^o$ its limit. It is straightforward to check that $\sigma^o$ and $u^o$ are related by

$$\left.\begin{array}{l} u^o \in V \ , \quad \sigma^o \in \mathbb{L}^2(\Omega) \ , \ \mathrm{div}\,\sigma^o - f = 0 \ \text{ in } \ \Omega \\ \sigma^o = a\,\varepsilon(u^o) \ \text{ in } \ \Omega^+ \cup \Omega^- \end{array}\right\}$$

and the only remaining point concerns the behavior of $\sigma^o$ and $u^o$ on $S$. $u^o$ may admit discontinuities on $S$, while $\sigma^o.n$ is uniquely defined in $H^{-1/2}(S)^N$ (by virtue of the equilibrium equations). For this purpose an intermediate result will be usefull : note that for a "smooth" stress field $\tau$ (in $W^{1,\infty}(\Omega)^{N^2}_S$ for instance) we have

$$\lim_{\delta\to 0} \int_{B^\delta} \tau\varepsilon(u^\delta)dx = \int_S \tau.n[\![u^o]\!]dx \tag{4.18}$$

Indeed

$$\lim_{\delta\to 0} \int_{B^\delta} \tau\,\varepsilon(u^\delta)dx = \lim_{\delta\to 0}\Big[\int_\Omega \tau\,\varepsilon(u^\delta)dx - \int_{\Omega^\delta} \tau\varepsilon(u^\delta)dx\Big]$$

$$= \lim_{\delta\to 0} \Big[-\int_\Omega \mathrm{div}\,\tau\, u^\delta dx - \int_{\Omega_+\cup\Omega_-} \chi_{\Omega^\delta}\tau\varepsilon(u^\delta)dx\Big]$$

According to (4.16)(4.17) and (4.6) we obtain

$$= -\int_\Omega \mathrm{div}\,\tau\, u^o dx - \int_{\Omega_+\cup\Omega_-} \tau\,\varepsilon(u^o)dx = \int_S \tau.n[\![u^o]\!]ds$$

which proves (4.18) .

<u>Contact law</u> It remains to prove that $\sigma^o.n$ and $[\![u^o]\!]$ are related by the contact law (4.5) . For this purpose, according to the maximal monotonicity of the operator $j'_p(z) = |z|^{p-2}z$ , it is sufficient to prove

$$\int_S (\sigma^o.n - Kj'_p([\![v]\!]))([\![u]\!] - [\![v]\!])ds \geq 0 \tag{4.19}$$

for every $v$ in $V$ . (4.19) is derived from the inequality

$$\int_{B^\delta} (\sigma^\delta - k^\delta\varphi'_p(\varepsilon(v^\delta)) \ , \ \varepsilon(u^\delta) - \varepsilon(v^\delta))dx \geq 0 \tag{4.20}$$

applied to a suitable approximation $v^\delta$ of $v$ in $V$ which we further describe.

Let $v$ be an element of $V$. By a density argument we can clearly assume that $v$ is regular $(C^\infty(\Omega^+)^N \times C^\infty(\Omega^-)^N)$ in $\Omega_+$ and $\Omega_-$ , and that its support is compact in $\Omega$ .

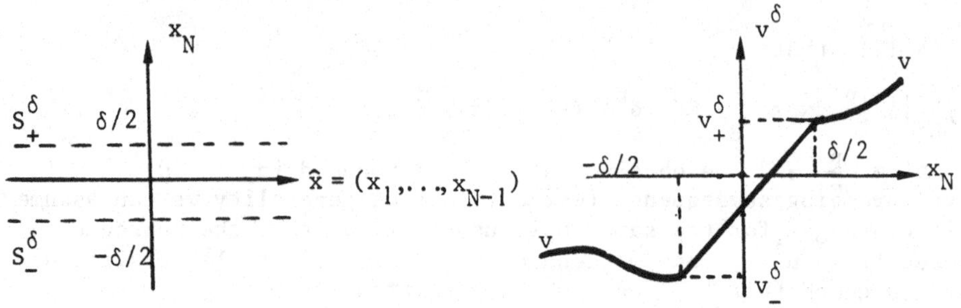


- Figure 4.2 Approximation $v^\delta$

We approximate v by a more regular vector field $v^\delta$ continuous across S :

$$v^\delta(x) = \begin{cases} v(x) & \text{if } |x| > \frac{\delta}{2} \\ \frac{1}{2}(v_+^\delta(\hat{x}) + v_-^\delta(\hat{x})) + \frac{x_N}{\delta} [\![v]\!]^\delta(\hat{x}) & \text{if } |x_N| \leq \frac{\delta}{2} \end{cases}$$

where $v_\pm^\delta(\hat{x}) = v(\hat{x}, \pm\frac{\delta}{2})$ , and $[\![v]\!]^\delta(\hat{x}) = v_+^\delta(\hat{x}) - v_-^\delta(\hat{x})$ . $v^\delta$ exhibits the following properties :

i) $v^\delta$ tends to v in $L^2(\Omega)^N$ strong.

ii) Trace operator on a surface $x_N = \pm\delta/2$ is a continuous function of the surface. Therefore

$$\lim_{\delta\to 0} [\![v]\!]^\delta(\hat{x}) = [\![v]\!](\hat{x}) \text{ in } H^{1/2}(S)^N (\text{hence } L^2(S)^N) \text{ strong} \quad (4.21)$$

and consequently, for every $r \in [1,2]$

$$\lim_{\delta\to 0} \delta^{-1} | [\![v]\!]^\delta \otimes_s n - [\![v]\!] \otimes_s n |^r_{\mathbb{L}^r(B^\delta)} = 0 \quad (4.22)$$

iii) Deformation associated with $v^\delta$ reads as

$$i \text{ and } j \neq N \quad \varepsilon_{ij}(v^\delta) = \frac{1}{2}(\varepsilon_{ij}(v_+^\delta) + \varepsilon_{ij}(v_-^\delta)) + \frac{x_N}{2\delta}(\varepsilon_{ij}(v_+^\delta) - \varepsilon_{ij}(v_-^\delta)) \quad (4.23)$$

$$i \neq N \quad j = N \quad \varepsilon_{iN}(v^\delta) = \frac{1}{2\delta}(v_{+i}^\delta - v_{-i}^\delta) + \frac{1}{4}\left(\frac{\partial v_{+N}^\delta}{\partial x_i} + \frac{\partial v_{-N}^\delta}{\partial x_i}\right) + \frac{x_N}{2\delta}\left(\frac{\partial v_{+N}^\delta}{\partial x_i} - \frac{\partial v_{-N}^\delta}{\partial x_i}\right)$$

$$i = N \quad j = N \quad \varepsilon_{NN}(v^\delta) = \frac{1}{\delta}(v_{+N}^\delta(\hat{x}) - v_{-N}^\delta(\hat{x}))$$

*Lemma 4.1 Let* v *be an element of* $C^\infty(\Omega_+ \cup \Omega_-)^N$ , *and* $v^\delta$ *defined as above. Then*

i) $\lim_{\delta\to 0} \int_{B^\delta} k^\delta \varphi_p'(\varepsilon(v^\delta))\varepsilon(v^\delta)dx = K \int_S j_p'([\![v]\!])[\![v]\!]ds$

ii) $\lim_{\delta\to 0} \int_{B^\delta} k^\delta \varphi_p'(\varepsilon(v^\delta))\varepsilon(u^\delta)dx = K \int_S j_p'([\![v]\!])[\![u^o]\!]ds$

iii) $\lim_{\delta\to 0} \int_{B^\delta} \sigma^\delta \varepsilon(v^\delta)dx = \int_S \sigma^o.n[\![v]\!]ds$

iv) $\lim_{\delta\to 0} \sup \int_{B^\delta} \sigma^\delta \varepsilon(u^\delta)dx \leq \int_S \sigma^o.n[\![u^o]\!]ds$ .

It is obvious from the lemma that (4.19) is a mere consequence of (4.20). Therefore the proof of lemma 4.1 will complete the proof of theorem 4.1.

<u>Proof of lemma 4.1</u> For every $r \in [1,2]$

$$\lim_{\delta\to 0} \int_{-\delta/2}^{+\delta/2} dx_N \int_S |\varepsilon(v^\delta) - \frac{1}{\delta}[\![v]\!]^\delta \otimes_s n|^r ds \leq \lim_{\delta\to 0} \int_{-\delta/2}^{+\delta/2} |c_1 + c_2 \frac{x_N}{\delta}|^r dx_N$$

(where $c_1$ and $c_2$ depend only on v) . Therefore

$$\lim_{\delta\to 0} |\varepsilon(v^\delta) - \frac{1}{\delta}[\![v]\!]^\delta \otimes_s n|^r_{\mathbb{L}^r(B^\delta)} = 0 \qquad (4.24)$$

Combining (4.22) and (4.24) yields

$$\lim_{\delta\to 0} \delta^{p-1} |\varepsilon(v^\delta) - \frac{[\![v]\!]}{\delta} \otimes_s n|^p_{\mathbb{L}^p(B^\delta)} = 0 \qquad (4.25)$$

Consequently

$$\lim_{\delta\to 0} K\delta^{p-1} \int_{B^\delta} \varphi'_p(\varepsilon(v^\delta))\varepsilon(v^\delta)dx = \lim_{\delta\to 0} p\, K\, \delta^{p-1} |\varepsilon(v^\delta)|^p_{\mathbb{L}^p(B^\delta)}$$

$$= |K \lim_{\delta\to 0} \delta^{-1}([\![v]\!] \otimes_s n|^p_{\mathbb{L}^p(B^\delta)} = K\int_S |[\![v]\!]\otimes_s n\,{}^p ds = K\int j'_p([\![v]\!])[\![v]\!]ds \qquad (4.26)$$

which proves point i) .

Moreover $\varphi'$ is Hölder continuous from $\mathbb{L}^p(B^\delta)$ into $\mathbb{L}^{p'}(B^\delta)$ (cf GLOWINSKI-MAROCCO [50])

$$|\varphi'_p(e_1) - \varphi'_p(e_2)|_{\mathbb{L}^{p'}(B^\delta)} \leq C|e_1 - e_2|^{p-1}_{\mathbb{L}^p(B^\delta)}$$

Therefore

$$|K\, \delta^{p-1} \int_{B^\delta} (\varphi'_p(\varepsilon(v^\delta)) - \varphi'_p(\frac{1}{\delta}[\![v]\!] \otimes_s n))\varepsilon(u^\delta)dx|$$

$$\leq C\, \delta^{p-1} |\varepsilon(v^\delta) - \frac{1}{\delta}[\![v]\!] \otimes_s n|^{p-1}_{\mathbb{L}^p(B^\delta)} |\varepsilon(u^\delta)|_{\mathbb{L}^p(B^\delta)}$$

$$\leq C(\delta^{p-1} |\varepsilon(v^\delta) - \frac{1}{\delta}[\![v]\!] \otimes_s n|^p_{\mathbb{L}^p(B^\delta)})^{p-1/p} (\delta^{p-1/p} |\varepsilon(u^\delta)|_{\mathbb{L}^p(B^\delta)})$$

We deduce from estimate (4.12) and from (4.25) that

$$\lim_{\delta\to 0} \int_{B^\delta} k\, \varphi'_p(\varepsilon(v^\delta))\varepsilon(u^\delta)dx = \lim_{\delta\to 0} K\delta^{p-1} \int_{B^\delta} \varphi'_p(\frac{[\![v]\!]}{\delta} \otimes_s n)\varepsilon(u^\delta)dx$$

$$= \lim_{\delta\to 0} K \int_{B^\delta} \tau\varepsilon(u^\delta)dx \quad , \quad \text{where} \quad \tau = \varphi_p'([\![v]\!] \otimes_s n)$$

according to (4.18) we obtain

$$= K \int_S \varphi_p'([\![v]\!] \otimes_s n).n[\![u^o]\!]ds = \int_S K\, j_p'([\![v]\!])[\![u^o]\!]ds$$

which proves point ii) .

Then we note that

$$\lim_{\delta\to 0} \int_{B^\delta} \sigma^\delta\varepsilon(v^\delta)dx = \lim_{\delta\to 0} \int_\Omega \sigma^\delta\varepsilon(v^\delta)dx - \int_\Omega \sigma^\delta \chi_{\Omega^\delta}\, \varepsilon(v)dx$$

(note that $\chi_{\Omega^\delta}\,\varepsilon(v^\delta) = \chi_{\Omega^\delta}\,\varepsilon(v)$)

$$= \lim_{\delta\to 0} (-\int_\Omega f\, v^\delta dx - \int_\Omega \sigma^\delta\chi_{\Omega^\delta}\, \varepsilon(v)dx)$$

$$= -\int_\Omega f\,v\,dx - \int_{\Omega_+\cup\Omega_-} \sigma^o\, \varepsilon(v)dx = \int_S \sigma^o n[\![v]\!]ds$$

which proves iii) .

By lower semi continuity we obtain

$$\int_{\Omega_+\cup\Omega_-} \sigma^o\varepsilon(u^o)dx = \int_{\Omega_+\cup\Omega_-} a\varepsilon(u^o)\varepsilon(u^o)dx \leqslant \liminf_{\delta\to 0} \int_{\Omega_+\cup\Omega_-} a(\chi_{\Omega^\delta}\varepsilon(u^\delta))(\chi_{\Omega^\delta}\varepsilon(u^\delta))dx$$

$$\leqslant \liminf_{\delta\to 0} \int_{\Omega^\delta} \sigma^\delta\varepsilon(u^\delta)dx$$

Using the equilibrium equations for both $\sigma^o$ and $\sigma^\delta$ yields

$$-\int_S \sigma^o n[u^o]ds \leqslant \liminf_{\delta\to 0} [-\int_{B^\delta} \sigma^\delta\varepsilon(u^\delta)dx]$$

(Note that $\int_\Omega f\,u^o dx = \lim_{\varepsilon\to 0} \int_\Omega f\,u^\delta dx = \int_\Omega \sigma^\delta\varepsilon(u^\delta)dx$

Thus

$$\int_S \sigma^o.n[u^o]ds \geqslant \limsup_{\delta\to 0} \int_{B^\delta} \sigma^\delta\varepsilon(u^\delta)dx$$

which ends the proof of lemma 4.1 and of theorem 4.1 .

## 4.3 COMMENTS, OPEN PROBLEMS

1. As noted in the introduction a more physically relevant problem would be that of a real viscoplastic layer. Equations of the problem read as

$$\left.\begin{array}{l} \sigma^\delta = a\,\varepsilon(u^\delta) \text{ in } \Omega^\delta \quad , \quad \sigma^\delta = k^\delta|\varepsilon(\dot u^\delta)|^{p-2}\,\varepsilon(\dot u^\delta) \text{ in } B^\delta \\ \operatorname{div}\sigma^\delta + f = 0 \text{ in } \Omega \; , \; u^\delta = 0 \text{ on } \partial\Omega \end{array}\right\}$$

2. Again a more physically relevant problem should include incompressibility of the layer

$$\operatorname{div} u^\delta = 0 \text{ in } B^\delta \quad , \quad \sigma^\delta = -\,p\,\mathbb{I}d + k^\delta|\varepsilon(u^\delta)|^{p-2}\,\varepsilon(u^\delta) \;.$$

It follows from this constraint that $[\![u^o]\!].n = 0$ on $S$ , and the resulting constitutive law of the interface should still be given by (4.5) with due account of the above equality. However the precise statement and the proof of this result have not been completed by the author.

3. Theorem 4.1 lets open the case $\alpha < p-1$ . One could think that the limit behavior is simply obtained by letting K tend to $+\infty$ in (4.5) which yields

$$[\![u^0]\!] = 0 .$$

This result would be incomplete as proved by CAILLERIE [51] for $p = 2$ . For negative $\alpha$ the interface behaves like a membrane or like a plate. The total energy of the system contains a contribution of S with first or second derivatives of $u^0$ on S . The extension of CAILLERIE's results and the determination of critical exponents $\alpha$ in the present context $(1 < p \leq 2)$ is an open problem to the author's knowledge. When a plate behavior of the interface is expected, BLANCHARD & PAULMIER's [52] modelling of viscoplastic plates should be useful.

4. The case of elasto-plastic layers is treated in DESTUYNDER & NEVEU [53] . It is proved that, assuming that the safety assumption is met *uniformly with respect to* $\delta$ , the interface has a limited strength characterized by a domain C which delimits the set of admissible stress vectors

$$\sigma.n \in C \quad , \quad ([\![u]\!],\sigma.n - T) \geq 0 \qquad \forall\, T \in C .$$

C is nothing else than the set of stress vectors defined in section 2 . This result can be proved by a very simple use of the epi-convergence theory (see next section) and of Mosco-convergence (see ATTOUCH [3]) . An interesting problem arises when discussing the limit load problem. Let us denote by $\lambda^\delta$ the limit load for a structure $\Omega$ made from two elastic parts bonded by a band of width $\delta > 0$ . When $\delta$ goes to 0 does $\lambda^\delta$ tends to $\lambda$ , obtained by considering an interface with the limited strength defined by C ? The answer is *yes* if the structure is not loaded on its boundary (but fixed), and *no* if the structure is loaded on its boundary. The next section will provide an example of a similar phenomenon.

5. Similar problems to those dealt with in the present section, are treated by BREZIS & al [54] , ACERBI & al [55] .

# 5. EPI-CONVERGENCE, DISCONTINUITIES AND PLASTICITY

## 5.1 ELEMENTS ON EPI-CONVERGENCE

The theory of *epi-convergence*(+) applies typically to the following problem. Let $(X,\tau)$ be a topological vector space (for simplicity $\tau$ is assumed to be metrizable), and $J^n$ a sequence of functionals mapping $X$ into $\mathbb{R} \cup \{+\infty\}$. Consider the variational problem

$$\lambda_n = \operatorname*{Inf}_{x \in X} J^n(x) \tag{5.1}$$

We define the $\tau$ epi-limits inf and sup of $J^n$ as follows

$$\forall\, x \in X \qquad \underline{J}(x) = \operatorname*{Inf}_{x_n \xrightarrow{\tau} x} \ \liminf_{n \to +\infty} J^n(x_n) \tag{5.2}$$

$$\forall\, x \in X \qquad \overline{J}(x) = \operatorname*{Inf}_{x_n \xrightarrow{\tau} x} \ \limsup_{n \to +\infty} J^n(x_n) \tag{5.3}$$

Note that $\overline{J} \geq \underline{J}$ . $J^n$ is said to be $\tau$ epi-convergent to $J$ if $\underline{J} = \overline{J} = J$ . A first important result significantly reduces the number of possible $\tau$ epi-limits.

*Theorem* 5.1 *With the above notations* $\underline{J}$ *and* $\overline{J}$ *are* $\tau$ *lower semi continuous functionals.*

We have already encountered epi-convergence (but we did not know it) as proved by the important corollary.

*Corollary* 5.1 *Take* $J^n = J$ *for all* $n$ . *Then* $J^n$ *epi converges to the* $\tau$ *closure of* $J$ .

Therefore epi-convergence *contains* the difficult problem of computing the lower semi continuous hull of a given function.

The following abstract theorem specifies the behavior of the infima $\lambda_n$ in (5.1) .

---

(+) a general and complete exposure on epi-convergence, together with numerous applications and references can be found in ATTOUCH [3] .

*Theorem 5.2 Let* $J^n$ *be a sequence of functionals* $\tau$ *epiconverging to* $J$ *, and let* $x_n$ *be such that :*

$$J^n(x_n) \leq \lambda_n + 1/n$$

*Then for every* $\tau$ *converging subsequence* $x_{n_k}$ *(limit* $\bar{x}$*) we have*

$$J(\bar{x}) = \underset{x\in X}{\text{Min}}\ J(x) = \lim_{k\to+\infty} J^{n_k}(x_{n_k})$$

Provided that we can extract from the approximate minimizers a $\tau$ converging subsequence, the above theorem ensures convergence of the infima :

$$\lambda = \underset{x\in X}{\text{Min}}\ J(x) = \lim_{x\to+\infty} \lambda_n \tag{5.4}$$

A very simple corollary of theorem 5.2 will be useful in the study of limit analysis problems :

*Corollary 5.2 In addition to the assumptions of theorem 5.2 let* $L$ *be a continuous linear form on* $(X,\tau)$ *and consider the variational problem*

$$\lambda_n = \text{Inf}\{J^n(x)\ ,\ L(x) = 1\} \tag{5.5}$$

*and let* $x_n$ *be such that* $L(x_n) = 1$. $J^n(x_n) \leq \lambda_n + 1/n$ . *Then for every* $\tau$ *converging subsequence* $x_{n_k}$ *(limit* $\bar{x}$*)*$^n$ *we have*

$$J(\bar{x}) = \text{Inf}\{J(x), L(x) = 1\} = \lim_{k\to+\infty} \lambda^{n_k} = \lim_{k\to+\infty} J^{n_k}(x_{n_k}) \tag{5.6}$$

In conclusion we recall a useful criterion in two steps in order to prove an epi-convergence result :

$J$ *is the* $\tau$ *epi-limit of* $J^n$ *if the two following conditions are met*

i) $(\forall x \in X)(\forall x_n \in X) \lim_{n\to+\infty} x_n = x$ *in* $(X,\tau), J(x) \leq \liminf_{n\to+\infty} J^n(x_n)$ (5.7)

ii) $(\forall x\in X)(\exists x_n \in X) \lim_{n\to+\infty} x_n = x$ *in* $(X,\tau)$ *and* $\limsup_{n\to+\infty} J^n(x_n) \leq J(x)$ (5.8)

When $\tau$ is not metrizable, i) and ii) define the sequential epi-limit of $J^n$ . Such is the case of weak $\star$ topologies on $M^1(\Omega)$ or $BD(\Omega)$. For simplicity we shall omit the word "sequential", which will be implicit as soon as a non metrizable topology is used.

## 5.2 A MODEL PROBLEM OF SLIPPING BLOCKS

The onset and the development of macroscopic plastic strains is physically related to dislocations, i.e. to defects of the atomic structure at the microscopic scale. Motion of dislocations is similar to a friction phenomenon and the present section is devoted to a modelling of the microscopic situation. We consider a periodic assembly of parallelepipedic blocks slipping at their interfaces (one could also think of a wall made from bricks), and our goal is to derive the limit (or macroscopic) constitutive law of this assembly when the parallepipeds' size goes to 0 . This procedure has been popularized under the term *homogenization*.

For simplicity we reduce our attention to 2-dim problems, and we assume that the domain $\Omega$ is the assembly of identical cells stuck together on their boundaries, the strength of which is limited :

$$|\sigma_{ij}n_j(x)| \leq k \qquad i = 1,2 \qquad \text{for all} \quad x \in \Gamma \tag{5.9}$$

where $\Gamma$ denotes the union of all cells boundaries while $\Omega^*$ denotes the remaining part of the body $\Omega^* = \Omega - \Gamma$ .

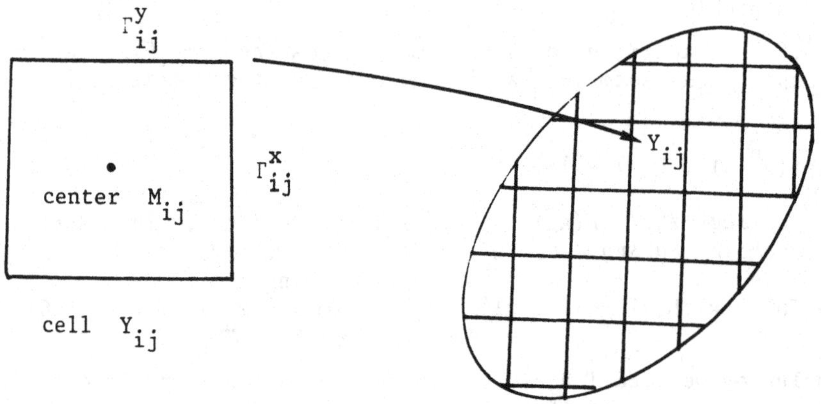


- Figure 5.1 -

Each open cell $Y_{ij}$ is centered on the node $M_{ij}$ of a periodic network extending on $\mathbb{R}^2$. The vertical right part, the horizontal upper part of $\partial Y_{ij}$ are denoted respectively $\Gamma^x_{ij}$ and $\Gamma^y_{ij}$ . We note that

$$\bigcup_{-\infty < i,j < +\infty} (Y_{ij} \cap \Omega) = \Omega^* , \qquad \bigcup_{-\infty < i,j < +\infty} (\Gamma^x_{ij} \cup \Gamma^y_{ij}) \cap \Omega = \Gamma \tag{5.10}$$

We shall only deal here with the limit analysis problem. Blocks can support an arbitrary load but the interfaces have a limited strength (both to shear and to normal stresses) described by (5.9) . For simplicity the body is loaded by body forces only ($\Gamma_o = \partial\Omega$) . The statical definition of the limit load is

$$\lambda^s = \text{Sup}\{\lambda \in \mathbb{R} \mid \exists \sigma \in \mathbb{L}^2(\Omega) \ , \ \text{div}\,\sigma + \lambda f = 0 \ \text{ a.e. in } \Omega \quad |\sigma.n| \leq k \ \text{ a.e. on } \Gamma\} \tag{5.11}$$

A kinematical definition can also be proposed. Let $u$ be a displacement rate field element of

$$V = \{u \in H^1(\Omega^*)^2 \ , \ u = 0 \ \text{ on } \ \partial\Omega\}^{(+)} \ ,$$

and $\lambda < \lambda^s$ , $\sigma$ associated with $\lambda$ by (5.11) . Then

$$\int_{\Omega^*} \sigma\, \varepsilon(u)\, dx = \sum_{i,j} \int_{Y_{ij}\cap\Omega} \sigma\, \varepsilon(u)\, dx$$

$$= \lambda \left[ \sum_{ij} \int_{Y_{ij}\cap\Omega} f\, u\, dx \right] + \int_{\partial Y_{ij}\cap\Omega} \sigma.n\, u\, ds \tag{5.12}$$

Adjacent cells have opposite normal vectors on their common boundaries, and a rearrangment of the second integral in (5.12) yields (according to (5.10))

$$\lambda \int_\Omega f\, u\, dx = \int_\Gamma \sigma.n\, [\![u]\!]\, ds + \int_{\Omega^*} \sigma\, \varepsilon(u)\, dx \tag{5.13}$$

Now we take the supremum on $\sigma$ satisfying (5.11) . Since we have no information on $\sigma|_{\Omega^*}$ the supremum is $+\infty$ except if

$\varepsilon(u) = 0$ in every $Y_{ij}$ , i.e. $u = \rho_{ij}$ in $Y_{ij}$

where $\rho_{ij}$ is a rigid displacement on $Y_{ij}$ . For those displacement rate fields we use the limited strength condition in the form

$$\sigma.n\, [\![u]\!] \leq \pi^s([\![u]\!]) = k(|[\![u_1]\!]| + |[\![u_2]\!]|) \tag{5.14}$$

and obtain

$$\lambda^k = \operatorname*{Inf}_{u \in V} \{\Pi(u) \ , \ L(u) = 1 \ , \ u = \rho_{ij} \ \text{ on } \ Y_{ij}\} \tag{5.15}$$

where

$$\Pi(u) = \int_\Gamma \pi^s([\![u]\!])\, ds \ , \ L(u) = \int_\Omega f u\, dx$$

Equality between $\lambda^s$ and $\lambda^k$ (left as an exercise to the reader) is a consequence of the abstract result recalled in section 2.3 .

We turn now to our main point of interest. The limit behavior of the assembly of cells with a size $1/n$ which goes to $0$ . Cells number is of the order of $n^2$ , cells boundaries $\Gamma_n$ and limit loads $\lambda^k_n$ are now indexed by $n$ . Our problem of variational convergence can be set in the general form (5.5) , $(X,\tau)$ being identified to $L^1(\Omega)^2$ endowed with its strong topology ($BD(\Omega)$ weak $*$ would lead to identical developments):

$$J^n(u) = \begin{cases} \int_{\Gamma^n} k|[\![u]\!]|\, ds & \text{if } \ u \in H^1(\Omega^*)^2, \ u = \rho_{ij} \ \text{ on } Y_{ij} \ , \ u = 0 \ \text{ on } \partial\Omega \\ +\infty & \text{otherwise} \end{cases} \tag{5.16}$$

(+) Note that if $u$ belongs to $H^1(\Omega^*)^2$ it can admit discontinuities along the cell boundaries.

*Theorem 5.3 Let* P *be defined as*

$$P = \{\sigma \in \mathbb{R}^4_s \ , \ |\sigma_{11}| \leq k \ , \ |\sigma_{12}| \leq k \ , \ |\sigma_{22}| \leq k\} \tag{5.17}$$

*and* $\pi$ *the support function of* P : $\pi(e) = k(|e_{11}| + 2|e_{12}| + |e_{22}|)$ , *then* $J^n$ (5.16) *epi-converges in* $L^1(\Omega)^2$ *strong to*

$$J(u) = \begin{cases} \int_\Omega \pi(\varepsilon(u)) + \int_{\partial\Omega} \pi^b(-u)ds & \text{if} \quad u \in BD(\Omega)^{(+)} \\ +\infty & \text{otherwise} \end{cases} \tag{5.18}$$

*Proof.*(++) The boundary term in (5.18) is obviously a relaxation term due to the boundary condition $u = 0$ in the definition of $J^n$ . This term, which is not our main point of interest, complicates the analysis, and we prove a weaker (but essential) form of theorem 5.3 , which does not account for the boundary conditions.

*Let* $\hat{J}^n$ *be defined as*

$$\hat{J}^n(u) = \begin{cases} \int_{\Gamma^n} k|[\![u]\!]|ds & \text{if} \quad u = \rho_{ij} \text{ on } Y_{ij},\ u \in H^1(\Omega^*)^2 \\ +\infty & \text{otherwise} \end{cases}$$

*Then* $\hat{J}^n$ *epi-converges in* $L^1(\Omega)^2$ *strong to*

$$\hat{J}(u) = \int_\Omega \pi(\varepsilon(u)) \quad \text{if} \quad u \in BD(\Omega) \quad , \quad +\infty \quad \text{otherwise.}$$

The result is attained with the help of the above described criterion (5.7)(5.8) for epi-convergence.

<u>First step</u> Let $u_n$ be a sequence in $L^1(\Omega)^2$ converging to u and such that $\liminf\limits_n \hat{J}^n(u_n) < +\infty$ . $u_n$ is piecewise smooth and belongs to BD(Ω) (cf(2.25)) with

$$|\varepsilon(u_n)|_{M^1(\Omega)^4_s} \leq \int_\Gamma |[\![u_n]\!]|ds \leq \frac{1}{k}\hat{J}^n(u_n) < C \tag{5.19}$$

Moreover $(u_n)$ is bounded in $L^1(\Omega)^2$ (since it is convergent) and with (5.19) $(u_n)^n$ is a bounded sequence in BD(Ω) . Therefore u belongs to BD(Ω) and $\hat{J}(u)$ is finite.

Consider σ in $C^o_c(\Omega)^4_s$ such that

$$(\forall\, x \in \Omega) \qquad \sigma(x) \in P \tag{5.20}$$

Then for $u_n$ in the domain of $\hat{J}^n$ (i.e. piecewise $C^\infty$) we have according to (2.25) and understanding integrals in the sense of the

(+) $\int_\Omega \pi(\varepsilon(u))$ is to be understood in the sense of functions of measure (§ 2.5) , $\pi^b$ is defined in (2.29) .

(++) This proof is adapted from BOUCHITTE [56] who treated the case of a scalar unknown u .

duality $C_c^o(\Omega)$ , $M^1(\Omega)$ :

$$\int_\Omega \sigma\varepsilon(u_n) \leq \int_{\Gamma^n} k\,|[\![u_n]\!]|ds \leq \hat{J}^n(u_n)$$

$\varepsilon(u_n)$ converges to $\varepsilon(u)$ in $M^1(\Omega)^4_S$ weak * and therefore

$$\sup_{\substack{\sigma \in C_c^o(\Omega)^4_s \\ \sigma(x) \in P}} \int_\Omega \sigma\,\varepsilon(u) \leq \liminf_{n\to+\infty} \hat{J}^n(u_n) \tag{5.21}$$

But the first member in (5.21) is the true definition of $\int_\Omega \pi(\varepsilon(u))$ according to section 2.5 . We have achieved the first step (5.7)

$$(\forall\, u \in L^1(\Omega)^2) \qquad \hat{J}(u) \leq \liminf_{n\to+\infty} \hat{J}^n(u_n)$$

Second step In order to prove (5.8) in our specific example, we can restrict attention to $u$ in $C^\infty(\Omega)^2$ which is dense in $L^1(\Omega)^2$ .

For $u$ in $C^\infty(\Omega)^2$ we define its vorticity by

$$\omega_{k\ell} = 1/2\,\left(\frac{\partial u_k}{\partial x_\ell} - \frac{\partial u_\ell}{\partial u_k}\right)$$

(Note that $\omega_{11} = \omega_{22} = 0$ , $\omega_{21} = -\omega_{12} = \frac{1}{2}\left(\frac{\partial u_2}{\partial x_1} - \frac{\partial u_1}{\partial x_2}\right))$ . Define a rigid displacement $\rho$ on $Y_{ij}$ by

$$\rho_1^{ij}(x) = u_1(M_{ij}) - \omega_{21}(M_{ij})(x_2 - x_2^{ij})\ ,\ \rho_2^{ij}(x) = u_2(M_{ij}) + \omega_{21}(M_{ij})(x_1 - x_1^{ij})$$

where $x_k^{ij} = x_k(M_{ij})$ .

Finally we propose to approximate $u$ in $L^1(\Omega)^2$ strong by

$$u_n(x) = \rho^{ij}(x) \quad \text{if} \quad x \in Y_{ij}\ .$$

The following (easy) inequality ensures the convergence of $u_n$ towards $u$ in $L^1(\Omega)^2$ strong

$$|u_n - u|_{L^\infty(\Omega)^2} \leq \frac{1}{n}\,|\text{grad } u|_{L^\infty(\Omega)^4}\ .$$

$\hat{J}^n(u_n)$ is the sum of elementary integrals on $\Gamma^x_{ij}$ and $\Gamma^y_{ij}$ (see notations on figure 5.1)

$$k\int_{\Gamma^x_{ij}} (|\rho_1^{i+1,j} - \rho_1^{i,j}| + |\rho_2^{i+1,j} - \rho_2^{i,j}|)ds \tag{5.22}$$

and a similar integral over $\Gamma^y_{ij}$ . In order to compute the above integral we use a Taylor development of $u$ and $\omega_{12}$ in the neighborhood of $M_{ij}$ :

$$u_1(M_{i+1,j}) - u_1(M_{ij}) = \frac{1}{n}\frac{\partial u_1}{\partial x_1}(M_{ij}) + 0\left(\frac{1}{n^2}\right)$$

$$u_2(M_{i+1,j}) - u_2(M_{ij}) = \frac{1}{n}\frac{\partial u_2}{\partial x_1}(M_{ij}) + 0\left(\frac{1}{n^2}\right)$$

$$= \frac{1}{n}(\varepsilon_{21}(u)(M_{ij}) + \omega_{21}(M_{ij})) + 0\left(\frac{1}{n^2}\right)$$

$$|\omega_{21}(M_{i+1,j}) - \omega_{21}(M_{ij})| \leq \frac{1}{n}\,|D^2u|_{L^\infty(\Omega)^8}$$

Therefore, recalling that on $\Gamma^x_{ij}$ we have $x_1 - x_1^{ij} = \frac{1}{2n}$, $x_1 - x_1^{i+1,j} = -\frac{1}{2n}$, and $x_2^{i+1,j} = x_2^{ij}$, $x_1^{i+1,j} = x_1^{ij} + \frac{1}{n}$, we obtain

$$|\rho_1^{i+1,j}(x) - \rho_1^{ij}(x)| = \frac{1}{n}\left|\frac{\partial u_1}{\partial x_1}(M_{ij})\right| + 0(\frac{1}{n^2})$$

$$|\rho_2^{i+1,j}(x) - \rho_2^{ij}(x)| = |u_2(M_{i+1,j}) - u_2(M_{ij}) - (\omega_{21}(M_{i+1,j}) + \omega_{21}(M_{ij}))\frac{1}{2n}|$$

$$= |u_2(M_{i+1,j}) - u_2(M_{ij}) - \frac{1}{n}\omega_{21}(M_{ij})| + 0(\frac{1}{n^2}) = \frac{1}{n}|\varepsilon_{21}(u)(M_{ij})| + 0(\frac{1}{n^2})$$

We also note that

$$\int_{Y_{ij}} |\varepsilon_{kh}(u(x))|dx = \frac{1}{n^2}|\varepsilon_{kh}(u)(M_{ij})| + 0(\frac{1}{n^3})$$

Therefore the integral (5.22) over $\Gamma^x_{ij}$ amounts to

$$k\int_{Y_{ij}} (|\varepsilon_{11}(u)| + |\varepsilon_{12}(u)|)dx + 0(\frac{1}{n^3}) \tag{5.23}$$

A similar argument shows that the integral over $\Gamma^y_{ij}$ amounts to

$$k\int_{Y_{ij}} (|\varepsilon_{12}(u)| + |\varepsilon_{22}(u)|)dx + 0(\frac{1}{n^3}) \tag{5.24}$$

Combining (5.23) and (5.24), achieving the sum over indices ij yields :

$$\hat{J}^n(u_n) = \int_\Omega k(|\varepsilon_{11}(u)| + 2|\varepsilon_{12}(u)| + |\varepsilon_{22}(u)|)dx + 0(\frac{1}{n}))$$

and therefore

$$\lim_{n\to+\infty} \hat{J}^n(u_n) = \hat{J}(u)$$

Proof of theorem 5.3 is complete.

*Corollary* 5.3 $\lim_{n\to+\infty} \lambda_n^k = \lambda^k$ *where*

$$\lambda_n^k = \operatorname*{Inf}_u \{J^n(u), L(u) = 1\} \tag{5.25}$$

$$\lambda^k = \operatorname*{Inf}_u \{J(u), L(u) = 1\} \tag{5.26}$$

To combine theorem 5.3 with corollary 5.2 it is sufficient to prove that approximate minimizers of (5.25) form a compact set in $L^1(\Omega)^2$, or are bounded in $BD(\Omega)$ (use the compact embedding of $BD(\Omega)$ into $L^1(\Omega)^2$ cf § 2.3). But this last estimate is a result of (5.19) and (2.19) with the following semi-norm

$$r(u) = \int_{\partial\Omega} |u|ds$$

which vanishes on $J^n$'s domain. The proof of corollary 5.3 is complete.

(5.26) has an important mechanical interpretation :

*An assembly of rigid blocks, the interfaces of which have a limited strength, behaves macroscopically* (i.e. when the blocks' size becomes so small that they cannot be distingued one from another) *as a rigid-plastic body with a strength domain* P *defined in* (5.17) .

## 5.3  ASSEMBLY OF MATERIALS WITH LIMITED STRENGTH I. NO LOADS ON THE BOUNDARY

This section deals with the homogenized properties of a periodic assembly of materials with a limited strength.

$$P(y) = \begin{cases} P^1 & \text{if } y \in Y_1 \\ P^2 & \text{if } y \in Y_2 \end{cases}$$

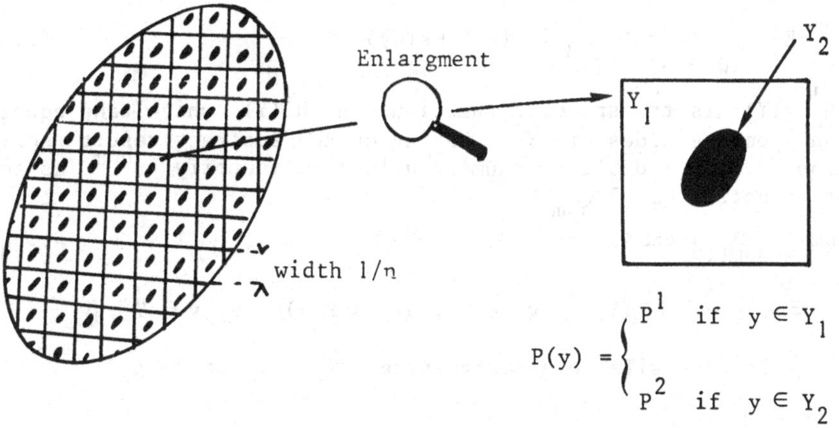

-  Figure  5.2  -

Main notations of the problem are displayed on figure  5.2 . Material properties are defined on the unit cell  $Y = ]0,1[^3$  and extended by periodicity to  $\mathbb{R}^3$ . We assume for simplicity that  P(y)  is simply defined by a ball of radius  k(y)  (assumption  (2.4))  and that  k(y)  takes two (non vanishing) values on two borelian sets  $Y_1$  and  $Y_2$  :

$$k(y) = k_1 \text{ if } y \in Y_1 \quad , \quad k(y) = k_2 \text{ if } y \in Y_2 , \tag{5.27}$$

thus  $\pi(y,e) = k(y)|e|$ .  k(y)  is extended by periodicity to  $\mathbb{R}^3$  and we set

$$k^n(x) = k(nx) \quad , \quad \pi^n(x,e) = k^n(x)|e|$$

$k^n$  is periodic with a period  1/n  in each direction.

The sequence of limit loads is defined by

$$\lambda_n = \text{Inf}\{\Pi^n(u) , u \in L^1(\Omega)^3 , L(u) = 1\} \tag{5.28}$$

$$\Pi^n(u) = \begin{cases} \int_\Omega \pi^n(x,\varepsilon(u))dx & \text{if } u \in H^1(\Omega)^3 , u = 0 \text{ on } \Gamma_o \quad (5.29) \\ \\ +\infty & \text{otherwise} \end{cases}$$

$$L(u) = \int_\Omega f\,u\,dx$$

(in this section we consider that $\Gamma_o = \partial\Omega$ , $\Gamma_1 = \emptyset$) .
The homogenized strength properties is given by theorem 5.4 due in its generality to BOUCHITTE [26] , extending a partial result of the author [57] . We need further notations to formulate it clearly : < > denotes the average symbol on the unit cell

$$\langle f\rangle = \frac{1}{|Y|}\int_Y f(y)dy \ .$$

We define a convex l.s.c. function, positively homogeneous of degree one, on $\mathbb{R}^9_s$

$$\pi^{hom}(E) = \operatorname*{Inf}_{w\in H^1_{per}(Y)^3} \langle\pi(y, E+\varepsilon(w))\rangle \tag{5.30}$$

where $H^1_{per}(Y)$ is the space of functions in $H^1(Y)$ which take equal values on opposite sides of $Y$ . Basing on the abstract result recalled in section 2.2 , a duality argument allows to identify $\pi^{hom}$ as the conjugate function of $I_{P^{hom}}$

$$P^{hom} = \operatorname{dom}((\pi^{hom})^*) \tag{5.31}$$

$$= \{\Sigma\in\mathbb{R}^9_s \,|\, \exists\sigma\in\mathbb{L}^2(Y),\ \operatorname{div}\sigma = 0\ ,\ \sigma(y)\in P(y) \text{ a.e. } y\in Y\ ,$$

$$\sigma.n \text{ opposite on opposite sides of } Y\ ,\ \langle\sigma\rangle = \Sigma\} \ .$$

Then we have

*Theorem 5.4 [26] Assume that assumption (2.1) is met. Then* $\pi^n$ *epi-converges in* $L^1(\Omega)^3$ *strong to* $\Pi^{hom}$

$$\Pi^{hom}(u) = \begin{cases} \int_\Omega \pi^{hom}(\varepsilon(u)) + \int_{\Gamma_o}(\pi^{hom})^b(-u)ds & \textit{if } u\in BD(\Omega) \\ +\infty & \textit{otherwise} \end{cases} \tag{5.32}$$

*Moreover*

$$\lim_{n\to+\infty}\lambda_n = \lambda = \operatorname{Inf}\{\Pi^{hom}(u)\ ,\ L(u) = 1\} \tag{5.33}$$

*Comment.* Theorem 5.4 states that when heterogeneities'size becomes small ($1/n \to 0$) , the limit load computed on the nonhomogeneous material is well approximated by the limit load of a fictitious homogenized material the strength domain of which is $P^{hom}$ (5.29) .

## 5.4 ASSEMBLY OF MATERIALS WITH LIMITED STRENGTH. LOADS ON THE BOUNDARY

When the assumption $\Gamma_o = \partial\Omega$ , $\Gamma_1 = \emptyset$ is abandonned, new interesting problems arise. Indeed the linear form $L$ is no more continuous on $BD(\Omega)$ (as we already noticed in section 2.5) , because of its boundary term

$$L(u) = \int_\Omega fu\,du + \int_{\Gamma_1} g\,u\,ds \tag{5.34}$$

Therefore Corollary 5.2 no more applies. More specifically, conter-examples to an extrapolation of (5.33) have been given in the recent litterature (DE BUHAN [58] , TURGEMAN-PASTOR [59] : consider the sequence $\lambda_n$ of limit loads for the nonhomogeneous materials, defined by (5.28) but where L is now defined by (5.34) . Then $(\lambda_n)$ *does not converge* to the natural guess

$$\lambda_\Omega = \mathrm{Inf}\{\Pi^{hom}(u)\ ,\ L(u) = 1\} \quad \Pi^{hom} \text{ given by (5.32))} \tag{5.35}$$

We describe an adaptation of DE BUHAN's example (in which $\partial\Omega$ was not $C^1$) which illustrates this surprising negative result.

Example Consider in $\mathbb{R}^2$ a stratified medium made from two homogeneous constituents obeying (5.27)

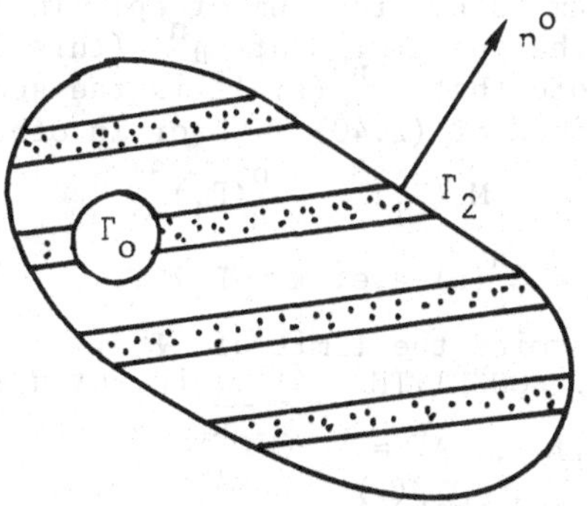


\- Figure 5.3 -

$P^{hom}$ (given by (5.31)) can be determined. It is contained in the ball of center 0 and radius $k_2$ , and *strictly* contains the ball of center 0 and radius $k_1$ (we assume $k_2 > k_1$) . Therefore

$$(\exists\ \Sigma^o \in P^{hom})(\exists n^o \in \mathbb{R}^3)\ ,\ |n^o| = 1\ ,\quad |\Sigma^o.n^o| > k_1$$

(By continuity $n^o$ can be chosen to be *not normal* to the stratification plane). $\Omega$ is chosen with a smooth boundary containing a flat part $\Gamma_2$ normal to $n^o$ (cf Figure 5.3) , and we set

$$g(x) = \Sigma^o.n(x) \qquad \forall\ x \in \Gamma_1\ ,\quad f = 0$$

In the nonhomogeneous material we have for an admissible $\lambda < \lambda^s$

$$|\lambda g(x)| \leqslant k(x) \qquad \text{a.e. } x \in \Gamma_1$$

i.e. on $\Gamma_2$

$$\lambda_n \leqslant \frac{k_1}{|\Sigma^o.n^o|} = \beta < 1$$

Therefore $\limsup \lambda_n \leqslant \beta < 1$ . But it can be checked that $\Sigma^o$ satisfies

$$\Sigma^o \in P^{hom} , \ \operatorname{div} \Sigma^o = 0 , \ \Sigma^o.n(x) = g(x) \quad \forall x \in \Gamma_1$$

and therefore that $\lambda_\Omega \geq 1$ . In this example $\limsup \lambda_n$ and $\lambda_\Omega$ cannot be equal.

The correct limit to the sequence $(\lambda_n)$ is provided by corollary 5.4 below. In order to understand it we recall that we introduced in section 2.5 an appropriate relaxed form of the limit analysis problem, for loaded boundaries (cf(2.30))

$$\lambda_n = \operatorname*{Inf}_{u,u^+} \{\hat{\Pi}^n(u,u^+) , \hat{L}(u,u^+) = 1 , (u,u^+) \in H^1(\Omega)^3 \times H^{1/2}(\Gamma_1)^3\}$$

$$\hat{\Pi}^n(u,u^+) = \int_\Omega \pi^n(\varepsilon(u))dx + \int_{\Gamma_o} \pi^{nb}(-u)ds + \int_{\Gamma_1} \pi^{nb}(u^+ - u)ds$$

$$= \Pi^n(u) + J^n(u^+ - u).$$

Although the epi-limit of a sum is not the sum of epi limits, it can be useful to compute separately the epi limits of $\Pi^n$ (this is done in theorem 5.4), and $J^n$ . We note that $\pi^{nb}(x,.)$ is the support function of set $C^n(x)$ previously defined cf (2.40) , or in other terms that

$$J^n = (I_{\mathbb{C}^n})^* \text{ in the duality } M^1(\Gamma_1)^3 - C^o_c(\Gamma_1)^3$$

$$\mathbb{C}^n = \{\varphi \in C^o(\Gamma_1)^3 , \varphi(x) \in C^n(x) \text{ a.e. } x \in \Gamma_1\} .$$

By duality we have now to determine the limit of $\mathbb{C}^n$ . It is given by the following result due to G. BOUCHITTE [4] . *Let us define*

$$\Omega^n_1 = \{x \in \Omega \quad P^n(x) = P_1\} , \quad A^n = \Gamma_1 \cap \Omega^n_1$$

*We assume that* (2.3) is met and that(+)

$$H^2(\partial A^n) = 0 , \quad \operatorname{int}(A^n) \to A , \quad \operatorname{int}(\Gamma_1 \setminus A) \to B$$

*Then* $I_{\mathbb{C}^n}$ *epi-converges in* $C^o(\Gamma_1)^3$ *strong to* $I_\mathbb{C}$ :

$$\mathbb{C} = \{\varphi \in C^o(\Gamma)^3 , \varphi(x) \in \mathbb{C}(x) \quad \forall x \in \Gamma_1\}$$

$$\mathbb{C}(x) = P_1 n(x) \ \text{ if } \ x \in A \setminus B , \quad \mathbb{C}(x) = P_2 n(x) \ \text{ if } \ x \in B \setminus A$$

$$\mathbb{C}(x) = P_1 n(x) \cap P_2 n(x) \ \text{ if } \ x \in A \cap B .$$

Let $\pi_{\Gamma_1}(x,.)$ denote the support function of $\mathbb{C}(x)$ which can be extended in a function of measure on $M^1(\Gamma_1)^3$ , and set

$$J(\mu) = \int_{\Gamma_1} \pi_{\Gamma_1}(x,\mu)$$

Then a duality result (ATTOUCH [3]) asserts that $J^n$ epi-converges (sequentially) in $M^1(\Gamma_1)^3$ weak $\star$ to $J$ . We are now in a position to state a precise result (BOUCHITTE & SUQUET [20]) .

---

(+) $H^2$ is the 2-dim Hausdorff measure

*Theorem* 5.5 $\hat{\Pi}^n$ *epi-converges (sequentially) in* $BD(\Omega) \times M^1(\Gamma_1)^3$ *weak* ✱ *to*

$$\hat{\Pi}^{hom}(u,u^+) = \Pi^{hom}(u) + \int_{\Gamma_o} \pi^{hom\, b}(-u)ds + \int_{\Gamma_1} \pi_{\Gamma_1}(u^+ - uds) \quad (5.36)$$

*Corollary* 5.4 *Assume that* g *belongs to* $C^o(\Gamma_1)^3$ . *Then* $\lambda_n$ *converges to* $\lambda_{hom}$ *where* :

i) $\lambda_{hom} = \operatorname{Inf}_{u,u^+} \{\Pi^{hom}(u,u^+) \ , \ \hat{L}(u,u^+) = 1 \ , \ u,u^+ \in BD(\Omega) \times M^1(\Gamma_1)^3\}$

ii) $\lambda_{hom} = \operatorname{Inf}(\lambda_\Omega \ , \Lambda)$ (5.37)

$\lambda_\Omega$ *is defined by* (5.35) *while* $\Lambda$ *is given by*

$$\Lambda = \inf\{\alpha \in \mathbb{R} \mid \alpha g(x) \in C^{hom}(x) \cap C(x) \quad \forall\, x \in \Gamma_1\} \quad (5.38)$$

*where* $C^{hom}(x) = P^{hom} \cdot n(x)$ .

*Comment.* Corollary 5.4 contains two statements
i) is a direct consequence of theorem 5.5 , since $\hat{L}$ *is continuous* on $BD(\Omega) \times M^1(\Gamma_1)^3$ weak ✱ (therefore corollary 5.2 applies).
ii) shows that the homogenized limit analysis problem is twofold. A first limit analysis problem set on an homogeneous $\Omega$ with strength domain $P^{hom}$ , is to be solved and yields $\lambda^\Omega$ . A second limit analysis problem, in which only the strength of $\Gamma_1$ is limited, is to be solved for $\Lambda$ . The real limit load is the infimum of these two values. It should be noted that $\Lambda$ cannot be derived from the homogenized behavior in $\Omega$ (open set) only : the homogenized behavior of $\Gamma_1$ is different. It should also be noted that "relaxation and epi-convergence do not always commute", in the following sense :

$$\Pi^n_R(u) = \Pi^n(u) + \Lambda^n \,|1 - L(u)|$$

which epi-converges to

$$\Pi^{hom}_R(u) = \Pi^{hom}(u) + \Lambda \,|1 - L(u)| \ .$$

However $\Pi^n$ epi-converges to $\Pi^{hom}$ , and the relaxation term for $\Pi^{hom}$ associated with the constraint $L(u) = 1$ is

$$\Pi^{hom}(u) + \Lambda^{hom}|1 - L(u)|$$

$$\Lambda^{hom} = \operatorname{Sup}_\alpha \{\alpha \in \mathbb{R} \mid \alpha g(x) \in C^{hom}(x) \quad \forall\, x \in \Gamma_1\}$$

$$\Lambda^{hom} \geqslant \Lambda \quad \text{(strict inequality can occur)} \ .$$

## 5.5 BIBLIOGRAPHICAL COMMENTS, OPEN PROBLEMS

1. Epi-convergence and Γ-convergence have been extensively studied by a very active Italian school (see BUTTAZO & al [60] , DE GIORGI & al [61] , and additional references in ATTOUCH [3]) .

2. The approach developped here is *essentially kinematical* (epi-convergence of functionals defined on displacement rates). By duality

arguments we can obtain informations on homogenized strength capacities. However a direct study of statical problems (formulated in terms of stresses only) would be interesting. For duality arguments we refer to AZE [62] , and for convergence of variational inequalities with obstacles to DAL MASO & al [63] ATTOUCH-PICARD [64] .

3. Theorem 5.4 is due to G. BOUCHITTE [65] who extended a previous (partial) result of the author [57] . This work includes results on the elasto-plastic problem. DEMENGEL & TANG QUI [66] extended BOUCHITTE's work (by a different method) to the case of Von Mises criterion, for which a constraint of incompressibility is to be added in definitions of $\Pi^n$ and $\Pi^{hom}$ . However the treatment of boundary loads in [66] does not let appear clearly a different homogenized behavior on $\Gamma_1$ . The present exposure of "loaded boundaries" follows BOUCHITTE & SUQUET [20] in which, to the author's knowledge, the necessity of two limit analysis problems (inside $\Omega$ and on $\Gamma_1$) was first established.

4. We have explicitely made regularity assumption : $\partial\Omega$ is $C^1$ , g is $C^0$ . However interesting problems arise when these assumptions are dropped, as illustrated by TURGEMAN & PASTOR's example [59] in which $\partial\Omega$ exhibits corners.

# REFERENCES


1. TEMAM R., STRANG, G. : "Duality and relaxation in Plasticity", *Journal de Mécanique*, 19, 1980, p. 493-527.
2. TEMAM R. : *Problèmes mathématiques en Plasticité*, Gauthier-Villars, Paris, 1983.
3. ATTOUCH H. : *Variational convergence for functions and operators*, Pitman, London, 1984.
4. BOUCHITTE G. : "Calcul des variations en cadre non réflexif...", Thesis. University of Perpignan. France. 1987.
5. SALENCON J. : *Calcul à la rupture et analyse limite*, Presses ENPC, Paris, 1983.
6. NAYROLES B. : "Essai de théorie fonctionnelle des structures rigides plastiques parfaites", *J. de Mécanique*, 9, 1970, 491-506.
7. ROCKAFELLAR R.T. : *Convex Analysis*. Princeton Univ. Press. Princeton. 1970.
8. EKELAND I., TEMAM R. : *Convex Analysis and Variational Problems*, North Holland, Amsterdam, 1976.
9. PANAGIOTOPOULOS P.D. : *Inequality problems in Mechanics and Applications*. Birkhaüser. Boston. 1985.
10. STRANG G. : "A family of model problems in Plasticity", *Proc. Symp. Comp. Meth. in Appl. Sc.* Ed. R. Glowinski and J.L. Lions, Lecture Notes in Math. n° 704, 1978, p. 292-308.
11. SUQUET P. : "Sur un nouveau cadre fonctionnel pour les équations de la Plasticité", *C.R. Acad. Sci. Paris*, A, 286, 1978, p. 1129-1132.
12. SUQUET P. : "Un espace fonctionnel pour les équations de la Plasticité", *Ann. Fac. Sci. Toulouse*, 1, 1979, p. 77-87.
13. BOURBAKI N. : *Intégration*. Hermann. Paris. 1954.
14. MATTHIES H., STRANG G. : "The saddle point of a differential program" in *Energy Methods in Finite Element Analysis*, Ed. Glowinski, Robin, Zienckiewicz. John Wiley. New-York. 1979.
15. GOFFMAN C., SERRIN J. : "Sublinear functions of measures and variational integrals", *Duke Math. J.*, 31, 1964, p. 159-178.
16. DEMENGEL F., TEMAM R. : "Convex functions of a measure and applications", *Indiana Univ. Math. J.*, 33, 1984, p. 673-709.
17. GIAQUINTA M., MODICA G., SOUCEK J. : "Functionals with linear growth in the calculus of variations", *Com. Math. Univ. Carolina*, 20, 1979, p. 143-171.
18. VALADIER M. : "Fonctions et opérateurs sur les mesures", *C.R. Acad. Sc. Paris*, I, 304, 1986, p. 135-138.

19. HADHRI T. : "Fonction convexe d'une mesure", *C.R. Acad. Sc. Paris*, I, 301, 1985, p. 687-690.
20. BOUCHITTE G., SUQUET P. : "Charges limites, plasticité et homogénéisation : le cas d'un bord chargé", *C.R. Acad. Sc. Paris*, I, 305, 1987, p. 441-444 .
21. SUQUET P. : "Existence and regularity of solutions for Plasticity problems" in *Variational Methods in the Mechanics of Solids*, Ed S. NEMAT-NASSER, Pergamon Press, Oxford, 1980, p. 304-309.
22. SUQUET P. : "Sur les équations de la Plasticité : existence et régularité des solutions". *J. de Mécanique*, 20, 1981, p. 3-39.
23. DEMENGEL F. : "Espaces de fonctions à dérivées mesures et application à l'analyse limite en plasticité", *C.R. Acad. Sc. Paris*, I, 302, 1986, p. 179-182.
24. HADHRI T. : "Sur le glissement à l'interface de deux milieux homogènes constituant un solide de Hencky", *C.R. Acad. Sc. Paris*, II, 302, 1986, p. 1181-1184.
25. DAL MASO G. : "Integral representation on BV(Ω) of Γ-limit of variational integrals", *Manuscripta Math.*, 30, 1980, p. 387-416.
26. BOUCHITTE G. : "Convergence et relaxation de fonctionnelles du calcul des variations à croissance linéaire. Application à l'homogénéisation en plasticité", *Ann. Fac. Sc. Toulouse*, 8, 1986-1987, p. 7-36.
27. MERCIER B. : "Une méthode pour résoudre le problème des charges limites", *J. Méca.*,16, 1977.
28. FREMOND M., FRIAA A. : "Les méthodes statique et cinématique en calcul à la rupture et en analyse limite", *J. Méca. Th. Appl.*, 1, 1982, p. 881-905.
29. JOHNSON C. : "Error estimates for some finite element methods for a model problem in perfect plasticity", Meeting "Problemi matematici della meccanica dei continui", Trento, Italy, 1981.
30. CHRISTIANSEN E. : "Computation of limit loads", *Int. J. Num. Meth. Eng.*, 17, 1981, p. 1547-1570.
31. NGUYEN DANG HUNG : "Sur la plasticité et le calcul des états limites par éléments finis", Thèse de Doctorat Spécial. Liège, 1985.
32. GAVARINI C. : "Plastic analysis of structures and duality in linear programming", *Meccanica* 1, 1966.
33. DUVAUT G., LIONS J.L. : *Les inéquations en Mécanique et en Physique*, Dunod, Paris, 1972.
34. NECAS J., HLAVACEK I. : *Mathematical theory of elastic and elastico-plastic bodies : an introduction*, Elsevier, Amsterdam, 1981.
35. KOHN R., TEMAM R. : "Dual spaces of stresses and strains with applications to Hencky Plasticity", *Appl. Math. Optim.*, 10, 1983, p. 1-35.
36. HADHRI T. : "Convex function of a measure and application to a problem of nonhomogeneous plastic material", Preprint 140, Ecole Polytechnique, Palaiseau, France, 1986.
37. BEN DHIA H., HADHRI T. : "Existence result and discontinuous finite element discretization for a plane stress Hencky problem", Preprint 146, Ecole Polytechnique, Palaiseau, France, 1986.
38. BEN DHIA H. : "Numerical analysis of a bidimensional Hencky problem approximated by a discontinuous finite element method", Preprint 161, Ecole Polytechnique, Palaiseau, France, 1987.

39. JOHNSON C. : "Existence theorems for Plasticity problems", *J. Math. Pures Appl.*, 55, 1976, p. 431-444.
40. SUQUET P. : *Plasticité et homogénéisation*, Thèse d'Etat, Paris, 1982. See also "A few mathematical aspects of incremental plasticity", Cours CIMPA, Monastir, 1986. To be published. Internal report 86-2, Montpellier, France.
41. BREZIS H. : "Un problème d'évolution avec contraintes unilatérales dépendant du temps", *C.R. Acad. Sc. Paris*, A, 274, 1972, p. 310-312.
42. MOREAU J.J. : "Evolution problèm associated with a moving convex set in a Hilbert space", *J. Diff. Eq.*, 26, 1977, p. 347-374.
43. ANZELOTTI : "On the existence of the rates of stress and displacement for Prandtl Reuss Plasticity", *Quart. Appl. Math.*, 46, 1983, p. 181-208.
44. PERZYNA P. : "Fundamental problems in viscoplasticity" in *Advances in Applied Mechanics*, 9, 1966, p. 241-258.
45. FRIAA A. : "Le matériau de Norton-Hoff généralisé et ses applications à l'analyse limite", *C.R. Acad. Sc. Paris*, A, 286, 1978, p. 953-956.
46. JOHNSON C. : "On plasticity with hardening", *J. Math. Anal. Appl.*, 62, 1978, p. 325-336.
47. NGUYEN QUOC SON : "Matériaux plastiques écrouissables. Distribution de la contrainte dans une évolution quasi-statique", *Ark. Mech.*, 25, 1973, p. 695-702.
48. JOHNSON C. : "A mixed finite element method for Plasticity with hardening", *SIAM J. Numer. Anal.*, 14, 1977, p. 575-583.
49. DJAOUA M. : "Analyse mathématique et numérique de quelques problèmes en mécanique de la rupture", Thèse de Doctorat d'Etat, Paris, 1983.
50. GLOWINSKI R., MAROCCO A. : "Sur l'approximation par éléments finis d'ordre un et la résolution par pénalisation-dualité d'une classe de problèmes de Dirichlet non linéaires", *RAIRO*, 1975; p. 41-76.
51. CAILLERIE D. : "The effect of a thin inclusion of high rigidity in an elastic body", *Math. Meth. Appl. Sc.*, 2, 1980, p. 251-270.
52. BLANCHARD D., PAUMIER J.C. : "Une justification de modèles de plaques viscoplastiques", *RAIRO Analyse Numérique*, 18, 1984, p. 377-406.
53. DESTUYNDER P., NEVEU D. : "Sur les modèles de lignes plastiques en Mécanique de la Rupture", *Math. Modelling and Numer. Anal.*, 20, 1986, p. 251-263.
54. BREZIS H., CAFFARELLI L., FRIEDMAN A. : "Reinforcement problems for elliptic equations and variational inequalities", *Ann. Mat. Pura. Appl.*, 123, 1980, p. 219-246.
55. ACERBI E., BUTTAZO G. : "Reinforcement problems in the calculus of variations", *Scuola Norm. Sup. Pisa*, 1984.
56. BOUCHITTE G. : Private communication, 1986.
57. SUQUET P. : "Analyse limite et homogénéisation", *C.R. Acad. Sc. Paris*, II, 296, 1983, p. 1355-1358.
58. DE BUHAN P. : "Approche fondamentale du calcul à la rupture des ouvrages en sols renforcés", Thèse d'Etat, Paris, 1986.
59. TURGEMAN S., PASTOR J. : "Comparaison des charges limites d'une structure hétérogène et homogénéisée", *J. Méca. Th. Appl.*, 6, 1987, 121-143.

60. BUTTAZO G., DAL MASO G. : "Γ limit of integral functionals", *J. Anal. Math.*, 37, 1980, p. 145-185.
61. DE GIORGI E., AMBROSIO L., BUTTAZO G. : "Integral representation and relaxation for functionals defined on measures", Scuola Norm. Sup. Pisa, 1986.
62. AZE D. : "Homogénéisation primale et duale par épi-convergence". Publications AVAMAC. Université de Perpignan, 1984.
63. DAL MASO G., LONGO P. : "Γ-limits of obstacles", *Ann. Mat. Pura Appl.*, 128, 1981, p. 1-50.
64. ATTOUCH H., PICARD C. : "Variational inequalities with varying obstacles", *J. Funct. Anal.*, 50, 1983, p. 329-386.
65. BOUCHITTE G. : "Homogénéisation sur BV(Ω) de fonctionnelles intégrales à croissance linéaire", *C.R. Acad. Sc. Paris*, I, 301, 1985, p. 785-788.
66. DEMENGEL F., TANG QI : "Homogénéisation en Plasticité", *C.R. Acad. Sc. Paris*, 303, I, 1986, p. 339-342.
67. DJAOUA M., SUQUET P. : "Evolution quasi-statique des milieux viscoplastiques de Maxwell-Norton", *Math. Meth. Appl. Sc.*, 6 , 1984, p. 192-205.
68. BLANCHARD D., LE TALLEC P. : "Numerical Analysis of the equations of small strains quasistatic elastoviscoplasticity", *Numer. Math.*, 50, 1986, p. 147-169.

# TOPICS ON UNILATERAL CONTACT PROBLEMS OF ELASTICITY AND INELASTICITY

J.J. Telega
Polish Academy of Sciences, Warsaw, Poland

ABSTRACT

These lectures present selected unilateral boundary and initial value problems of elasticity and inelasticity. Particularly, for elastic Signorini's problem with general subdifferential sliding rule two dual problems are derived as well as the bidual one. Some historical and modern views on friction phenomenon are briefly presented. The next chapter is concerned with unilateral frictionless and frictional problems for viscoelastic, plastic and viscoplastic bodies. In the last chapter homogenization of microfissured elastic solids and plates is studied. It is assumed that microfissures behave unilaterally.

## 1. INTRODUCTION

These lectures are focussed on variational approach to unilateral contact problems for elastic and inelastic bodies. Non-smooth effects are imposed by unilateral conditions at the boundary. In the case of perfect plasticity /Chapter 3/ and fissured solids and plates /Chapter 4/ certain functions like displacements and velocities are also non-smooth /discontinuous/.

In my opinion it is covenient to classify contact problems of solid and structural mechanics as follows.

/i/ External and internal contact problems. In the first case the body considered is in contact with another rigid or deformable body /elastic or inelastic/, or with a system of such bodies. Internal contact problems are posed by fracture and damage mechanics, see Chapter 4 of these lectures.

/ii/ Contact of bodies made of: inorganic materials, for instance metal - metal contact;organic materials, for example cartilage - cartilage contact in human and animal joints, see Refs [1,9,10,22,26,27,36,37;38,40,52] and Fig.1;

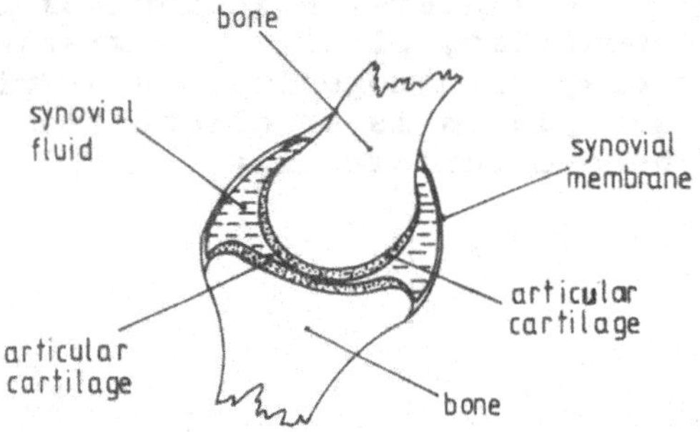


Fig.1, Diagrammatic representation of a human joint

organic and inorganic materials /in inplants, for instance bone-polymethyl methacrylate cement, bone-metal contact/, cf. Refs [2,6,12,13,16,17,28-31,48,49,57] and Fig.2.

/iii/ Dependence upon friction: frictionless contact, contact with friction but without lubrication /dry contact/, lubricated contact , see Refs [4,15,20,32,42,45,46,53,56].

/iv/ Unilateral or non-unilateral conditions.

/v/ Presence or absence of adhesion /see the lectures by Frémond/ and wear.

Obviously, in real situations various combinations of points /i/ - /v/ occur.

Now I shall briefly present the subject of my lectures. Since I largely deal with unilateral contact problems with friction, therefore some historical and modern

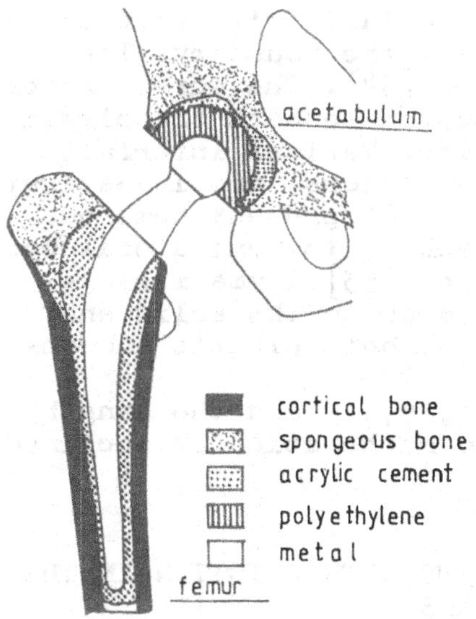


**Fig.2, Schematic illustration of a Charnley total hip prosthesis /acetabular and femoral component/ fixated in the bone with acrylic cement**

concepts of the friction phenomenon are first succintly presented. Various phenomenological friction laws or sliding rules are next introduced. Both associated and non--associated laws are discussed. Of main concern in these notes are subdifferential laws which are very convenient in variational formulations. Having formulated the elastic Signorini's contact problem with friction I pass to the study of dual and bidual problems. Toward this end the theory of duality in the sense of Mosco-Capuzzo Dolcetta-Matzen /M-CD-M theory of duality/ is used. Existence problems are also discussed.

Chapter 3 deals with unilateral contact problems for inelastic solids: viscoelastic, rigid-perfectly plastic, elasto-plastic and viscoplastic. Large deformations are also mentioned.

In the last chapter homogenization or averaging of periodically microfissured elastic solids and plates is considered. I assume that microfissures behave unilaterally. Therefore such problems leads naturally to homogenization of variational inequalities posed on domains depending on a small parameter . After homogenization we arrive at equations since microfissures are "smeared out". In the case of frictional behaviour of microfissures implicit variational inequality has to be homogenized.

Before ending this introductory chapter I would like to comment briefly on some relevant references. The reader interested in classical contact problems should refer to the books [23,24,25,32,47,51]. The classical, non-variational contact mechanics deals essentially with problems involving no unilateral conditions at the boundary. The founder of such approach was Hertz [54]. The survey paper [33] may serve as a good introduction to both the classical and unilateral contact mechanics. Various interface problems, including unilateral conditions, are investigated in [50]. The books [14,18,34,44,46] present numerous results on the variational approach to the unilateral contact mechanics. The extensive paper [55] summarizes results obtained by using such approach to the solid and structural mechanics in the case of both elastic and inelastic materials.

The books [3,5,7,11,19,20,21,35,41] and the lengthy paper [8] are strongly advised to mathematically inclined reader.

## 2. DUALITY FOR SIGNORINI'S PROBLEMS WITH FRICTION IN THE CASE OF LINEAR ELASTIC MATERIALS

### 2.1 Origins of friction as a science

Before passing to the presentation of phenomenological friction conditions and laws it seems worthwile to know something about the origins of friction as the science. In this regard confusion exists in the current literature, particularly in textbooks. An excellent historical account of friction up to modern times presents the book by Dowson [93].

First important contributions Dowson [93] attributes to Leonardo da Vinci /1452-1519/. However, Leonardo's investigations of friction remained unfamiliar since, as Dowson writes: "Valuable notebooks /of Leonardo da Vinci - J.J.T./ containing records in the form of sketches and notes of studies of bearings, friction and wear were found in Madrid as recently as 1967... His experimental approach was essentially the same as that employed by Charles-Augustin de Coulomb some three centuries later...".

Guillaume Amontons /1663-1705/. Due to studies by Dowson we now know that Amontons investigated not "dry" friction as is commonly believed but the frictional characteristics of surfaces greased with old pork fat under the conditions nowadays described as boundary lubrication. Amontons' observations embodied thefirst and second laws of friction, that is: 1. the friction force is proportional to the applied load, 2. the friction force is independent of the apparent area of contact. Amontons attributed friction to rigid and elastic asperities.

These two empirical laws of friction can in fact be deduced from Leonardo's studies.

Phillippe de la Hire /1640-1718/ introduced what is now called the concept of permanent surface deformation and shearing of the interface /asperities/. He pointed out the possibility that resistance to motion would depend upon the number of asperities, and hence the size of the surfaces.

Gottfried Wilhelm von Leibniz /1646-1716/ made a distinction between sliding and rolling friction.

John Theophilus Desaguliers /1683-1744/ initiated an alternative view of the friction phenomena, namely he considered the influence of adhesion between contacting bodies /and not cohesion, as Dowson misleadingly writes, see [78]/.

Leonard Euler /1707-1783/. An importance of Euler's contribution is due to the development of an analytical approach to friction, the introduction of the symbol $\mu$ for the "coefficient of friction" as well as the distinction between static and kinetic friction.

Isaac Newton /1642-1727/. In Section IX of Book II of his Principia Mathematica Philosophiae Naturalis /1687/, which deals with "Circular Motion of Fluids", the term "want of lubricity" was introduced. It corresponds to what we now describe as internal friction or viscosity.

Charles Augustin Coulomb /1736-1806/. He studied experimentally such materials as oak, green oak, guaiac wood, elm and fir in various combinations in both dry and lubricated conditions. Coulomb found that in most cases friction was almost proportional to load and independent of the size of the contacting surfaces. In the case of frictional resistance to sliding of horizontal surfaces the limiting friction force was written as

$$F = a + P/\mu , \qquad /2.1/$$

where $\mu$ is the inverse (!) of the coefficient of friction, while $a$ was attributed to adhesion. For Coulomb, however, adhesion had only a very small influence upon friction. It is woth noticing that he was the first to use the two-term expression /2.1/ for friction, where the second term is attributed to deformation or ploughing action of asperities. Coulomb investigated also kinetic and rolling friction.

## 2.2 On friction mechanisms and coefficient of friction

Nowadays it is commonly assumed that friction between sliding surfaces is caused by three phenomena: 1/ deformation of sperities, 2/ ploughing by wear particles and hard surfaces asperities and, 3/ adhesion; see also Fig.3. Thus the coefficient of friction expresses as follows

$$\mu = F/P = f(\mu_d, \mu_p, \mu_a), \qquad /2.2/$$

where $\mu_d$, $\mu_p$ and $\mu_a$ denote the asperity deformation

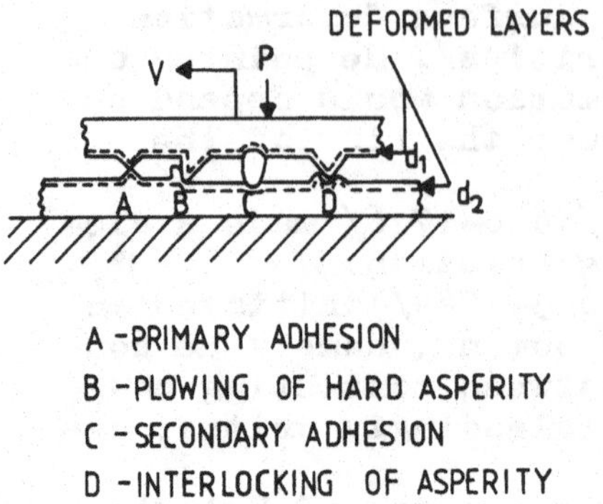


Fig.3, Typical asperity and wear particle at a sliding interface [137]

component, the ploughing component and adhesion component, respectively; F is the friction force while P denotes the applied load. The relative contribution of these components depends on the condition of the sliding interface, which is affected by the history of sliding, the surfaces topography, the specific materials used and the enviroment. Suh and Sin [163] affirm that the coefficient of friction is not an inherent material property.
The static coefficient of friction is determined primarily by asperities deformation.

Figures below exemplify influences of various factors on the coefficient of friction.

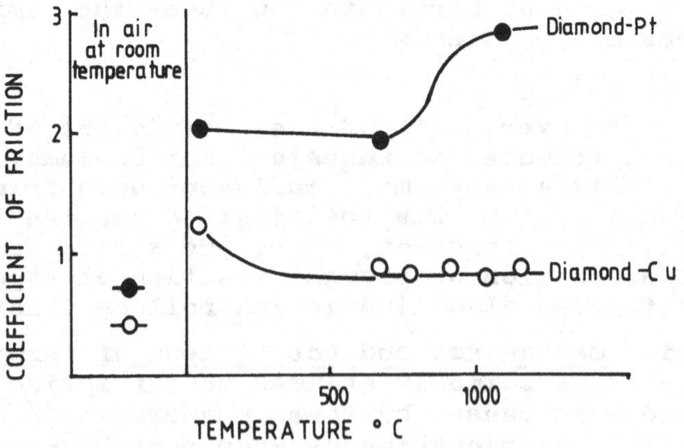


Fig.4, Friction of diamond on metal after cleaning in vacuum. High friction, especially with platinum, shows that very strong interfacial adhesion can occur [78]

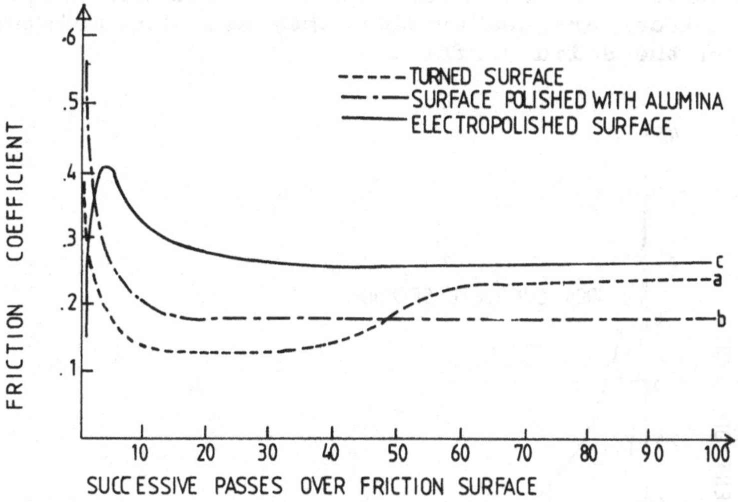


**Fig.5, Effect of surface condition of aluminium on friction coefficient [78]**

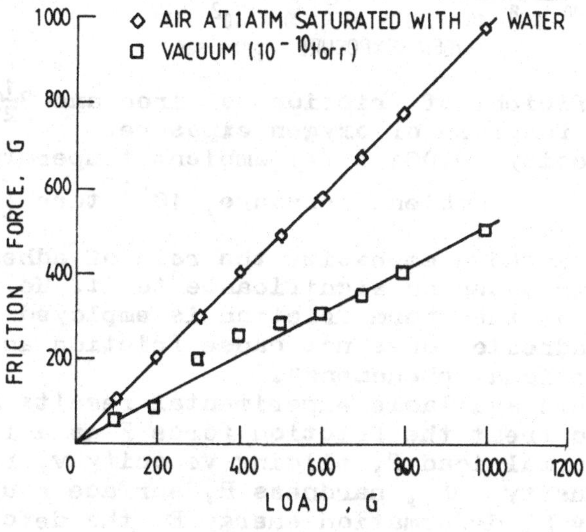


**Fig.6, Friction force as function of load for glass sliding on glass. Sliding velocity 30 cm/min; temperature 23°C [78]**

**We observe that glass is one of the few materials exhibiting higher friction in moist air than it does in a**

vacuum environment in the clean state. In the clean state the bonding forces are weaker than they are when moisture is present on the solid surface.

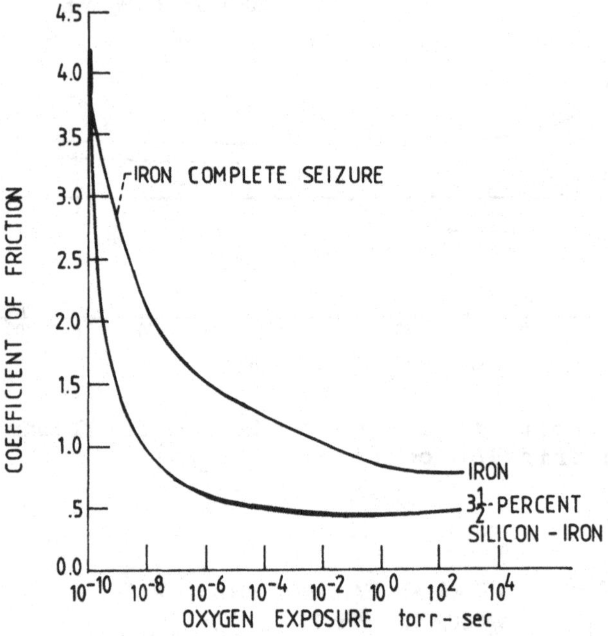


Fig.7, Coefficient of friction for iron and $3\frac{1}{2}\%$ silicon-iron as function of oxygen exposure.
Sliding velocity, 0.001 cm/s; ambient temperature 20°C; ambient pressure, $10^{-10}$torr [78]

Though many authors emphasize the role of adhesion yet Bikerman [75] attributes no significance to it. He writes /p.13/: "As long as the term friction is employed in its common meaning, adhesion does not cause friction and has no influence on frictional phenomena".

Basing on some available experimental results Madakson [137] proposed to treat the friction force F as a function of the applied normal load P, sliding velocity v, real area of contact A, density $\rho$, hardness H, surface roughness R, surface energy S, deformation energy E, the deformed area Q, temperature T, the environment D and the experimental system Y. Thus we can write

$$F = \hat{F}(P,v,A,\rho,H,R,S,E,Q,T,D,Y) . \qquad /2.3/$$

Dimensional analysis yields

$$F = \hat{F}\left(P,\ \frac{\rho v^2 A}{P},\ \frac{S}{Hv^2},\ \frac{EQ}{PAR}\right) K \qquad /2.4/$$

where K is a constant depending on temperature, environment and experimental system. The formula /2.4/ applies to two identical materials. In the case of dissimilar materials $\rho$ is the density of the softer one.
We observe that in Madakson's formula /2.4/ the temperature does not enter explicitly. It seems, however, that in a general formula the temperature should be explicitly present according to experimental results, see [76,79,134] and Figs 8,9.

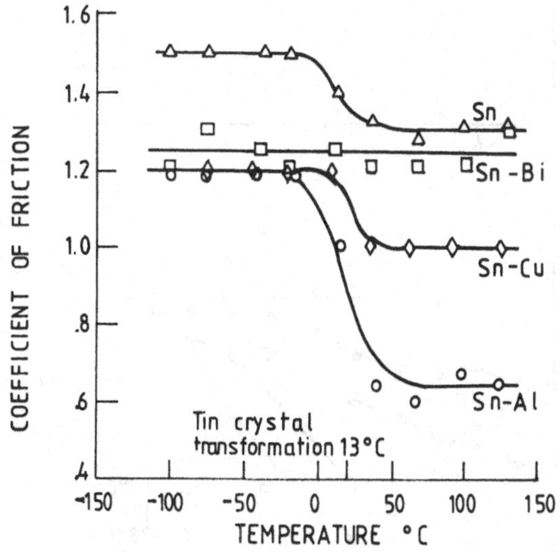


Fig.8, Coefficient of friction for polycrystalline tin and tin alloys at various temperatures. Sliding velocity, 0.7 mm/min; load 10 G;
pressure $1,33\times10^{-8}$ $N/m^2$/$10^{-10}$torr/ [78]

For instance, Bereznyakov et al. [71] proposed the following formula for temperature dependence of the friction force

$$F = 0,15\left[p_c\left(1+\alpha T\ /HB_o\right]^{1/4} P + 5,94\cdot10^{-2}\rho\, L\ln\left(T_m/T\right) \qquad /2.5/$$

where $\alpha$ is a small parameter, $\alpha \leqslant 0,01$; $HB_o$ denotes hardness in the sense of Brinell at T = 293K, $p_c$ is the contour pressure while $T_m$ and L stand for melting temperature and specific heat, respectively. Hence we infer that $\mu$ also depends on T.

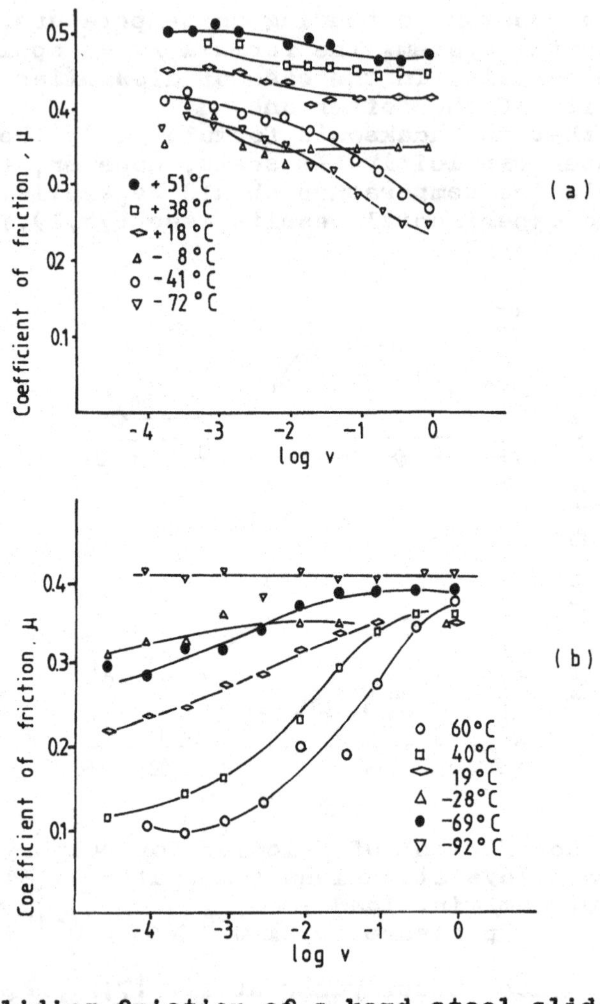


Fig.9, Sliding friction of a hard steel slider over a polymer surface as a function of speed and temperature. /a/ Nylon 6,6; /b/ Low-molecular weight branched polyethene [135]

Specific form of /2.4/ is [137]

$$\mu = F/P = KF_1\left(\frac{\rho A v^2}{P}\right)F_2\left(\frac{S}{Hv^2}\right)F_3\left(\frac{EQ}{PAR}\right). \qquad /2.6/$$

If the sliding process is in equilibrium then we can assume

that the functions $F_2$ and $F_3$ are equal to constants. Hence

$$\mu = K_1 F_1 \left( \frac{\varrho A v^2}{P} \right) . \qquad /2.7/$$

An approximate form of /2.7/ is

$$\mu = K_1 \left( \frac{\varrho A v^2}{P} \right)^c , \qquad /2.8/$$

where c is a constant.

If a loaded hemispherical slider rests on a flat surface of an identical material then $A=K_2P^n$, where $K_2$ and n are constants; n=1 for plastic deformation and $n<1$ for elastic deformation.

Fig.10 represents experimental and theoretical results, the latter being obtained from /2.8/ for c=0,12 and $K_1$=2,4 [137]

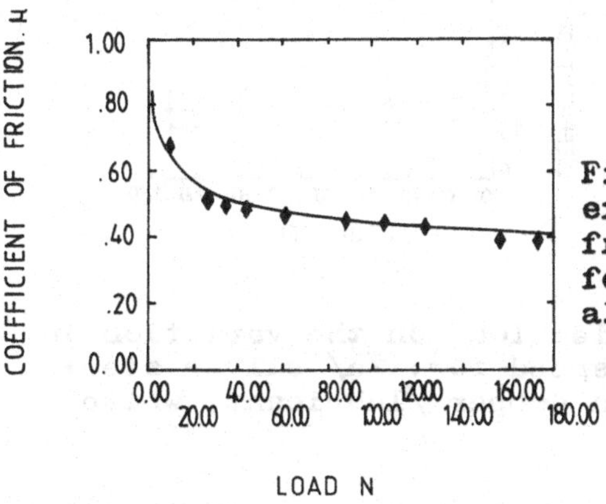


Fig.10, Theoretical /—/ and experimental /◆/ variations of friction coefficient with load for Duraluminium sliding on aluminium at 154 mm/s [137]

Further theoretical results derived from /2.8/ are presented in Fig.11.

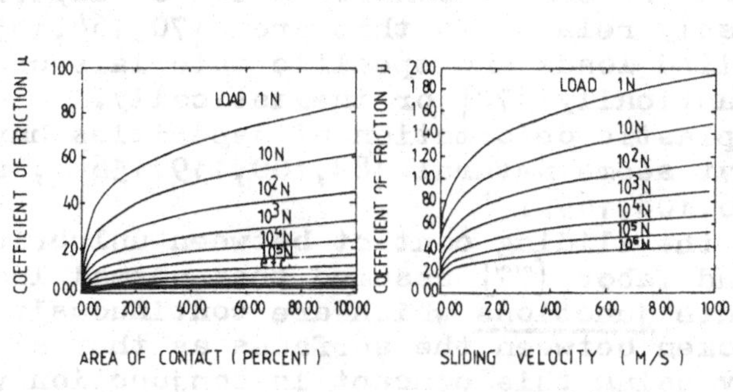


Fig.11, Plots of friction coefficient against:/a/area of contact,/b/sliding velocity for Duraluminium on aluminium

From the last figure we infer that with increasing loading the coefficient of friction becomes almost independent of velocity and area of contact. Thus the classical observations by Leonardo da Vinci and Coulomb are verified.

Experimental results of the dependence of $\mu$ on the sliding speed are presented in Figs 12 and 13 [137].

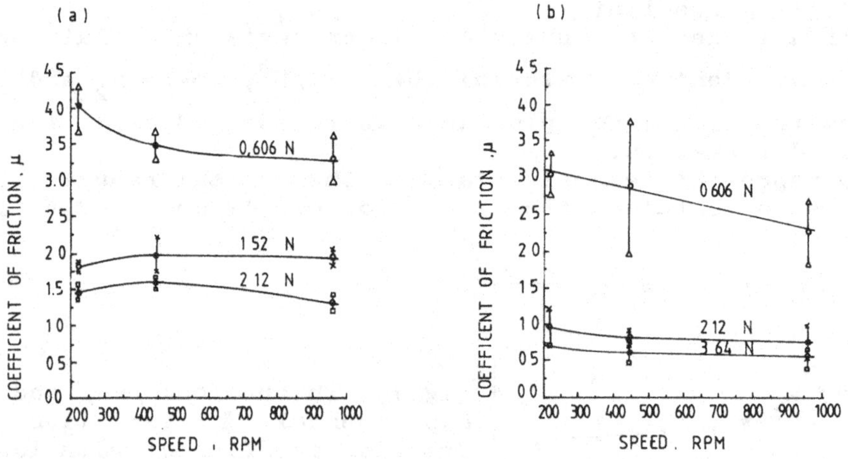


Fig.12, The effect of applied load on the variation in friction coefficient with speed for: /a/ silver steel on aluminium, /b/ copper on copper /lubricant, white spirit/

We observe that a rapid increase in load during sliding leads to a sharp decrease in the coefficient of friction, see also [103].

Earlier I have already mentioned about the importance of the real area of contact. Deformation of asperities is obviously closely related to this area [70,164,165]. Depending upon applied loads and specific materials used asperities deform elastically [72] or inelastically.
The study of plastic deformation of asperities have been preferred, what seems natural [83,100,119,156] ; see also [32,56,102,106,108,169].

To study the sliding contact between unlubricated metals Bowden and Tabor [77] assumed that normal load is supported by minute junctions which are continuously being formed and broken between the surfaces as they slide over each other. By using this concept in conjunction with Griffith theory of fracture and Rice J-integral De Celis [90] estimated the size of junctions. Thus for brittle ma-

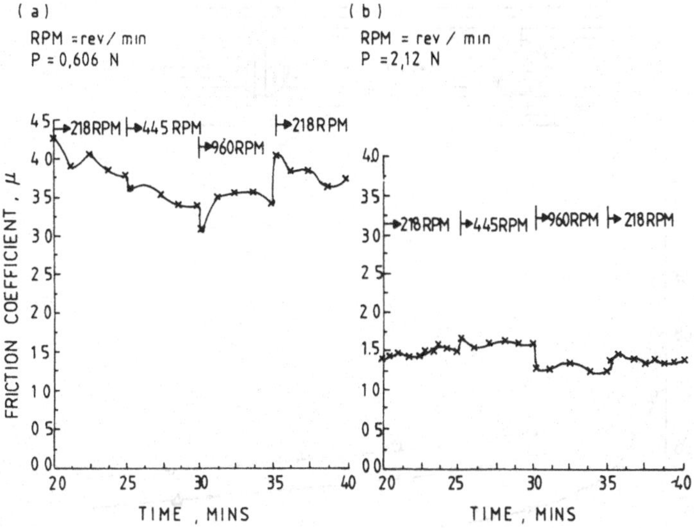


Fig.13, The effect of rapid change in sliding speed on the friction coefficient for silver steel on aluminium /lubricant, white spirit/

terials they are of the order of $10^{-9}$ m while for ductile materials their size is about $10^{-4}$- $10^{-5}$m.

With sliding friction at low speeds is closely connected the complex but very important phenomenon of "stick--slip", see [60,67,155]. The study of this process have even raised some doubts about the validity of the concept of static friction.

Apart from metals of practical interest are frictional characteristics of many other materials like wood [64,165], polymeric materials [56,65,69,135,165], rubber [68,165] and ceramics [165].

In my subsequent presentation of boundary value problems, generally anisotropic friction conditions and sliding rules will be taken into account.
Directional effects occuring during sliding have already been reported in the available literature [59,78,1 6,173], see also the following figure 14.

For a study of a structure-soil interaction frictional characteristics play an important role. A detailed experimental study for the determination of the coefficient of friction between various sands and steel was carried out in the papers [124,170] , see also the references cited therein and [118]. The coefficient of friction was found to be

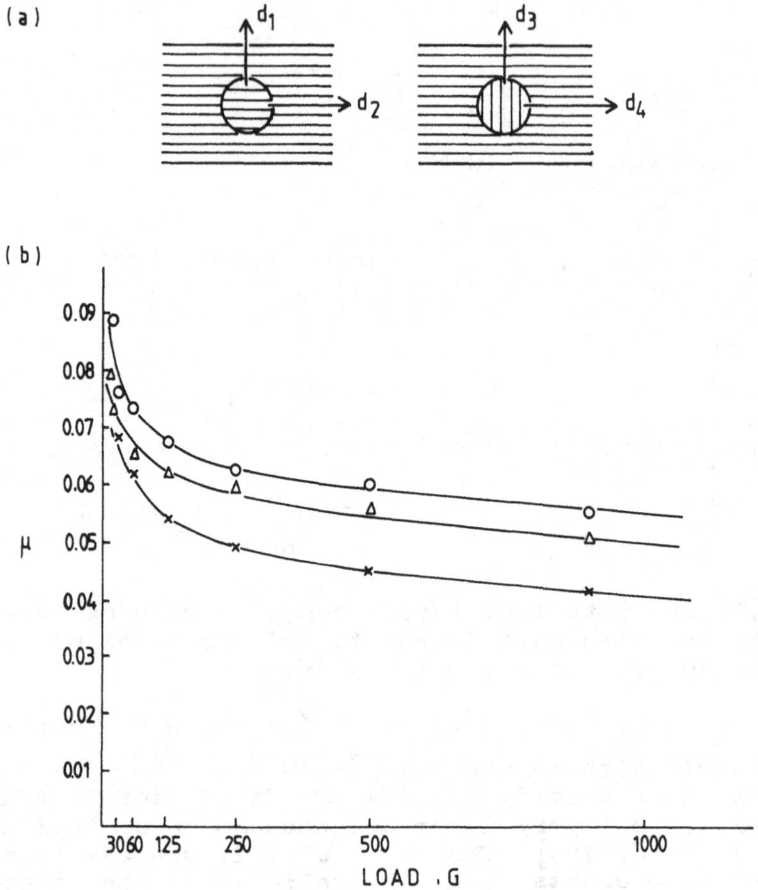


Fig. 14, /a/ The four possible directions of sliding for an oriented PTFE/polytetrafluor thylene/ slider on an oriented PTFE surface, /b/ the variation of the coefficient of friction between an oriented PTFE hemispherical slider and an oriented PTFE surface with load. o, direction $d_1$; x, direction $d_2$; Δ , directions $d_3$ and $d_4$ [166]

strongly influenced by sand type and surface roughness of steel while influences of normal stresses and mean grain size were of little significance.

Last but not least are frictional characteristics of human and animal joints, see [22,94,112,144]. In this case three types of joints must be distinguished: normal, pathological and artificial ones. Fig. 15, taken from [61], exhibit that the coefficient of friction in normally functioning synovial joints is extremely low.

(a)

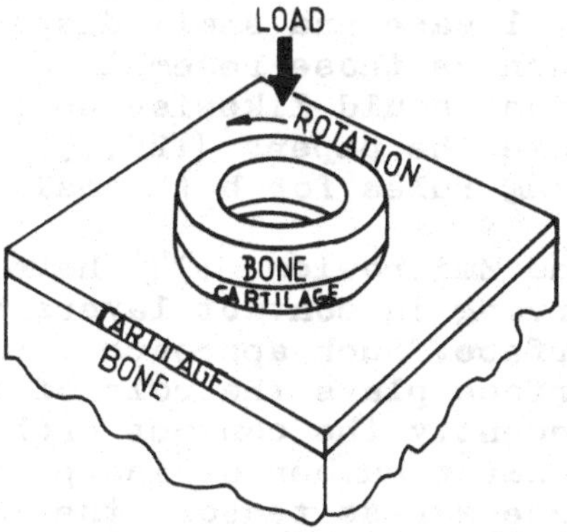


(b)

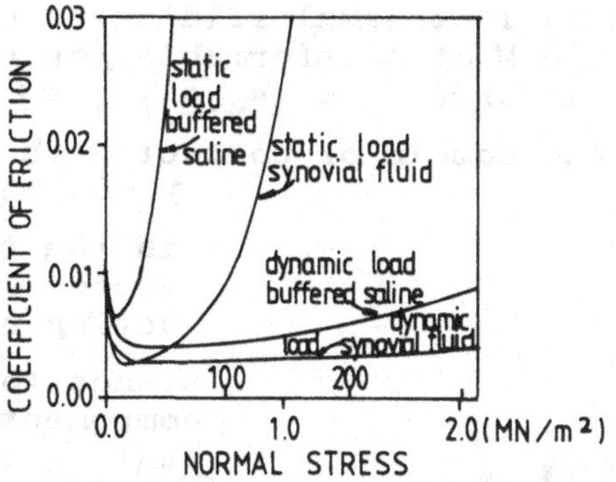


Fig.15, /a/ Experimental configuration, /b/ variation of friction coefficient for articular cartilage with load

Dowson et al. [94] reported the results of an experiment in which a severely damaged rheumatoid hip was investigated at loads up to 1500 N. The coefficient of friction was as high as 0.4, thus indicating a fifteen-fold increase over the average value of 0,025 recorded for healthy hip joints.

## 2.3 Phenomenological friction conditions and sliding rules

Physical considerations like those presented in the preceding section are very important in our understanding of friction phenomenon yet are insufficient for the study of boundary and initial value problems. For instance, friction laws or sliding conditions do not follow from such considerations, so they must be postulated.

A broad class of sliding rules can be obtained from the corresponding constitutive equations of the mechanics of solids. Such point of view is now briefly presented, see also [88]. Though I restrict myself, for the sake of simplicity, to isothermal case and small displacements yet non--isothermal laws such as those resembling thermoplastic constitutive equations could likewise be proposed. For instance, one could use the papers [129,154] to formulate thermoplastic sliding rules for both small and large sliding displacements.

Felder [99] and Zmitrowicz [174] have proposed to model phenomena occurring in contact layers by introducing of a material interface. Such approach seems to be promissing, yet the interface plays the role of third, two-dimensional body. Consequently the contact problem becomes more complicated due to large number of independent physical fields like interface stress tensor; the latter may even be non-symmetric [174]. Signorini's contact conditions, which are of main interest in these lectures, were not taken into account in [99,174].

Let us pass to isothermal sliding rules. Suppose that a material particle M of a deformable /or rigid/ body $\Omega$ is acted upon by a force $\underline{s} = (s_N, \underline{s}_T)$, $M \in S$, see Fig.16. Here S is the plane domain of contact with rigid plane $y_3=0$, and $\underline{s}_T=(s_{T\alpha})$ $(\alpha=1,2)$ is the tangential force vector or friction force acting on M while $s_N$ stands for the normal component of $\underline{s}$. Let $\underline{t}=(t_{ij})$ be a stress tensor, dev$\underline{t}$ its deviatoric part and $p=(\mathrm{tr}\underline{t})/3$. Further, let $\underline{n}$ be the exterior unit normal to $y_3=0$. and $\underline{e}=(e_{ij})$ the strain tensor; i,j=1,2,3.

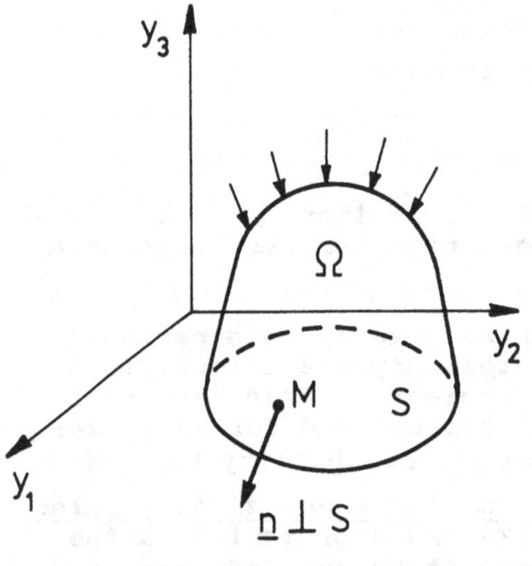


Fig.16

Under the following correspondences

$$\underline{e} = \underline{e}^{e} + \underline{e}^{p} \longleftrightarrow \underline{u} = \underline{u}^{e} + \underline{u}^{p} ,$$

$$\underline{\dot{e}} = \underline{\dot{e}}^{e} + \underline{\dot{e}}^{p} \longleftrightarrow \underline{\dot{u}} = \underline{\dot{u}}^{e} + \underline{\dot{u}}^{p} , \qquad /2.9/$$

$$\underline{t} = \mathrm{dev}\,\underline{t} + p\underline{I} \longrightarrow \underline{s} = s_N\underline{n} + \underline{s}_T ,$$

various sliding rules can readily be formulated; the superscripts "e" and "p" mean "elastic", "plastic" /or generally inelastic/, respectively. Here $\underline{u}$ is the displacement vector of M and $\underline{\dot{u}}$ its velocity.

For sufficiently small loads the material element M deforms elastically, thus $\underline{u}^p = 0$. Then we can assume

$$u_N^e = u_N = ks_N , \quad u_{T\alpha} = E_{\alpha\beta}\, s_\beta , \qquad /2.10/$$

in the linear case. Here $k > 0$ , $\underline{E} = \underline{E}^T$ and

$$E_{\alpha\beta}\, s_\alpha s_\beta \geqslant k_1 s_\alpha s_\alpha \quad , k_1 > 0 \ ,\forall \underline{s} = (s_\alpha) \qquad /2.11/$$

Suppose now that M deforms plastically, see [58]. Obviously if the body considered is rigid then the term "plastic" has no meaning yet plasticity type sliding rules still apply. Let a friction condition be given by

$$g(s_N, \underline{s}_T) \leqslant 0 \quad , \qquad /2.12/$$

where g is not necessarily isotropic.
Assume that the set

$$K = \{ \underline{s} = (s_N, \underline{s}_T) \mid g(s_N, \underline{s}_T) \leqslant 0 \} \qquad /2.13/$$

is convex and closed. The associated subdifferential sliding rule has the form

$$(\dot{u}_N^p , -\underline{\dot{u}}_T^p) \in \partial I_K(\underline{s}) , \qquad /2.14/$$

where $I_K$ is the indicator function of K

$$I_K(\underline{s}) = \begin{cases} 0 , & \text{if } \underline{s} \in K \\ + & , \text{otherwise} \end{cases} \qquad /2.15/$$

Such friction law is not very reasonable physically since it yields the normal component of the velocity $\dot{u}_N \neq 0$. In the case of Coulomb's friction condition

$$|\underline{s}_T| - \mu |s_N| \leqslant 0 , \qquad /2.16/$$

we obtain $\dot{u}_N^p = -\lambda \operatorname{sgn} s_N = \lambda\mu \geqslant 0$ since $\lambda \geqslant 0$ and

$s_N \leq 0$ /non-physical situation/.
If instead of /2.14/ we take

$$\left(-\dot{u}_N^p , -\dot{\underline{u}}_T^p\right) \in \partial I_K(\underline{s}), \qquad /2.17/$$

then we get $\dot{u}_N^p = -\lambda\mu \leq 0$ and M would be forced to leave the plane $y_3 = 0$. The friction laws /2.14/ and /2.17/ correspond to constitutive equations of associated plasticity of compressible materials. Such laws could probably be used in the case of a deformable foundation.

Of practical interest are the sliding rules corresponding to non-associated incompressible plasticity. Now the friction condition is still given by /2.12/ yet the assumption concerning convexity may be weakened. We set [167]

$$K(s_N) = \left\{ \underline{t} = (t_\alpha) \mid g(s_N, \underline{t}) \leq 0 \right\}, \qquad /2.18/$$

and assume that for each $s_N$ the set $K(s_N)$ is closed and convex. Thus the set K, defined by /2.13/, is not necessarily convex too.
As is known Coulomb's friction law expresses as follows

$$\begin{aligned} |\underline{s}_T| < \mu |s_N| &\Longrightarrow \dot{\underline{u}}_T^p = 0 , \\ |\underline{s}_T| = \mu |s_N| &\Longrightarrow \dot{\underline{u}}_T^p = -\lambda \underline{s}_T , \quad \lambda \geqslant 0 . \end{aligned} \qquad /2.19/$$

Coulomb's friction /sliding/ potential h is thus given by

$$h(\underline{s}_T) = |\underline{s}_T| + p , \qquad /2.20/$$

where p is a parameter, since potential is in fact a one parameter family. The parameter p is defined from the condition $h(\underline{s}_T) = g(s_N, \underline{s}_T)$. The potential /2.20/ yields

$$\dot{\underline{u}}_T^p = -\tilde{\lambda}\, \partial h/\partial s_T = -\lambda\, \underline{s}_T , \quad \lambda = \tilde{\lambda} / \underline{s}_T \qquad /2.21/$$

or

$$-\dot{\underline{u}}_T^p \in \partial I_{K(s_N)}(\underline{s}_T) , \text{ where } K(s_N) = \left\{ \underline{t} \,\middle|\, |\underline{t}| - \mu |s_N| \leq 0 \right\} .$$

Hence the friction law /2.19/ is not associated with Coulomb's friction condition /2.16/.
We observe that subdifferential approach to contact rules was initiated by Moreau, see [140].

I pass now to the general case. The sliding rule is assumed in the form

$$-\dot{\underline{u}}_T^p \in \partial I_{K(s_N)}(\underline{s}_T) , \qquad /2.22/$$

which is equivalent to

$$\left(\underline{t} - \underline{s}_T\right) \cdot \underline{\dot{u}}_T^p \geqslant 0 \quad , \quad \forall\ \underline{t} \in K(s_N) \cdot \qquad /2.23/$$

Denoting by $\underline{t}$ the conjugate variable, the support function $d(s_N, \underline{t}^*)$ of the convex set $K(s_N)$ is given by

$$d(s_N, \underline{t}_T^*) = \sup_{\underline{t}}\left(\underline{t}^* \cdot \underline{t} - I_{K(s_N)}(\underline{t})\right) =$$

$$= \sup\left\{\underline{t}^* \cdot \underline{t} \mid \underline{t} \in K(s_N)\right\} . \qquad /2.24/$$

The function $d(s_N, \cdot)$ is positively homogeneous of degree one [157].

Duality of the indicator function $I_{K(s_N)}(\cdot)$ with the support function $d(s_N, \cdot)$ results in

$$\underline{s}_T \in \partial_2 d(s_N, -\underline{\dot{u}}_T^p) , \qquad /2.25/$$

where $\partial_2 d(s_N, -\underline{\dot{u}}_T^p)$ denotes the subdifferential of the function $d(s_N, \cdot)$. Hence we infer that $\underline{s}_T^* = -\underline{\dot{u}}_T^p$. Either of sliding rules /2.22/ or /2.25/ is equivalent to

$$I_{K(s_N)}(\underline{s}_T) + d(s_N, -\underline{\dot{u}}_T^p) = \underline{s}_T \cdot (-\underline{\dot{u}}_T^p) , \qquad /2.26/$$

or, since $\underline{s}_T \in K(s_N)$,

$$j(s_N, \underline{\dot{u}}_T^p) \overset{df}{=} d(s_N, -\underline{\dot{u}}_T^p) = \underline{s}_T \cdot (-\underline{\dot{u}}_T^p) . \qquad /2.27/$$

The function j is the friction dissipation density. We observe that due to the nonassociated character of the friction law /2.22/ or /2.29/, the friction dissipation depends explicitly on normal stresses and not only on $\underline{\dot{u}}_T^p$.

By using Rockafellar's Th.13.5 we get

$$j(s_N, \underline{\dot{u}}_T^p) = \inf_{\lambda > 0} \lambda\ g^*(s_N, -\lambda^{-1}\underline{\dot{u}}_T^p) \text{ ,for } \underline{\dot{u}}_T^p \neq 0 . \qquad /2.28/$$

Obviously, for $\underline{\dot{u}}_T^p = 0$ we have $j(s_N, 0) = 0$.

If $K(s_N)$ is closed but nonconvex then the friction law /2.28/ remains valid, provided that $\partial I_{K(s_N)}(\underline{s}_T)$ stands for the subdifferential in the sense of Clarke [85] or rather Clarke-Rockafellar. Unfortunately, in the nonconvex case the definition of the subdifferential is more complicated, particularly in the case of functions which are not locally Lipschitzian, see also the lectures by Panagiotopoulos.

I think that even in such nonconvex case one can still consider convex sliding rules by simple application of the duality; that is we take /2.25/ as the subdifferential friction law since the function $d(s_N, \cdot)$ is always convex [97] /and lower semicontinuous/. Also the formulae /2.26/ and /2.27/ hold.

Example 2.1. Suppose that g is symmetric /isotropic/ in the following sense:

$$g(A_1 s_N, \underline{A}_2 \underline{s}_T) = g(s_N, \underline{s}_T) \quad , \qquad /2.29/$$

where $A_1$ is the orthogonal transformation which reverses the sign of the normal component, whereas $\underline{A}_2$ is an orthogonal transformation. Then we have, see [157, $\bar{p}$, 110]

$$g(s_N, \underline{s}_T) = G\left(|s_N|, |\underline{s}_T|\right). \qquad /2.30/$$

Hence we obtain

$$j\left(s_N, \underline{\dot{u}}_T^p\right) = |\underline{s}_T| \, |\underline{\dot{u}}_T^p| \quad . \qquad /2.31/$$

In the specific case of Coulomb's friction the last relation immediately yields

$$j(s_N, \underline{\dot{u}}_T^p) = \mu \, |s_N| \, |\underline{\dot{u}}_T^p| \, . \qquad /2.32/$$

Example 2.2 [167]. As a natural generalization of Coulomb's friction condition the following anisotropic one is proposed

$$g(s_N, \underline{s}_T) = \frac{1}{2} N_{\alpha\beta} \, s_{T\alpha} \, s_{T\beta} - H(s_N) \, . \qquad /2.33/$$

The tensor $\underline{N}$ can be called anisotropic friction tensor. We assume that $\underline{N} = \underline{N}^T$. Now we have

$$K(s_N) = \left\{ \underline{t} = (t_\alpha) \,\middle|\, \frac{1}{2} N_{\alpha\beta} \, t_\alpha t_\beta - H(s_N) \leq 0 \right\} . \qquad /2.34/$$

The set $K(s_N)$ is convex if and only if $\underline{N}$ is positive semi-definite. The nonassociated friction law takes on the form

$$\underline{\dot{u}}_T^p = -\lambda \, \underline{N} \, \underline{s}_T \quad , \qquad \lambda \geqslant 0 \, , \qquad /2.35/$$

where

$$\lambda = \frac{\left(M_{\alpha\beta} \, \dot{u}^p_{T\alpha} \, \dot{u}^p_{T\beta}\right)^{1/2}}{\left(2H(s_N)\right)^{1/2}} \quad , \quad \underline{M} = \underline{N}^{-1} \qquad /2.36/$$

provided that $\underline{N}$ is invertible.
The friction dissipation density is expressed as follows

$$j(s_N, \underline{\dot{u}}_T^p) = \left(2M_{\alpha\beta}\, \dot{u}^p_{T\alpha}\, \dot{u}^p_{T\beta}\, H(s_N)\right)^{1/2} . \qquad /2.37/$$

Non-associated sliding rules corresponding to compressible non-associated plastic flow laws can also be studied.

In the case of purely static considerations the velocity should obviously be replaced by displacement, see the next section. Then friction laws correspond to the deformational theory of plasticity sometimes called Hencky plasticity. For instance, the subdifferential sliding rule takes the form

$$-\underline{u}_T^p \in \partial I_{K_1(s_N)}(\underline{s}_T) \qquad /2.38/$$

Let us pass to sliding rules corresponding to incompressible non-associated plasticity accounting for hardening and/or softening, see also [140]. Now the friction condition is given by $g(s_N, \underline{s}_T, \underline{Q}) \leq 0$ , where $(\underline{s}, \underline{Q})$ are generalized forces. We set

$$K_1(s_N) = \{ (\underline{t}, R) \mid g(s_N, \underline{t}, \underline{R}) \leq 0 \} \qquad /2.39/$$

and assume that $K_1(s_N)$ is convex.

The generalized sliding rule is given by

$$(-\underline{\dot{u}}_T^p, -\underline{\dot{q}}) \in \partial I_{K_1(s_N)}(\underline{s}_T, \underline{Q}) . \qquad /2.40/$$

Thus $\underline{\dot{q}}$ is the generalized velocity associated with $\underline{Q}$. The friction dissipation density is now given by

$$j(s_N, \underline{\dot{u}}_T^p, \underline{\dot{q}}) := d(s_N, -\underline{\dot{u}}_T^p, -\underline{\dot{q}}) =$$

$$= \sup \{ \underline{t} \cdot (-\underline{\dot{u}}_T^p) - \underline{R} \cdot \underline{q} \mid (\underline{t}, \underline{R}) \in K_1(s_N) \} . \qquad /2.41/$$

Example 2.3. By using anisotropic yield conditions of plasticity with hardening it seems natural to generalize the friction condition studied in Example 2.2 as follows

$$g(s_N, \underline{s}_T, \underline{Q}_1, Q_2) = a + \frac{1}{2} N_{\alpha\beta} ( s_{T\alpha} - Q_{1\alpha})(s_{T\beta} - Q_{1\beta}) -$$

$$- H(s_N, Q_2) , \qquad /2.42/$$

where, for instance,

$$q_2(t) = q_2(0) + \int_0^t | \dot{\underline{u}}_T^p \cdot \dot{\underline{u}}_T^p | \, d\tau \quad , \; t \text{ - time}, \qquad /2.43/$$

and a stands for adhesion.
The sliding rule /2.40/ gives

$$\dot{\underline{u}}_T^p = - \lambda N_{\alpha\beta} s_T \quad , \; \dot{q}_{1\alpha} = \lambda N_{\alpha\beta} Q_{1\beta} \; , \; \dot{q}_2 = \lambda \partial H / \partial Q_2 , \quad \lambda \geqslant 0$$

provided that $\underline{N} = \underline{N}^T$ is positive semi-definite and $H(s_N)$ is concave.

Remark 2.1

Dynamical equations of discrete systems with Coulomb friction and Signorini's type conditions have been investigated in [114,152], see also the paper by Moreau in [14].

Moreau [141] has formulated a subdifferential sliding rule not requiring the decomposition of the contact force into normal and tangential parts.

Contact conditions are also discussed in [50,89,116, 125,126,147,148].

## 2.4. Signorini's contact problem with friction: kinematic and stress approach

In the sequel of the present Chapter I shall study in detail the static Signorini's contact problem for a linear elastic body. In this case the sliding rule is of deformational plasticity type /2.38/ and $u_N^p = 0$ , $\underline{u}_T^e = 0$ /on the surface of contact/.

Let $\Omega \subset R^3$ be a bounded sufficiently regular domain. Its boundary $\partial\Omega$, denoted by S, consists of three nonoverlapping parts: $S_0, S_1, S_2$ and $S = \bar{S}_0 \cup \bar{S}_1 \cup \bar{S}_2$, where $S_0$ denotes the closure of $\bar{S}_0$, etc. By $\underline{n} = (n_i)$ we denote the outer unit normal to S. Let $\underline{u} = (u_i)$ be a displacement vector and $e_{ij}(\underline{u}) = (u_{i,j} + u_{j,i})/2$ the strain tensor. Stress tensor is denoted by $\underline{s} = (s_{ij})$. Latin indices take values 1,2,3 while Greek indices 1,2.

Since $S_2$ is in general a surface and not a plane domain therefore the sliding rule /2.38/ have to be reconsidered. Let $\omega \subset R^2$ be a two-dimensional domain and $S_2 = \Phi(\omega)$ where $\Phi$ is of class $C^1(\bar{\omega})$. The plane $R^2$ is referred to coordinates $(y^\alpha)$ while $R^3$ to $(x_i)$. For $x = \Phi(y)$ we set $\underline{a}_{,\alpha} = \partial\Phi / \partial y^\alpha$, or $x_{i,\alpha} = \partial \Phi_i / \partial y^\alpha$. Now we have, on S

$$u_i = u_N n_i + u_{Ti} \quad , \quad s_{ij} n_j = s_N n_i + s_{Ti} \quad , \qquad /2.44/$$

where $s_N = s_{ij} n_i n_j \quad , \quad s_{Ti} = s_{ij} n_j - s_N n_i$ .

Let $\underline{A} = [A_{ia}] = [\underline{a}_\alpha, \underline{n}] = [x_{i,\alpha}, n_i]$ . Obviously, the matrix $\underline{A}$ is invertible. Hence we can write

$$u_{T\alpha} = x_{i,\alpha} u_{Ti} \; , \; u_{Ta} = A_{ia} u_{Ti} \; , \; (u_{Ta}) = (u_{T\alpha}, u_{T3})$$

$$s_{Ti} = x_{i,\alpha} s_{T\alpha} = A_{ia} s_{Ta} \; , \; \underline{s}_T = s_{T\alpha} \underline{a}_\alpha + s_{T3} \underline{n} \; .$$

Obviously we have $u_{T3} = n_i u_{Ti} = 0$ , $s_{T3} = 0$ , $s_{Ti} u_{Ti} = s_{T\alpha} u_{T\alpha}$.

Let at a point $x = \Phi(y)$ the continuous friction condition be given by $g_o(s_N, s_{T\alpha}) \leq 0$ . Then, always at the same point,

$$g(s_N, s_{Ti}) = g_o(s_N, \underline{A}^{-1} \underline{s}_T) \; , \; \underline{s}_T = (s_{Ti}) . \qquad /2.45/$$

The function $g(s_N, \cdot)$ remains convex if such is $g_o(s_N, \cdot)$ We set

$$K_0(s_N) = \{ \underline{t} = (t_\alpha) \mid g_o(s_N, \underline{t}) \leq 0 \} \; , \qquad /2.46/$$

$$K(s_N) = \{ \underline{t} = (t_i) \mid g(s_N, \underline{t}) \leq 0 \} . \qquad /2.47/$$

Hence

$$I_{K(s_N)}(s_T) = I_{K_0(s_N)}(A^{-1} \underline{s}_T) , \; \underline{s}_T = (s_{Ti}) ,$$

and

$$I_{K_0(s_N)}(s_{Ta}) = I_{K_0(s_N)}(s_{T\alpha}) = I_{K(s_N)}(A_{ia} s_{Ta}) \qquad /2.48/$$

The subdifferential sliding rule has the form

$$(-u_{T\alpha}) \in \partial I_{K_0(s_N)}(s_{T\alpha}) . \qquad /2.49/$$

The following elementary theorem is very useful [157,pp.38] and [225].

Theorem 2.1. Let A be a linear transformation from $R^n$ to $R^m$. Then for each convex function f on $R^m$, the function fA defined by

$$(fA)(z) = f(Az) \; , \qquad , \qquad /2.50/$$

is convex on $R^n$. Moreover, if f is a proper lower semicontinuous /=l.s.c./ convex function and there exists a point in the range of A at which f is finite and continuous then

$$\partial (fA)(z) = A^T \partial f(Az) , \qquad /2.51/$$

where $A^T$ is the transpose of A.

By using this theorem, from /2.48/ and /2.49/ we readily get

$$\left(-u_{T\alpha},0\right) = \left(-A_{ia}u_{Ti}\right) \in \partial I_{K_o(s_N)}\left(s_{Ta}\right) = \partial I_{K(s_N)}\left(A_{ia}s_{Ta}\right).$$

Hence

$$-\underline{A}^T\underline{u}_T \in \underline{A}^T \partial I_{K(s_N)}(\underline{s}_T) \quad , \quad \underline{s}_T = (s_{Ti}) \quad .$$

Since $\underline{A}$ is invertible, so we finally obtain

$$-\underline{u}_T \in \partial I_{K(s_N)}(\underline{s}_T) \; , \; \underline{u}_T = (u_{Ti}). \qquad /2.52/$$

In the above considerations the function $g_o(s_N,\underline{s}_T)$ was identified with a certain function $g_1(s_N,(\underline{s}_T,0))$.

After these preliminaries we can pass to the Signorini's contact problem with friction. Its strong /displacement/ formulation reads:

Problem $P_1$

Find a displacement field $\underline{u} = \underline{u}(x)$, $x \in \Omega$ such that

$$s_{ij,j} + b_i = 0 \; , \text{ in } \Omega, \qquad /2.53/$$

$$s_{ij}(\underline{u}) = a_{ijkl}e_{kl}(\underline{u}) \; , \qquad /2.54/$$

$$\underline{u} = 0 \; , \text{ on } S_o \; , \qquad /2.55/$$

$$s_{ij}n_j = F_i \; , \quad \text{on } S_1 \qquad /2.56/$$

$$u_N \leqslant 0, \; s_N \leqslant 0, \quad s_Nu_N = 0 \; , \text{ on } S_2 \; , \qquad /2.57/$$

$$-\underline{u}_t \in \partial I_{K(s_N)}(\underline{s}_T) \; , \qquad \text{on } S_2. \qquad /2.58/$$

Here $\underline{b}$ and $\underline{F}$ stand for the body forces and surface traction, respectively. The elasticity tensor $(a_{ijkl})$ possesses the usual symmetry properties: $a_{ijkl} = a_{klij} = a_{jikl}$ ; moreover

$$a_{ijkl}e_{kl}e_{ij} \geqslant c\, e_{ij}e_{ij} \; , \text{ for each } \underline{e} \in \mathbb{M}^3_s(\mathbb{R}), \qquad /2.59/$$

where c is a positive constant while $\mathbb{M}^3_s(\mathbb{R})$ denotes the space of symmetric 3x3 real matrices. The three conditions entering /2.57/ are the Signorini's condition or complementarity ones.

We assume that

$$a_{ijkl} \in L^\infty(\Omega) \; , \qquad (\underline{b},\underline{F}) \in V^*, \qquad /2.60/$$

where

$$V = \{ \underline{v} = (v_i) \mid \underline{v} = 0 \quad \text{on } S_o \} \qquad /2.61/$$

Particularly, we can assume that $b_i \in L^2(\Omega)$, $F_i \in L^2(S_1)$.
The choice of the space of displacements V is discussed in Section 2.7.

Let us put

$$K_d = \{ \underline{v} \mid \underline{v} \in V \ , \ v_N \leq 0 \quad \text{on} \quad S_2 \}, \qquad /2.62/$$

$$a(\underline{u},\underline{v}) = \int_\Omega a_{ijkl} e_{ij}(\underline{u}) e_{kl}(\underline{v}) dx \ ; \ \underline{u},\underline{v} \in V, \qquad /2.63/$$

$$J(\underline{u},\underline{v}) = \int_{S_2} j(s_N \ \underline{u} \ , \underline{v}_T) dS \ ; \quad \underline{u},\underline{v} \in V, \qquad /2.64/$$

$$\langle \underline{L},\underline{v} \rangle = L(\underline{v}) = \int_\Omega b_i v_i dx + \int_{S_1} F_i v_i dS \ , \ \underline{v} \in V. \qquad /2.65/$$

Problem $P_1$ is now formulated in the variational, or weak form as

<u>Problem $P_d$</u>

Find $\underline{u} \in K_d$ such that

$$a(\underline{u},\underline{v}-\underline{u}) + J(\underline{u},\underline{v}) - J(\underline{u},\underline{u}) \geq L(\underline{v}-\underline{u}), \quad \forall \underline{v} \in K_d. \qquad /2.66/$$

We observe that /2.66/ is valid regardless of whether the closed set $K(s_N)$ is convex or not.

The proof of /2.66/ is straightforward.
Multiplying /2.53/ by $\underline{v}-\underline{u}$, $\underline{v} \in K_d$ and performing the integration by parts we obtain

$$a(\underline{u},\underline{v}-\underline{u}) + \int_{S_2} (-\underline{s}_T)\cdot(\underline{v}_T-\underline{u}_T) dS = L(\underline{v}-\underline{u}) + \int_{S_2} s_N (v_N - u_N) dS \qquad /2.67/$$

The subdifferential friction law /2.52/ is equivalent to

$$j(s_N,\underline{v}_T) - j(s_N,\underline{u}_T) \geq (-\underline{s}_T)\cdot(\underline{v}_T-\underline{u}_T) \ \forall \underline{v}_T .$$

Taking account of the Signorini's conditions /2.57/ and the last inequality in /2.67/ we arrive at the <u>implicit variational inequality</u> = I.V.I. /2.66/.

Let us study now the stress formulation of the same frictional problem. Toward this end we set $\underline{b} = \underline{a}^{-1}$ and

$$K_s(s_N) = \{ \underline{t} = (t_{ij}) \in V_s \mid t_N \leq 0,\ \underline{t}_T \in K_s(s_N) \text{ on } S_2 \} \qquad /2.68/$$

where

$$V_s = \{ \underline{t} = (t_{ij}) \in H_s \mid t_{ij,j} + b_i = 0, \text{ in } \Omega \ ; t_{ij}n_j = F_i \text{ on } S_1 \}. \qquad /2.69/$$

The function spaces are introduced in Section 2.7.
Let us multiply the constitutive equation

$b_{ijkl}s_{kl} - e_{ij}(\underline{u}) = 0$ by $(\underline{t}-\underline{s})$, $\underline{s},\underline{t} \in K_s(s_N)$,
and integrate over $\Omega$. We readily obtain

$$0 = b(\underline{s},\underline{t}-\underline{s}) - \int_{S_o} (t_{ij}-s_{ij}) n_j u_i dS - \int_{S_2} \{(t_N-s_N) u_N - (\underline{t}_T-\underline{s}_T)\cdot \underline{u}_T\} dS \ , \forall \underline{t} \in K_s(s_N) \qquad /2.70/$$

where

$$b(\underline{s},\underline{t}) = \int_\Omega b_{ijkl}s_{ij}t_{kl}dx \ ; \ \underline{t},\underline{s} \in H_s . \qquad /2.71/$$

We know that $\underline{u}=0$ on $S_o$. Taking account of /2.57/ we infer that the integral over $S_2$ is non-negative. Thus we can formulate

**Problem $P_s$**

Find $\underline{s} \in K_s(s_N)$ such that

$$b(\underline{s},\underline{t}-\underline{s}) \geqslant 0 \qquad \forall\ \underline{t} \in K_s(s_N). \qquad /2.72/$$

We observe that if $\underline{u} = \underline{u}_o$ on $S_o$ then /2.70/ yields

$$b(\underline{s},\underline{t}-\underline{s}) - \int_{S_o} (t_{ij}n_j - s_{ij}n_j) u_{oi} dS \geqslant 0 \ , \forall \underline{t} \in K_s(s_N) \qquad /2.73/$$

The inequality /2.72/ or /2.73/ is the <u>principle of virtual stresses</u> for the Signorini problem with friction.
It is worth noticing that /2.72/ is the <u>quasi-variational inequality</u> = Q.V.I., since the set $K_s(s_N)$ of statically admissible stresses depends on the solution $\underline{s}$.

## 2.5. Mosco-Capuzzo Dolcetta-Matzeu theory of duality for implicit variational problems

Mechanicians are usually convinced that for variational problems like $P_d$ and $P_s$ no dual formulation is available. Such statement is false. To prove it, I shall employ the so called Mosco-Capuzzo Dolcetta-Matzeu theory of duality for I.V.Ps. For short, here I call it M-CD-M theory of duality.

In the present section the general theory is presented, since the original paper [82] avoids important details and is restricted to reflexive Banach spaces.

Let V and $V^*$ be locally convex topological vector spaces = LCTVS put in separating duality by a bilinear form $\langle \cdot,\cdot \rangle : V^* \times V \to R$. Thus for each $v \in V$, $v \neq 0$, there exists $v^* \in V^*$ such that $\langle v^*, v \rangle \neq 0$ and for each $v^* \in V^*$, $v^* \neq 0$, there exists $v \in V$ such that $\langle v^*, v\rangle \neq 0$. Let Y be another LCTVS and

$(H_1)$ | $A: V \to Y$ a linear and continuous operator.

The implicit variational problem to be studied involves functionals $\varphi$ and g such that

$(H_2)$ | $w \to \varphi(Av, w)$ is, for every $v \in V$, a convex l.s.c. function on V not identically equal to $+\infty$.

$(H_3)$ | $w \to g(v,w)$ is, for every $v \in V$, a real-valued convex function which is continuous for $w = v$.

$(H_4)$ | $w \to g(v,w)$ has, for every $v \in V$, the Gâteaux derivative with respect to the second variable $D_2 g(v,w)$ at $w=v$, such that for every $w^* \in V^*$ the set $\{v \in V \mid D_2 g(v,v) = w^*\}$ contains at most one element $(D_2 g)^{-1}(w^*)$.

I recall that the aforementioned Gâteaux derivative is defined as follows

$$\langle D_2 g(v,w), z\rangle = \lim_{\lambda \to 0^+} \frac{g(v, w+\lambda z) - g(v,w)}{\lambda}, \forall z \in V, \qquad /2.74/$$

The primal I.V.P. reads:

(P) | find $u \in V$ such that
$$\varphi(Au,u) + g(u,u) \leq \varphi(Au,w) + g(u,w), \ \forall w \in V. \qquad /2.75/$$

The Fenchel conjugate of $\varphi(Av,\cdot)$ is defined by

$$\varphi^*(Av, w^*) = \sup\{\langle w^*, w\rangle - \varphi(Av,w) \mid w \in V\}, w^* \in V^* \qquad /2.76/$$

The functional $g^*(v,w^*)$ is defined similarly.

The subdifferential of the convex function $g(v,\cdot)$ is

$$\partial_2 g(v,w) = \{w^* \in V^* \mid g(v,z) - g(v,w) \geq \langle w^*, z-w\rangle, \ \forall z \in V\}. \qquad /2.77/$$

Now we formulate the dual problem $P^*$ as follows

(P) | find $u \in V$, $u^* \in V^*$ such that
$-u^* \in \partial_2 g(u,u)$,
$\varphi^*(Au,u^*) - \langle u^*,u\rangle \leq \varphi^*(Au,w^*) - \langle w^*,u\rangle, \ \forall w^* \in V^*$.

The problems P and $P^*$ are interrelated by

Theorem 2.2. Let the assumptions $(H_1)-(H_3)$ be satisfied. If u is a solution of P, then there exists $u^*\in V^*$ such that $(u,u^*)$ solves the dual problem $P^*$. Conversely, if $(u,u^*)$ solves $P^*$ then u is a solution of the problem P. Moreover, the extremality condition

$$\varphi(Au,u) + \varphi^*(Au,u^*) = \langle u^*,u \rangle = -\big(g(u,u)+ g^*(u,-u^*)\big) \qquad /2.78/$$

is satisfied.

Before proving this theorem I first recall some indispensable notions of the convex analysis [97,131,157,158,228].

By $\Gamma_o(V)$ we denote the set of proper lower semi-continuous convex functions on V, or $\Gamma_o(V)=\{f\,|\,f:V\to R\cup\{+\infty\},\ f\not\equiv+\infty,\ f$ - convex and l.s.c.$\}$. Now several useful lemmas are formulated.

Lemma 2.1. Let $f\in\Gamma_o(V)$. The following properties are equivalent

/i/ $f(u) + f^*(u^*) = \langle u^*,u\rangle$ ,

/ii/ $u^*\in\partial f(u)$ ,

/iii/ $u\in\partial f^*(u^*)$ .

Lemma 2.2. Let $f_1,f_2\in\Gamma_o(V)$. Suppose that there exists a point $v_o\in V$ such that $f_1$ and $f_2$ take finite values at this point and one of them is continuous at $v_o$. Then

$$\partial(f_1+f_2)(v) = \partial f_1(v)+\partial f_2(v),\ \forall v\in V. \qquad /2.79/$$

Lemma 2.3. Let $f:V\to\overline{R} = R\cup\{-\infty\}\cup\{+\infty\}$.
An element $u\in V$ satisfies

$$f(u) = \min_{v\in V} f(v) \quad , \quad \text{/with } f(u) \text{ finite/}$$

if and only if $0\in\partial f(u)$ .

Definition 2.1. Let $f_1$ and $f_2$ be two functions, $f_\alpha : V\to\overline{R}$. The function

$$f(v):= \inf\{f_1(v_1)+ f_2(v_2)\,|\,v_1+v_2 = v,\ v_\alpha\in V\} =$$
$$= \inf\{f_1(v_1) + f_2(v-v_1)\,|\,v_1\in V\} \qquad /2.80/$$

is called the inf-convolution /infimal convolution/ and will be denoted by $f = f_1\nabla f_2$.

Some properties of inf-convolution are now listed.

/a/ If $f_1$ and $f_2$ are convex, then $f_1\nabla f_2$ is convex too.

/b/ $(f_1\nabla f_2)^* = f_1^* + f_2^*$ ,

provided that $f_\alpha\not\equiv +\infty$.

/c/ Let $f_1,f_2\in\Gamma_o(V)$. Suppose that there exists $v^*\in V^*$ such that $f_1^*(v^*)$, $f_2^*(v^*)$ are finite and one of them is continuous at $v^*$. Then $f_1\nabla f_2\in\Gamma_o(V)$. Moreover, the inf-convolution is exact.

Definition 2.2. We say that $f_1 \nabla f_2$ is exact if for each $v \in V$ the infimum in the definition of inf-convolution is attained, i.e. for each $v \in V$

$$(f_1 \nabla f_2)(v) = \min\{f_1(v_1) + f_2(v) \mid v_1+v_2 = v\} .$$

Thus there exist $v_1, v_2 \in V$, such that $v_1+v_2 = v$ and $(f_1 \nabla f_2)(v) = f_1(v_1) + f_2(v_2)$.

/d/ Let $f_1, f_2 : V \to \overline{R}$, $f_\infty \not\equiv +\infty$. If the inf-convolution $f_1 \nabla f_2$ is exact at $v = v_1+v_2$ then one has

$$\partial(f_1 \nabla f_2)(v) = \partial f_1(v_1) \cap \partial f_2(v_2) .$$

/e/ Under the assumptions of Lemma 2.2 one has

$$(f_1+f_2)^* = f_1^* \nabla f_2^* ,$$

and the inf-convolution is exact.

Proof of Th.2.2. Suppose that u solves the primal problem P. Then

$$\varphi(Au,w) + g(u,w) - (\varphi(Au,u)+ g(u,u)) \geq \langle 0, w-u\rangle, \forall w \in V$$

Hence, by using $(H_3)$ and Lemmas 2.2 and 2.3

$$0 \in \partial_2(\varphi(Au,u)+ g(u,u)) = \partial_2 \varphi(Au,u) + \partial_2 g(u,u) ,$$

where $0 \in V^*$. Lemma 2.1 and the properties of the inf-convolution result in

$$0 \in \partial_2(\varphi + g)(Au,u) \Longleftrightarrow u \in \partial_2(\varphi + g)^*(Au,0) =$$
$$= \partial_2(\varphi^* \nabla g^*)(Au,0) = \partial_2 \varphi^*(Au,u^*) \cap \partial_2 g^*(Au,-u^*) ,$$

since $0 = u_1^* + u_2^*$, or $u^* = u_1^* = -u_2^*$.

Thus we can write

$$u \in \partial_2 \varphi^*(Au,u^*) \text{ and } u \in \partial_2 g^*(u,-u^*),$$

or, equivalently

$$u^* \in \partial_2 \varphi(Au,u) \text{ and } -u^* \in \partial_2 g(Au,u). \qquad /2.81/$$

By using the property /i/ of Lemma 2.1 we infer

$$\varphi(Au,u) + \varphi^*(Au,u^*) = \langle u^*,u\rangle , \qquad /2.82/$$
$$g(u,u) + g^*(u,u^*) = \langle -u^*,u\rangle . \qquad /2.83/$$

Further, we have, see [97,131]

$$\varphi(Au,u) = \varphi^{**}(Au,u) = \sup\{\langle w^*,u\rangle - \varphi^*(Au,w^*) \mid w^* \in V^*\}.$$

The last relation and /2.82/ yield

$$\varphi^*(Au,u^*) - \langle u^*,u\rangle \leq \varphi^*(Au,w^*) - \langle w^*,u\rangle \quad \forall w^* \in V.$$

The last inequality combined with $/2.81/_2$ exhibit that $(u,u^*)$ solves the dual problem $P^*$.

The second part of the thesis of theorem results by an inversion of the above reasoning. Finally, the extremality relation /2.78/ follows immediately from /2.82/ and /2.83/. Thus the theorem has been proved.

To eliminate the primal variable u from the dual problem $P^*$ we use the assumption $(H_4)$. Consequently we arrive at

Theorem 2.3. Under the assumptions $(H_1)-(H_4)$ an element $u \in V$ is a solution of the primal problem P if and only if $u^* = -D_2 g(u,u)$ solves the following dual problem

$$(P^*)\left|\begin{array}{l}\text{find } u^* \in V^* \text{ such that}\\ \varphi^*\big(A[(D_2 g)^{-1}(-u^*)],u^*\big) - \langle u^*,(D_2 g)^{-1}(-u^*)\rangle \leq\\ \leq \varphi^*\big(A[(D_2 g)^{-1}(-u^*)],w^*\big) - \langle w^*,(D_2 g)^{-1}(-u^*)\rangle,\ \forall w^* \in V^* \end{array}\right. \qquad /2.84/$$

Moreover, the extremality relation /2.78/ holds.

Bidual problem $P^{**}$. Treating the problem $P^*$ as the primal one the duality operation can be iterated. In this manner we shall obtain the bidual problem $P^{**}$, provided that the assumption $(H_4)$ remains valid. Toward this end we set

$$\tilde{\varphi}_A(v^*,w^*) = \varphi^*\big(A[(D_2 g)^{-1}(-v^*)],w^*\big), \qquad /2.85/$$

$$\tilde{g}(v^*,w^*) = -\langle w^*,(D_2 g)^{-1}(-v^*)\rangle . \qquad /2.86/$$

for all $v^*, w^* \in V^*$. The bidual problem reads

$$(P^{**})\left|\begin{array}{l}\text{find } u \in V \text{ such that}\\ \tilde{\varphi}^*_A\big([(D_2\tilde{g})^{-1}(-u)],u\big) - \langle (D_2\tilde{g})^{-1}(-u),u\rangle \leq\\ \leq \tilde{\varphi}^*_A\big([(D_2 g)^{-1}(-u)],w\big) - \langle (D_2 g)^{-1}(-u),w\rangle,\ \forall w \in V \end{array}\right.$$

The functions $g$ and $\tilde{g}$ are interrelated by

Lemma 2.4. The function $\tilde{g}$ satisfies $(H_4)$, and for each $v,w \in V$ one has

$$(D_2\tilde{g})^{-1}(-w) = -D_2 g(w,w), \qquad /2.87/$$

$$\tilde{\varphi}^*_A\big([(D_2\tilde{g})^{-1}(-v)],w\big) = \varphi(Av,w). \qquad /2.88/$$

Proof. The relation /2.86/ gives

$$D_2\tilde{g}(w^*,w^*) = -(D_2 g)^{-1}(-w^*).$$

Thus for any $w \in V$ there exists a uniquely defined $w^* \in V^*$ such that

$$D_2\tilde{g}(w^*,w^*) = -w ,$$

$w^*$ being obviously given by $w^* = -D_2 g(w,w)$.
Hence /2.87/ is proved.

To prove /2.88/ we observe that /2.85/ and /2.87/ yield

$$\varphi_A^*\left(\left[(D_2\tilde{g})^{-1}(-v)\right],w\right) =$$
$$= \sup\left\{\langle w^*,w\rangle - \varphi^*\left(A\left[(D_2 g)^{-1}(-D_2 g^{-1}(-v)\right],w^*\right) \mid w^* \in V^*\right\} =$$
$$= \sup\left\{\langle w^*,w\rangle - \varphi^*(Av,w^*) \mid w^* \in V^*\right\} = \varphi^{**}(Av,w) =$$
$$= \varphi(Av,w) .$$

Due to Lemma 2.4, the problem $P^{**}$ can be formulated as follows

$(P^{**})$ | find $u \in V$ such that
$\langle Dg(u,u),w-u\rangle \geq \varphi(Av,u) - \varphi(Au,w), \forall w \in V$.

A simple consequence of the above considerations is

Theorem 2.4. Suppose that the assumptions $(H_1)$ - $(H_4)$ are satisfied. Then $u \in V$ is a solution of $(P)$ if and only if $u$ is a solution of $(P^{**})$.

Proof. From Th.2.2 we infer that u is a solution of $(P)$ if and only if $u^* = -Dg(u,u)$ solves $(P^*)$, hence if and only if

$$-D\tilde{g}(u^*,u^*) = (Dg)^{-1}(-u^*) = u$$

is a solution of the Problem $P^{**}$.

Corollary 2.1. The variational problems P and $P^{**}$ coincide if and only if

$$g(v,w) = \langle (D_2\tilde{g})^{-1}(-v),v-w\rangle . \qquad /2.89/$$

Proof. Suppose that $(P)$ coincides with $(P^{**})$. Then

$$g(v,w) - g(v,v) = \langle D_2 g(v,v),w-v\rangle = \langle (D_2 g)^{-1}(-v),v-w\rangle ,$$

by virtue of /2.87/.

Now let us assume that /2.89/ holds. Then

$$g(v,w) - g(v,v) = \langle (D_2\tilde{g})^{-1}(-v),v-w\rangle =$$
$$= \langle D_2 g(v,v),w-v\rangle .$$

Thus we finally have $(P) = (P^{**})$.

Remark 2.2. In the original paper by Capuzzo Dolcetta and Matzeu [82], A = identity and V is a reflexive Banach space. We observe that A could even be a nonlinear operator.

Remark 2.3. For $\varphi(v,w)$ independent of v and $g(v,w) = \langle N(v),w-v\rangle$, where N is an injective map, the dual variational inequality was introduced by Mosco [142], see also

[101,143]. In this case the problems P and $P^*$ have the following form, respectively

$$(P)\ |\ u\in D(N):\ \langle N(u),v-u\rangle \geqslant \varphi(u)-\varphi(v),\ \forall\ v\in V, \qquad /2.90/$$

$$(P^*)\ |\ u^*\in D(N'):\ \langle N'(u^*),v^*-u^*\rangle \geqslant \varphi^*(u^*)-\varphi^*(v^*),\ \forall v^*\in V^* \qquad /2.91/$$

provided that $D(N)$ is the domain of N and $N' = -N^{-1}-: v^* \longrightarrow -A^{-1}(-v^*)$, $D(N') =$ range $(N)$, while $\varphi$ is convex and l.s.c.

Problem /2.91/ results directly from equivalences

$$(P) \Longleftrightarrow \{u\in D(N) : -N(u)\in \partial\varphi(u)\} \Longleftrightarrow \{u^*\in D(N') : -N'(u)\in\partial\varphi^*(u^*)\}$$

Mosco [142] has also generalized such theory of duality to the case when N is a multivalued mapping.

It is worth noticing that Mosco's theory of duality and more generally, M-CD-M theory, apply to problems involving non-potential operators for which extremum principles are not available. Problems described by potential operators are certainly easier, see for instance [136].

## 2.6. Application of M-CD-M theory of duality to Signorini's problem with friction

In the present section, by using M-CD-M theory of duality problems $P_d^*$ and $P_s^*$ will be derived. It will also be shown that $(P_s^*) = (P_d)$ and $(P_d^{**}) = (P_d)$.

Let us pass to the derivation of the problem $P_d^*$. We set

$$g(u,w) = a(\underline{u},\underline{w}-\underline{u}) - L(\underline{w}-\underline{v}), \qquad /2.92/$$

$$\varphi(A\underline{v},w) = J(\underline{v},\underline{w}) + I_{K_d}(\underline{w}), \qquad /2.93/$$

where $A: V \longrightarrow s_N(\underline{v})$; V is defined by /2.61/.

Thus space Y coincides with $H^{-1/2}(S_2)$ in the general case or is equal to $L^2(S_2)$, see the next section.

The assumptions $(H_1)-(H_4)$ being obviously verified we pass to proving $(H_4)$. Let us first find conjugate functionals. The following elementary lemma is useful for the derivation of $\varphi^*(A\underline{v},\underline{w}^*)$.

Lemma 2.5. If V is a reflexive Banach space and $C\subset V$ a closedconvex cone such that $0\in C$, then

$$(I_C)^*(v^*) = I_{C^*}(v^*), \qquad /2.94/$$

where $C^*$ is the polar cone of C, that is

$$C^* = \{v^*\ |\ v^*\in V^*, v^*\leqslant 0\}, \qquad /2.95/$$

Here $v^*\leqslant 0$ means that $\langle v^*,v\rangle \leqslant 0$, $\forall v\in C$.

$$= a(\underline{v},\underline{v}) - L(\underline{v}) + \begin{cases} 0, \text{ if } \underline{w}^* = B\underline{v}-\underline{L} , \\ +\infty , \text{ otherwise} \end{cases} \qquad /2.105/$$

where $\underline{L} = (\underline{b},\underline{F})$.

The Gâteaux derivative $D_2 g$ expresses as follows

$$\langle D_2 g(\underline{v},\underline{w}),\underline{h}\rangle = a(\underline{v},\underline{h}) - L(\underline{h}) = \langle B\underline{v}-\underline{L},\underline{h}\rangle . \qquad /2.106/$$

Hence we have

$$D_2 g(\underline{u},\underline{u}) = B\underline{u}-\underline{L} , \text{ or } B\underline{u} = \underline{L} + D_2 g(\underline{u},\underline{u}), \qquad /2.107/$$

and

$$\underline{u} = GB\underline{u} = G(\underline{b},\underline{F}) + GD_2 g(\underline{u},\underline{u}). \qquad /2.108/$$

Comparing /2.104/ with /2.108/ we infer that

$$\underline{\Sigma} = D_2 g(\underline{u},\underline{u}) , \text{ on } S_2 . \qquad /2.109/$$

Hence we can write $(D_2 g)^{-1}(-\underline{u}^*) = \underline{u} = \underline{u} = G(\underline{\Sigma}) + \hat{\underline{u}}$ , and the property $(H_4)$ is satisfied.

Taking account of /2.84/ and /2.109/ we finally formulate the dual problem of $(P_d)$ as

Problem $P_d^*$

Find $\underline{\Sigma} = (s_N,\underline{s}_T) \in (-K_N^*) \times C(s_N)$ such that

$$\langle \underline{t}-\underline{\Sigma}, G(\underline{\Sigma}) + \hat{\underline{u}}\rangle \geq 0, \ \forall \underline{t} = (t_N,\underline{t}_T) \in (-K_N^*) \times C(s_N) , \qquad /2.110/$$

or

$$\langle t_N - s_N, [G(\underline{\Sigma})]_N + \hat{u}_N\rangle + \langle \underline{t}_T - \underline{s}_T, [G(\underline{\Sigma})]_T + \hat{\underline{u}}_T\rangle \geq 0 \qquad /2.111/$$
$$\forall t_N \in -K_N^* , \ \forall \underline{t}_T \in C(s_N).$$

The extremality condition /2.78/ gives

$$J(\underline{u},\underline{u}) = -\langle \underline{s}_T,\underline{u}_T\rangle - \langle s_N,u_N\rangle = -a(\underline{u},\underline{u}) + L(\underline{u}) . \qquad /2.112/$$

Hence

$$J(\underline{u},\underline{u}) = -\langle \underline{s}_T,\underline{u}_T\rangle , \quad \langle s_N,u_N\rangle = 0 , \text{ on } S_2 \qquad /2.113/$$

and

$$a(\underline{u},\underline{u}) - L(\underline{u}) - \langle \underline{s}_T,\underline{u}_T\rangle - \langle s_N,u_N\rangle = 0 \qquad /2.114/$$

Obviously, in /2.110/ - /2.114/ the duality pairings $\langle .,. \rangle$ are defined over the part $S_2$ of the boundary. Further, it is worth noticing that /2.110/, or equivalently /2.111/, cannot be uncoupled into two inequalities for an independent determination of the normal $s_N$ and tangential $\underline{s}_T$ stresses over the contact surface $S_2$. The coupling is implied by $\underline{s}_T \in C(s_N)$, since /2.110/ is the <u>quasi-variational inequality</u> posed on $S_2$ only. Such Q.V.I. is useful for the determination of contact pressure and tangential /frictional/ stresses provided that the Green's matrix $[G_{ik}(x,y)]$, or the Green's operator G is known. Unfortunately, this is not generally the case and approximation procedures are mostly indispensable.

To apply Lemma 2.5 we observe that

$$I_{K_d}(\underline{w}) = I_{K_N}(\gamma_N \underline{w}) = I_{K_N}(w_N),\ \underline{w} \in V ,$$

where

$$K_N = \{z \in V_N \mid z \leq 0 ,\quad z = (\gamma \underline{z})|_{S_2},\ \underline{z} \in V\} \qquad /2.96/$$

and $\gamma$ is the trace operator; $w_N = \gamma_N(\underline{w}) = (\gamma \underline{w})_N$.
We can assume that $V_N = H^{1/2}(S_2)$. If $s_N \in H^{-1/2}$, then $\langle s_N, v_N \rangle$ is to be understood in the sense of duality. On the other hand, if $s_N \in L^2(S_2)$ then

$$\langle s_N, v_N \rangle = \int_{S_2} s_N v_N dS ,$$

in the sense of Lebesque.

Having introduced indispensable preliminaries, we find

$$\varphi^*(Av, w^*) = \sup\{\langle w^*, w\rangle - J(\underline{v}, \underline{w}) - I_{K_d}(\underline{w}) | \underline{v} \in V\} =$$

$$= \sup\Big\{[\langle w_N^*, w_N\rangle - I_{K_N}(w_N)] + [\langle \underline{w}_T^*, \underline{w}_T\rangle -$$

$$- \int_{S_2} j(s_N(\underline{v}), \underline{w}_T)\, dS] \;\Big|\; (\gamma \underline{w})|_{S_2} = (w_N, \underline{w}_T)|_{S_2},\ \underline{w} \in V\Big\}$$

Taking account of /2.64/, we conclude that

$$\varphi^*(A\underline{v}, \underline{w}^*) = I_{K_N^*}(w_N^*) + I_{C(s_N(\underline{v}))}(\underline{w}_T^*) , \qquad /2.97/$$

where

$$K_N^* = \{z^* \in V_N^* \mid \langle z^*, z\rangle_{V_N \times V_N} \leq 0,\ \forall z \in K_N\} , \qquad /2.98/$$

while

$$C(s_N(\underline{v})) = \{\underline{t}_T \mid \forall\ \underline{w}_T = (\gamma \underline{w})_T \text{ , on } S_2, \langle \underline{t}_T, \underline{w}_T\rangle \leq J(\underline{v}, \underline{w})\} \qquad /2.99/$$

The last relation follows directly from Corollary 13.2.1 of Rockafellar [157], which for l.s.c. functions reads: "Let f be any positively homogeneous l.s.c. convex function which is not identically $+\infty$. Then f is the support function of a certain closed convex set C, namely

$$C = \{v^* \mid \forall v,\ \langle v^*, v\rangle \leq f(v)\}".$$

Although Rockafellar considers the finite-dimensional case only, yet the above corollary readily extends to infinite-dimensional case, such as studied in the preceding section.

The closed and convex set $C(s_N)$ is the global one. If

$s_N \in L^2(S_2)$ then $C(s_N)(x) = K(s_N(x))$, almost everywhere on $S_2$. The relation /2.97/ yields

$$\varphi^*(A\underline{v},\underline{w}^*) = \begin{cases} 0, \text{ if } w_N^* \in K_N^* \text{ and } \underline{w}_T^* \in C(s_N(v)), \\ +\infty, \text{ otherwise.} \end{cases} \qquad /2.100/$$

To find an explicit form of $g^*(\underline{v},\underline{w}^*)$, we introduce the linear operator $B, B: V \to V^*$, defined as follows

$$\langle B\underline{v},\underline{w}\rangle = \int_\Omega s_{ij}(\underline{v})e_{ij}(\underline{v})dx = a(\underline{v},\underline{w}); \ \underline{v},\underline{w} \in V. \qquad /2.101/$$

Hence

$$\langle Bv,v\rangle = a(\underline{v},\underline{v}) \geq c\|\underline{v}\|_V^2 \quad , c>0, \qquad /2.102/$$

provided that meas $S_o > 0$; c is a constant.
As a norm on V we can take any of equivalent norms, for instance [18,43]

$$\int_\Omega e_{ij}(\underline{v})e_{ij}(\underline{v})\,dx.$$

The operator B is continuous, since

$$\langle B\underline{v},\underline{w}\rangle \leq c_1\|\underline{v}\|_V\|\underline{w}\|_V \quad ; \ \underline{v},\underline{w} \in V.$$

Hence we infer that an inverse operator $G=B^{-1}$ exists, $G: V^* \to V$. This operator is obviously continuous and $V^*$-elliptic, that is

$$\langle \underline{v}^*, G\underline{u}^*\rangle \leq c_2\|\underline{u}^*\|_{V^*}\|\underline{v}^*\|_{V^*} \quad ,$$

$$\langle \underline{u}^*, G\underline{u}^*\rangle \geq c_3\|\underline{u}^*\|^2_{V^*} \quad ,$$

where $c_2$ and $c_3$ are positive constants. The linear operator G is nothing else as the Green's operator for the mixed boundary value problem of the elastostatics.

Let us put $\underline{\Sigma} = (s_{ij}n_j)$, on $S_2$. Now we can write, see [146,p.135]

$$u_k(x) = [G(\underline{b},\underline{F},\underline{\Sigma})]_k = \int_\Omega b_i(y)G_{ik}(x,y)\,dS(y) + \qquad /2.103/$$
$$+\int_{S_1} F_i(y)G_{ik}(x,y)\,dS(y) + \int_{S_2} \Sigma_i(y)G_{ik}(x,y)\,dS(y).$$

Due to the linearity of G we can write

$$\underline{u} = G(\underline{b},\underline{F}) + G(\underline{\Sigma}) = \hat{\underline{u}} + G(\underline{\Sigma}), \qquad /2.104/$$

where $\hat{\underline{u}} = G(\underline{b},\underline{F})$.

Taking account of /2.92/ and /2.101/ we obtain

$$g^*(\underline{v},\underline{w}^*) = \sup\{\langle \underline{w}^*,\underline{w}\rangle - a(\underline{v},\underline{w}-\underline{v}) + L(\underline{w}-\underline{v}) \mid \underline{w} \in V\} =$$
$$= a(\underline{v},\underline{v}) - L(\underline{v}) + \sup\{\langle \underline{w}^*,\underline{w}\rangle - a(\underline{v},\underline{w}) + L(\underline{w}) \mid \underline{w} \in V\} =$$

The proof that $(P_d^{**}) = (P_d)$ is straightforward and results from Corollary 2.1 and Lemma 2.4. In our case we have

$$g(\underline{v},\underline{w}) = a(\underline{v},\underline{w}-\underline{v}) - L(\underline{w}-\underline{v}) = \langle B\underline{v}-L,\underline{w}-\underline{v}\rangle =$$

$$= \langle D_2 g(\underline{v},\underline{v}),\underline{w}-\underline{v}\rangle = \langle (D_2\tilde{g})^{-1}(-\underline{v}),\underline{v}-\underline{w}\rangle \text{ ; similarly } P_s^{**}=P_s.$$

Proof that $(P_s^*) = (P_d)$. Toward this end we set

$$g_1(\underline{s},\underline{t}) = b(\underline{s},\underline{t}-\underline{s}) \ ; \quad \underline{s},\underline{t} \in \left[L^2(\Omega)\right]^6 , \qquad /2.115/$$

$$\varphi_1(s_N,\underline{t}) = I_{K_s(s_N)}(\underline{t}). \qquad /2.116/$$

Hence we obtain

$$g_1^*(\underline{s},\underline{t}^*) = \sup_{\underline{t}}\left\{\langle \underline{t}^*,\underline{t}\rangle - b(\underline{s},\underline{t}-\underline{s})\right\} =$$

$$= b(\underline{s},\underline{s}) + \sup_{\underline{t}}\left\{\int_\Omega (t_{ij}^* - b_{ijkl}s_{kl})\, t_{ij}\, dx\right. =$$

$$= \begin{cases} b(\underline{s},\underline{s}) \ , & \text{if } t_{ij}^* = b_{ijkl}s_{kl} , \\ +\infty \ , & \text{otherwise} \end{cases} \qquad /2.117/$$

Thus $\underline{t}^*$ is the strain tensor.
The Gâteaux derivative is

$$\langle D_2 g_1(\underline{s},\underline{t}),\underline{h}\rangle = b(\underline{s},\underline{h}).$$

Hence

$$D_2 g_1(\underline{s},\underline{t}) = (b_{ijkl}s_{kl}) , \qquad /2.118/$$

and thus

$$(D_2 g_1)^{-1}(-\underline{s}^*) = \underline{s} = -\underline{a}\,\underline{s} . \qquad /2.119/$$

Let us pass to the derivation of the conjugate functional $\varphi_1^*(s_N,\underline{t}^*)$. We may take $\underline{t}^* = \underline{e}(\underline{w})$, $\underline{w} \in V$, where V is defined by /2.61/. Then one has

$$\varphi_1^*(s_N,\underline{e}(\underline{w})) = \sup\left\{\langle \underline{t},\underline{e}(\underline{w})\rangle - \varphi_1(s_N,\underline{t}) \mid \underline{t} \in H_s\right. =$$

$$= \sup\left\{\int_\Omega t_{ij}e_{ij}(\underline{w})\,dx \mid \underline{t}\in K_s(s_N)\right\} =$$

$$= \sup\left\{-\int_\Omega t_{ij,j}w_i\,dx + \int_S t_{ij}n_jw_i\,dS \mid \underline{t} \in K_s(s_N)\right\} =$$

$$= L(\underline{w}) + \sup_{t_N \leqslant 0}\int_{S_2} t_N w_N\,dS + \sup\left\{\int_{S_2} \underline{t}_T\cdot\underline{w}_T\,dS \mid \underline{t}_T \in C(s_N)\right\} =$$

$$= \begin{cases} L(\underline{w}) + \tilde{J}(s_N,\underline{w}_T) \ , & \text{if } w_N \geqslant 0 \text{ on } S_2 , \\ +\infty \ , & \text{otherwise} \end{cases} \qquad /2.120/$$

where

$$\tilde{J}(s_N, \underline{w}_T) = \sup \left\{ \int_{S_2} \underline{t}_T \cdot \underline{w}_T dS \,\middle|\, \underline{t}_T \in C(s_N) \right\} \qquad /2.121/$$

In /2.121/ the roles of supremum and integration may be intechanged provided that $\underline{t}_T \cdot \underline{w}_T$ is a measurable function belonging to a so called <u>decomposable</u> space, for instance to $L^1(S_2)$ [158,p.185].

Taking account of Th.2.3, /2.119/ and /2.120/, the dual problem $P_s^*$ reads

$$(P_s^*) \left| \begin{array}{l} \text{find } \tilde{\underline{u}} \in -K_d \quad \text{such that} \\ L(\tilde{\underline{u}}) + \tilde{J}(-s_N(\tilde{\underline{u}}), \tilde{\underline{u}}_T) + a(\underline{u},\underline{u}) \leq L(\underline{w}) + \tilde{J}(-s_N(\tilde{\underline{u}}), \underline{w}_T) + \\ \qquad + a(\underline{w}, \tilde{\underline{u}}) \;, \; \forall \, \underline{w} \in -K \, . \end{array} \right. \qquad /2.122/$$

Now we set $\underline{u} = -\tilde{\underline{u}}$ and $\underline{v} = -\underline{w}$. Then the problem $P_s^*$ takes on the form

$$(P_s^*) \left| \begin{array}{l} \text{find } \underline{u} \in K_d \quad \text{such that} \\ -L(\underline{u}) + J(\underline{u},\underline{u}) + a(\underline{u},\underline{u}) \leq - L(\underline{v}) + J(\underline{u},\underline{v}) + \\ \qquad + a(\underline{v},\underline{u}), \quad \forall \, \underline{v} \in K_d \, , \end{array} \right. \qquad /2.123/$$

where $J(\underline{u},\underline{v}) = \tilde{J}(s_N(\underline{u}), -\underline{v}_T)$ , see also /2.64/. Thus we see that $(P_s^*) = (P_d)$.

The extremality condition /2.78/ takes now the form

$$-L(\underline{u}) + J(\underline{u},\underline{u}) = -\langle \underline{s}, \underline{e}(\underline{u}) \rangle = -b(\underline{s},\underline{s}) \, . \qquad /2.124/$$

where $\langle \underline{s}, \underline{e}(\underline{u}) \rangle = \int_\Omega s_{ij} e_{ij}(\underline{u}) dx$ .

Summarizing, the main results of this section can be represented by the following diagram

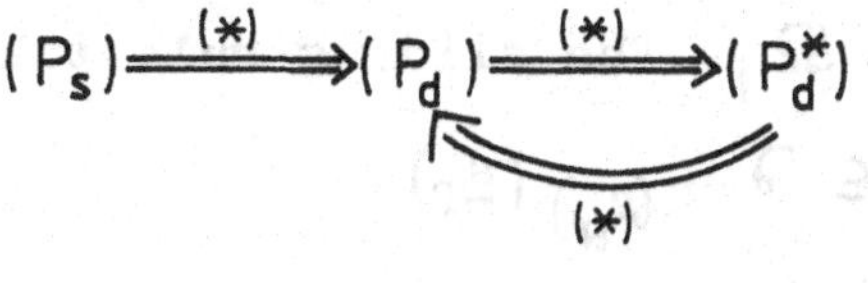


$$(P_s^*) = (P_d) \quad , \quad (P_d^{**}) = (P_d)$$

Fig.17

provided that $(*)$ stands for the duality operation.

Remark 2.4. Suppose that in /2.111/ $\underline{s}_T=\underline{t}_T=0$, on $S_2$. Then the Q.V.I. /2.111/ simplifies to the following variational inequality defined, as previously, on $S_2$

$$(P_V^*)\quad \begin{array}{|l} \text{find } s_N \in -K_N^* \text{ such that} \\ \langle t_N-s_N, [G(s_N)]_N + \hat{u}_N\rangle \geq 0, \ \forall\, t_N \in -K_N^*. \end{array} \qquad /2.125/$$

The latter V.I. has already been derived by Kikuchi [120] and called the "reciprocal variational inequality". This author does not employ the Mosco's theory of duality, see Remark 2.2. We observe that /2.125/ follows directly from /2.91/. To solve the problem $P_V^*$ a numerical procedure has been proposed in [120], see also the lectures by Haslinger and [91,121,123]. Reciprocal formulation of the Signorini's problem with Coulomb's friction and prescribed contact pressure together with an approximation by Ritz-Galerkin method is studied in [110]. Similar problem in the dynamical case is investigated in [162].

Various numerical methods for solving elastic unilateral contact problems with friction have been proposed in [81,105,109,111,117,122,123,125,126,128,130,138,147,149, 151,162].

An a alytical approach employing complex variable method to dynamical frictional problems without Signorini's conditions is used in [160].

Geometrically nonlinear problems are studied in [98, 150,159,161] ; the frictionless case is considered in [84].

Mechanically important frictional problems for plates are investigated in [62,63,113]. The sliding rule employed in [62,63] is motivated numerically and not physically.

Remark 2.5. M-CD-M theory of duality readily applies to two elastic solids $\Omega_\alpha$ $(\alpha=1,2)$, clamped along $S_0^\alpha$. Now we have $S_2=S_2^1=S_2^2$ and $\underline{s}_T=\underline{s}_T^1=-\underline{s}_T^2$, $[\![u_N]\!] = u_N^{(1)} - u_N^{(2)} \leq 0$, $s_N=s_N^{(1)}= -s_N^{(2)}$, $s_N[\![u_N]\!] = 0$ /on $S_2$/, provided that the unit normal $\underline{n}$ is taken as exterior to $\Omega_1$. The sliding rule takes the form

$$[\![\underline{u}_T]\!] \in \partial I_{K(s_N)}(\underline{s}_T). \qquad /2.126/$$

Let us set $\underline{u} = (\underline{u}^{(1)},\underline{u}^{(2)})$, $\underline{b} = (\underline{b}^{(1)},\underline{b}^{(2)})$, $\underline{F} = (\underline{F}^{(1)},\underline{F}^{(2)})$ and

$$a(\underline{u},\underline{v}) = \sum_\alpha a^{(\alpha)}(\underline{u}^{(\alpha)}; \underline{v}^{(\alpha)}), \quad L(\underline{v}) = \sum_\alpha L^{(\alpha)}(\underline{v}^{(\alpha)}), \qquad /2.127/$$

and, for instance

$$L^{(1)}(\underline{v}^{(1)}) = \int_{\Omega_1} b_i^{(1)} v_i^{(1)}\, dx + \int_{S_1} F_i^{(1)} v_i^{(1)}\, dS \qquad /2.128/$$

The set of kinematically admissible displacements is

$$K_d = \{ \underline{v} = (\underline{v}^{(1)}, \underline{v}^{(2)}) \in V \,|\, [\![ v_N ]\!] \leq 0 \quad \text{on } S_2 \} , \qquad /2.129/$$

where

$$V = V_1 \times V_2 = \{ (\underline{v}^{(1)}, \underline{v}^{(2)}) \,|\, \underline{v}^{(\alpha)} \in V_\alpha \,,\; \underline{v}^{(\alpha)} = 0 \text{ on } S_o^\alpha \} \qquad /2.130/$$

With obvious identifications, the Signorini's problem with the sliding rule /2.126/ takes on the same form as the problem $P_d$, studied previously. I denote it by $(P_{d2})$.

Now we have

$$\underline{u}^{(1)} = \hat{\underline{u}}^{(1)} + G_1(\underline{\Sigma}) \,,\; \underline{u}^{(2)} = \hat{\underline{u}}^{(2)} - G_2(\underline{\Sigma}) \,, \qquad /2.131/$$

where

$$\hat{\underline{u}}^{(\alpha)} = G_\alpha(\underline{b}^{(\alpha)}, \underline{F}^{(\alpha)}) \quad , \text{ /no summing!/.}$$

The dual problem $P^*_{d2}$ reads:

$$(P^*_{d2}) \left| \begin{array}{l} \text{find } \underline{\Sigma} \in -K_N^* \times C(s_N) \quad \text{such that} \\ \langle \underline{t} - \underline{\Sigma},\, G_1(\underline{\Sigma}) + G_2(\underline{\Sigma}) + \hat{\underline{u}}^{(1)} + \hat{\underline{u}}^{(2)} \rangle \geqslant 0, \;\forall \underline{t} \in -K_N^* \times C(s_N) \end{array} \right. \qquad /2.132/$$

where

$$K_N^* = \{ z^* = (z^{(1)*}, z^{(2)*}) \in V_N^* = V_{1N}^* \times V_{2N}^* \,|\, \langle z^*, z \rangle \leq 0, \; \forall\, z \in K_N \} \qquad /2.133/$$

and

$$K_N = \{ z = (z^{(1)}, z^{(2)}) \in V_N = V_{1N} \times V_{2N} \,|\, z^{(1)} - z^{(2)} \leqslant 0 \,,$$

$$z^{(1)} = [\gamma_1(\underline{z}^{(1)})]_{|S_2} \,,\; z^{(2)} = [\gamma_2(\underline{z}^{(2)})]_{|S_2} \,,$$

$$\underline{z} = (\underline{z}^{(1)}, \underline{z}^{(2)}) \in V . \qquad /2.134/$$

Of particular interest is the case where one of the domains, say $\Omega_2$, is <u>unbounded</u>. For instance, let $\Omega_2$ be a half-space:

$$\Omega_2 = \{ x = (x_i) \in R^3 \quad x_3 < 0 \} .$$

Further, let $S_o^1$ be empty. Then the bilinear form $a^{(1)}(\cdot,\cdot)$ is not coercive, yet the Green operator $G_2$ still exists [146] for that traction /second/ boundary value problem of elasticity.

We observe that the numerical procedure proposed by Kikuchi [120] for frictionless case extends to problems, where $S_o^1$ is empty by introducing of an artificial part of the boundary with prescribed displacements.

2.7. Existence problems

The existence of a solution for the problem $P_d$ is solved only partially, even in the case of Coulomb friction [86,92,95,115,123,130,145,171]. Particularly, Jarušek [115] studies the case where $\mu \in C^1(S_2)$ and supp$\mu$ is compact. Thus the coefficient of friction is equal to zero in the vicinity of $\partial S_2$, what is not very realistic.

Duvaut [95] introduced the notion of non-local friction which has been exploited in many papers by Oden and his coworkers, see [55,92,123,138,147,148,149,151,168]. This notion proves also useful from the numerical point of view.

Cocu and Radoslovescu [87] have proved the following regularity theorem for the problem $P_d$ in the case of non--local Coulomb friction.

Theorem 2.5. Let $\Omega$ be of class $C^3$ in all $x \in S_2$. Then for every open set U, $\overline{U} \subset \Omega \subset S_2$, we have

$$\underline{u} \in \left[H^2(U)\right]^3 .$$

Existence problems for elastic bodies in the presence of viscous sliding rules /parabolic problem/, but without Signorini's conditions have been studied in [96,133].

Problem $P_{d2}$ for both linear elastic and viscoelastic bodies is investigated by Boucher [183,279], see also [192, 220]. However, friction is neglected, but $\Omega_2$ is a half space, see Remark 2.5. The space $V_2$ is a Barros-Neto space [278], equivalently defined as follows

$$V_2 = \{\underline{v} = (v_i) \mid v_i \in L^6(\Omega_2) \ , \ v_{i,j} \in L^2(\Omega_2)\} . \qquad /2.135/$$

Such space is larger then $H^1(\Omega_2)$ and both spaces coincide if $\Omega_2$ is bounded.

It can also be shown that there is a continuous imbedding of the trace space $K^{1/2}(R^2) = K^{1/2}(\partial\Omega_2)$ of functions of $V_2$ into $L^4(\partial\Omega_2)$. Thus we see that the linear form $L^{(2)}(v^{(2)})$ makes sense if $b_i \in L^{6/5}(\Omega_2)$, $F_i \in L^{4/3}(S_1^2)$. Rigid body motions of the body $\Omega_1$ are also studied by Boucher [183, 279].

Oden and Martins [147] proposed new contact conditions with a rough surface. I will now show that those conditions correspond to associated rigid plasticity with hardening and/or softening.

Let the initial gap between the body and the foundation be $g \geq 0$ along $S_2$. According to the paper [147], the contact conditions are:

/i/ normal interface response

$$s_N = -c_N(u_N-g)_+^{m_n} \quad , \quad \text{on } S_2 \qquad /2.136/$$

/ii/ friction conditions

$$u_N \leq g \Rightarrow \underline{s}_T = 0$$

$$u_N > g \Rightarrow \begin{cases} |\underline{s}_T| \leq c_T(u_N-g)_+^{m_T} , \\ |\underline{s}_T| < c_T(u_N-g)_+^{m_T} \Rightarrow \underline{u}_T = \underline{\dot{U}}_T^o , \\ |\underline{s}_T| = c_T(u_N-g)_+^{m_T} \Rightarrow \exists \lambda \geq 0, \ \underline{\dot{u}}_T - \underline{\dot{U}}_T^o = -\lambda \underline{s}_T , \end{cases} \qquad /2.137/$$

where $c_n, c_T, m_n$ and $m_T$ are material parameters, and $\underline{\dot{U}}_T^o$ is the tangential velocity of the subgrade with which $S_2$ comes in contact; $(u_N-g)_+$ stands for the positive part of $(u_N-g)$.

We observe that Coulomb's friction law is recovered if $m_n = m_T$; then $\mu = c_T/c_n$ is the usual coefficient of friction.

Oden-Martins formulae imply that the friction condition is given by

$$f(s_N, \underline{s}_T, u_N^+) = |\underline{s}_T| - \mu(u_N^+)|s_N| , \qquad /2.138/$$

where $\mu(u_N^+) = c_T(u_N-g)_+^{m_T}/c_n(u_N-g)_+^{m_n}$ is the coefficient of friction, which depends on the normal displacement $u_N$. In the case considered $u_N^e=0$, $\underline{u}_T^e=0$ . The sliding rule takes on the form

$$(-\dot{u}_N, -\underline{\dot{v}}_T) \in \partial I_{K(u_N^+)}(s_N, \underline{s}_T) , \qquad /2.139/$$

where $\underline{\dot{v}}_T = \underline{\dot{u}}_T - \underline{\dot{U}}_T^o$ , and

$$K(u_N^+) = \{ \underline{t} = (t_N, \underline{t}_T) \mid f(t_N, \underline{t}_T, u_N^+) \leq 0 \} . \qquad /2.140/$$

Hence the density of dissipation produced by normal and tangential /frictional/ stresses is given by

$$j(u_N^+, \dot{u}_N, \underline{\dot{u}}_T) = \sup\{ t_N \cdot (-\dot{u}_N) + t_T \cdot (-\underline{\dot{v}}_T) \mid (t_N, \underline{t}_T) \in K(u_N^+) \} =$$

$$= c_n(u_n-g)_+^{m_n} |\dot{u}_N| + c_T(u_n-g)_+^{m_T} |\underline{\dot{u}}_T - \underline{\dot{U}}_T^o| \qquad /2.141/$$

The coefficient of friction $\mu(u_N^+)$ may be seen as a specific case of Madakson's formula /2.6/.

Static case is studied in [153], where existence and uniqueness of variational /weak/ solutions are proved. Results are established either for arbitrary external forces and sufficiently small coefficient of friction or for arbitrary coefficient of friction and sufficiently small external forces. For this purpose the implicit function theorem is used.

Now I pass to proving that a solution of the problem $P_d$ exists in the general case of a nonlocal sliding rule. Such rules are introduced as follows.

We assume that

$$V = \{ \underline{v} = (v_i) \mid v_i \in H^1(\Omega),\ \underline{v} = 0 \text{ on } S_o \}, \qquad /2.142/$$

$$H_s = \{ \underline{t} = (t_{ij}) \mid t_{ij} \in L^2(\Omega),\ t_{ij,j} \in L^2(\Omega) \}. \qquad /2.143/$$

Then the Green formula shows that $t_{ij}n_j \in H^{-1/2}(S)$, and particularly $s_N(\underline{u}) \in H^{-1/2}(S_2)$. This is a source of mathematical troubles with the problem $P_d$. One way of overcoming them is to replace $s_N$ by more regular normal stresses $s_N^h$. For instance, we can take

$$s_N^h(x) = \int_{S_2} \omega_h(|x-y|\ s_N(y)\,dS(y)\ ; \quad x,y \in S_2 .$$

The last operation is called mollification. The function $\omega_h(r)$ is of class $C^\infty$ and has the following properties

/i/ $\omega_h(r) > 0, \quad r < h,$

/ii/ $\omega_h(r) = 0, \quad r \geqslant h,$

/ii/ $\int_{r<h} \omega_h(r)\,dy = \int_{r<h} \omega_h(r)\,dx = 1 .$

Let us put

$$J_h(\underline{u},\underline{v}) = \int_{S_2} j(s_N^h(\underline{u}), \underline{v}_T)\,dS . \qquad /2.144/$$

Instead of the problem $P_d$ we consider the following one

$$(P_d^h) \left| \begin{array}{l} \text{find } \underline{u} \in K_d \text{ such that} \\ a(\underline{u},\underline{v}-\underline{u}) + J_h(\underline{u},\underline{v}) - J_h(\underline{u},\underline{u}) \geqslant L(\underline{v}-\underline{u}),\ \forall \underline{v} \in K_d \end{array} \right. \qquad /2.145/$$

Now let us recall a general existence result [143], here formulated for reflexive spaces only.

Theorem 2.6. Let $V$ be a reflexive Banach space and $g: C\times C \to R$, with $g(\underline{v},\underline{v}) \geq 0$ $\forall v \in C$. We assume that
/a/ $C \subset V$ is a closed convex subset.
/b/ For each $v \in C$, $g(v,\cdot)$ is convex.
/c/ For each $w \in C$, $g(\cdot,w)$ is weakly upper semicontinuous /u.s.c./.
/d/ There exists a bounded subset $B \subset V$ and an element $w_0 \in B \cap C$ such that for all $v \in C\setminus B$ we have $g(v,w_0) < 0$.
Then there exists at least one solution $\underline{u} \in C \cap B$ of the inequality

$$g(u,v) \geq 0 \qquad \forall v \in C. \qquad /2.146/$$

To employ this theorem we shall also need the following one, proved in [66], see also [80].

Theorem 2.7. Let $F: \Theta \times R^s \times R^t \to R \cup \{+\infty\}$ $(R \cup \{-\infty\})$ satisfy the following properties:
/i/ $F(\cdot,z,v): \Theta \to R \cup \{+\infty\}$ $(R \cup \{-\infty\})$ is measurable for every $(z,v) \in R^s \times R^t$,
/ii/ $F(x,\cdot,\cdot): R^s \times R^t \to R \cup \{+\infty\} (R \cup \{-\infty\})$ is continuous for almost all $x \in \Theta$,
/iii/ $F(x,z,\cdot): R^t \to R \cup \{+\infty\}(R \cup \{-\infty\})$ is convex /concave/ for almost all $x \in \Theta$ and all $z \in R^s$.

Let $z^{(n)}, z: R^s \to R \cup \{+\infty\} (R \cup \{-\infty\})$ be measurable functions such that $z^{(n)} \to z$ almost everywhere and let $v^{(n)} \to v$ in $[L^1(\Theta)]^t$ as $n \to \infty$. Suppose further that there exists $f \in L^1(\Theta)$ such that

$$F(x,z^{(n)}(x),v^{(n)}(x)) \geq f(x), \quad F(x,z(x),v(x)) \geq f(x),$$
$$(F(x,z^{(n)}(x),v^{(n)}(x)) \leq f(x), \quad F(x,z(x),v(x)) \leq f(x),$$

for all n and almost all $x \in \Theta$. Then

$$\int_\Theta F(x,z(x),v(x))\,dx \leq \underline{\lim}_{n\to\infty} \int_\Theta F(x,z^{(n)}(x),v^{(n)}(x))\,dx \quad ,/2.147/$$

$$\left( \int_\Theta F(x,z(x),v(x))\,dx \geq \overline{\lim}_{n\to\infty} \int_\Theta F(x,z^{(n)}(x),v^{(n)}(x))\,dx \right). \quad /2.148/$$

The next lemma has been proved in [92].

Lemma 2.6. Let $\underline{u}_n \to \underline{u}$ weakly in $V$, where $V$ is defined by /2.142/. Then

/a/ $s_N(\underline{u}_n) \to s_N(\underline{u})$ in $H^{-1/2}(S)$ weakly, that is

$$\langle s_N(\underline{u}_n),z\rangle \to \langle s_N(\underline{u}),z\rangle \qquad \forall z \in H^{1/2}(S).$$

/b/ $s_N^h(\underline{u}_n) \rightarrow s_N^h(\underline{u})$ in $L^2(S_2)$ strongly.

Let us set $C = K_d$ , and

$$g(\underline{v},\underline{w}) = a(\underline{v},\underline{w}-\underline{v}) + J_h(\underline{v},\underline{w}) - J_h(\underline{v},\underline{v}) - L(\underline{w}-\underline{v}). \qquad /2.149/$$

The function F involved in Th.2.7 is d. As we know, the function $d(x,z,.)$ is convex. For a fixed $\underline{v}$, $d(x,z,\underline{v})$ is trivially concave with respect to $\underline{v}$.
Let us assume that $d(x,.,.)$ is continuous while $d(.,z,\underline{v})$ measurable. The assumpion concerning continuity can be weakened [80]. In the case of Coulomb's friction continuity is obvious while measurability is satisfied if $\mu \in L^{\infty}(S_2)$.
Our next assumption is /growth condition/

$$J_h(\underline{v},\underline{w}) \leqslant c(\underline{w}) \| \underline{v} \|_V^{2-\varepsilon} , \quad \varepsilon > 0 , \ \underline{v} \in V, \qquad /2.150/$$

provided that $\underline{w} \in V$ is held fixed. In the case of Coulomb friction $\varepsilon = 1$.

The function $g(\underline{v},.)$ defined by /2.149/ is obviously convex.

The functional $g(.,\underline{w})$ is weakly u.s.c. for each $\underline{w} \in K_d$. To corroborate this statement we observe that $a(\underline{v},\underline{v})$ is weakly l.s.c., hence $-a(\underline{v},\underline{v})$ is weakly u.s.c. Further, the functionals $a(.,\underline{v})$ and $L(.)$ are linear and continuous on V. To prove that $J_h(\underline{v},\underline{v})$ is weakly l.s.c., we emply Th.2.7 and Lemma 2.6. The latter implies that at least for a subsequence, still denoted by $\{\underline{u}_n\}$, $s_N^h(\underline{u}_n) \rightarrow s_N^h(\underline{u})$ almost everywhere. Since $J_h(\underline{v},\underline{w})$ represents the work of the frictional stresses, hence it is reasonable to assume that $d(x,z,\underline{v}) \geqslant f(x)$ $x \in S_2$, or even that $f(x) \equiv 0$. Thus we see that $J_h(\underline{u},\underline{u})$ is l.s..c., and $-J_h(\underline{u},\underline{u})$ u.s.c. Now we need the upper semicontinuity of $J_h(.,\underline{w})$, which results from our assumptions and the "concavity part" of Th.2.7.
Thus the condition /c/ of Th.2.6 is satisfied. It is worth noticing that, apart from other rather classical properties, all we need to prove the condition /c/ is the sequentially weak continuity of $J_h(.,\underline{w})$.

To prove the condition /d/, let us take $\underline{w}_o \in K_d$ and define B by

$$B = \{ \underline{v} \in K_d \mid g(\underline{v},\underline{w}_o) \geqslant 0 \} .$$

The set B is bounded. In order to prove it let us suppose that D is unbounded. Then a sequence $\{\underline{v}_n\} \subset K_d$ exists, $\| \underline{v}_n \|_V \rightarrow \infty$ as $n \rightarrow \infty$ , such that

$$g(\underline{v}_n,\underline{w}_o) \geqslant 0 \,, \qquad /2.151/$$

for all n.

Consider the functional $-g(\underline{v}_n,\underline{w}_o)$. By using Korn's inequality, /2.150/ and knowing that $J_h(\underline{v}_n,\underline{v}_n)$ is bounded from below we arrive at

$$-g(\underline{v}_n,\underline{w}_o) \geqslant c_1 \| \underline{v}_n \|_V^2 - c(\underline{w}_o) \| \underline{v}_n \|_V^{2-\varepsilon} - c_2 \| \underline{v}_n \|_V + c_3 \,,$$

where $c, c_1, c_2$ and $c_3$ are constants and $c_1 > 0$. Hence, $-g(\underline{v}_n,\underline{w}_o) \longrightarrow \infty$ or $g(\underline{v}_n,\underline{w}_o) \to -\infty$ ; a contradiction with /2.151/.

Thus Th.2.6 yields finally the existence of $\underline{u} \in K_d$ solving the nonlocal frictional problem $P_d^h$.

Such a solution is <u>unique</u> under additional assumptions, see [86,Th.3.1].

Let H be a real Hilbert space, $C \subset H$ a non-empty closed convex subset and let $V: C \to H$ be an operator, such that

$$\begin{aligned} &(B(u) - B(v), u-v) \geqslant m \| u-v \|^2 , \quad \forall u,v,w \in C \\ &(B(u) - B(v), w) \leqslant M \| u-v \| \, \| w \| , \end{aligned} \qquad /2.152/$$

where m and M are positive constants, $(\cdot,\cdot)$ stands for the inner product on H and $\| \cdot \|^2 = (\cdot,\cdot)$.

In our case the operator B is defined by /2.101/ and then /2.152/ certainly holds.

<u>Theorem 2.8.</u> Let us assume that /2.152/ is satisfied and let $J: C \times C \to R \cup \{+\infty\}$ be a proper functional such that $J(v,\cdot)$ is convex and l.s.c. on C. Let L be a linear and continuous functional on H and suppose that J satisfies the following inequality

$$|J(u_1,v_1) + J(u_2,v_1) - J(u_1,v_1) - J(u_2,v_2)| \leqslant c \| u_1 - u_2 \| \, \| v_1 - v_2 \|$$
$$\forall \, u_1,u_2,v_1,v_2 \in C \,, \qquad /2.153/$$

where $c > 0$ is a constant, $m > c$.

<u>Then there exists a unique</u> solution of the following problem

$$(P_B) \left| \begin{array}{l} \text{find } u \in C \quad \text{such that} \\ (B(u), v-u) + J(u,v) - J(u,u) \geqslant L(v-u) \quad \forall v \in C . \end{array} \right. \qquad /2.154/$$

To apply Th.2.8 to the Problem $P_d^h$, I assume that the functional $J_h$ satisfies /2.153/ and the requirements of this theorem. <u>Then $\underline{u} \in K_d$ solving the problem $P_d^h$ is unique.</u>

In the case of Coulomb's friction uniqueness is implied by a "smallness" of $\mu$ [86,92], provided that $\mu \in L^\infty(S_2)$ and

$0 \leq \mu(x) \leq \mu_1$. Cocu [82] proves that in such a case

$$|J_h(\underline{u}_1, \underline{v}_2) + J_h(\underline{u}_2, \underline{v}_1) - J_h(\underline{u}_1, \underline{v}_1) - J_h(\underline{u}_2, \underline{v}_2)| \leq$$
$$\leq c_o \, \|\mu\|_{L^\infty(S_2)} \| \underline{u}_1 - \underline{u}_2 \| \| \underline{v}_1 - \underline{v}_2 \| \quad . \qquad /2.155/$$

Taking $\mu_1 < mc_o$ we conclude that /2.153/ and $m > c$ are satisfied if $c = c_o \mu_1$.

## 3. INELASTIC UNILATERAL CONTACT PROBLEMS

As one could expect, mostly studied are unilateral elastic problems. For instance, mathematically rigorous study of unilateral contact problems for plastic bodies and structures is at the very begining.

In this chapter I will summarize the variational approach to the study of unilateral problems of viscoelasticity and plasticity. Some new results are also presented as well as open problems.

### 3.1. Impact of a viscoelastic rod with an obstacle

In the papers [189,190] an impact of a viscoelastic rod with rigid or deformable /linear/ obstacles has been studied. Omitting the dissipation and shock waves during a shock between the rod and the obstacle, the problem is in fact considered as the dynamic Signorini's problem without friction. In the case of a rigid obstacle the variational formulation results in a variational equation and a V.I., which are coupled. It seems that this dynamic situation for the rod is typical for many unilateral problems involving the Signorini's conditions. Surprisingly, but this simple conclusion seems to have been overlooked in the relevant literature. Formally the same problem arises in the quasi--static case involving the Signorini's conditions and the flow law type sliding rule /2.22/, see Section 3.2.

Let us pass to the study of the aforementioned rod problem. Suppose that in the undeformed configuration the rod is identified with $\Omega = (0,l)$, $l > 0$, see Fig.18. The density is denoted by $\rho$, while $s(x,t)$ stands for the normal stress in a section $x$ at time $t$. Viscoelastic constitutive equation has the form

$$s = a \quad + b\dot{\varepsilon} \quad , \qquad a > 0 \quad , \; b > 0 \; , \qquad /3.1/$$

where $\varepsilon(u) = u_{,x} = \partial u/\partial x$, $\dot{\varepsilon} = \partial\varepsilon/\partial t$ . The case $b=0$ corresponds to purely elastic material. Let us put

$$U(t) = u(0,t), \; R(t) = -s(0,t), \; t \in [0,T], \; T > 0. \qquad /3.2/$$

Thus U denotes the displacement of the lower section of the rod while R is the reaction of the obstacle. The Signorini's conditions take on the form

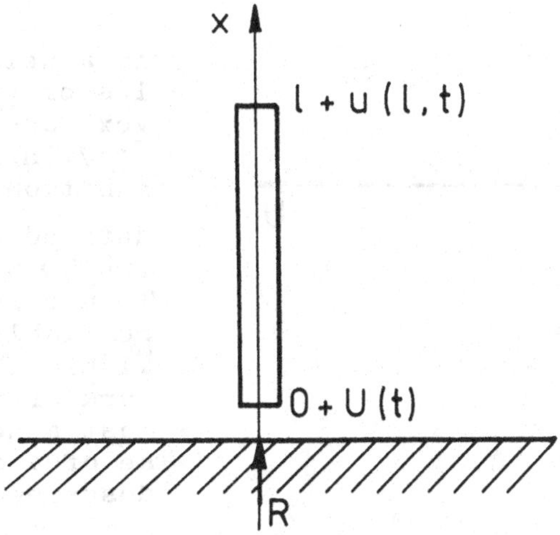


Fig. 18

$$U \geqslant 0 \ , \quad R \geqslant 0 \ , \quad UR = 0 \ , \qquad /3.3/$$

and are equivalent to

$$U \geqslant 0 \ , \quad R(W-U) \geqslant 0 \ \ \forall W \in K \ . \qquad /3.4/$$

The convex set K of kinematically admissible displacements for the lower section is defined by

$$K = \{ W(t) = W(0,t) \mid W(t) \geqslant 0 \ , \ t \in (0,T) \} \ . \qquad /3.5/$$

Obviously, $W(t)$ stands for the virtual displacement of the lower section of the rod and will be precised in the sequel.

From the viewpoint of the existence theory and numerical calculations it is convenient to replace the rigid obstacle by a deformable one /penalty method/.

Elastic obstacle. Denoting by $U^+, U^-$ the positive and negative parts of U $(U = U^+ - U^-)$, we have

$$R = kU^- \Longleftrightarrow R = \begin{cases} 0 \ , & \text{if } U \geqslant 0 \ , \\ -kU, & \text{if } U \leqslant 0 \ , \end{cases} \qquad /3.6/$$

where k stands for the rigidity of the obstacle. The rigid case is recovered in the limit as $k \rightarrow \infty$ .

More general obstacle behaviour can be described as follows. For instance, we may take a nonlinear dependence

$$R = \hat{R}(U^-) \ , \qquad /3.7/$$

where $\hat{R}$ is a continuous and non-decreasing function, not necessarily differentiable, see Fig. 19. In such a case the function

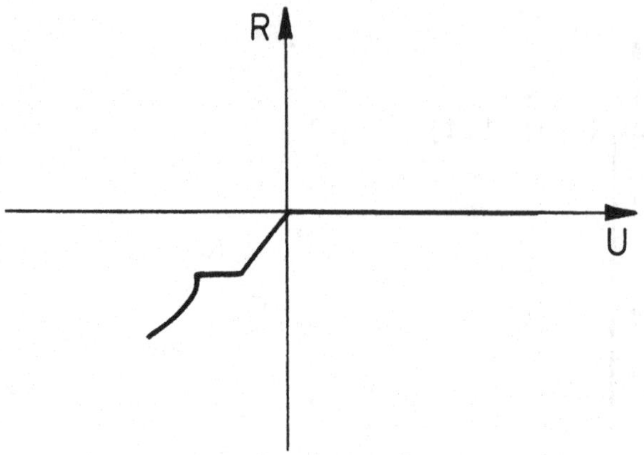


Fig. 19

$$r(U^-) = \int_0^{U^-} \hat{R}(Z)\, dZ \qquad /3.8/$$

is a well defined l.s.c. proper convex function, see [157,Th.24.2]. The function r can be defined on $R^1$, since for $U \geq 0$ $r(U) \equiv 0$. Then r is subdifferentiable on $R^1$. Within the framework of covexity most general response of a subgrade is described by maximally monotone graphs [257] or equivalently complete non-decreasing curves [157,p.232].

<u>Viscoelastic obstacle</u>. If the behaviour of the obstacle is linear then

$$R = kU^- + \nu \dot{U}^-, \qquad /3.9/$$

where $\nu$ is the coefficient of viscosity and $\dot{U}^- = \overline{\dot{U}}^-$.

Similarly as previously, both elastic and viscous behaviour can be nonlinear.

Let us pass to the variational formulation. The equation of motion is

$$s_{,x} + f = \varrho\, \partial^2 u/\partial t^2 = \varrho\, \ddot{u}\,, \qquad /3.10/$$

where f stands for density of body forces. Assume that $\varrho \equiv 1$ while the initial conditions are

$$u(0) = u_o\,, \quad \dot{u}(0) = u_1\,. \qquad /3.11/$$

Now a function $v(x,t)$ is treated as a vector function $v\colon (0,T) \to B$, where B is a Banach space.

A <u>virtual velocity</u> v is such that $v(T) = 0$. Multiplying /3.10/ by v and integrating over $(0,1)$ we get

$$(\ddot{u},v) + (s, \varepsilon(v)) = (f,v) + R(t)\, V(t)\,, \quad V(t) = v(0,t)\,, \qquad /3.12/$$

where

$$(v,w) = \int_0^1 vw\,dx\,.$$

Let us set

$$a(u,v) = \int_0^1 a\, \varepsilon(u)\, \varepsilon(v)\, dx\,, \qquad /3.13/$$

$$b(\dot{u},v) = \int_0^1 b\,\varepsilon(\dot{u})\,\varepsilon(v)dx . \qquad /3.14/$$

Integrating /3.4/ and /3.12/ over the time interval, next taking account of /3.1/ and /3.11/ we readily obtain the variational formulation of the impact problem considered.

**Problem P**

Find a function $u(t)$, $t \in [0,T]$, such that

$$\int_0^T \left\{-(\dot{u}(t),\dot{v}(t)) + a(u(t),v(t)) + b(\dot{u}(t),v(t))\right\} dt - (u_1,v(0)) =$$

$$= \int_0^T \left\{(f(t),v(t)) + R(t)V(t)\right\} dt , \quad \forall v,\ v(T)=0 \qquad /3.15/$$

$$U(t) = u(0,t) \geqslant 0 , \qquad /3.16/$$

$$\int_0^T R(t)\left(W(t) - U(t)\right) dt \geqslant 0 \qquad \forall W \in K , \qquad /3.17/$$

$$u(0) = u_0 . \qquad /3.18/$$

As one can see, Problem P consists of the variational equation /3.15/ and the V.I. /3.17/. The latter involves only displacements whereas the former both displacements and velocities.

Substituting /3.9/ into /3.12/ we obtain the penalized problem, provided that $k>0$ and $\nu \geqslant 0$ are given.

**Problem $P_1$**

Find a function $u(t)$, $t \in [0,T]$, such that

$$(\ddot{u}(t),v) + a(u(t),v) + b(\dot{u}(t),v) = (f(t),v) +$$

$$+ kU^-(t)V + \nu\dot{U}^-(t)V , \quad \forall v , \qquad /3.19/$$

and which satisfies /3.11/.

Let us pass to discussion of results concerning existence and uniqueness. The case $\nu=0$ is studied.

**Theorem 3.1.** Let $k>0$, $T>0$, $u_0 \in H^1(0,1)$, $u_1 \in L^2(0,1)$ and $f \in L^2(0,T;L^2(0,1))$. Then there exists a unique function $u_k$ which satisfies /3.19/ ($\nu=0$) for almost every $t\in(0,T)$ and every $v \in H^1(0,1)$ and such that

$$u_k \in L^\infty(0,T;H^1(0,1)) , \quad \dot{u}_k \in L^\infty(0,T;L^2(0,1)) \cap L^2(0,T;H^1(0,1)) \qquad /3.20/$$

$$u_k(0) = u_0 , \qquad \dot{u}_k(0) = u_1 . \qquad /3.21/$$

<u>Sketch of the proof</u>. The notation $u_k$ for a function solving /3.19/ with $k>0$ is used to stress that u depends on k.

To prove uniqueness we use Gronwall's lemma [257,p.156]: "Let $m\in L^1(0,T;R^1)$ be a function such that $w\geq 0$ a.e. on $(0,T)$ and let a be a constant $\geq 0$. Let $w\in C([0,T])$ be a function such that

$$w(t)\leq a+\int_0^t m(s)\,w(s)\,ds \quad \text{for each } t\in[0,T].$$

Then

$$w(t) = a e^{\int_0^t m(s)\,ds} \text{"}. \qquad /3.22/$$

More familar form of this lemma is for m being a constant, see [283].

It is known that for given $\lambda>0$, there exists $\alpha>0$ ($\beta>0$) such that [18]

$$\begin{aligned} &a(v,v)+\lambda|v|^2\geq\alpha\|v\|^2,\ \forall v\in H^1(0,1),\\ &(b(v,v)+\lambda|v|^2\geq\beta\|v\|^2,\ \forall v\in H^1(0,1).\end{aligned} \qquad /3.23/$$

Here $\|\cdot\|=\|\cdot\|_{H^1(0,1)}$, $|\cdot|=|\cdot|_{L^2(0,1)}$.

<u>Uniqueness</u>. Suppose that $u_{k(1)}$ and $u_{k(2)}$ are two solutions. We put $u=u_{k(1)}-u_{k(2)}$ for $\lambda>0$ given and fixed; we set

$$z_\alpha(t)=e^{-\lambda t}u_{k(\alpha)}, \quad z(t)=e^{-\lambda t}u(t), \quad \alpha=1,2,$$

For the sake of simplicity we take k=1.
Writing /3.19/ ($\nu=0$) for $u_{k(1)}$ and $u_{k(2)}$, next subtracting and taking $v=\dot z$ we arrive at

$$\frac{1}{2}\frac{d}{dt}\{|\dot z|^2+\lambda^2|z|^2+a(z)+\lambda b(z)\}+2\lambda|\dot z|^2+b(\dot z)\leq|Z||\dot Z|,$$

since $|Z_1^- - Z_2^-|\leq|Z_1-Z_2|$; here $a(z)=a(z,z)$, $b(z)=b(z,z)$, $Z(t)=z(0,t)$. Knowing that $z(0)=0$, $\dot z(0)=0$, using /3.23/ and Hölder's inequality we finally obtain

$$\alpha\|z\|^2\leq c\int_0^t\|z\|^2ds,$$

where c is a positive constant. Applying Gronwall's lemma we conclude that z=0. Hence, the solution is unique.

<u>Existence</u> is proved by using the Faedo-Galerkin method and compacity arguments. For details, the reader should refer to [190].

The results for the original problem P are the following.

**Theorem 3.2.** Let the data on $T, u_0, u_1$ and f be as previously. Further, let $U_o = u(0,0) > 0$. Then, as $k \to \infty$ we have, at least for a subsequence

$$u_k \to u \quad \text{weak star in } L^\infty(0,T;H^1(0,1)) , \qquad /3.24/$$

$$\dot{u}_k \to u \quad \text{weak star in } L^\infty(0,T;L^2(0,1)) \text{ and}$$

$$\text{weakly in } L^2(0,T;H^1(0,1)) , \qquad /3.25/$$

$$U_k \to U \quad \text{uniformly on } [0,T] , \qquad /3.26/$$

$$R_k = kU_k^- \to R \text{ in the weak topology of } M([0,T]). \qquad /3.27/$$

Moreover, the following relations are satisfied

$$\int_0^T \{-(\dot{u}(t),v(t)) + a(u(t),v(t)) + b(\dot{u}(t),v(t))\} - (u_1,v(0)) =$$

$$= \int_0^T (f(t),v(t))\,dt + \langle R,V\rangle, \ \forall v \in L^2(0,T;H^1(0,1)) \text{ with } \dot{v} \in L^1(0,T;L^2(0,1)) , \ V \in C^o([0,T]) , v(T)=0 , \qquad /3.28/$$

$$u(0) = u_0 , \qquad /3.29/$$

$$U \geq 0, \quad \text{on } (0,T) , \qquad /3.30/$$

$$\langle R, W-U\rangle \geq 0, \ \forall W \in C^o([0,T]) , \ W \geq 0 \quad \text{on } (0,T). \qquad /3.31/$$

A detailed proof is given in [190] for $\nu = 0$. It starts from taking $v = \dot{u}_k(t)$ in /3.19/, and next integration over $(0,T)$ is performed. A priori estimates result in /3.24/-/3.26/. The second a priori estimates are deduced by taking in /3.19/ for the virtual motion a rigid one, i.e. $v(x) = V < 0$, V being a constant. After integration over $(0,T)$ we arrive at the estimation

$$\int_0^T kU_k^-(t)\,dt \leq \text{const., or } kU_k^- \text{ is bounded in } L^1(0,T). \qquad /3.32/$$

Hence /3.27/ results. In fact at this moment Banach-Alaoglu theorem is used. It states that if X is a Banach space then the unit ball in $X^*$ is compact in the topology $\sigma(X^*,X)$. To prove /3.31/ one proceeds in the following way. Let $W \in C^o([0,T])$. The function

$$h(y) = \begin{cases} y^2/2 & \text{if } y \leq 0 , \\ 0 & \text{if } y > 0 , \end{cases}$$

is obviously convex. Subdifferentiability furnishes

$$\tfrac{1}{2}\left[(W^-)^2 - (U_k^-)^2\right] \geq -U_k^-(W-U_k) , \text{ on } [0,T] .$$

Taking now $W \geq 0$ on $[0,T]$, we obtain

Hence
$$U_k^-(W-U_k) \geqslant 0 , \quad \text{on} \quad [0,T] ,$$
$$\int_0^T kU_k^-(W-U_k)dt \geqslant 0 .$$

If $k \to \infty$, then by taking account of /3.32/ we arrive at /3.31/.

<u>Remark 3.1</u>. Do [189] has formulated a theorem similar to Th.3.2 above, yet for $\nu \geqslant 0$. During the passage with k to infinity $\nu$ is held fixed. Obviously, then u depends on $\nu$ i.e. $u=u_\nu$. According to [189] such $u_\nu$ also satisfies /3.24/--/3.31/. This result, given by Do without proof, is not clear.

We also observe that u solving the problem P is not necessarily unique, or at least uniqueness remains, as far as I know, an open problem.

## 3.2. General variational formulations for a class of inelastic bodies

The model problem P of the previous section can be generalized in such a way that the flow law type sliding rule is readily included.

Let $\Omega \subset R^3$ be a sufficiently regular domain while $\overline{\Omega}$ represents an undeformed configuration of a not necessarily elastic body. The equations of motion are

$$\rho \frac{\partial^2 u_i}{\partial t^2} = s_{ij,j} + b_i , \text{ in } \Omega . \qquad /3.33/$$

Now all functions may depends on $x \in \Omega$ and $t \in [0,T]$, t -time. The initial conditions are

$$\underline{u}(x,0) = \underline{u}_0(x) , \qquad \dot{\underline{u}}(x,0) = \underline{u}_1(x). \qquad /3.34/$$

In the quasi-static case the dynamic term in /3.33/ and /3.34/$_2$ must be omitted. Suppose that the boundary conditions on $S_2$ are given by /2.22/, /2.55/-/2.57/. Due to /2.55/, $\dot{\underline{u}}=0$ on $S_2$. Multiplying /3.33/ by $\underline{v}-\dot{\underline{u}}(t)$, where $\underline{v}$ is a virtual field, and integrating we arrive at

$$\int_\Omega \rho\ddot{\underline{u}}(t)\left(\underline{v}-\dot{\underline{u}}(t)\right) dx = - \int_\Omega s_{ij}(t)\, e_{ij}\left(v-\dot{\underline{u}}(t)\right) dx + \int_{S_2} s_N(t)\left(v_N-\dot{u}_N(t)\right) dS$$
$$+ \int_{S_2} \underline{s}_T(t)\left(\underline{v}_T-\dot{\underline{u}}_T\right) dS + L\left(\underline{v}-\dot{\underline{u}}(t)\right) , \quad t \in [0,T], \ \forall v \in V, \qquad /3.35/$$

where V is a space of virtual fields

$$V = \left\{\underline{v} = (v_i) \,\middle|\, \underline{v} = 0 \quad \text{on } S_0\right\}. \qquad /3.36/$$

It is assumed that all integrals make sense.

Taking account of /2.22/ in /3.35/ we readily get

$$\int_\Omega [\rho \ddot{\underline{u}}(t)(\underline{v}-\dot{\underline{u}}(t)) + s_{ij}e_{ij}(\underline{v}-\dot{\underline{u}}(t))]\,dx + J_1(s_N,\underline{v}) - J_1(s_N,\dot{\underline{u}}(t)) - \int_{S_2} s_N(t)(v_N-\dot{u}_N(t)\,dS \geq L(\underline{v}-\dot{\underline{u}}(t)),\ t\in[0,T], v\in V, \quad /3.37/$$

where

$$J_1(s_N,\dot{\underline{u}}(t)) = \int_{S_2} d(s_N,-\dot{\underline{u}}_T(t))\,dS.$$

The global form of the Signorini's condition is

$$u_N \in K_N : \int_{S_2} s_N(t)(w-u_N(t))dS \geq 0,\quad t\in[0,T], \forall w\in K_N, \quad /3.38/$$

where $K_N$ is a closed cone in a suitably chosen space of vector valued functions defined over $S_2$; for instance $K_N = \{w \in L^2(0,T;H^{1/2}(S_2)) \mid w \leq 0\}$ in the case of linear elasticity or viscoelasticity.
Thus we see that /3.36/ and /3.38/ constitute the coupled system of two inequalities. The coupling is produced by the Signorini's conditions and flow law type sliding rule.

We can formulate the following variational problem.

Problem $P_2$

Find $\underline{u}(t)$, $t\in[0,T]$, such that /3.34/, /3.37/ and /3.38/ are satisfied.

It should be remembered that now $\underline{s}=\underline{s}(\underline{e}(\underline{u}))$, or $\underline{s}(\underline{e}(\dot{\underline{u}}))$, or $\underline{s}(\underline{e}(\underline{u}),\underline{e}(\dot{\underline{u}}))$ depending upon the body considered.

It is worth noticing that such principle of virtual fields holds irrespective of the material of the body considered.

Examples of constitutive relations allowable by /3.37/:
/a/ $\underline{s} \in \partial D(\underline{e}(\underline{u}))$ or nonlinear elastic material with subdifferentiable strain potential.
/b/ Viscoelastic materials with long or short memory.
/c/ $$s^*_{ij}(t) = e^d_{ij}(t) - \int_0^t R_1(t+p(x),\tau+p(x))\,e^d_{ij}(\tau)\,d\tau,$$

$$s^*(t) = e(t) - \int_0^t R_2(t+p(x),\tau+p(x))\,e(\tau)\,d\tau, \quad .$$

where $\underline{s} = \underline{s}^d + \frac{1}{3}s\underline{I}$, $\underline{e} = \underline{e}^d + \frac{1}{3}e\underline{I}$, and

$$s^*_{ij} = \frac{s_{ij}(t)}{2G(t+p(x))}, \qquad s^*(t) = \frac{s(t)}{E(t+p(x))} .$$

The functions $p,G,E,R_1$ and $R_2$ are given.
Such constitutive equations describe the so called ageing isotropic materials. Viscoelastic materials constitute a

particular class, where $R_\alpha(t+p, \tau+p) = R_\alpha(t-\tau)$. Obviously, anisotropic ageing materials may also be considered [177]. The function $p(x)$ characterizes nonhomogeneity of the process of ageing.

/d/ Rigid-perfectly plastic materials, where

$$\underline{s} \in \partial D(e(\underline{\dot{u}})) .$$

For more details, see Section 3.5.

Remark 3.2. Quasi-static Signorini's problem without friction for ageing materials is studied in [177,178,181,182,202], see also [204,240]. Various problems related to unilateral contact problems for linear viscoelastic bodies are investigated in [18,183,220,221,230]. All those problems are simpler than the Signorini's problem with friction.

An alternative approach to the study of the Signorini problem with prescribed normal contact stresses for a class of inelastic materials has been proposed by Raous [241], see also [239,240]. I think that more general setting including the sliding rule /2.38/ is readily available. Unfortunately, Raous' approach precludes the flow law type sliding rule /2.22/. Let $\underline{e} = (e_{ij})$ denote the strain tensor and $e^{in}$ its inelastic part, i.e. $\underline{e} = \underline{e}^e + \underline{e}^{in}$. We can write

$$s_{ij} = a_{ijkl}(e_{kl} - e^{in}) . \qquad /3.39/$$

Let us consider the quasi-static formulation. I recall that time dependent functions are treated as vector functions mapping the time interval into a Banach space. Therefore I write $\underline{b}(t)$, $\underline{F}(t)$, $\underline{s}(t)$, etc. The convex set $K_d$ of kinematically admissible displacements is given by /2.62/.

The constitutive relation is assumed in the following form

$$\underline{\dot{e}}^{in} = \mathcal{F}(\underline{e}, \underline{e}^{in}), \quad \underline{e}^{in}(0) = \underline{e}_o , \qquad /3.40/$$

where $\underline{e}_o$ is a given function of $x \in \Omega$.

For instance, in the case of Maxwell model of viscoelastic behaviour of isotropic materials we have [233]

$$\underline{\dot{e}}^{in} = \underline{\dot{e}}^{v} = \eta^{-1}\underline{s} = \eta^{-1}\underline{a}(\underline{e} - \underline{e}^{v}) , \qquad /3.41/$$

where $\underline{a} = a_{ijkl}$ stands for the isotropic tensor; $a_{iikl} = 0$ since $\operatorname{div}\underline{\dot{e}}^{v} = 0$.

In the case of elastic perfectly-plastic materials /3.40/ is written in the form

$$\underline{\dot{e}}^{in} = \underline{\dot{e}}^{p} \in \partial I_K(\underline{s}) = \partial I_K[\underline{a}(\underline{e} - \underline{e}^{p})] , \qquad /3.42/$$

where K denotes the so called elasticity convex determined by the yield locus. The inelastic part $e^{in}$ of the strain tensor may also stand for viscoplastic deformation.

Hardening and/or softening effects can readily be taken into account.

Let us pass to the variational, or rather mixed /variational-strong/, formulation of the quasi static evolution problem for materials obeying /3.40/. The Signorini's conditions /2.57/ and the sliding rule /2.38/ are present on $S_2$. Multiplying /2.53/ by $(\underline{v}-\underline{u}(t))$ and taking account of /3.39/, /3.40/ as well as of the boundary conditions /2.55/-/2.58/ one can formulate the following problem, where $e_{ij}(\underline{u}(t)) = \partial u_i(t)/\partial t$.

Problem $P_2$

Find $\underline{u}(t) \in K_d$ and $\underline{e}^{in}(t)$, $t \in [0,T]$ such that

$$a(\underline{u}(t), \underline{v}-\underline{u}(t)) - \int_\Omega a_{ijkl} e^{in}_{kl}(t)\, e_{ij}(\underline{v}-\underline{u}(t))\, dx +$$

$$+ J(\underline{u}(t), \underline{v}) - J(\underline{u}(t), \underline{u}(t)) \geq L(\underline{v}-\underline{u}(t)), \quad \forall v \in K_d, \qquad /3.43/$$

$$\dot{\underline{e}}^{in} = \mathcal{F}(\underline{e}, \underline{e}^{in}), \qquad /3.44/$$

$$\underline{e}^{in}(0) = \underline{e}_o. \qquad /3.45/$$

Obviously, the problem $P_2$ is not purely variational.

Existence of a solution of the problem $P_2$ is not a simple matter and as such the problem remains open. Nevertheless in [240,241] Raous presented implicit and semi-implicit methods for numerical solving of some particular cases of the problem $P_2$. Results obtained exhibit that a formulation such as /3.43/-/3.45/ is not useless.

### 3.3. Steady-state rolling contact of rigid cylinder traversing a viscoelastic half-space

Consider the viscoelastic half-space $\{(x_\alpha, z) \mid z \geq 0\}$, $(x = (x_\alpha))$, traversed by a rigid roller moving with a constant velocity V, see Fig.20. This figure shows that relative to the symmetric position x=0 /occupied by the center of the contact region for an elastic solid/, the contact region is displaced in the direction of rolling. Thus, even in the absence of surface shear tractions /frictionless rolling/, the net force on the cylinder possesses a nonvanishing component opposing the motion. Such frictionless case is investigated in this section. The model of a behaviour of a material of the half-space is given by the generalized Voigt's model with one additional spring, for which $G_D$ stands for the dynamic shear modulus, see Fig.21. For this model the creep function is given by

$$\varphi(t) = G_D^{-1}\Big[1+\sum_{r=1}^{n} f_r(1-e^{t/\tau_r})\Big], \qquad /3.46/$$

where $f_r = G_r^{-1}/G_D^{-1}$, while $\tau_r\ (r=1,\dots,n)$ are relaxation ti-

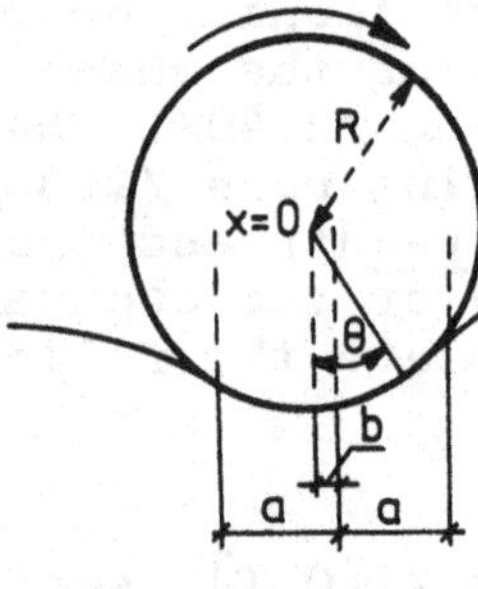


Fig.20, Schematic diagram of rolling cylinder

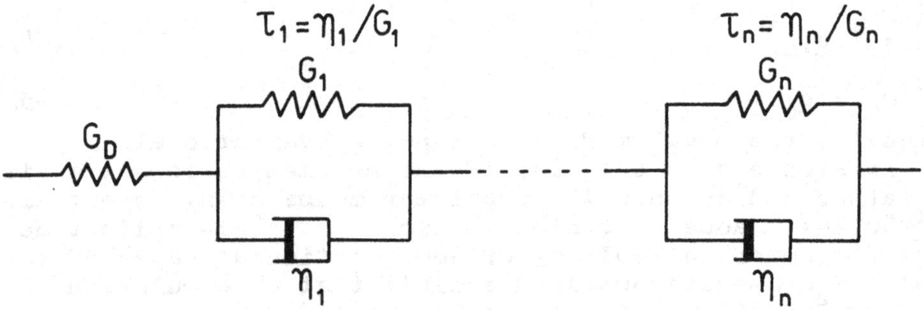


Fig.21

mes. Specifically, the creep-strain behaviour of the material in shear , $\gamma(t)$, in response to a constant stress s, applied at time t=0 is expressed by

$$\gamma(t)= s\,\Psi(t)\,H(t) \quad , \qquad /3.47/$$

where H(t) is Heaviside's unit step function.

Hunter found that in the coordinate system moving with the roller and having the direction of a motion as the positive $x_1$-axis, the normal surface displacement is [203], see also[1] [236]

$$w(x)= u_e(x)+ \sum_r f_r \int_0^\infty e^{-\lambda} u_e(x_1+V\tau_r\lambda, x_2)d\lambda , \qquad /3.48/$$

where $u_e(x)$ is the instantaneous elastic displacement. This elastic displacement is given by the Boussinesq equation

$$u_e(x)=\frac{1-\nu}{2\pi G_D}\int_{R_2}\frac{p(y)}{|x-y|}\,dy \quad , \qquad /3.49$$

The inverse Fourier transform $F^{-1}$ is expressed by

$$F^{-1}(\tilde{f}) = f(x) = \frac{1}{(2\pi)^2}\int_{R^2} \tilde{f}(\xi)\, e^{-i(x,\xi)}\, d\xi . \qquad /3.52/$$

A linear functional f defined and continuous on $S(R^2)$ is called a tempered distribution /in $R^2$/. The totality of such tempered distributions is denoted by $S'(R^2)$.

The Fourier transform of the tempered distribution $f \in S'(R^2)$ ( f-functional on $S(R^2)$) is defined as the generalized function $\tilde{f} \in S'(R^2)$ such that

$\tilde{f}(\tilde{\varphi}) = \langle \tilde{f}, \tilde{\varphi}\rangle = (2\pi)^2 \langle f, \varphi\rangle$, $\forall \varphi \in S(R^2)$ .

Below we shall use the following relations

$$F\left(\frac{1}{|x|}\right) = 2\pi \frac{1}{|\xi|} , \qquad /3.53/$$

$$F\big(f(x-a)\big) = e^{i(a,\xi)}\, \tilde{f}(\xi),\quad a = (a_1, a_2) \in R^2 \qquad /3.54/$$

Applying now formally the Fourier transform to /3.49/ and taking account of /3.53/ and /3.54/, we obtain

$$\tilde{u}_e(\xi) = \int_{R^2} u_e(x)\, e^{i(x,\xi)}\, dx = \frac{1-\nu}{2\pi G_D}\int_{R^2\times R^2} \frac{p(y)}{|x-y|} e^{i(x,\xi)}\, dx dy =$$

$$= \frac{1-\nu}{G_D}\, \frac{1}{|\xi|}\, \tilde{p}(\xi) . \qquad /3.55/$$

Employ now the Fourier transform to /3.48/; we have

$$\tilde{w}(\xi) = \tilde{u}_e(\xi) + \sum_r f_r \int_{R^2}\left(\int_0^\infty e^{-\lambda}\, u_e(x_1 + V\tau_r\lambda, x_2)\, d\lambda\right) e^{i(x,\xi)} d\xi =$$

$$= \tilde{u}_e + \sum_r f_r \tilde{u}_e(\xi) \int_0^\infty e^{-(1+\xi_1 V\tau_r i)\lambda}\, d\lambda .$$

Performing the integration in the last relation we arrive at

$$\tilde{w}(\xi) = \tilde{u}_e(\xi) + \sum_r f_r \frac{\tilde{u}_e(\xi)}{1+\xi_1 V\tau_r i} \qquad /3.56/$$

Eliminating $\tilde{u}_e(\xi)$ by applying the inverse Fourier transform to /3.56/ we can write

$$w = Bp = \frac{1}{(2\pi)^2}\int_{R^2} \tilde{w}(\xi)\, e^{-i(x,\xi)}\, d\xi =$$

$$= \frac{1-\nu}{4\pi^2 G_D} \sum_r \int_{R^2} \frac{1}{|\xi|}\, \frac{(1+f_r) + V\tau_r\xi_1 i}{1+V\tau_r\xi_1 i}\, \tilde{p}(\xi)\, e^{-i(x,\xi)} d\xi . \qquad /3.57/$$

The inverse Fourier transform $F^{-1}$ is expressed by

$$F^{-1}(\tilde{f}) = f(x) = \frac{1}{(2\pi)^2}\int_{R^2}\tilde{f}(\xi)\, e^{-i(x,\xi)}\, d\xi . \qquad /3.52/$$

A linear functional f defined and continuous on $S(R^2)$ is called a tempered distribution /in $R^2$/. The totality of such tempered distributions is denoted by $S'(R^2)$.

The Fourier transform of the tempered distribution $f \in S'(R^2)$ ( f-functional on $S(R^2)$) is defined as the generalized function $\tilde{f} \in S'(R^2)$ such that

$\tilde{f}(\tilde{\varphi}) = \langle \tilde{f}, \tilde{\varphi}\rangle = (2\pi)^2\langle f, \varphi\rangle, \ \forall \varphi \in S(R^2)$ .

Below we shall use the following relations

$$F\left(\frac{1}{|x|}\right) = 2\pi\,\frac{1}{|\xi|} , \qquad /3.53/$$

$$F\big(f(x-a)\big) = e^{i(a,\xi)}\,\tilde{f}(\xi), \quad a = (a_1, a_2) \in R^2 \qquad /3.54/$$

Applying now formally the Fourier transform to /3.49/ and taking account of /3.53/ and /3.54/, we obtain

$$\tilde{u}_e(\xi) = \int_{R^2} u_e(x)\, e^{i(x,\xi)}\, dx = \frac{1-\nu}{2\pi G_D}\int_{R^2\times R^2}\frac{p(y)}{|x-y|}\, e^{i(x,\xi)}\, dx dy =$$

$$= \frac{1-\nu}{G_D}\,\frac{1}{|\xi|}\,\tilde{p}(\xi) . \qquad /3.55/$$

Employ now the Fourier transform to /3.48/; we have

$$\tilde{w}(\xi) = \tilde{u}_e(\xi) + \sum_r f_r \int_{R^2}\left(\int_0^{\infty} e^{-\lambda}\, u_e(x_1 + V\tau_r\lambda, x_2)\, d\lambda\right) e^{i(x,\xi)} d\xi =$$

$$= \tilde{u}_e + \sum_r f_r \tilde{u}_e(\xi)\int_0^{\infty} e^{-(1+\xi_1 V\tau_r i)\lambda}\, d\lambda .$$

Performing the integration in the last relation we arrive at

$$\tilde{w}(\xi) = \tilde{u}_e(\xi) + \sum_r f_r \frac{\tilde{u}_e(\xi)}{1+\xi_1 V\tau_r i} \qquad /3.56/$$

Eliminating $\tilde{u}_e(\xi)$ by applying the inverse Fourier transform to /3.56/ we can write

$$w = Bp = \frac{1}{(2\pi)^2}\int_{R^2}\tilde{w}(\xi)\, e^{-i(x,\xi)}\, d\xi =$$

$$= \frac{1-\nu}{4\pi^2 G_D}\sum_r \int_{R^2}\frac{1}{|\xi|}\,\frac{(1+f_r) + V\tau_r\xi_1 i}{1+V\tau_r\xi_1 i}\,\tilde{p}(\xi)\, e^{-i(x,\xi)} d\xi . \qquad /3.57/$$

The fundamental properties of the pseudodifferential operator B are given below. Beforehand however, the functional setting is introduced. $H^s(R^2)$ $(s \in R)$ is the usual Sobolev's space with the norm

$$\|v\|_s^2 = \int_{R^2} \left(1+|\xi|^2\right)^s |\tilde{v}(\xi)|^2 d\xi . \qquad /3.58/$$

Let D be some fixed, sufficiently regular domain of $R^2$. Fundamental for the study of the operator B above are the following spaces:

$$\mathring{H}^s(D) = \{v \in H^s(R^2) \mid \operatorname{supp} v \subset \bar{D}\} , \qquad /3.59/$$

which is the closure of $C_0^\infty(D)$ in $H^s(R^2)$; next is the space $H^s(D)$ or the space of functions /distributions for $s<0$/ in D admitting an extension on $R^2$ belonging to $H^s(R^2)$. The norm of $v \in H^s(D)$ is

$$\|v\|_{s,D} = \inf \left\{ \|Lv\|_s \mid Lv \in H^s(R^2) \text{ and } Lv|_D = v \right\} .$$

We have $\mathring{H}^{1/2}(D) \subset H^{1/2}(D)$, and the inclusion is strict, so that the norms $\|\cdot\|_s$ and $\|\cdot\|_{s,D}$ are different.

The notion of a pseudodifferential operator (P.D.O) A is introduced after Eskin [195].

Let $A(\xi) \in L^1_{loc}(R^n)$ satisfy the following condition

$$|A(\xi)| \le c\left(1+|\xi|\right)^\alpha . \qquad /3.60/$$

The class of such functions is denoted by $S_\alpha^o$. Obviously, in our case n=2.

<u>Definition 3.1</u>. The operator A is called pseudodifferential if it is defined on $S(R^n)$ according to the formula

$$Au = \frac{1}{(2\pi)^n} \int_{R^n} A(\xi)\, \tilde{u}(\xi)\, e^{-i(x,\xi)}\, d\xi . \qquad /3.61/$$

The function $A(\xi)$ is called the symbol of the operator A.

If $A(\xi)$ is a polynomial with respect to $\xi$, i.e. $A(\xi) = \sum_{k \le m} a_k \xi^k$, then

$$Au = \frac{1}{(2\pi)^n} \int_{R^n} \sum_{|k| \le m} a_k \xi^k \tilde{u}(\xi) e^{-i(x,\xi)}\, d\xi = \sum_{|k| \le m} a_k D^k u(x) .$$

Thus A is then the differential operator.

Comparing /3.57/ with /3.61/ we conclude that <u>B is the pseudodifferential operator</u>.

The symbol $B(\xi)$ of B is

$$B(\xi) = \frac{1-\nu}{(4\pi)^2 G_D} \sum_r \frac{1}{|\xi|} \frac{(1+f_r)+V\tau_r\xi_1 i}{1+V\tau_r\xi_1 i} . \qquad /3.62/$$

Lemma 3.1. Let $A(\xi) \in S^o$. Then the P.D.O. A with the symbol $A(\xi)$ satisfies, for each $s \in R^1$, the inequality

$$\|Au\|_{s-\alpha} \leq c_s \|u\|_s , \quad \forall u \in S(R^n). \qquad /3.63/$$

Consequently, by continuity it can be extended to the bounded operator from $H^s(R^n)$ to $H^{s-\alpha}(R^n)$.

Proof. We have

$$\|Au\|^2_{s-\alpha} = \int_{R^n} (1+|\xi|^2)^{(s-\alpha)} |A(\xi)\tilde{u}(\xi)|^2 d\xi \leq$$
$$\leq \tilde{c}_s \int_{R^n} (1+|\xi|)^{2s} |\tilde{u}(\xi)|^2 d\xi \leq c_s \|u\|^2_s .$$

Properties of the operator B

1. B is positive, that is $Bq \geq 0$ for all $q \in C_0^\infty(R^2), q \geq 0$. This property follows directly from /3.48/.

2. $|B(\xi)| \leq c|\xi|^{-1}$, where $c = c(f_r, V\tau_r)$; see /3.62/.

3. The previous property and Lemma 3.1 imply that the operator B is bounded from $\overset{\circ}{H}{}^s(D)$ to $H^{s+1}(D)$ for every $s \in R$ $(\alpha=-1)$. Hence

$$(Bp,q) = \int_{R^2} (Bp)(x)\, q(x)\, dx \quad ; \quad p,q \quad \overset{\circ}{H}{}^{-1/2}(D) , \qquad /3.64/$$

is well defined by the continuity for all $p,q$ $\overset{\circ}{H}{}^{-1/2}(D)$.

4. The operator B is coercive on $\overset{\circ}{H}{}^{-1/2}(D)$, that is

$$(Bq,q) \geq C \|q\|^2_{-1/2} , \quad \forall q \quad \overset{\circ}{H}{}^{-1/2}(D) . \qquad /3.65/$$

where C is a positive constant.

To prove /3.65/ let us decompose B into the sum: $B=B_1+B_2$, which is obtained in the following way, We may write

$$Bq = \frac{1-\nu}{4\pi^2 G_D} \sum_r \int_{R^2} \frac{[(1+f_r)+V\tau_r\xi_1 i](1-V\tau_r\xi_1 i)}{|\xi|(1+V\tau_r\xi_1 i)(1-V\tau_r\xi_1 i)} \tilde{q}(\xi) e^{-i(x,\xi)} d\xi .$$

Hence

$$B_1 q = \frac{1-\nu}{4\pi^2 G_D} \sum_r \int_{R^2} \frac{1}{|\xi|} \frac{(1+f_r)+(V\tau_r)^2\xi_1^2}{1+(V\tau_r)^2\xi_1^2} \tilde{p}(\xi) e^{-i(x,\xi)} d\xi \qquad /3.66/$$

$$B_2 q = -\frac{1-\nu}{4\pi^2 G_D} \sum_r \int_{R^n} \frac{i}{|\xi|} \frac{f_r V\tau_r\xi_1}{1+(V\tau_r)^2\xi_1^2} \tilde{p}(\xi) e^{-i(x,\xi)} d\xi . \qquad /3.67/$$

We see that $B_2=-B_2^*$, where $B_2^*$ is the adjoint of $B_2$. Thus $(B_2q,q)=0$ for $q\in \mathring{H}^{-1/2}(D)$.

In physical situations we always have $\nu<1$. Then from /3.66/ we obtain $(B_1q,q)\geq C\|q\|^2_{-1/2}$, since $|\xi|^{-1}\geq(1+|\xi|^2)^{-1/2}$, and

$$(B_1q,q)=\frac{1-\nu}{4\pi^2G_D}\sum_r\int_{R^2}\frac{1}{|\xi|}\frac{(1+f_r)+(v\tau_r)^2\xi_1^2}{1+(v\tau_r)^2\xi_1^2}|\tilde{q}(\xi)|^2d\xi .$$

The latter relation follows from the Parseval's formula

$$\int_{R^2}f(x)g(x)\,dx=\frac{1}{(2\pi)^2}\int_{R^2}\tilde{f}(\xi)\bar{\tilde{g}}(\xi)\,d\xi ,$$

where $g(x)$ is a real function. Thus the coerciveness property /3.65/ of B has been proved.

Having introduced the P.D.O. B we can pass to the variational formulation. In a steady motion the rigid roller immerses at a fixed depth $\Delta$. Let $G\subset R^2$ be the unknown domain of contact; therefore the rolling problem considered belongs to so called free boundary problems /as for the rest the unilateral problems studied previously/.

For $z=0$ we can write

$$w(x)=\Delta+h(x)\ ,\ p(x)\geq 0\ ,\ x\in\bar{G}\ ,$$
$$w(x)\geq\Delta+h(x)\ ,\ p(x)=0,\ \ x\notin G\ . \qquad /3.68/$$

Taking account of /3.57/ we arrive at the following complementarity conditions

$$p(x)\geq 0,\ Bp(x)\geq\Delta+h(x),\ p(x)Bp(x)=p(x)(\Delta+h(x))\ , \qquad /3.69/$$

for all $x\in R^2$, provided that h has been extended on $R^2$. Such an extension must preserve /3.69/$_2$.

A known and fixed domain D is now chosen, $G\subset D$. For instance, the projection of the roller on x-plane may be suitable. We put

$$H_Q^+=\{q\in\mathring{H}^{-1/2}(D)\,|\,q\geq 0\ \text{and}\ (1,q)=Q\}. \qquad /3.70/$$

The set $H_Q^+$ is closed and convex in $\mathring{H}^{-1/2}(D)$ and Q has the meaning of the total pressure. We see that at least for $p\in L^2(D)$ /3.69/ is equivalent to

$$\left|\begin{aligned}&(Bp,q)\geq(\Delta+h,q),\ \forall q\in H_Q^+\ ,\\&(Bp,p)=(\Delta+h,p).\end{aligned}\right. \qquad /3.71/$$

Subtraction yields

$$(Bp,q-p)\geq(\Delta,q-p)+(h,q-p)=\Delta(1,q-p)+(h,q-p).$$

But $(1,q-p)=(1,q)-(1,p)=0$, since $p,q\in H_Q^+$.

In this manner we eventually obtain the following V.I. for the determination of a distribution of the contact pressure p:

$$(P_3)\quad \left|\begin{array}{l}\text{find } p\in H_Q^+ \text{ such that}\\ (Bp,q-p)\geqslant(h,q-p),\ \forall q\in H_Q^+ .\end{array}\right. \qquad /3.72/$$

To solve this problem, the weight Q of the roller must be prescribed.

The next problem arises for $\Delta > 0$ prescribed. Toward this end let us define the closed and convex cone $H^+ \subset \mathring{H}^{-1/2}(D)$, i.e.

$$H^+ = \{q\in \mathring{H}^{-1/2}(D)\mid q\geqslant 0\}. \qquad /3.73/$$

From /3.69/ one readily obtains the following variational problem:

$$(P_4)\quad \left|\begin{array}{ll}\text{find } p\in H^+ \text{ such that} & \\ (Bp,q)\geqslant(\Delta+h,q),\ \forall q\in H^+ & /3.74/\\ \text{and } (Bp,p)=(\Delta+h,p). & /3.75/\end{array}\right.$$

The relations /3.74/ and /3.75/ are equivalent to the following variational ineqality

$$\mid p\in H^+:\ (Bp,q-p)\geqslant(\Delta+h,q-p),\ \forall q\in H^+ . \qquad /3.76/$$

Now we can formulate [179]

Theorem 3.3. a/ The variational problem $P_3$ possesses a unique solution $p_Q$ for each $h\in H^{1/2}(D)$, $Q\geqslant 0$.

b/ The variational problem $P_4$ has a unique solution $p_\Delta$ for each $h\in H^{1/2}(D)$, $\Delta\in R^+=[0,\infty)$.

c/ $p_\Delta = p_Q$ if $Q=(1,p_\Delta)$ and $p_Q=p_\Delta$ if $\Delta=(Bp_Q-h,p_Q)/Q$.

d/ The correspondence $\Delta\longmapsto Q=(1,p_\Delta)$ is a continuous, strictly increasing function from $R^+$ onto $R^+$. The same holds for its inverse: $Q\longmapsto\Delta=(Bp_Q-h,p_Q)/Q$.

e/ Suppose that $p_1$ and $p_2$ are solutions of $(P_3)$ with $Q=Q_1\neq 0$ and $Q=Q_2\neq 0$, respectively. Then

$$|Q_1-Q_2|\geqslant \|Q_1p_2-Q_2p_1\|_{-1/2} .$$

Points a/ and b/ result from the general theory of variational inequalities due to /3.65/ and the continuity of the respective linear functionals [35]. The remaining points are proved in [179]. Finite element approximations are also studied in [179]. Elastic problem was earlier investigated in [180].

Remark 3.3. It would be interesting to apply Mosco's theory of duality to variational problems $P_3$ and $P_4$.

Remark 3.4. L'vov [223] developed an incremental procedure for the study of a frictionless contact of a shallow creeping shell with a rigid stamp. Geometrically linear case was investigated.

### 3.4. A finitely deforming viscoelastic cylinder in steady-state rolling contact with a rough foundation

An inverse problem, in a certain sense, to that studied in the preceding section will be presented in this section. Now a cylinder rolling with a constant velocity $v_o$ is finitely deformable while the foundation is rigid but rough /rolling with friction/. The physical situation may be describe in the following way: a viscoelastic cylinder undergoing finite deformations is rolling on a rough foundation at a constant angular velocity $\omega$, see Fig.22 for an example. Such problem has been intensively studied in [234,235], see also [150]. By $\underline{v}_o$ we denote the velocity of the foundation /e.g. roadway or belt/ relative to the axis of the cylinder /in the absence of slipping, $v_o=R_o\omega$ for rigid bodies/. The cylinder can be attached to a rigid cylindrical axle of radius $R_i$, and this axle spins at a constant angular velocity $\omega$. An observer riding on the axle will see the same geometry of the deformed cylinder at all times t. The situation is also equivalent to a cylinder spinning about a fixed axis and in contact with a moving belt. A key consideration is the kinematics: we compare the geometry of the deformed cylinder in its current configuration C with that of a rigid cylinder spinning at the same angular velocity $\omega$, the latter

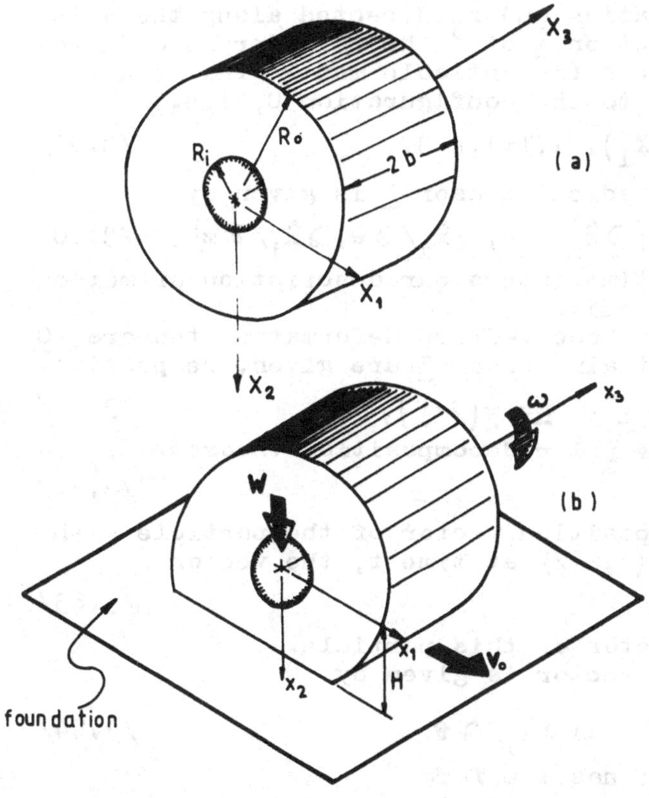


Fig.22,/a/ Geometry of a rolling cylinder. /b/ Rolling cylinder under applied load

characterizing the reference configuration $C_o$. Polar cylindrical coordinates $r,\theta,z$ of a particle with labels $(R,\Theta,Z)$ at some arbitrary reference time t=0 are defined by

$$r=R, \quad \theta=\Theta+\omega t, \quad z=Z, \quad \text{for } t\geqslant 0. \qquad /3.77/$$

Alternatively, one can employ the Cartesian reference coordinates

$$X_1=r\cos\theta, \quad X_2=r\sin\theta, \quad X_3=z. \qquad /3.78/$$

Let us denote by $(x_i)$ $(i=1,2,3)$ the spatial coordinates; these are the Cartesian coordinates of points in space relative to a moving coordinate frame with origin fixed on the axes of the moving cylinder: $x_1$ parallel to the foundation, $x_2$ normal to the foundation and $x_3$ directed along the axis of the cylinder. The motion $\hat{\underline{x}}$ of the cylinder is defined by an invertible, twice differentiable map $\chi$ that takes the configuration $C_o$ into the configuration C; i.e.,

$$x_i=\hat{x}_i(r,\theta,z)=\chi_i(X_I), \quad i,I=1,2,3. \qquad /3.79/$$

The deformation gradient tensor $\underline{F}$ is given by

$$\underline{F}=\nabla\hat{\underline{x}}, \text{ or } \underline{F}(r,\theta,z)=[\partial\hat{x}_i/\partial r, \partial\hat{x}_i/\partial\theta, \partial\hat{x}_i/\partial z]. \qquad /3.80/$$

Thus we conclude that time enters our description of motion only implicity as $\theta=\Theta+\omega t$.

The right and left Cauchy-Green deformation tensors, $\underline{C}$ and $\underline{B}$, and the Green strain tensor $\underline{E}$ are given, respectively, by

$$\underline{C}=\underline{F}^T\underline{F}=\underline{U}^2, \quad \underline{B}=\underline{F}\,\underline{F}^T=\underline{V}^2, \quad \underline{E}=\tfrac{1}{2}(\underline{C}-\underline{1}), \qquad /3.81/$$

where, according to the polar decomposition theorem,

$$\underline{F}=\underline{R}\,\underline{U}=\underline{V}\,\underline{R}. \qquad /3.82/$$

If $\underline{r}$ denotes the position vector of the particle with reference coordinates $(r,\theta,z)$ at time t, the vector

$$\underline{u}=\hat{\underline{x}}(r,\theta,z)-\underline{r}, \qquad /3.83/$$

is the displacement vector of this particle.
The reference velocity vector is given by

$$v_i=\frac{\partial\hat{x}_i(r,\theta,z)}{\partial t}=\omega\,\partial\hat{x}_i/\partial\theta. \qquad /3.84/$$

The acceleration vector has the form

$$a_i=\frac{\partial v_i}{\partial t}=\omega^2\,\partial^2\hat{x}_i/\partial\theta^2. \qquad /3.85/$$

We observe that since the reference coordinates are time dependent, hence

$$\partial u_i/\partial t=v_i-\partial r_i/\partial t=\omega(\partial\hat{x}_i/\partial\theta-\partial r_i/\partial\theta), \qquad /3.86/$$

where $r_i$ are the Cartesian components of $\underline{r}$.

In our case we have

$$\partial X_\beta/\partial t=\omega\,\varepsilon_{\alpha\beta}X_\alpha, \quad \partial X_3/\partial t=0 \qquad /3.87/$$

where Greek indices take values 1 and 2 while $(\varepsilon_{\alpha\beta})$ is the permutation symbol.
Further, /3.80/ furnishes

$$\dot{\underline{F}}=\omega \frac{\partial}{\partial\theta}\left[\partial\hat{x}_i/\partial r, \partial\hat{x}_i/\partial\theta, \partial\hat{x}_i/\partial z\right] = \omega\partial\underline{F}/\partial\theta . \qquad /3.88/$$

Taking account of /3.87/, from /3.88/ we obtain

$$\dot{F}_{\alpha\beta} = \omega X_{\gamma\lambda}\varepsilon_{\gamma\lambda}\hat{x}_{\alpha,\lambda\beta} \quad , \dot{F}_{3j}=\dot{F}_{j3}=0 ,$$

where $\hat{x}_{\alpha,\beta\gamma} = \partial^2\hat{x}_\alpha/\partial X_\beta\partial X_\gamma$. Hence

$$\dot{C}_{\alpha\beta} = \omega X_\gamma\varepsilon_{\gamma\lambda}\left(\hat{x}_{\mu,\lambda\alpha}\hat{x}_{\mu,\beta} + \hat{x}_{\mu,\lambda\beta}\hat{x}_{\mu,\alpha}\right), C_{j3}=0 . \qquad /3.89/$$

Let $\underline{s}$ denote the Cauchy stress tensor and $\underline{S}$ the second /symmetric/ Piola-Kirchhoff stress tensor, related to one another according to

$$\underline{S}= (\det\underline{F})\underline{F}^{-1}\underline{s}\,\underline{F}^{-T}. \qquad /3.90/$$

The first /unsymmetric/ Piola-Kirchhoff stress tensor $\underline{T}$ is expressed by

$$\underline{T} = \underline{F}\,\underline{S} . \qquad /3.91/$$

A viscoelastic material of the rolling cylinder obeys the following constitutive equation

$$\underline{S}=\underline{G}^{(1)}(\underline{X},\underline{C}) + \underline{G}^{(2)}(\underline{X},\dot{\underline{C}}) , \qquad /3.92/$$

where $\underline{G}^{(1)}$ and $\underline{G}^{(2)}$ are response functions.

In the case considered, taking account of /3.89/, we can write

$$\underline{S}(\underline{X}, \nabla\hat{\underline{x}}, \nabla^2\hat{\underline{x}}, \omega) = \underline{G}^{(1)}(\underline{X}, \hat{x}_{i,j}\hat{x}_{j,k}) +$$
$$+ \underline{G}^{(2)}(X, \omega X_\gamma\varepsilon_{\gamma\varrho}\hat{x}_{\mu,\varrho\alpha}\hat{x}_{\mu,\beta} + \hat{x}_{\mu,\varrho\beta}\hat{x}_{\mu,\alpha}) . \qquad /3.93/$$

Thus rate dependence enters /3.93/ as a dependence of $G^{(2)}$ on $\nabla^2\underline{x}$ and $\omega$.
The equation of motion is expressed by

$$\mathrm{Div}\underline{T}^T + \varrho\underline{b} = \varrho\ddot{\underline{x}} , \qquad /3.94/$$

where Div stands for the divergence operator corresponding to the reference configuration, $\underline{b}$ is the body force per unit mass in the reference configuration and $\ddot{\underline{x}}=(\ddot{x}_i)$, $\ddot{x}_i=a_i$, cf. the formula /3.85/.

Let $S_0 = \{(r,\theta,z)\ r=R_i\}$. We assume that if the cylinder is fixed to the rigid spinning axle, then

$$\hat{\underline{x}}(R_i,\theta,z) = (R_i,\theta,z) \quad \text{on } S_0 . \qquad /3.95/$$

Since the foundation is rough, so contact and friction conditions have to be imposed on $S_2$. Here $S_2$ denotes the candidate contact surface in the reference configuration /$S_2$ could even be taken as the entire outside surface of the cylinder/. Let $\underline{n}_o$ denote a unit exterior normal to $S_2$ while $\underline{n}$ stands for a unit outward normal to the deformed surface, see Fig.23. The surface traction vector $\underline{t}$ at a point on $S_2$ is expressed by

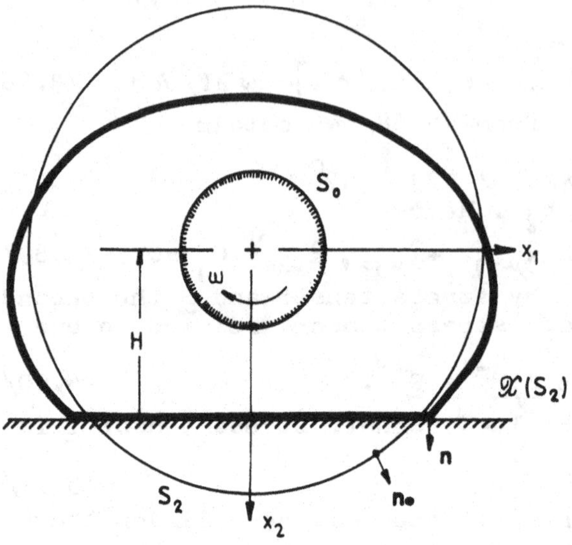


Fig.23

$$\underline{t} = \underline{T}^T \underline{n}_o \qquad /3.96/$$

The vector $\underline{t}$ can be decomposed into the normal and tangential parts in the following way

$$t_n = \underline{t} \cdot \underline{n} = t_J \, \delta_{Ji} n_i, \qquad /3.97/$$

$$\underline{t}_n = t_n \underline{n}, \quad \underline{t}_T = \underline{t} - \underline{t}_n .$$

A good presentation of the kinematics of finite deformations as well as of stress measures is given in the book [285].

The slip velocity $\underline{w}_T$ is defined by

$$\underline{w}_T = \underline{v} - \underline{v}_o \ , \text{ on } S_2 \ , \qquad /3.98/$$

where $\underline{v}$ denote the particle velocity on $S_2$.

The Signorini's conditions take the form

$$\hat{x}_2 - H \leq 0, \quad t_n \leq 0, \quad t_n(\hat{x}_2 - H) = 0, \text{ on } S_2 \ , \qquad /3.99/$$

where H is the distance from the axis of the cylinder to the foundation , see Fig.23.

The friction law is assumed to be of the flow law type, cf./2.22/. Now we have

$$-\underline{w}_T \in \partial I_{K(t_n)}(\underline{t}_T) \ , \text{ on } S_2 \ , \qquad /3.100/$$

where

$$K(t_n) = \{ \underline{z} \mid g(t_n, \underline{z}) \leq 0 \} \ , \text{ on } S_2 .$$

We note that in [234,235] such general law is not considered.

The set of admissible motions is defined by

$$V = \{ \underline{\hat{x}} = \underline{\hat{x}}(r, \theta, z), \underline{\hat{x}} \in B \mid \underline{x} \text{ is prescribed on } S_o \text{ and the energy is well defined} \}, \qquad /3.101/$$

where B is in general a Banach space, for instance $B = [W^{2,p}(\Omega_o)]^3$ and $\Omega_o$ is the interior of the cylinder in the reference configuration.

The set K of admissible motions has the form

$$K = \{\hat{\underline{x}} \in V \,|\, \hat{x}_2 - H \leq 0 \quad \text{on } S_2\}. \qquad /3.102/$$

Let us also introduce a special velocity - motion space

$$V_M = \{\underline{\eta} = \tilde{\nabla}_\theta \hat{\underline{x}} \,|\, \hat{\underline{x}} \in V\} , \qquad /3.103/$$

and

$$K_M = \{\underline{\eta} \in V_M \,|\, \underline{\eta} = \tilde{\nabla}_\theta \hat{\underline{x}},\ \hat{\underline{x}} \in K\} , \qquad /3.104/$$

where $\tilde{\nabla}_\theta \hat{\underline{x}} = (\omega \partial_\theta \hat{x}_1, \hat{x}_2, \omega \partial_\theta x_3)$. The vector $\underline{\eta} \in V_M$ seems to be artificial since its second component has the dimension of a displacement whereas the first and third of speed. Therefore it would be better to present the problem at the very begining in a non-dimensional form to avoid such ambiquity.

It is convenient to introduce a collection of nonlinear forms:

$$A(\hat{\underline{x}}, \underline{\eta}) = \int_{\Omega_o} \mathrm{tr}\left[\underline{F}\ \underline{S}(X, \nabla\hat{\underline{x}}, \nabla^2\hat{\underline{x}}, \omega)\nabla\underline{\eta}\right] d\Omega_o , \qquad /3.105/$$

/virtual power of the stresses/;

$$B(\hat{\underline{x}}, \underline{\eta}) = \int_{\Omega_o} \varrho\,(\partial_\theta \hat{\underline{x}}) \cdot \partial_\theta \underline{\eta}\, d\Omega_o , \qquad /3.106/$$

/the power developed by inertial forces/;

$$L(\underline{\eta}) = \int_{\Omega_o} \varrho\, \underline{b} \cdot \underline{\eta}\, d\Omega_o \qquad /3.107/$$

/the virtual power of body forces/;

$$J(\hat{\underline{x}}, \underline{\eta}) = \int_{S_o} d\left(t_n (\nabla\hat{\underline{x}}, \nabla^2\hat{\underline{x}}, \omega), -\underline{w}_T\right) dS ,$$

/virtual power of the frictional forces/,

where tr stand for trace of tensors.

The variational formulation is obtained in the usual way: we multiply the equations of motion /3.94/ by $(\underline{\eta} - \tilde{\nabla}_\theta \hat{\underline{x}})$, integrate over $\Omega_o$, perform the integration by parts and next use /3.97/ and /3.100/. Finally we arrive at the following <u>implicit variational inequality</u>:

(P) Find $\hat{\underline{x}} \in K$ such that

$$A(\hat{\underline{x}}, \underline{\eta} - \tilde{\nabla}_\theta \hat{\underline{x}}) + J(\hat{\underline{x}}, \underline{\eta}) - J(\hat{\underline{x}}, \tilde{\nabla}_\theta \hat{\underline{x}}) \geq$$
$$\geq \omega^2 B(\hat{\underline{x}}, \underline{\eta} - \tilde{\nabla}_\theta \hat{\underline{x}}) + L(\underline{\eta} - \tilde{\nabla}_\theta \hat{\underline{x}}), \ \forall \underline{\eta} \in K_M . \qquad /3.108/$$

IVI /3.108/ is not the usual one since $\hat{\underline{x}} \in K$ while $\underline{\eta} \in K_M$. Moreover, the sets K and $K_M$ are not necessarily convex and closed due to a dependence of the set V of admissible motions on the energy. Such an example is given by the condition $\det \underline{F} > 1$.

We note that formulation such as /3.108/ is available for a very limited class of problems. It seems that even for geometrically and kinematically the same problem but in the case of transient process, where time enters explicitly, IVI similar to /3.108/ cannot be derived. As in Section 3.2, however, we can always obtain a variational formulation in the form of coupled inequalities. The general procedure is similar to that employed in Section 3.2. Let us apply this method to the same steady-state rolling problem of finitely deforming cylinder.

For this purpose by $V_{ad}$ is denoted the space of sufficiently regular virtual velocity fields; in general $V_{ad} \neq V_M$. Multiplying /3.94/ by $(\underline{\eta} - \underline{v})$, $\underline{v} = \dot{\underline{\hat{x}}}$, $\underline{\eta} \in V_{ad}$, next integrating over $\Omega_0$, in the standard manner we obtain the IVI /3.109/ below. The Signorini's conditions /3.99/ are represented in the global form /3.110/ below where $V(S_2)$ is a properly chosen space of functions defined on $S_2$; for instance $V(S_2) = [W^{2-1/p,p}(S_2)]^* = W^{-2+1/p,q}(S_2)$, $1/p+1/q=1$.

The variational formulation of the steady-state rolling of the cylinder considered reads:

$$(P_{nat}) \quad \begin{cases} \text{Find } \underline{\hat{x}} \in V \text{ such that} \\ A(\underline{\hat{x}}, \underline{\eta} - \underline{v}) + J(\underline{\hat{x}}, \underline{\eta}) - J(\underline{\hat{x}}, \underline{v}) - \int_{S_2} t_n(\eta_n - v_n)\, dS \geq \\ \quad \geq \omega^2 B(\underline{\hat{x}}, \underline{\eta} - \underline{v}) + L(\underline{\eta} - \underline{v}), \quad \forall \underline{\eta} \in V_{ad}, \qquad /3.109/ \\ \text{and} \\ \hat{x}_2 - H \leq 0 \; : \int_{S_2} t_n(z - \hat{x}_2)\, dS \geq 0, \quad \forall z \in K_n. \qquad /3.110/ \end{cases}$$

/$P_{nat}$ - natural formulation/. Obviously $t_n$ depends on $\nabla \underline{\hat{x}}$, $\nabla^2 \underline{\hat{x}}$ and $\omega$; moreover

$$K_n = \{ z \in V(S_2) \mid z - H_2 \leq 0 \}. \qquad /3.111/$$

As I have already mentioned, formulation such as $(P_{nat})$ can be used to non-stationary problems, yet the form of inequalities /3.109/ and /3.110/ will depend on the particular case considered.

A great number of numerical examples is given in [150, 234,235]. Particularly, in [234] to the hyperelastic law /see $\underline{G}^{(1)}$ in /3.92/ / viscoelastic aspects of the response of the cylinder are appended in the form of the following constitutive equation

$$\underline{G}^{(2)}(\tau)_{\tau=-\infty}^{t} = \nu_g \left\{ \underline{E}(t) - \frac{1}{T_r} \int_{-\infty}^{t} \exp\{-(t-\tau)/T_r\} \underline{E}(\tau)\, d\tau \right. \qquad /3.112/$$

where $\nu_g$ is the glassy modulus of the material and $T_r$ is the relaxation time. Such law can describe viscous behaviour of elastomers.

To discretize the problem P, the non-smooth functional J has to be regularized. In the case of Coulomb's friction one can take

$$\underline{t}_T = -\mu |t_n| \varphi_\varepsilon(|\underline{w}_T|)\, \underline{w}_T/|\underline{w}_T| , \qquad /3.113/$$

where $\varphi_\varepsilon$ is a sufficiently smooth function, see Fig.24. It is a continuous, monotone, real-valued function of the non-negative real numbers $|\underline{w}_T|$, depending on a parameter $\varepsilon > 0$ such that

/a/ $0 \leq \varphi_\varepsilon(|\underline{w}_T|) \leq 1$ ;

/b/ $\lim_{\varepsilon \to 0} \varphi_\varepsilon(|\underline{w}_T|) = 1, \forall |\underline{w}_T| > 0$;

/c/ $\lim_{|\underline{w}_T| \to \infty} \varphi_\varepsilon(|\underline{w}_T|) = 1, \ \forall \varepsilon > 0$ .

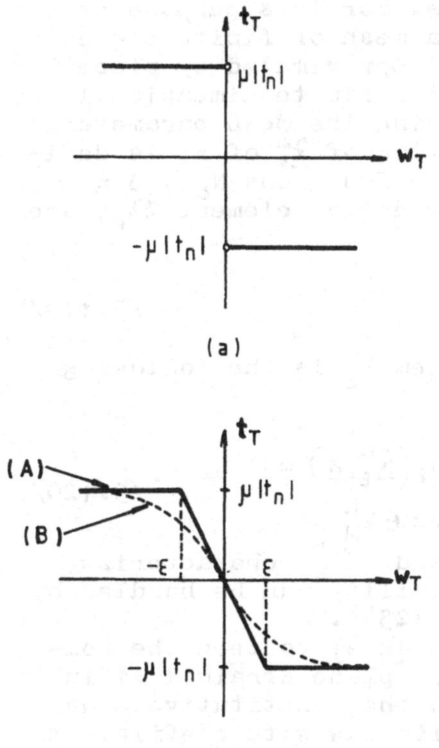


Fig.24, /a/ Classical Coulomb's friction law; /b/ regularized Coulomb's friction law

Examples of functions satisfying these properties are

$$\varphi_\varepsilon(|\underline{w}_T|) = \begin{cases} 1, \text{ if } |\underline{w}_T| > \varepsilon , \\ |\underline{w}_T|/\varepsilon , \text{if } |\underline{w}_T| \leq \varepsilon , \end{cases} \qquad /3.114/$$

and

$$\varphi_\varepsilon(|\underline{w}_T|) = \tanh(|\underline{w}_T|/\varepsilon \qquad /3.115/$$

Instead of J we take the regularized functional $J_\varepsilon$ defined by

$$J_\varepsilon(\hat{\underline{x}}, \underline{\eta}) = = \int_{S_2} d\left(t_n, \varphi_\varepsilon(|\underline{\eta}|)\right) ds, \ \varepsilon > 0 \qquad /3.116/$$

provided that the density d depends on $t_n$ and $\underline{w}_T$. As previously, $t_n$ depends on $\nabla\hat{\underline{x}}$, $\nabla^2\hat{\underline{x}}$ and $\omega$.

The rigid foundation is replaced by a deformable one, characterized, for instance, by the functional, see the formula /2.136/,

$$D(\hat{\underline{x}}, \underline{\eta}) = \int_{S_2} c_n(\hat{x}_2 - H)_+^{m_n} \eta_2 dS. \qquad /3.117/$$

Regularization of the problem P signifies that for each fixed $\varepsilon > 0$ we consider the problem:

$$(P_\varepsilon) \left| \begin{array}{l} \text{find } \hat{\underline{x}}_\varepsilon \in V \text{ such that} \\ A(\hat{\underline{x}}_\varepsilon, \underline{\eta}) - \omega^2 B(\hat{\underline{x}}, \underline{\eta}) + D(\hat{\underline{x}}_\varepsilon, \underline{\eta}) + J_\varepsilon(\hat{\underline{x}}_\varepsilon, \underline{\eta}) = \\ \qquad = L(\underline{\eta}) \quad \forall \underline{\eta} \in V_M . \end{array} \right. \qquad /3.118/$$

The variational equation /3.118/ is to be used as the basis of discretizations . Problem $P_\varepsilon$ can be discretized by using Lagrangian finite elements. For this purpose we construct a partition of $\Omega_o$ into a mesh of finite elements $\Omega_e$ over which the motion $\hat{\underline{x}}_\varepsilon$ is approximated by piecewise polynomials in such a way that a finite dimensional subspace $V^h$ of V is formed, $h > 0$ being the mesh parameter. Over each element $\Omega_e$, the restriction of $\hat{\underline{x}}_\varepsilon^h$ of $\hat{\underline{x}}_\varepsilon$ is defined as a linear combination of shape functions $N_\Delta$ , $\Delta = 1,\dots,N_e$; $N_e$ being the number of nodes of element $\Omega_e$, see Fig.25. Thus we have

$$\hat{\underline{x}}_\varepsilon^h = \sum_{\Delta=1}^{N_e} N_\Delta(\xi,\eta)\, \hat{\underline{x}}^\Delta . \qquad /3.119/$$

The discrete form of the problem $P_\varepsilon$ is the following problem:

$$(P_\varepsilon^h) \left| \begin{array}{l} \text{find } \hat{\underline{x}}_\varepsilon^h \in V^h \text{ such that} \\ A(\hat{\underline{x}}_\varepsilon, \underline{z}) - \omega^2 B(\hat{\underline{x}}_\varepsilon^h, \underline{z}) + D(\hat{\underline{x}}_\varepsilon^h, \underline{z}) + J_\varepsilon(\hat{\underline{x}}_\varepsilon^h, \underline{z}) = \\ \qquad = L(\underline{z}) \quad \forall z \in V_M^h . \end{array} \right. \qquad /3.120/$$

Formulations such as (P), $(P_\varepsilon)$ and $(P_\varepsilon^h)$ characterize compressible materials. Incompressibility can be handled by the method of Lagrange multipliers [234].

Numerical results presented in [234] concern the rolling cylinder which is in a state of plane strain; $R_i = 1$ in. and $R_o = 2$ in. The elastic term $\underline{G}^{(1)}$ in the constitutive equation is characterized by Mooney-Rivlin law with coefficients of the stored elastic energy function of $C_1 = 80$ psi and $C_2 = 20$ psi. I recall that the Mooney-Rivlin strain energy function W is expressed by /incompressible material/

$$W(I_1, I_2) = C_1(I_1 - 3) + C_2(I_2 - 3),$$

where $I_1, I_2$ are the principal invariants of the tensor $\underline{C}$ or

$$I_1 = \text{tr}\underline{C}, \quad I_2 = (\text{tr}\underline{C})^2/2 - (\text{tr}\underline{C}^2)/2 ,$$

since for incompressible materials $I_3 = (\det\underline{F})^2 = 1$, cf.[285].

Further data are: $\omega = 10$ rad/s, DISP=$R_o$-H=0.2 in, $v_o = \omega H$, $c_n = 10^5$, $m_n = 1$, $\mu = 0.3$ /coefficient of friction/,

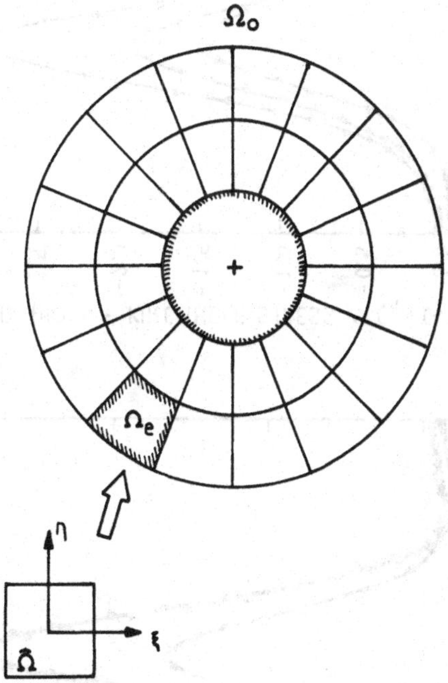


Fig.25

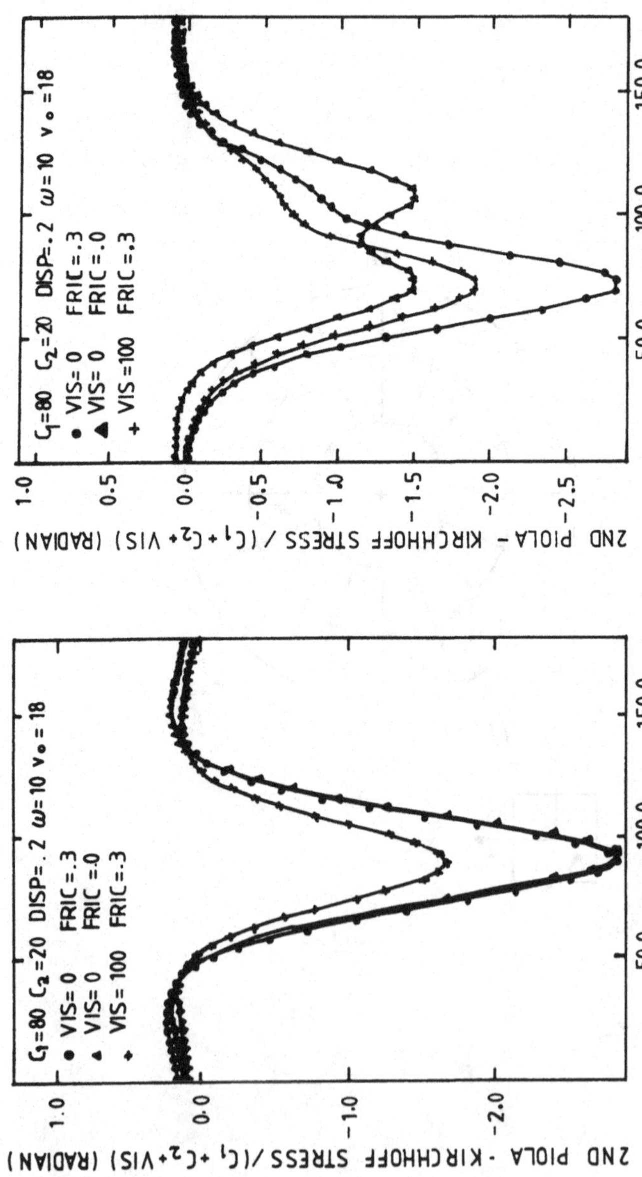


Radial stress near bonded surface

Circumferential stress near bonded surface

Fig.26

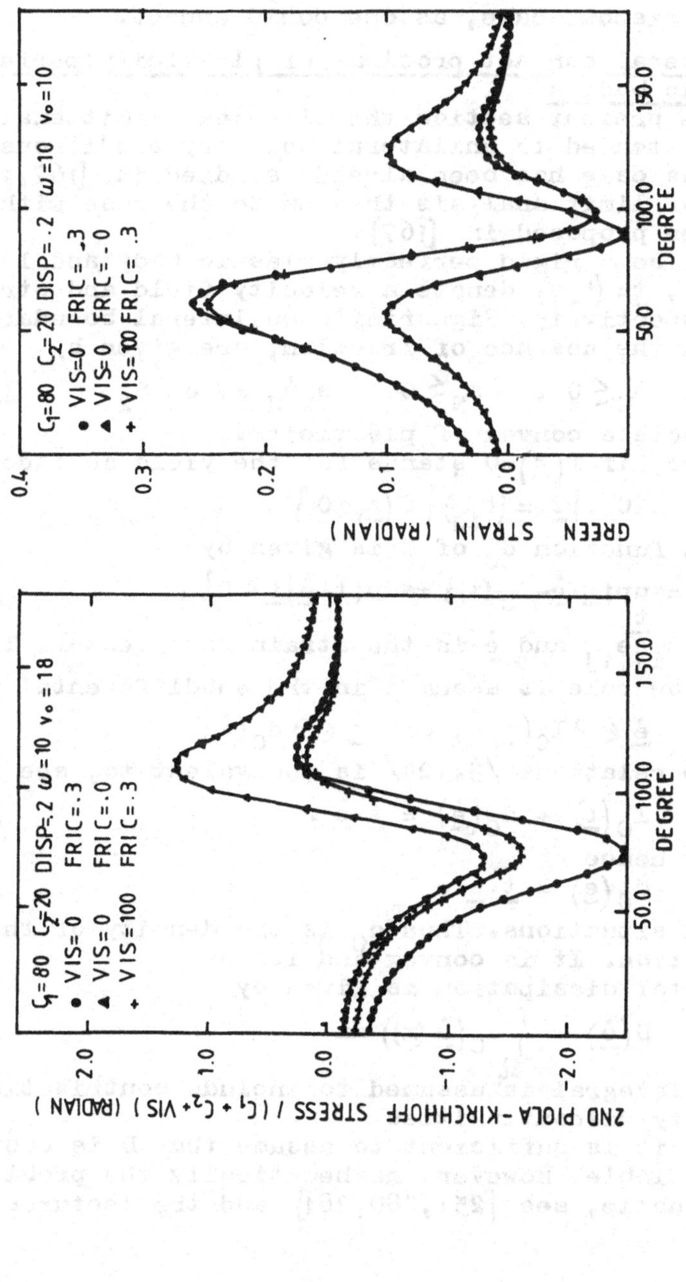


Shear stress near bonded surface

Radial strain near bonded surface

Fig. 27

$\varepsilon$ =0.25 in/s, VIS= $\gamma_c$=100 psi, $T_r$=0.1 s. Some of the numerical results are represented in Figs 26 and 27 after [234]. We conclude that the inclusion of viscoelastic effects produces a softer material: peak values of the dimensionless radial stress, circumeferential stress, and contact pressure are seen to be smaller for the viscous case than for the same elastic stress. All strain components increase slightly for the viscous case, as one could expect.

## 3.5. Unilateral contact problems of plasticity: perfectly plastic bodies

In the present section the classical limit analysis theory is extended to unilateral boundary conditions. The frictionless case has been already studied in [167,198,249]. Extension of limit analysis theorem to the case with friction has been proposed in [167].

Let $\Omega$ be a rigid perfectly-plastic body and let $\underline{\dot{u}}=(\dot{u}_i)$ and $\underline{s}=(s_{ij})$, $\underline{t}=(t_{ij})$ denote a velocity field and stress fields, respectively. Signorini's unilateral boundary conditions, in the absence of friction, are given by

$$\underline{s}_T=0 \; , \quad \dot{u}_N \leq 0 \; , \quad s_N \leq 0 \; , \quad s_N \dot{u}_N = 0 \text{ on } S_2 \; . \qquad /3.121/$$

By C we denote a convex of plasticity.
For instance, if $f(\underline{t})=0$ stands for the yield surface, then

$$C = \{ \underline{t} = (t_{ij}) \mid f(\underline{t}) \leq 0 \} \; . \qquad /3.122/$$

The support function $d_C$ of C is given by

$$d_C(\underline{\dot{e}}) = \sup_{\underline{t}} \{ \underline{t} \cdot \underline{\dot{e}} - I_C(\underline{t}) \} = \sup \{ \underline{t} \cdot \underline{\dot{e}} \mid \underline{t} \in C \} \; , \qquad /3.123/$$

where $\underline{t} \cdot \underline{\dot{e}} = t_{ij} \dot{e}_{ij}$ and $\underline{\dot{e}}$ is the strain rate tensor. The associated flow rule is assumed in the subdifferental form

$$\underline{\dot{e}} \in \partial I_C(\underline{t}) \; , \text{ or } \; \underline{t} \in \partial d_C(\underline{\dot{e}}) \; . \qquad /3.124/$$

Each of the relations /3.124/ is equivalent to, see Lemma 2.1

$$I_C(\underline{t}) + d_C(\underline{\dot{e}}) = \underline{t} \cdot \underline{\dot{e}} \; . \qquad /3.125/$$

Since $\underline{t} \in C$, hence

$$d_C(\underline{\dot{e}}) = \underline{t} \cdot \underline{\dot{e}} \; , \qquad /3.126/$$

in physical situations. Thus $d_C$ is the density of the plastic dissipation. It is convex and l.s.c.

The total dissipation is given by

$$D(\underline{\dot{e}}) = \int_{\Omega} d_C(\underline{\dot{e}}(x)) \, dx \; . \qquad /3.127/$$

The volume integral is assumed to include contributions from velocity discontinuities.
Physically, it is sufficient to assume that D is convex and subdifferentiable. However, mathematically the problem is much more subtle, see [251,280,281] and the lectures by Suquet.

For this purpose I recall the notion of <u>convex functionals of a measure</u>.

Let f be a convex function on $R^n$ satisfying the following property

$$\exists\, c_0 \geq c_1 > 0, \quad \exists\, k_0 \;, \; c_1|z| - k_0 \leq f(z) \leq c_0(1+|z|) \quad \forall z \in R^n \qquad /3.128/$$

The condition /3.128/ implies continuity of f. Then K = $= \operatorname{dom} f^* = \{z \mid f^*(z) < +\infty\}$ is a bounded convex set in $R^n$. The support function of K is defined by

$$f_\infty(z) = \sup\{z\cdot z^* \mid z^* \in K = \lim_{\lambda \to 0^+} \lambda f(z/\lambda) \;.$$

It is known that $f_\infty$ is convex, positively homogeneous and satisfies /3.128//with $k_0=0$/.

Let $\Omega \subset R^n$ be a bounded domain and $M^1(\Omega, R^n)$ the space of bounded Radon measures on $\Omega$ with values in $R^n$. The norm of $\mu \in M^1(\Omega, R^n)$ is denoted by $\int_\Omega |\mu|$ /the total variation/ and is expressed by

$$\|\mu\| = \int_\Omega |\mu| = \sup\{|\langle\mu, g\rangle| \mid g \in C_0(\Omega, R^n);\ \sup_x |g(x)| \leq 1\}. \qquad /3.129/$$

We have

$$M^1(\Omega, R^n) = [C_0(\Omega, R^n)]^*.$$

<u>Definition 3.2</u>. To every convex function satisfying /3.129/ corresponds the functional F denoted by $\int_\Omega f(\mu)$ defined on $M^1(\Omega, R^n)$ by

$$F(\mu) = \sup\{\int_\Omega v\mu - \int_\Omega f^*(v(x))dx \mid v \in C_0(\Omega, R^n)\}. \qquad /3.130/$$

Notation $\int_\Omega f(\mu)$ is justified by the following integral representation theorem.

<u>Theorem 3.4</u>.

/i/ $F(\mu) = \int_\Omega f(\mu)$ is the weak lower semicontinuous regularized functional /weak l.s.c. envelope/ in $M^1(\Omega, R^n)$ /in the sense of duality $(M^1(\Omega, R^n), C_0(\Omega, R^n))$ of the functional F defined by

$$\tilde{F}(\mu) = \begin{cases} \int_\Omega f(h(x))\, dx \;, & \text{if } \mu = h(x)\, dx \\ +\infty \;, & \text{otherwise.} \end{cases}$$

F is convex and strongly continuous in $M^1(\Omega, R^n)$.

/ii/ If $\mu = hdx + \mu^s$ $(h \in L^1(\Omega, R^n))$ represents the Lebesque decomposition of $\mu$, then

$$\int_{\Omega} f(\mu) = \int_{\Omega} f(h(x))\, dx + \int_{\Omega} f_{\infty}\left(\frac{d\mu^s}{d|\mu|^s}\right)|\mu^s| \qquad /3.131/$$

The theorem just formulated is profound and some comments are necessary.

By l.s.c. envelope of $\tilde{F}$ is meant the greatest lower semicontinuous functional F, such that $F \leq \tilde{F}$, see [286]. $\frac{d\mu^s}{d|\mu^s|}$ is the Radon-Nikodym derivative of $\mu^s$ with respect to the total variation $|\mu^s|$.

Let us continue the study of the limit analysis. In this case we have $f=d_C$, $k_o=0$ and $d_C(\dot{\underline{e}}) \leq c_o(1+|\dot{\underline{e}}|)$, provided that C is bounded. The dissipation density is non-negative. We assume that l.h.s. inequality in /3.128/ is also satisfied. Then dom $f^*=C$ and $f_\infty$ is the bidual of f, that is $f_\infty = d_C$.

A proper space for velocity fields is the space BD defined by

$$BD(\Omega) = \{\underline{u} \in [L^1(\Omega)]^3 \mid e_{ij}(\underline{u}) \in M^1(\Omega)\}. \qquad /3.132/$$

Thus we have $\underline{e}(\underline{u}) \in M^1(\Omega, \mathbb{M}_s(R^3))$, where $(\mathbb{M}_s(R^3))$ is the space of symmetric 3x3 real matrices /2x2 in the two-dimensional case/.

If a velocity fields $\dot{\underline{v}} \in BD(\Omega)$, then the total dissipation $D(\underline{e}(\dot{\underline{v}}))$ has the precise sense resulting from /3.131/, or D is the convex functional of the measure $\underline{e}(\dot{\underline{v}})$.

Now I pass to the formulation of limit analysis theorems in the presence of Signorini's boundary conditions without friction.

A stress field $\underline{t}$ defined over $\Omega$ is said to be statically admissible if

$$t_{ij,j} + b_i = 0 \;, \quad \text{in } \Omega, \qquad /3.133/$$

$$\underline{t}(\underline{x}) \in C(\underline{x}) \;, \quad \text{in } \Omega, \qquad /3.134/$$

$$t_{ij} n_j = m^s p_i^o \;, \quad \text{on } S_1, \qquad /3.135/$$

$$\underline{t}_T = 0 \;, \quad t_N \leq 0 \;, \quad \text{on } S_2 . \qquad /3.136/$$

Here $m^s$ is the static loed multiplier, whereas $\underline{p}^o$ is prescribed. The conditions /3.136/ may be called "non-classical" in the theory of limit analysis. We also note that $m^s$ does not affect body forces. Such an approach seems to me reasonable, since the usually considered "body forces" $m^s\underline{b}$ are physically unrealizable within the theory considered.

A velocity field $\underline{v}$ defined over $\Omega$ is said to be <u>kinematically admissible if</u>

$$\dot{\underline{v}} = 0 \ , \text{ on } S_0 \ , \tag{3.137}$$

$$\dot{v}_N \leq 0 \ , \text{ on } S_2 \ , \tag{3.138}$$

$$\int_{S_1} \underline{p}^o \cdot \dot{\underline{v}} \, dS > 0 \ . \tag{3.139}$$

For incompressible materials such a field must additionally satisfy the incompressibility condition

$$\dot{v}_{i,i} = 0 \ , \text{ in } \Omega . \tag{3.140}$$

<u>The kinematically admissible load multiplier $m^k$</u> is defined as follows

$$m^k(\dot{\underline{v}}) = \frac{D\left(e(\dot{\underline{v}})\right) - \int_\Omega \underline{b} \cdot \underline{v} dx}{\int_{S_1} \underline{p}^o \cdot \underline{v} dS} \ . \tag{3.141}$$

If the load affects also body forces /usual approach/ then, instead of /3.141/, we have

$$\tilde{m}^k(\dot{\underline{v}}) = \frac{D\left(\underline{e}(\dot{\underline{v}})\right)}{\int_{S_1} \underline{p}^o \cdot \underline{v} dS + \int_\Omega \underline{b} \cdot \underline{v} dx} \ . \tag{3.142}$$

The difference between $m^k(\dot{\underline{v}})$ and $\tilde{m}^k(\dot{\underline{v}})$ is obvious.

By a <u>complete solution</u> of the limit analysis problem we mean a triple $(m, \underline{s}, \dot{\underline{u}})$ such that the stress fields $\underline{s}$ is statically admissible, the velocity field $\dot{\underline{u}}$ is kinematically admissible whereas m is the associated load multiplier, that is $m = m^s(\underline{s}) = m^k(\dot{\underline{u}})$. Moreover $\underline{s}$ and $\underline{e}(\dot{\underline{u}})$ are interrelated by the flow rule $/3.124/_1$ or $/3.124/_2$.

<u>Lower bound theorem</u> states that

$$m^s(\underline{t}) \leq m \ , \tag{3.143}$$

where $\underline{t}$ is statically admissible. In other words the lower bound theorem reads

$(P_s)$ | find $\sup\{ m^s(\underline{t}) \mid \underline{t} \in K_s \}$ ,

where

$$K_s = \{ \underline{t} = (t_{ij}) \mid \underline{t}(x) \in C(x) \ , \ x \in \Omega , \ t_{ij,j} + b_i = 0, \Omega ;$$

$$t_{ij} n_j = m^s p_i^o , \ S_1 ; \ t_N \leq 0 , \ \underline{t}_T = 0 , \ S_2 \} , \tag{3.144}$$

is the set of statically admissible fields.

To prove the inequality /3.143/ we take

$$\int_\Omega \left(t_{ij}-s_{ij}\right)_{,j} \dot{u}_i dx = 0 \;, \; \forall \underline{t} \in K_s . \qquad /3.145/$$

Integrating by parts and taking account of the Signorini's condition $s_N\dot{u}_N=0$, we readily arrive at

$$0 \leqslant \int_\Omega (t_{ij}-s_{ij})\, e_{ij}(\underline{\dot{u}})\,dx = (m^s-m)\int_{S_1} \overset{o}{p}_i \dot{u}_i dS + \int_{S_2} t_N \dot{u}_n dS .$$

The inequalities $t_N \leqslant 0$ and $\dot{u}_N \leqslant 0$ imply $t_N\dot{u}_N \geqslant 0$. Hence we eventually obtain /3.143/.

<u>Upper bound theorem</u> states that

$$m \leqslant m^k(\underline{\dot{v}}) \;, \; \forall \underline{\dot{v}} \in K_v \;, \qquad /3.146/$$

where $K_v$ is the set of kinematically admissible velocity fields, i.e.

$$K_v=\left\{\underline{\dot{v}} = (\dot{v}_i) \,\middle|\, \underline{\dot{v}} =0,\; S_0 \;;\; v_N \leqslant 0 \;,\; S_2\right\}, \qquad /3.147/$$

or in the case of incompressible materials

$$K_v=\left\{\underline{\dot{v}} \,|\, \dot{v}_{i,i}=0, \Omega;\; \underline{\dot{v}}=0,\; S_0;\; \dot{v}_N \leqslant 0,\; S_2\right\}. \qquad /3.148/$$

To prove /3.146/ we assume that D is subdifferentiable, then

$$D\left(\underline{e}(\underline{\dot{v}})\right) - D\left(\underline{e}(\underline{\dot{u}})\right) \geqslant \int_\Omega s_{ij} e_{ij}(\underline{\dot{v}}-\underline{\dot{u}})\,dx \;, \; \forall \underline{\dot{v}} \in K_v .$$

According to /3.141/, we substitute

$$m^k \int_{S_1} \underline{\overset{o}{p}} \cdot \underline{\dot{v}} dS + \int_\Omega \underline{b} \cdot \underline{\dot{v}} dx \;, \quad m \int_{S_2} \underline{\overset{o}{p}} \cdot \underline{\dot{u}} dS + \int_\Omega \underline{b} \cdot \underline{\dot{u}} dx \;,$$

for $D\left(\underline{e}(\underline{\dot{v}})\right)$ and $D\left(\underline{e}(\underline{\dot{u}})\right)$, respectively. Thus we readily arrive at

$$m^k \int_{S_1} \underline{\overset{o}{p}} \cdot \underline{\dot{v}} dS \geqslant m \int_{S_1} \underline{\overset{o}{p}} \cdot \underline{v} dS + \int_{S_2} s_N \dot{v}_N dS .$$

Since $s_N\dot{v}_N \geqslant 0$ and the relation /3.139/ holds, we finally obtain /3.146/.

Upper and lower bound theorem remain valid if, instead of the relation /3.141/, the following definition of the kinematic load multiplier is assumed:

$$m_1^k(\underline{\dot{v}}) = \frac{\int_\Omega \left(\underline{t} \cdot \underline{e}(\underline{\dot{v}}) - \underline{b} \cdot \underline{\dot{v}}\right) dx - \int_{S_2} t_N \dot{v}_N dS}{\int_{S_1} \underline{\overset{o}{p}} \cdot \underline{v} dS} \;, \qquad /3.149/$$

where $\underline{t} \in \partial d_C\left(\underline{e}(\underline{\dot{v}})\right)$. The formula /3.149/ results directly from the principle of virtual velocities. We observe that $m^k(\underline{\dot{u}}) = m_1^k(\underline{\dot{u}})$, provided that $\underline{\dot{u}}$ enters a complete solution, since then $s_N\dot{u}_N=0$.

Such theory of limit analysis where the load multiplier affects only some components of loadings can readily be extended to perfectly plastic plates and shells. In the existing mathematically rigorous studies of plastic plates the load multiplier always affects all components of an applied loading [185,188,250].

A rigorous proof of the relation

$$m = \sup_{\underline{t} \in K_s} m^s(\underline{t}) = \inf_{\underline{\dot{v}} \in K_v} m_1^k(\underline{\dot{v}}) , \qquad /3.150/$$

is still not available.

However, one can prove that

$$\tilde{m}_1 = \sup_{\underline{t} \in \tilde{K}_s} \tilde{m}^s \; \underline{t} = \inf_{\underline{\dot{v}} \in K_v} \tilde{m}^k(\underline{\dot{v}}) \qquad /3.151/$$

where

$$K_s = \{\underline{t} \mid \underline{t}(x) \in C(x), x \in \Omega , \; t_{ij,j} + \tilde{m}^s b_i = 0, \Omega ;$$
$$t_{ij}n_j = \tilde{m}^s \overset{o}{p}_i, S_1; \; t_N \leq 0, \; \underline{t}_T = 0, S_2\} . \qquad /3.152/$$

For this purpose Christiansen's min-max theorem [184] will be applied.

<u>Theorem 3.5</u>. Assume that X and Y are normed real vector spaces with a continuous bilinear form $a(\cdot,\cdot)$ on $X \times Y$. Let $B \subset X$ and $C \subset Y$ be convex sets such that B has nonempty interior, and C is closed. Assume the following conditions:

/i/ If $y^* \in Y^*$ satisfies $\sup\{y^*(y) \mid y \in C\} < \infty$, then there exists $x_o \in X$ such that $y^*(y) = a(x_o, y)$ for each $y \in Y$.

/ii/ If $x^* \in X^*$ satisfies

$\inf\{x^*(x) \mid x \in B\} > -\infty$ and

$a(x,y) = 0 \; \forall y \in Y \Longrightarrow x^*(x) = 0$

then there exists $y_o \in Y_o$ such that

$x^*(x) = a(x,y) \; \forall x \in X$ .

Under these conditions

$$\inf_{x \in B} \sup_{y \in C} a(x,y) = \max_{y \in C} \inf_{x \in B} a(x,y) ,$$

provided that the left hand side is finite.

$(\exists x_o \in \mathring{B} : \sup_{y \in C} a(x_o, y) < +\infty$ ; we can take $x_o = 0$; $\mathring{B}$ = interior of B) .

The detailed proof is given in [184].

To apply Th.3.5 we first define the bilinear form $a(\cdot,\cdot)$ and spaces X,Y. Let us confine to incompressible materials. In this case the yield locus is unbounded.

We set

$$\underline{t}^D = \underline{t} - \frac{1}{3}(\mathrm{tr}\,\underline{t})\underline{I}\ , \qquad \mathrm{tr}\,\underline{t} = t_{ii}\ ,\ \underline{I} = (\delta_{ij})\ ,$$

and $X = \Sigma$, $\Sigma$ being the stress space. The yield set $K_p$ is assumed to be of the form

$$K_p = \{\underline{t} \in \Sigma \mid \underline{t}^D(x) \in C \quad \text{a.e. in } \Omega\}\ , \qquad /3.153/$$

where $C \subset R^6$ is bounded, closed and convex and $0 \in \mathring{C}$.

Let $3 < p < \infty$ /in the three-dimensional case/. The stress space $\Sigma$ consists of tensors $\underline{t} \in \mathbb{M}_s(R)$ with the properties, see [184],

$$t_{ij} \in L^2(\Omega),\quad t^D_{ij} \in L^\infty(\Omega)\ ,\quad t_{ij,j} \in L^p(\Omega), \qquad /3.154/$$

$$t_{ij}n_j \in W^{1-(1/p),p}(S_1)\ ,\quad t_N \in L^\infty(S_2)\ . \qquad /3.155/$$

The norm of $\underline{t} \in \Sigma$ is clearly the sum of norms.

The space $Y=U$ of displacement rates consists in pairs

$$(\dot{\underline{v}}^{\Omega}, \dot{\underline{v}}^{S_1}) \in BD(\Omega) \times \left[\left(W^{1-(1/p),p}(S_1)\right)^*\right]^3\ , \qquad /3.156/$$

which satisfy the condition that the map defined for $t_{ij} \in W^{1,p}(\Omega)$ by

$$\underline{t} \to -\int_{\Omega}(\nabla\cdot\underline{t})\cdot\dot{\underline{v}}^{\Omega} + \int_{S_1}(\underline{n}\cdot\underline{t})\cdot\underline{v}^{S_1} \qquad /3.157/$$

is continuous with respect to the maximum norm on $\underline{t}$. I recall that $BD(\Omega) \subset [L^{n^*}(\Omega)]^n$, $n^* = n/(n-1)$ in the general case, and $W^{1,p}(\Omega) \subset C(\Omega)$. Moreover, the trace of a function $\underline{u} \in BD(\Omega)$ belongs to $[L^1(S)]^3$. Under our assumptions, /3.157/ is well defined.

U is the normed space /though incomplete/ in the norm

$$\|\underline{u}\|_U = \|\underline{u}^{\Omega}\|_{BD(\Omega)} + \|\underline{u}^{S_1}\|_{(W^{1-(1/p),p}(S_1))^*}\ . \qquad /3.158/$$

To simplify notations we put $\underline{u} = \underline{u}^{\Omega}$, ${}_1\underline{u} = \underline{u}^{S_1}$ and $\underset{\sim}{u} = (\underline{u}, {}_1\underline{u})$. Thus $\underset{\sim}{u} \in U$ if and only if there is a symmetric tensor of measures $\underset{\sim}{\mu} \in M(\bar{\Omega}, \mathbb{M}_s(R))$, such that for all $\underline{t} \in C^1(\Omega, \mathbb{M}_s(R))$ we have

$$\langle \underset{\sim}{u}, \underline{t} \rangle = -\int_{\Omega}(\nabla\underline{t})\cdot\underline{u} + \int_{S_1}(\underline{n}\cdot\underline{t})\cdot{}_1\underline{u}\ .$$

The principle of virtual velocities is defined in the following way

$$a(\underline{t}, \dot{\underline{u}}) = \int_{\Omega} t_{ij}e_{ij}(\dot{\underline{u}}) - \int_{S_2} t_N\dot{u}_N = -\int_{\Omega}(\nabla\underline{t})\cdot\underline{u} + \int_{S_1}(\underline{n}\cdot\underline{t})\cdot{}_1\underline{u} = L(\dot{\underset{\sim}{u}}) \qquad /3.159/$$

where

$$L(\dot{u}) = \int_{\Omega} b \cdot \dot{u} + \int_{S_1} p^o \cdot {}_1u , \qquad /3.160/$$

and $t \in \Sigma$, $\dot{u} \in U$.

By definition, the limit load multiplier $\tilde{m}_1$ is the maximal value of load multipliers for which there exists a stress field $s \in K_s^1 = K_p \cap \{ t \in \Sigma \mid t_N \leq 0 \text{ on } S_2 \}$, which is in equilibrium with $(\tilde{m}_1 b, \tilde{m}_1 p^o)$:

$$\tilde{m}_1 = \sup\{\tilde{m} \mid \exists s \in K_s^1 : a(s,\dot{u}) = \tilde{m}L(\dot{u}) \ \forall \dot{u} \in C_N\} =$$

$$= \sup_{t \in K_s^1} \inf_{\substack{L(\dot{u})=1 \\ \dot{u} \in K_v}} a(t,\dot{u}). \qquad /3.161/$$

The kinematical formulation is obtained by changing the sequence of inf and sup operations.

By applying Th.3.5 we conclude that

$$\sup_{t \in K_s^1} \inf_{\substack{L(\dot{u})=1 \\ \dot{u} \in K_v}} a(t,\dot{u}) = \min_{\substack{L(\dot{u})=1 \\ \dot{u} \in \tilde{K}_v}} \sup_{t \in K_s^1} a(t,\dot{u}) . \qquad /3.162/$$

The proof of /3.162/ follows along the same lines as in [187], and is omitted here.

To prove the existence of a maximum for /3.161/ we proceed in the following way. Let us define the subspace $U_o$ of U /since we deal with incompressibility/

$$U_o = \{ u \in U \mid tr(\mu) = 0 \text{ in } M(\bar{\Omega}, M_s(R)) \}. \qquad /3.163/$$

The convex of plasticity or the yield set is

$$K_0 = K_p / \{ \varphi I \mid \varphi \in L^2(\Omega) \} .$$

Thus $K_0$ contains only deviatoric parts of stress tensors defining the yield locus.

Take now a sequence $\tilde{m}_k \to \tilde{m}_1$ /from below/and let $\{\tilde{t}^k\}$ satisfy

$$\{t^k\} \in K_0 \cap K_s^1 , \quad a(\tilde{t}^k, \dot{u}) = \tilde{m}_k L(\dot{u}) \ \forall \dot{u} \in U_o . \qquad /3.164/$$

There exists a subsequence, still denotes by $\{\tilde{t}^k\}$, such that the trace representatives are weak-* convergent in $L^\infty(\Omega)$ /Banach-Alaoglu theorem/. If $s \in K_s^1$ is the limit, then it satisfies the equilibrium equations and the boundary condition on $S_1$.

Mathematical aspects of the theory of limit analysis in the absence of unilateral boundary conditions are studied in [184-188,250,251].

Let us pass to constructing of counterparts of the limit analysis theorems in the case of Signorini's conditions with friction. The friction law is given by /2.22/, where $\dot{\underline{u}}^p_T=\dot{\underline{u}}_T$. The counterparts mentioned were proposed in [167]. It is worth noticing that if the lower and upper bound theorems are applied in the usual manner to the boundary value problems with friction then it may happen that $m^s > m^k$ /or $\tilde{m}^s > \tilde{m}^k$. The approach which will now be presented excludes inherently such a possibility.

Let us set

$$K_s(\theta_N) = \{\underline{t}=(t_{ij}) \mid t_{ij,j}+b_i=0,\ \underline{t} \in K\ ,\ \text{in}\ \Omega;$$
$$t_{ij}n_j=m^s \overset{o}{p}_i, \text{on}\ S_1;\ t_N \leq 0,\ \underline{t}_T \in K(\theta_N), \text{on}\ S_2\} \qquad /3.165/$$

where K is the global convex of plasticity and $K(\theta_N)$ is defined by /2.18/.

The static approach is the following one. Let $\theta_N$ be defined over $S_2$, $\theta_N \leq 0$ and $g(\theta_N,\underline{t}_T) \leq 0$ makes sense. Then we consider the convex maximization problem

$$m^s_1(\theta_N,\underline{s}) = \sup_{\underline{t}\in K_s(\theta_N)} m^s(\underline{t}). \qquad /3.166/$$

A stress field $\underline{s}$ solving /3.166/ clearly depends on $\theta_N$, that is $\underline{s}=\underline{s}(\theta_N)$.

For a tensor field $\underline{t}$ belonging to an appropriately chosen space we define $T_N(\underline{t})$ in the following way:

$$\int_{S_2} T_N(\underline{t})\, w_N dS = \int_{\Omega} \left(t_{ij}e_{ij}(w) - b_i w_i\right) dx\ ,\ \forall w \ \text{such that}$$
$$\underline{w}_{S\setminus S_2}=0,\ \underline{w}_T|_{S_2}=0. \qquad /3.167/$$

From /3.167/ we conclude that if $\underline{t}$ satisfies the equilibrium equations /2.53/, then $T_N(\underline{t})=t_N$, on $S_2$. In this manner a nonlinear operator $N:\theta_N \to T_N(\underline{t})$ has been implicity defined.

Let $\underline{s}$ be a solution of the problem /3.166/ for some $\theta_N$. Suppose that $\theta_N=T_N(\underline{s})$. Then the stress field $\underline{s}$ is called a static solution of the limit analysis in the case of Signorini's conditions with friction.

Such a definition of the static solution is correct since then $\theta_N=T_N(\underline{s})=N(\theta_N)$. Hence we infer that $\theta_N=s_N$ is a fixed point of the operator N. We observe that if $s_N=N(s_N)$ then we have the following counterpart of the lower bound theorem

$$m^s(\underline{t}) \leq m^s_1(s_N,\underline{s}) \qquad \forall \underline{t} \in K_s(s_N), \qquad /3.168/$$

where $m^s(\underline{t})$ enters the definition of the convex set $K_s(s_N)$, see /3.165/. A field $\underline{t} \in K_s(s_N)$ is statically admissible.

If $\tilde{m}(\tilde{\underline{s}})$ is a solution of the maximization problem

$$m^s(\tilde{\underline{s}}) = \sup_{\underline{t}\in K_s} m^s(\underline{t}),$$

where

$$K_s=\{t=(t_{ij})\mid t_{ij,j}+b_i=0,\ \underline{t}\in K,\ \text{in } \Omega;$$

$$t_{ij}n_j=m^s p_i^o,\ \text{on } S_1;\ t_N\leq 0, g(t_N,\underline{t}_T)\leq 0, S_2\},\qquad /3.169/$$

then obviously we have $m_1^s(s_N,\underline{s})\leq m^s(\tilde{\underline{s}})$.

Thus non-associated friction laws diminish limit loads in comparison with corresponding associated laws.

If $s_N$ is such that $s_N=N(s_N)$, then we define a kinematically admissible velocity field as an arbitrary field $\dot{\underline{v}}\in K_v$, provided that the kinematical load multiplier is defined by

$$m^k(s_N,\dot{\underline{v}})=\frac{D(\underline{e}(\dot{\underline{v}}))+J(s_N,\dot{\underline{v}})-\int_\Omega \underline{b}\cdot\dot{\underline{v}}dx-\int_{S_1} s_N v_N dS}{\int_{S_1}\underline{p}^o\cdot\dot{\underline{v}}dS}\qquad /3.170/$$

where $J(s_N,\dot{\underline{v}})$ stands for the total power of frictional forces.

A counterpart of the upper bound theorem is formulated a minimization problem

$$\inf_{\dot{\underline{v}}\in K_v} m^k(s_N,\dot{\underline{v}})\qquad /3.171/$$

Obviously, we have $m_1^s(s_N,\underline{s})\leq m^k(s_N,\dot{\underline{v}}),\forall\dot{\underline{v}}\in K_v$.

It seems reasonable to define a limit load multiplier m in the general case of friction, as follows:

$$m=m_1^s(s_N,\underline{s})=m^k(s_N,\dot{\underline{u}}).\qquad /3.172/$$

where $\dot{\underline{u}}$ is a solution of /3.171/.

As one can see, the kinematical approach is now coupled with the static approach. Such conclusion is not surprising since the dissipation produced by friction forces depends not only on the tangential component of a velocity vector but also explicitly on $s_N$. From the viewpoint of applications, the method of the construction of a static field $\underline{s}$ is not effective. However, an iterative procedure overcomes this drawback. For this purpose let us take a function $t_N^{(0)}$ defined over $S_2$ and such that $t_N^{(0)}\leq 0$ and $K(t_N^{(0)})$ makes sense. In the next step we solve the convex problem

$$\sup_{\underline{t}\in K_s(t_N^{(0)})} m_1^s(\underline{t}).\qquad /3.173/$$

A solution $\underline{t}^{(1)}$ of /3.173/ yields $t_N^{(1)}$ on $S_2$ and hence also on $S_2$. In this manner we obtain a sequence of maximization problems

$$\sup_{\underline{t}\in K_s\left(t_N^{n-1}\right)} m_n^s(\underline{t}), \quad n=1,2,\ldots \qquad /3.174/$$

where

$$K_s\left(t_N^{(n-1)}\right)=\{\underline{t}\,|\,t_{ij,j}+b_i=0,\ \underline{t}\in K,\ \text{in}\ \Omega;\ t_{ij}n_j=m_n^s p_i^o,$$

$$\text{on}\ S_1;\ t_N\leq 0,\ \underline{t}_T\in K\left(t_N^{(n-1)}\right),\ \text{on}\ S_2\}, \qquad /3.175/$$

and

$$K\left(t_N^{(n-1)}\right)=\{\underline{t}_T\,|\,g\left(t_N^{(n-1)},\underline{t}_T\right)\leq 0,\ \text{on}\ S_2\}. \qquad /3.176/$$

Hence we have a sequence $\{\underline{t}^{(n)}\}$ $n=1,2,\ldots$ In the limit, if it exists, we obtain $\lim_{n\to\infty}\underline{t}^{(n)}=\underline{s}$.

Mathematically rigorous study of the limit analysis in the presence of friction remains open.

Remark 3.5. A large class of unilateral problems with friction can be studied in the form of an infinite sequence of minimum, maximum or mini-max problems. For instance, problem /2.66/ furnishes the following sequence of minimization problems

$$\min\{F_n(\underline{v})\,|\,\underline{v}\in K_d\}, \qquad /3.177/$$

where

$$F_n(\underline{v}) = \frac{1}{2}\,a(\underline{v},\underline{v})+J\left(s_N(\underline{u}_{n-1}),\underline{v}_T\right)-L(\underline{v}),$$

and $\underline{u}_{n-1}$ is a minimizer of the functional $F_{n-1}$. In connection with such sequence of functionals /or saddle functionals in the case of the limit analysis/ there arises a problem of an application of the theory of $\Gamma$-convergence, see [176, 252]. The problem remains open.

Remark 3.6. Hill [201] writes /p.23/:"Idealized rigid/plastic response, by contrast, has not been shown to admit dual potentials of any kind". It seems that Hill rediscovers an old problem solved already some fifteen years ago by Moreau and the French School of Mechanics /Débordes, Nayroles, Salençon/.

## 3.6. Unilateral contact problems of plasticity: elastic-plastic bodies

For such bodies no mathematically rigorous studies of the Signorini's contact problem with friction seems to exist in the available literature.

In [200,205,211-217,231] frictionless problems are studied for materials obeying Hencky's law. In fact such laws resemble small strain, nonlinear elasticity.

Duvaut [19] studies frictional contact problems without Signorini's conditions provided that normal stresses are a priori prescribed on $S_2$.

Kuz'menko [215-219] studies unilateral contact problems without and with friction for materials obeying the following constitutive relation

$$\dot{t}_{ij} = A_{ijkl}(\underline{r})\dot{e}_{kl} , \qquad /3.178/$$

where $\underline{r}$ is an internal parameter. He is interested in an evolution problem on a time interval, see Section 3.2 of the present lectures. However, he ignores the evolution of the parameter $\underline{r}$, and thus his variational approach seems to be doubtful.

Most popular are numerically motivated incremental formulations [194,197,207,208,242,243,248]. In the case of the geometrically linear theory such approach can be briefly presented in the following way.

Let us assume that the rate equations are given by

$$\dot{s}_{ij,j} + \dot{b}_i = 0, \quad \text{in } \Omega , \qquad /3.179/$$

$$e_{ij}(\dot{\underline{u}}) = \dot{u}_{(i,j)}, \text{ in } \Omega , \qquad /3.180/$$

$$\dot{s}_{ij}(\dot{\underline{u}}) = f_{ij}(\underline{r}, \underline{e}(\dot{\underline{u}})) , \text{ in } \Omega , \qquad /3.181/$$

$$\dot{\underline{u}} = 0, \text{ on } S_o , \qquad /3.182/$$

$$\dot{s}_{ij}n_j = \dot{F}_i, \text{ on } S_1, \qquad /3.183/$$

$$\dot{\underline{u}} \in K;\ \underline{s}_T = 0,\ \dot{s}_N(\dot{v}_N - \dot{u}_N) \geq 0, \text{ on } S_2 , \qquad /3.184/$$

where $\Omega$ is the known domain occupied by the body considered at a given instant of time. In practice rates are replaced by increments; for instance we take $\Delta\underline{u}$ instead of $\dot{\underline{u}}$. K is a convex set, possibly depending on internal parameter $\underline{r}$ and $\underline{u}$, which are known. we observe that /3.181/ includes also a class of non-associated flow laws /non-standard materials/.

In the usual way we arrive at the principle of virtual velocities:

$$\left| \begin{array}{l} \text{find } \dot{\underline{u}} \in K \text{ such that} \\ \int_\Omega s_{ij}(\dot{\underline{u}})\, e_{ij}(\dot{\underline{v}} - \dot{\underline{u}})dx \geq L(\dot{\underline{v}} - \dot{\underline{u}}) \quad \forall \dot{\underline{v}} \in K , \end{array} \right. \qquad /3.185/$$

where

$$L(\dot{\underline{v}}) = \int_\Omega \dot{b}_i \dot{v}_i dx + \int_{S_1} \dot{F}_i \dot{v}_i dS . \qquad /3.186/$$

Such an approach allows to include the following sliding rule

$$\underline{s}_T = \underline{G}(s_N, \underline{q}, \underline{u}_T, \dot{\underline{u}}_T) , \qquad /3.187/$$

where $\underline{q}$ are internal parameters describing the process of friction. For instance, we can consider the sliding rule of the form

$$\underline{s}_T = -\underline{G}(s_N, \underline{q}, \underline{u}_T)\, \dot{\underline{u}}_T . \qquad /3.188/$$

The variational formulation of the problem /3.179/--/3.184/ reads:

find $\dot{\underline{u}} \in K$ such that

$$\int_{\Omega} \dot{s}_{ij}(\dot{\underline{u}}) e_{ij}(\dot{\underline{v}}-\dot{\underline{u}}) dx - \int_{S_2} \dot{\underline{s}}_T(\dot{\underline{u}})(\dot{\underline{v}}_T-\dot{\underline{u}}_T) ds \geq L(\dot{\underline{v}}-\dot{\underline{u}}) \quad \forall \dot{\underline{v}} \in K \quad /3.189/$$

provided that the condition $\underline{s}_T=0$ /on $S_2$/, is replaced by /3.187/.

To rate problems formulated above one can apply the M-CD-M theory of duality provided that constitutive relation has the form /3.178/, say, while the sliding rule is given by /3.188/. Consider the frictionless case and set $K=K_d$, where $K_d$ is /2.62/; clearly now instead of displacements we take velocity vectors. The bilinear form /2.63/ must be replaced by

$$a(\dot{\underline{u}},\dot{\underline{v}}) = \int_{\Omega} A_{ijkl}(\underline{r},\underline{e})\, e_{ij}(\dot{\underline{u}})\, e_{kl}(\dot{\underline{v}})\, dx . \qquad /3.189/$$

Assuming that the following inequality is satisfied

$$A_{ijkl} e_{ij} e_{kl} \geq c_o e_{ij} e_{ij} , \quad \forall \underline{e} = (e_{ij}) \in \mathbb{M}_s(R) , \qquad /3.190/$$

we can construct the Green's operator G similarly as in Section 2.6. The variational inequality holds provided that instead of $s_N, t_N, \hat{u}_N$ we take their rates.

In the case of geometrically nonlinear problems various formulations are possible due to a flexibility in the choice of stress and strain measures and incremental procedures, see [207,242,243,247,248]. Consider a two-body frictionless unilateral contact problem [207]. One of them may be rigid. The times at the starting and finishing points in a load increment are denoted by t and t+ Δt, respectively. We assume that meas $S_o^{(2)}>0$ while $S_o^{(1)}=\emptyset$ is not excluded. The motions of the bodies are considered in a fixed Cartesian coordinate system in which all kinematic and static variables are defined. Body forces are neglected. A vector $(q_i)$ i=1,2,3 describes the unknown increment in the rigid body displacement of $\Omega_1$ from time t to t+ Δt. A superscript $\alpha$ is used to designate quantities associated with the body $\Omega_\alpha$ ($\alpha$=1,2). Comma denotes differentiation with respect to the space coordinates ${}^t x_j$. The coordinates of a point of the body at time t are $({}^t x_i)$ while at time t+ Δt we denote them by $({}^{t+\Delta t} x_i)$. The increments in the displacements from time t to t+ Δt are defined by

$$\Delta \underline{u} = {}^{t+\Delta t}\underline{x} - {}^t\underline{x} . \qquad /3.191/$$

Obviously, the configuration of the bodies at time t+ Δt is not known. Therefore the quantities are referred to a known configuration at time t.

The Cartesian components of the Cauchy stress tensor at time t+ Δt are denoted by ${}^{t+\Delta t}\underline{s}$ and those of the second Piola-Kirchhoff stress tensor corresponding to the configuration at time t+ Δt but measured in configuration at time t are denoted by ${}^{t+\Delta t}\underline{S}$. The relationship between these two tensors is as follows

$$ {}^{t+\Delta t}s_{ij} = \det\left[\frac{\partial\, {}^{t}x_p}{\partial\, {}^{t+\Delta t}x_q}\right] {}^{t+\Delta t}x_{i,m}\; {}^{t+\Delta t}x_{j,n}\; {}^{t+\Delta t}S_{mn}, \qquad /3.192/ $$

$$ {}^{t+\Delta t}\underline{S} = {}^{t}\underline{s} + \Delta\underline{S}\ , \qquad /3.193/ $$

since ${}^{t}\underline{s} = {}^{t}\underline{S}$. Here $\Delta\underline{S}$ is the /second/ Piola-Kirchhoff stress increment tensor.

The Green-Lagrange strain tensor constructed from $\Delta\underline{u}$ has the form

$$ E_{ij}(\Delta\underline{u}) = \frac{1}{2}\left[(\Delta u_i)_{,j} + (\Delta u_j)_{,i} + (\Delta u_k)_{,i}(\Delta u_k)_{,j}\right]. \qquad /3.194/ $$

The compressive contact pressure p at the current time t+ $\Delta$t is defined on $S_2 = S_2^{(1)} = S_2^{(2)}$ by the equation

$$ p = -\,{}^{t+\Delta t}s^{(1)}_{ij}\; {}^{t+\Delta t}n^{(1)}_i\; {}^{t+\Delta t}n^{(1)}_j\ , \qquad /3.195/ $$

where ${}^{t+\Delta t}\underline{n}^{(1)}$ is the unit outward normal vector to the boundary of the body 1 at time t+ $\Delta$t.

Since the body 1 may undergo rigid body displacements, and may even be rigid, therefore the equilibrium of the external forces acting on the boundary of the body 1 should be satisfied. We have

$$ \int_{S_1^{(1)}} F_i h_{ij}\; {}^{t+\Delta t}dS = \int_{S_2} p\,\alpha_{ij}\; {}^{t+\Delta t}n^{(1)}_i\; {}^{t+\Delta t}dS\ , \qquad /3.196/ $$

where $h_{ij}$ and $\alpha_{ij}$ represent rigid body displacements of a point on $S_1^{(1)}$ and $S_2$, respectively, in the i-th coordinate direction due to a unit displacement in the j-th rigid body degree of freedom. They are determined only by kinematic relations of the body 1. $F_i$ are the components of the surface traction at time t+ $\Delta$t and ${}^{t+\Delta t}dS$ stands for a surface element at time t+ $\Delta$t.

Internal equilibrium equations for each body are

$$ (\Delta S_{ij})_{,j} + \left[(\Delta u_i)_{,j}\; {}^{t+\Delta t}S_{jk}\right]_{,k} = 0 \qquad /3.197/ $$

since ${}^{t}s_{ij,j} = 0$.

The following constitutive relations are assumed for each body

$$ S_{ij} = C^{e-p}_{ijkl} E_{kl}, \qquad /3.198/ $$

where $C^{e-p}_{ijrs}$ is the elasto-plastic matrix at time t and is

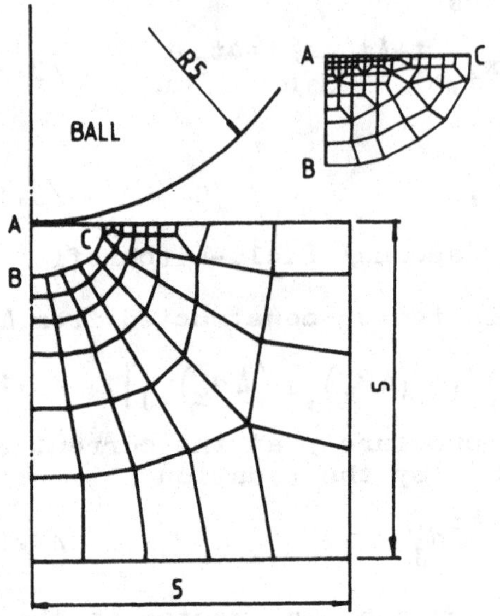


Fig.28, Finite element mesh for ball indentation problem. Dimensions in mm

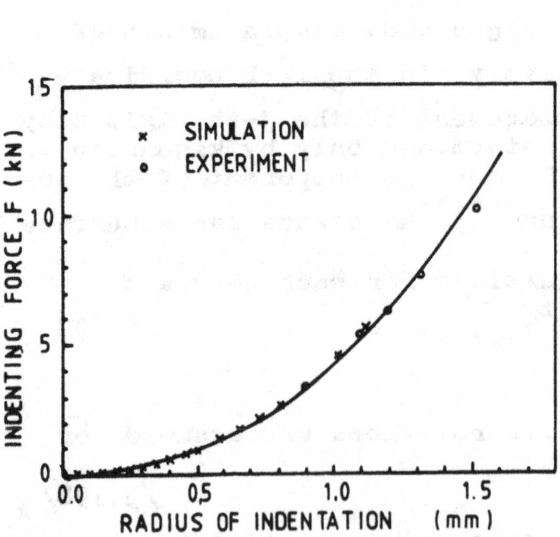


Fig.29, Load vs radius of indentation for ball indentation problem

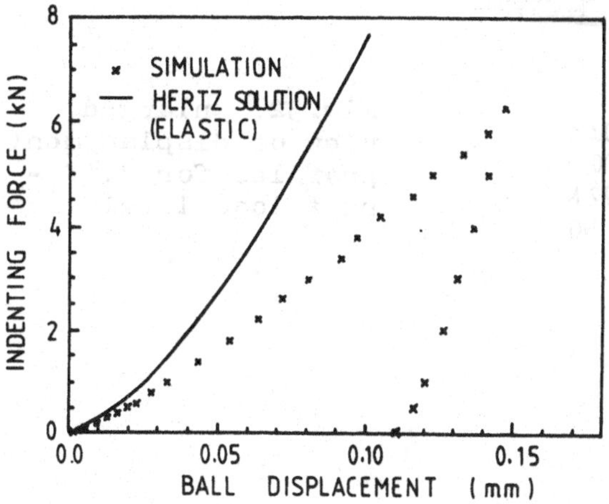


Fig.30, Load vs ball displacement for ball indentation problem

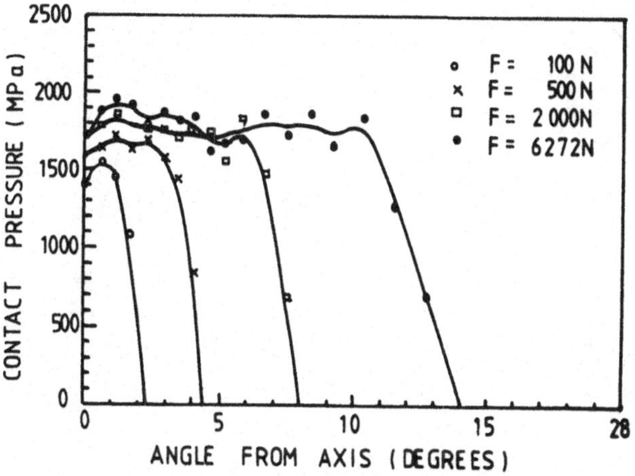


Fig.31, Contact pressure distribution for ball indentation problem

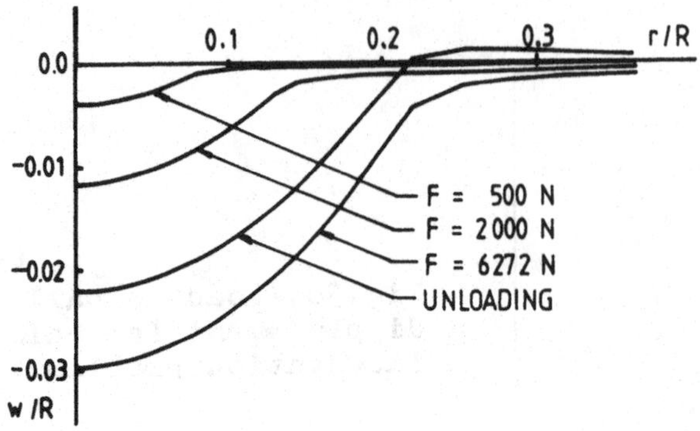


Fig.32, Enlarged view of displacement profiles for different load levels

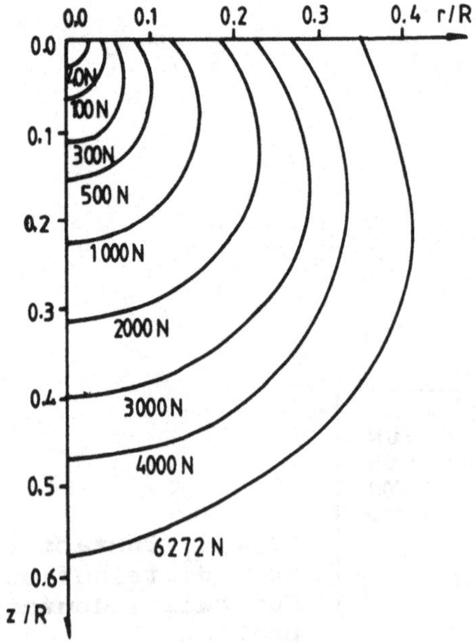


Fig.33, Plastic enclaves at various loads

determined by the yield function, flow law and hardening rule.

Boundary conditions:

$$\Delta\underline{u}^{(\alpha)} = 0 \quad \text{on} \quad S_o^{(\alpha)} , \qquad /3.199/$$

$$^{t+\Delta t}s_{ij} \; ^{t+\Delta t}n_j^{(\alpha)} = F_i^{(\alpha)} \quad \text{on } S_1 , \qquad /3.200/$$

$$p \geqslant 0 , \quad \Phi \geqslant 0 , \quad \Phi p = 0 \quad \text{on } S_2 , \qquad /3.201/$$

where $\Phi$ is the gap in the potential contact zone $S_2$ after incremental deformation and is given by

$$\Phi = \Phi_t - \left(\Delta u_i^{(1)} \; ^{t+\Delta t} - \Delta u_i^{(2)} \; ^{t+\Delta t} + \alpha_{ij} \; ^{t+\Delta t}q_j\right) n_i^{(1)} . \qquad /3.202/$$

Here $\Phi_t$ stands for the gap at time t.

In [207] the following minimization problem is formulated:

$$\left|\begin{array}{ll} \min F(\Delta\underline{u}^{(1)}, \Delta\underline{u}^{(2)}, q_i) & /3.203/ \\ \text{subject to } \Phi \geqslant 0 \text{ on } S_2 \text{ and } /3.199/ \text{ on } S_o^{(\alpha)} & /3.204/ \end{array}\right.$$

where

$$F = \int_{\Omega_1} \left({}^t s_{ij}^{(1)} + \tfrac{1}{2}\Delta s_{ij}^{(1)}\right) E_{ij}^{(1)} dV_t + \int_{\Omega_2} \left({}^t s_{ij}^{(2)} + \tfrac{1}{2}\Delta s_{ij}^{(2)}\right) E_{ij}^{(2)} dV_t -$$

$$- \int_{S_1^{(1)}} F_i^{(1)} \left(\Delta u_i^{(1)} + h_{ij} q_j\right) {}^{t+\Delta t} dS - \int_{S_1^{(2)}} F_i^{(2)} \Delta u_i^{(2)} \; {}^{t+\Delta t} dS \qquad /3.205/$$

where $dV_t$ stands for a volume element at time t.

Using a discretization technique, say FEM, one can obtain from /3.203/ and /3.204/ a nonlinear programming problem. Such approach has been applied in [207] to solve the axisymmetric ball indentation problem provided that the material obeys the Prandtl-Reuss equation, the von Mises yield criterion and the isotropic hardening rule. The ball itself is assumed to be rigid. The elastic modulus and Poisson's ratio used are 72.500 MPa and 0.33, respectively. In the effective stress-effective plastic strain relation which is expressed by $\bar{\sigma} = \sigma_y + \sigma_o \bar{\varepsilon}_p^n$ ; $\sigma_y$, $\sigma_o$ and n are 392 MPa, 301 MPa and 0.283, respectively.

Results of calculations are represented in Figs 28-34.

Remark 3.7. The reader interested in inelastic unilateral contact problems of structural mechanics should refer to [199,222-226,247].

In [244,245] a possibility of a generalization of the shakedown theorems [210,238] to frictional boundary conditions has been conjectured. However, no precise statement was given and the problem remains open.

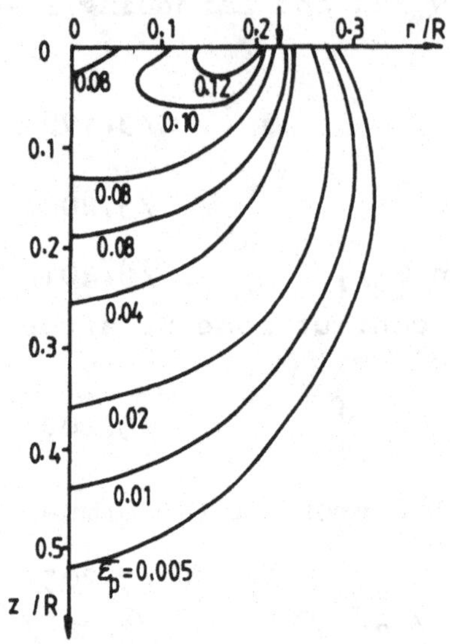


Fig.34, Effective plastic strain distribution at load 6272 N

4. HOMOGENIZATION OF FISSURED ELASTIC SOLIDS AND PLATES

In the case of nonhomogeneous solids and structures, like fissured elastic plates, it is desirable to know the overall properties. To determine such properties the methods of homogenization have proved to be very useful, see [252,254,255,256,260.261,262,266,271,272-275]. Yet these methods are most effective in the case of periodicity or non-uniform periodicity.

In this chapter I shall present the problem of homogenization of a geometrically linear, fissured elastic solid. Homogenization of Kirchhoff and Reissner-like plates will also be discussed.

4.1. Homogenization of a fissured elastic solid

Primarily the problem was posed and solved by Sanchez-Palencia [272], see also [265]. He employed the two-scale method of asymptotic expansions. Such an approach is formal, though very useful. Next Attouch and Murat [253] delivered a rigorous proof of convergence. Yet they considered only the scalar case. In the sequel the vector case is considered. The study follows the main lines of the approach proposed in [253].

Let us consider an elastic solid weakened by periodically distributed microfissures, see Fig.35, where two-dimensional case is schematically represented.

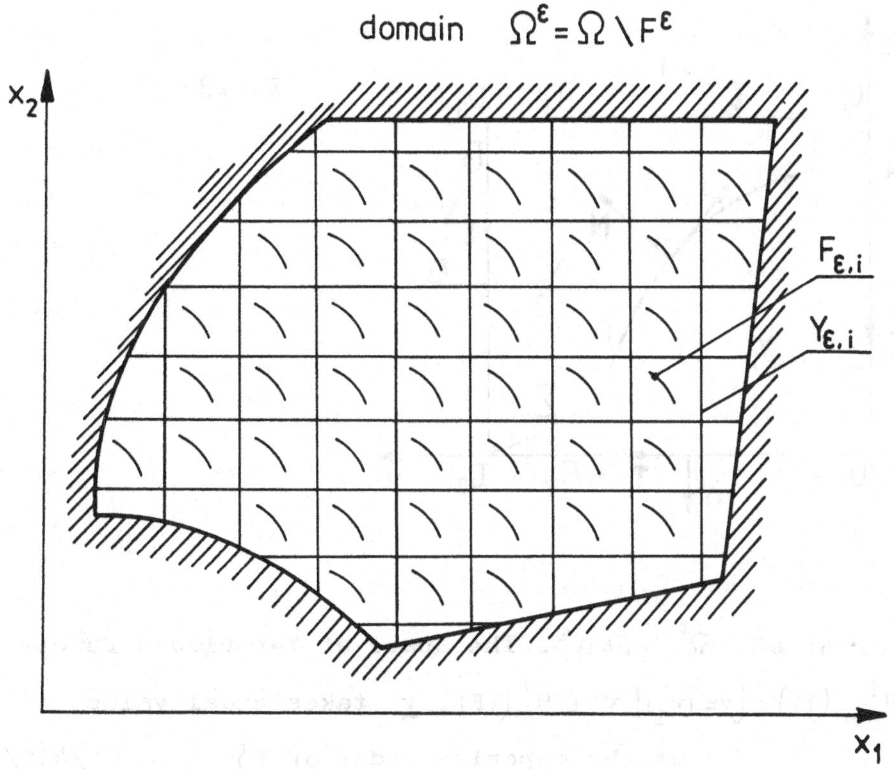


Fig.35

We assume that the body is clamped at the boundary while the microfissures $F_{\varepsilon,i}$ are distributed periodically and do not intersect the boundary $\partial\Omega$. Every cell $Y_{\varepsilon,i}$ is homothetic to the so called basic cell Y, see Fig.36. The latter is damaged by a fissure F. We assume that F is of class $C^1$ and $\overline{F}=F\subset Y$. Moreover we assume that $YF=Y\setminus F$ is connected /F may be a sum of disjoint fissures/. The parameter $\varepsilon$ is positive and homogenization signifies the passage to a limit with this parameter $/\varepsilon\rightarrow 0/$. The homogenized solid has no fissures, but elastic properties are different from those of the original material.

The following notation is introduced for the sum of microfissures such that the corresponding $\varepsilon Y$ - cells are contained in the domain $\Omega$

$$F^\varepsilon = \bigcup_{i\in I(\varepsilon)} F_{\varepsilon,i} \quad . \qquad /4.1/$$

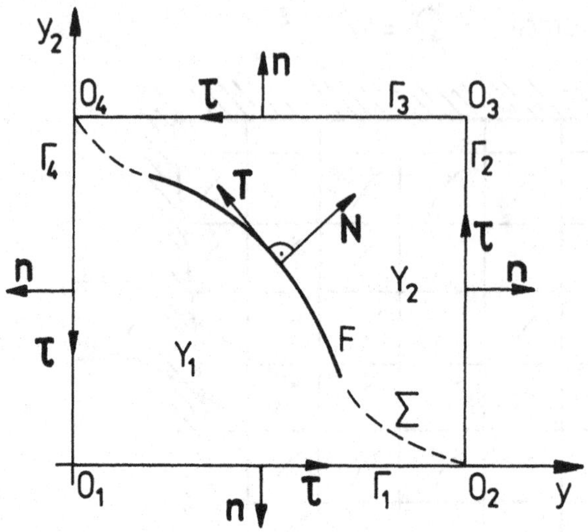


Fig.36

Further we set $\Omega^\varepsilon = \Omega \setminus F^\varepsilon$. The space of Y-periodic functions

$$H^1_{per}(YF) = \{\underline{v} = (v_i) \mid v_i \in H^1(YF),\ \underline{v} \text{ takes equal values at the opposite sides of } Y\} \qquad /4.2/$$

is of fundamental importance.

The Sobolev space $H^1(YF)$ may be defined in two equivalent manners [263]

$$H^1(YF) = \{v \in L^2(YF) \mid u_{,i} \in L^2(YF)\} =$$
$$= \{v \in H^1(Y_1) \cup H^1(Y_2) \mid \gamma_1(v) = \gamma_2(v) \text{ on } \Sigma \setminus F\}, \qquad /4.3/$$

where $\gamma_\alpha$ $(\alpha=1,2)$ is the trace operator for functions defined on $Y_\alpha$. Clearly, the choice of $\Sigma$ is arbitrary, not necessarily such as represented in Fig.36.

For $v \in H^1(YF)$ we define the jump across F in the following way

$$[\![v]\!] = \gamma_2(v) - \gamma_1(v) \quad , \text{ on } F. \qquad /4.4/$$

We assume that on $F^\varepsilon$ the Signorini's conditions are satisfied

$$\underline{s}^\varepsilon_T = 0 \ , \ [\![u^\varepsilon_N]\!] \geqslant 0 \ , \ s^\varepsilon_N \leqslant 0 \ , \ s^\varepsilon_N [\![u^\varepsilon_N]\!] = 0 \text{ on } F^\varepsilon, \qquad /4.5/$$

where $s^\varepsilon_N = s^\varepsilon_{N|1}$ on $F^\varepsilon$.

We put

$$a^{\varepsilon}(\underline{u},\underline{v}) = a_{ijkl} e_{ij}(\underline{u})\, e_{kl}(\underline{v})\, dx; \underline{u},\underline{v} \in \left[H^1(\Omega^{\varepsilon})\right]^3, \quad /4.6/$$

where $a_{ijkl} \in L^{\infty}(\Omega)$. One could also assume that the elasticities $a_{ijkl}$ are Y-periodic.

We assume that there exists constants $c_o, c_1$ such that

$$c_o e_{ij} e_{ij} \leqslant a_{ijkl} e_{ij} e_{kl} \leqslant c_1 e_{ij} e_{ij} \quad \forall \underline{e} \in \mathbb{M}_s(R). \quad /4.7/$$

Let us introduce the closed and convex sets of constraints

$$K^{\varepsilon} = \{\underline{v} \in [H^1(\Omega^{\varepsilon})]^3 \mid \underline{v}=0 \text{ on } \partial\Omega,\ [\![v_N]\!] \geqslant 0 \text{ on } F^{\varepsilon}\}. \quad /4.8/$$

Then the equilibrium problem of the fissured body $\Omega^{\varepsilon}$ may be formulated as the minimization problem

$$\min\{\tfrac{1}{2} a^{\varepsilon}(\underline{v},\underline{v}) - \int_{\Omega^{\varepsilon}} \underline{b}\cdot\underline{v}\, dx \mid \underline{v} \in K^{\varepsilon}\}, \quad /4.9/$$

or equivalently in the form of the VI:

$$(P^{\varepsilon}) \left| \begin{array}{l} \text{find } \underline{u}^{\varepsilon} \in K^{\varepsilon} \text{ such that} \\ a^{\varepsilon}(\underline{u}^{\varepsilon}, \underline{v}-\underline{u}^{\varepsilon}) \geqslant L(\underline{v}-\underline{u}^{\varepsilon}) \quad \forall \underline{v} \in K^{\varepsilon}, \end{array} \right. \quad /4.10/$$

where

$$L(\underline{v}) = \int_{\Omega^{\varepsilon}} \underline{b}\cdot\underline{v}\, dx . \quad /4.11/$$

Hence we infer that the problem $P^{\varepsilon}$ represents an internal Signorini's problem. In the homogenization we are interested in a passage with $\varepsilon$ to zero. One can prove the following results.

<u>Theorem 4.1</u>. Let $\underline{u}^{\varepsilon}$ solves $(P^{\varepsilon})$ and $s^{\varepsilon}_{ij} = a_{ijkl} e_{kl}(\underline{u}^{\varepsilon})$. Then we have

$$\underline{u}^{\varepsilon} \longrightarrow \underline{u} \quad \text{strongly in } L^2(\Omega, R^3),$$

$$\underline{s}^{\varepsilon} \longrightarrow \underline{s} \quad \text{weakly in } L^2(\Omega, \mathbb{M}_s(R)),$$

where $\underline{u}$ is the minimizer of the total potential energy of the homogenized solid

$$\min\{\tfrac{1}{2} a(\underline{v},\underline{v}) - \int_{\Omega} \underline{b}\cdot\underline{v}\, dx \mid \underline{v} \in [H^1_o(\Omega)]^3\}, \quad /4.12/$$

and

$$a(\underline{u},\underline{v}) = \int_{\Omega} a_{ijkl} e_{ij}(\underline{u})\, e_{kl}(\underline{v})\, dx; \quad \underline{u},\underline{v} \in [H^1_o(\Omega)]^3 \quad /4.13/$$

The <u>homogenized solid is hyperelastic</u> and its potential is given by

$$W^h(\tilde{\underline{e}}) = \min\{\tfrac{1}{2} \int_{YF} a_{ijkl} (\tilde{e}_{ij} + e_{ij}(\underline{w}))(\tilde{e}_{kl} + e_{kl}(\underline{w})) \mid \underline{w} \in K_{per}\}, \quad /4.14/$$

**Proof.** Such an operator may be constructed as follows. Let $\mathcal{R}$ denote the space of rigid displacements. Then each $\underline{v} \in [H^1(Y \setminus F_\eta)]^3$ can be decomposed according to

$$\underline{v} = \underline{v}_1 + \underline{r}, \qquad /4.18/$$

where $\underline{r} \in \mathcal{R}$ and $\underline{v}_1 \perp \mathcal{R}$ in $[L^2(Y \setminus F_\eta)]^3$.
To extend $\underline{v}$ on Y we extend $\underline{v}_1$ continuously. This operator is denoted by $\mathbb{P}$. Thus

$$\mathbb{P}\underline{v} = \mathbb{P}\underline{v}_1 + \underline{r}.$$

Let us prove /a/. We have

$$\|\mathbb{P}\underline{v}\|_{0,Y} = \|\mathbb{P}\underline{v}_1 + \underline{r}\|_{0,Y} = \|\mathbb{P}(\underline{v}_1 + \underline{r})\|_{0,Y} =$$

$$\leqslant c\|\underline{v}_1 + \underline{r}\|_{0,Y\setminus F_\eta} = c\|\underline{v}\|_{0,Y\setminus F_\eta}.$$

The first inequality in /b/ has been proved by Léné [266]. The second one is obvious since $Y \setminus F_\eta \subset YF$.

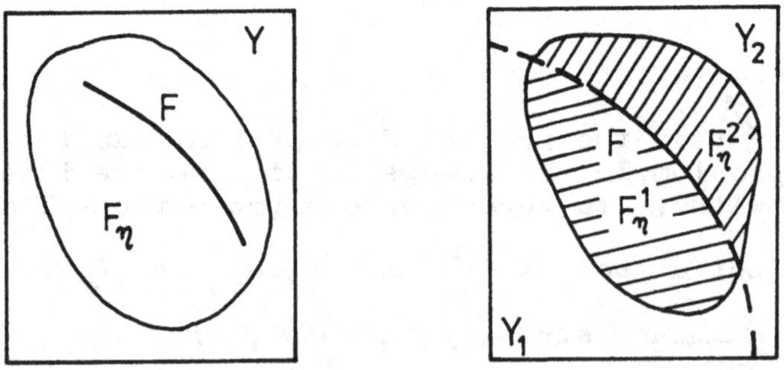


$F = \bar{F} \subset F_\eta$

$Y = Y_1 \cup Y_2 \cup \Sigma, \quad Y_1 \cap Y_2 = \emptyset$

$F = F_\eta^1 \cup F_\eta^2 \cup (\Sigma \cap F_\eta)$

$F_\eta^\alpha = F_\eta \cap Y_\alpha, \quad \alpha = 1, 2$

Fig.37

where

$$K_{per}=\{\underline{w}\in H^1_{per}(YF) \quad [\![w_N]\!]\geqslant 0 \text{ on } F\}, \qquad /4.15/$$

and $\tilde{\underline{e}} \in \mathbb{M}_s(R)$.

Moreover

$$\underline{s} = \partial W^h/\partial \tilde{\underline{e}} \quad , \qquad /4.16/$$

and

$$a^\varepsilon(\underline{u}^\varepsilon,\underline{u}^\varepsilon) \longrightarrow \int_\Omega W^h(\underline{e}(\underline{u}))\, dx \,. \qquad /4.17/$$

The next theorem characterizes the potential $W^h$.

Theorem 4.2. The elastic potential $W^h$ is of class $C^1$, positive, strictly convex and satisfies the following properties

/i/ $\quad c_2 e_{ij}e_{ij} \leqslant W^h(\underline{e}) \leqslant c_1 e_{ij}e_{ij}$ , $\forall \underline{e} \in \mathbb{M}_s(R)$,

for some $c_2 > 0$.

/ii/ For each $\underline{e} \in \mathbb{M}_s(R)$ we have

$$\underline{s} = \partial W^h/\partial \underline{e} \,.$$

The proofs of the above theorems are lengthy. Therefore I shall only sketch the main points. The scalar case is studied in [253]. The proof of Th.4.2. is given in [272]. The convergence in the vector case is implied by the results obtained in [277] for Reissner-like plates /two-dimensional case/.

The proof of Th.4.1 is divided into several steps: 1/ the construction of an extension operator $\mathbb{Q}^\varepsilon$, 2/ boundednes of the sequence $\{\underline{u}^\varepsilon\}$, 3/ localization, 4/ identification of $\underline{s}$ and $\underline{u}$.

We observe that the domain $\Omega^\varepsilon$ depends on $\varepsilon$. According to the approach of Attouch and Murat [253] we must construct an extension operator $\mathbb{Q}^\varepsilon$ such that $\mathbb{Q}^\varepsilon \underline{u}^\varepsilon \in [H^1_0(\Omega)]^3$. For this purpose we "enlarge" the fissure F to a hole $F_\eta$, see Fig.37.

The parameter $\eta > 0$ is kept fixed, $\eta = \eta_0$ and $F \subset F_\eta$. The boundary of $F_\eta$ is sufficiently smooth. To construct the operator $\mathbb{Q}^\varepsilon$ we proceed as follows.

We shall first construct the extension operator $\mathbb{Q}$.

Lemma 4.1. There exists an extension operator

$$\mathbb{P} : [H^1(Y\backslash F_\eta)]^3 \longrightarrow [H^1(Y)]^3 \quad \text{such that}$$

/a/ $\quad \|\mathbb{P}\underline{v}\|_{0,Y} \leqslant c\|\underline{v}\|_{0,Y\backslash F_\eta}$ ,

/b/ $\quad \sum_{i,j}\|e_{ij}(\mathbb{P}(\underline{v}))\|_{0,Y} \leqslant c \sum_{i,j}\|e_{ij}(\underline{v})\|_{0,Y\backslash F_\eta} \leqslant c\|\underline{e}(\underline{v})\|_{0,YF}$ .

The operator $\mathbb{Q}$ is defined as follows.

Definition 4.1. The extension operator

$$\mathbb{Q} : [H^1(YF)]^3 \to [H^1(Y)]^3$$

is equal to

$$\mathbb{Q} = \mathbb{P} \circ \mathbb{R}, \qquad /4.19/$$

where $\mathbb{R} : [H^1(YF)]^3 \to [H^1(Y\setminus F_\eta)]^3$ is the restriction operator.

The operator $\mathbb{Q}$ is characterized by

Lemma 4.2. The operator $\mathbb{Q}$ has the following properties

/i/ $\mathbb{Q}\,\underline{v} = \underline{v}$ on $Y\setminus F_\eta$.

/ii/ $\|\mathbb{Q}\,\underline{v}\|_{0,Y} \leqslant c\|\underline{v}\|_{0,Y\setminus F_\eta} \leqslant c\|\underline{v}\|_{0,YF} = c\|\underline{v}\|_{0,Y}$.

/iii/ $\|\underline{e}(\mathbb{Q}\,\underline{v})\|_{0,Y} \leqslant c\,\|\underline{e}(\underline{v})\|_{0,Y\setminus F_\eta} \leqslant c\,\|e(\underline{v})\|_{0,YF}$.

/iv/ $\|\mathbb{Q}\,\underline{v}-\underline{v}\|_{1,YF} \leqslant c\,\|\underline{e}(\underline{v})\|_{0,YF}$.

The proof is given in [277].

Having defined and examined the operator $\mathbb{Q}$ we pass to the extension operator $\mathbb{Q}^\varepsilon$ acting on functions determined on $\Omega^\varepsilon$. We know that $F_{\varepsilon,i} \subset Y_{\varepsilon,i}$ for every $i \in I(\varepsilon)$ and $Y_{\varepsilon,i}$ is the $\varepsilon Y$ cell corresponding to "i" or

$$Y_{\varepsilon,i} = \varepsilon Y + \underline{\xi}_{i,\varepsilon}, \qquad \underline{\xi}_{i,\varepsilon} \in R^3$$

Next operators $\mathbb{Q}^{\varepsilon,i}$ are constructed, see the scheme below. The global operator $\mathbb{Q}^\varepsilon$ is derived from $\mathbb{Q}^{\varepsilon,i}$ by the method of stitching of the operators $\mathbb{Q}^{\varepsilon,i}$. The global operator $\mathbb{Q}^\varepsilon$ is obtained in the following way

$\underline{z} \in [H^1(\Omega)]^3$

↓ restriction

$\underline{z}^{\varepsilon,i} \in [H^1(Y_{\varepsilon,i}\setminus F_{\varepsilon,i})]^3 = [H^1(\varepsilon YF + \underline{\xi}_{i,\varepsilon})]^3$

↓ translation and change of scale

$\underline{z}_1 \in [H^1(YF)]^3$ or $\underline{z}_1(y) = \underline{z}^{\varepsilon,i}(\varepsilon y + \underline{\xi}_i)$,

↓ restriction and extension

$\underline{z}_2 = \mathbb{Q}\underline{z}_1 \in [H^1(Y)]^3$

↓ translation and change of scale

$\underline{z}_3 = \mathbb{Q}^{\varepsilon,i}\underline{z}_2 \in [H^1(Y_{\varepsilon,i})]^3$ or $\underline{z}_3(x) = \underline{z}_2\left(\dfrac{x-\underline{\xi}_{i,\varepsilon}}{\varepsilon}\right)$

↓ stitching with respect to $i \in I(\varepsilon)$

$\underline{z}_4 = \mathbb{Q}^\varepsilon \underline{z}$.

We observe that $x = y + \underline{\xi}_{i,\varepsilon}$ implies $y = (x - \underline{\xi}_{i,\varepsilon})/\varepsilon$. According to our earlier assumptions the operator $\mathbb{Q}^\varepsilon$ may be set equal to the identity near the boundary S of $\Omega$, see Lemma 4.3 below. Then

$$\underline{z} \in [H_1^1(\Omega^\varepsilon)]^3 \Longrightarrow Q^\varepsilon \underline{z} \in [H_0^1(\Omega)]^3 .$$

where

$$H_1^1(\Omega^\varepsilon) = \{\underline{v} \in [H^1(\Omega^\varepsilon)]^3 \mid \underline{v} = 0 \quad \text{on } S\} .$$

The basic properties of the operator $Q^\varepsilon$ are given by

Lemma 4.3. For each $\varepsilon > 0$ the operator $Q^\varepsilon : [H^1(\Omega^\varepsilon)]^3 \longrightarrow [H^1(\Omega)]^3$ is linear and continuous. Moreover we have

$$Q^\varepsilon \underline{z} = \underline{z} \quad \text{on} \quad \Omega \setminus F_\varepsilon , \quad F_\varepsilon^\eta = \cup F_{\varepsilon,i}^\eta ,$$

$$\| Q^\varepsilon \underline{z} \|_{0,\Omega} \leq c \| \underline{z} \|_{0,\Omega} ,$$

$$\| \underline{e}(Q^\varepsilon \underline{z}) \|_{0,\Omega} \leq c \| \underline{e}(\underline{z}) \|_{0,\Omega} ,$$

$$\| Q^\varepsilon \underline{z} - \underline{z} \|_{0,\Omega} \leq c\,\varepsilon \| \underline{e}(\underline{z}) \|_{0,\Omega^\varepsilon} ,$$

$$\| \nabla (Q^\varepsilon \underline{z} - \underline{z}) \|_{0,\Omega^\varepsilon} \leq c \| \underline{e}(\underline{z}) \|_{0,\Omega^\varepsilon} .$$

By noting that $\| Q^\varepsilon \underline{z} - \underline{z} \|_{1,\Omega^\varepsilon} \leq (c_1 \varepsilon + c) \| \underline{e}(\underline{z}) \|_{0,\Omega^\varepsilon}$ we can formulate

Theorem 4.3. /Korn's inequality for $\Omega^\varepsilon$ /. For each $\underline{z} \in [H_1^1(\Omega^\varepsilon)]^3$ the Korn's inequality

$$\| \underline{z} \|_{1,\Omega^\varepsilon} \leq (c\,\varepsilon + c_1) \| \underline{e}(\underline{z}) \|_{0,\Omega^\varepsilon} . \qquad /4.20/$$

is satisfied. Here $0 < \varepsilon < \varepsilon_0$ and $\varepsilon_0$ is held fixed.

Remark 4.1. For the scalar case a similar role is played by the Poincaré inequality in which the parameter $\varepsilon$ also enters explicitly [253].

Now we can formulate

Theorem 4.4. For any sequence $\{\underline{v}^\varepsilon\}_{\varepsilon \to 0}$ satisfying $\sup_{\varepsilon > 0} \| \underline{v}^\varepsilon \|_{1,\Omega^\varepsilon} < \infty$ there exists a sequence $\{Q^\varepsilon \underline{v}^\varepsilon\}_{\varepsilon \to 0}$ bounded in $[H^1(\Omega)]^3$ and such that

$$\| Q^\varepsilon \underline{v}^\varepsilon - \underline{v}^\varepsilon \|_{0,\Omega} \longrightarrow 0 \quad \text{as } \varepsilon \longrightarrow 0.$$

By using /4.9/, /4,20/ and Th. 4.4 we prove that

$$\underline{u}^\varepsilon \longrightarrow \underline{u} \quad \text{strongly in} \quad [L^2(\Omega)]^3 ,$$

$$s_{ij} \rightharpoonup s_{ij} \quad \text{weakly in} \quad L^2(\Omega),$$

where $s_{ij}^\varepsilon = a_{ijkl} e_{kl}(\underline{u}^\varepsilon)$. Next we consider the inequality

$$\int_{\Omega^\varepsilon} \varphi(x) \left( a_{ijkl} e_{kl}(\underline{u}^\varepsilon) - a_{ijkl} e_{kl}(\underline{v}_{\underline{\varepsilon}}^\varepsilon) \right) e_{ij}(\underline{u}^\varepsilon - \underline{v}_{\underline{\varepsilon}}^\varepsilon)\, dx \geq 0 \qquad /4.21/$$

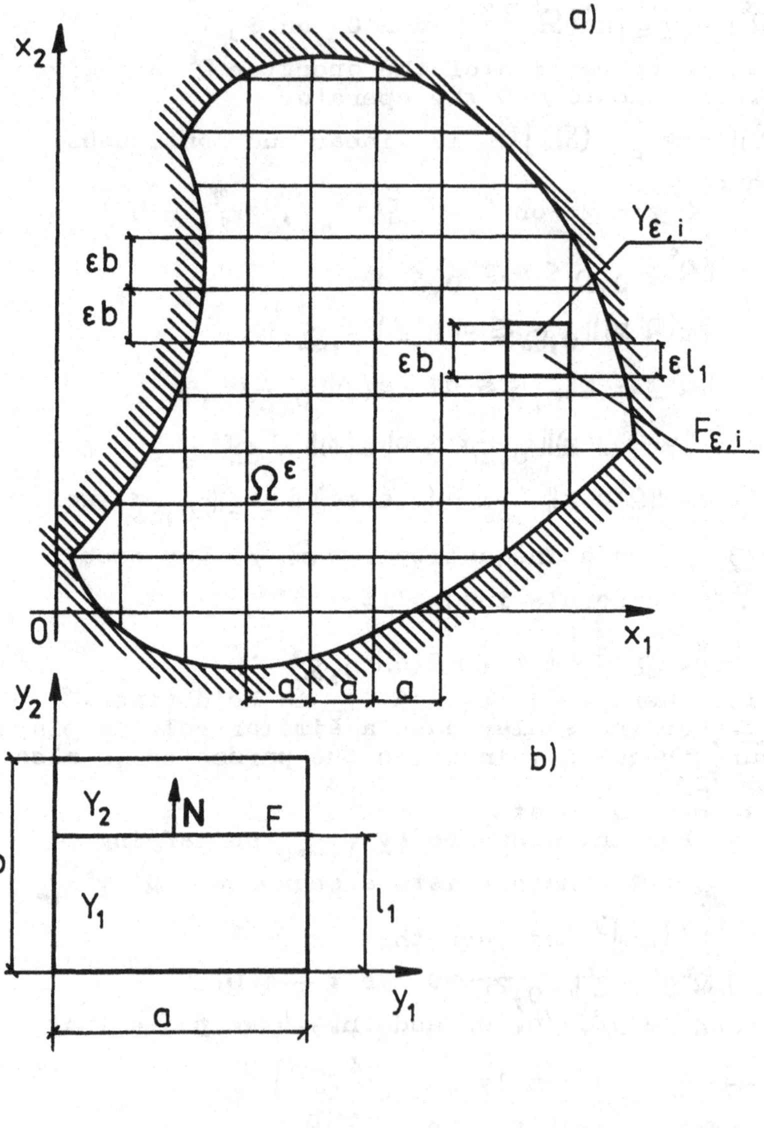
a)
$x_2$
$\varepsilon b$
$\varepsilon b$
$Y_{\varepsilon,i}$
$\varepsilon b$
$\varepsilon l_1$
$F_{\varepsilon,i}$
$\Omega^\varepsilon$
0
$x_1$
a
a
a
$y_2$
b)
$Y_2$
N
F
b
$Y_1$
$l_1$
$y_1$
a

Fig.38

where $\varphi \in \mathcal{D}^+(\Omega)$, $\underline{\tilde{e}} \in \mathbb{M}_s(\mathbb{R})$ and

$$\underline{v}^{\varepsilon}_{\tilde{e}} = \varepsilon \underline{v}(x/\varepsilon) = \underline{P}^1(x) + \varepsilon \underline{v}_{\tilde{e}}(x/\varepsilon), \quad P^1_i = \tilde{e}_{ij}x_j . \qquad /4.22/$$

Here $\underline{v}_{\tilde{e}}(y)$ is a solution of the local problem /4.14/. Performing the integration by parts in /4.21/ and localizing we arrive at the inequality

$$\langle \underline{s}(x) - \partial W^h(\underline{\tilde{e}}),\ \underline{e}(\underline{u}(x)) - \underline{\tilde{e}} \rangle_{R^3 \times R^3} \geqslant 0 \quad \forall \underline{\tilde{e}} \in \mathbb{M}_s(\mathbb{R}).$$

The maximal monotonicity of the subdifferential $\partial W^h$ implies, see [258, p.22]

$$\underline{s}(x) = \partial W^h(\underline{e}(x)) \quad \text{for almost every } x \in \Omega.$$

To prove /4.17/ one must employ the methods of $\Gamma$-convergence, see [253].

Remark 4.2. As a by-product, in [253] the authors show that $H^1(YF) \subset BV(Y)$, where BV stands for the space of /scalar/ integrable functions v such that $v_{,i} \in M^1(Y)$. Similarly it can be demonstrated that $[H^1(YF)]^3 \subset BD(Y)$.

Remark 4.3. Obviously, if friction occurs on $F^{\varepsilon}$ then the homogenization problem becomes much more difficult. Rigorous results are not available, yet heuristic approach is used in [265,276]. For instance, in [276] the local problem has the form of the IVI:

$$(P_{loc}) \quad \left| \begin{array}{l} \text{for } \underline{\tilde{e}} \text{ and } \underline{\alpha} \text{ given find } \underline{w} \in K_{per} \\ \text{such that} \\ \displaystyle\int_{YF} a_{ijkl}\left(e_{ij}(\underline{w}) + \tilde{e}_{ij}\right) e_{kl}(\underline{v}-\underline{w})\,dy \; + \\ \qquad + J(\underline{w}, \underline{v}-\underline{\alpha}) - J(\underline{w}, \underline{w}-\underline{\alpha}) \geqslant 0 \quad \forall \underline{v} \in K_{per}, \end{array} \right.$$

where J is the total work of frictional forces. The macroscopic potential is defined as follows

$$W_{s_N}(\underline{\tilde{e}}, \underline{\alpha}) = \frac{1}{|Y|}\frac{1}{2} \int_{Y\setminus F} a_{ijkl}\left(e_{kl}(\underline{w}) + \tilde{e}_{kl}\right)\left(e_{ij}(\underline{w}) + \tilde{e}_{ij}\right) dy \; + $$
$$+ \frac{1}{|Y|} \int_F d\left(s_N(\underline{w}),\ [\![\underline{w}_T]\!] - \underline{\alpha}\right) ds ,$$

and $s_N(\underline{w}(\underline{\tilde{e}}, \underline{\alpha}))$ is treated as a parameter in the process of differentiation.

4.2. Comments on homogenization of fissured elastic plates

So far, only fissured Kirchhoff plates in bending [267,268] and Reissner-like plates [269,277] were studied. In the case of the Kirchhoff plates in bending we have five

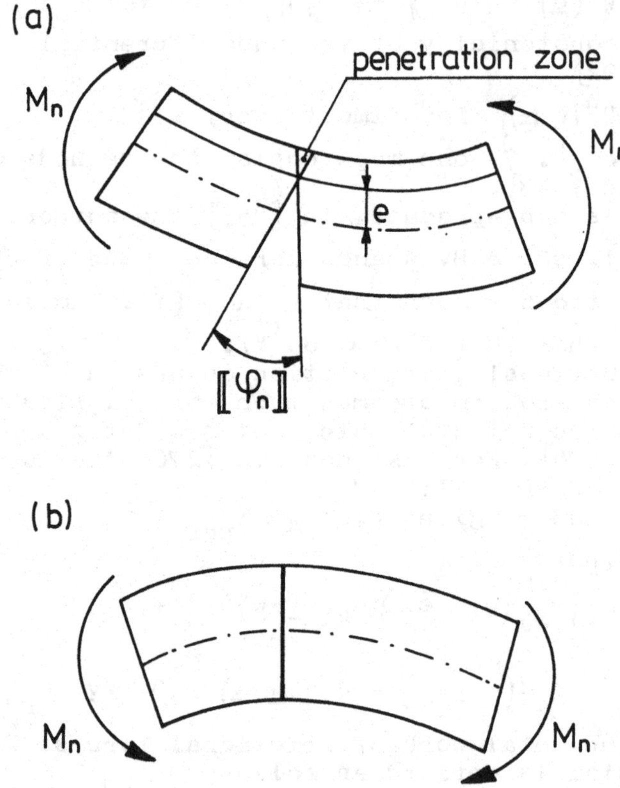


Fig. 39

different sets of unilateral constraints

$$K_\varepsilon^1 = \{v \in H_1^2(\Omega^\varepsilon) \mid [\![v]\!] \leqslant 0 \ , \text{ on } F^\varepsilon\},$$

$$K_\varepsilon^2 = \{v \in H_1^2(\Omega^\varepsilon) \mid [\![v]\!] \leqslant 0, \ [\![\partial v/\partial \underline{n}]\!] = 0, \text{on } F^\varepsilon\} \ ,$$

$$K^3 = \{v \in H_1^2(\Omega^\varepsilon) \mid [\![v]\!] = 0, \ [\![\partial v/\partial \underline{n}]\!] \leqslant 0, \text{ on } F^\varepsilon\} \ ,$$

$$K_\varepsilon^4 = \{v \in H_1^2(\Omega^\varepsilon) \mid [\![\partial v/\partial \underline{n}]\!] \leqslant 0 \ , \text{ on } F^\varepsilon\} \ ,$$

$$K_\varepsilon^5 = \{v \in H_1^2(\Omega^\varepsilon) \mid [\![v]\!] \leqslant 0 \ , [\![\partial v/\partial \underline{n}]\!] \leqslant 0, \text{ on } F^\varepsilon \ ,$$

where

$$H_1^2(\Omega^\varepsilon) = \{v \in H^2(\Omega^\varepsilon) \mid \gamma(v) = 0, \ \gamma\left(\frac{\partial v}{\partial \underline{n}}\right) = 0, \text{on } S\} .$$

Obviously now the domain $\Omega$ stands for the mid-plane of the undamaged plate.
Hence after homogenization we obtain five hyperelastic plates without fissures. In [267,268] the method of two-scales asymptotic expansions has been used. It is interesting to note that it was necessary to postulate the following expansion for the transverse displacement $w^\varepsilon$ of the microfissured plate

$$w^\varepsilon(x) = w^{(0)}(x) + \varepsilon^2 w^{(2)}(x,y) + \dots, \ y = x/\varepsilon \ ,$$

where $w^{(0)} \in H_0^2(\Omega)$ and $w^{(2)}$ is defined on $\Omega \times YF$..

The results obtained were applied to the plate in bending weakened by fissures parallel to the $x_1$-axis, see Fig.38. Clearly, then the assumption that fissures should not meet the boundary is violated. Consequently, the strict convexity of the homogenized potential is lost.

Still more complicated are fissured Reissner-like plates. In this case 211 modes of unilateral fissures are admitted by the model. To have a greater flexibility the mid-plane is shifted by e, see Fig.39.

We observe that both Kirchhoff and Reissner-like plates admit penetration zone/within the model considered/.

To avoid interpretation zones more refined models of plates are required(or we just have to solve three-dimensional problems!).

Acknowledgement. I am indebted to Professors J.J.Moreau and P.D. Panagiotopoulos for their invitation to deliver these lectures at CISM.

## REFERENCES

### INTRODUCTION


1. Armstrong, G.C., V.C.Mow: Biomechanics of Normal and Osteoarthrotic Articular Cartilage, in: Clinical Trends in Orthopaedics /Ed. P.D.Wilson and L.R.Straub/, Thieme-Stratton, New York 1982, 189-197.
2. Atkinson, J.R., D.Dowson, G.H.Isaac and B.M.Wroblewski: Laboratory wear tests and clinical observations of the penetration of femoral heads into acetabular cups in total replacement hip joints. II: A microscopical study of the surfaces of Charnley polyethylene acetabular sockets. III: The measurement of internal volume changes in explanted Charnley sockets after 2-16 years in vivo and determination of wear factors, Wear, 104/1985/, 217-224, 225-244.
3. Baiocchi, C. and A.Capelo: Variational and Quasivariational Inequalities: Applications to Free-Boundary Problems, John Wiley and Sons, Chichester 1984.
4. Bayada, G. and M.Chambat: On the Various Aspects of the Thin Film Equation in Hydrodynamic Lubrication when the Proughness Occurs, in: Application of Multiple Scaling in Mechanics /Ed. P.G.Ciarlet and E.Sanchez-Palencia/, Masson, Paris 1987.
5. Bensoussan, A. and J.L.Lions: Application des Inéquations Variationelles et Contrôle Stochastique, Dunod, Paris 1978.
6. Bernadou, M., P.Christel and Y.M.Crolet: Simulation Numérique des Contraintes aux Interfaces Prothèse de Hanche-Os, in: Computing Methods in Applied Sciences and Engineering VI /Ed. R.Glowinski and J.L.Lions/, Elsevier Science Publishers, North Holland 1984, 381-400.
7. Bossavit, A., A.Damlamian and M.Frémond /Eds/: Free Boundary Problems: Applications and Theory, Vol.III, and IV, Pitman, London 1985.
8. Brézis, H.: Problèmes unilatéraux, J.Math.Pures Appl., 51 /1972/, 1-168.
9. Brown, T.D. and A.M.Gigioia III: A contact-coupled finite element analysis of the natural adult hip, J.Biomechanics, 17/1984/, 437-448.
10. Chand, R., E.Haug and K.Rim: Stresses in the human knee joint, J.Biomechanics, 9/1976/, 417-422.
11. Chipot, M.: Variational Inequalities and Flow in Porous Media, Springer-Verlag, New York-Berlin 1984.
12. Clech, J.P., L.M.Keer and J.L.Lewis: A model of tension and compression cracks with cohesive zone at a bone-cement interface, J.Biomechanical Eng., 107/1985/, 175-182.

13. Crolet, J.-M.: L'ancrage du composant cotyloïdien dans les prothèses totales de hanche: simulation exploratoire Thése de Doctorat d'Etat en Sciences, Université de Technologie de Compiégne, 1985.
14. Del Piero, G. and F.Maceri /Eds/: Unilateral Problems in Structural Analysis, Springer-Verlag, Wien 1985.
15. Dowson, D. and G.R.Higgison: Elasto-Hydrodynamic Lubrication, Pergamon Press, Oxford 1977.
16. Dowson, D. and N.C.Wallbridge: Laboratory wear tests and clinical observations of the penetration of femoral heads into acetabular cups in total replacement hip joints. I: Charnley prostheses with polytetrafluoroethylene acetabular cups, Wear, 104/1985/, 203-215.
17. Dowson, D., B.J.Gillis and J.R.Atkinson: Penetration of Metallic Femoral Components into Polymeric Tibial Components Observed in a Knee Joint Simulator, in: ACS Symposium Series No 287, Polymer Wear and Its Control, /Ed. Lieng-Huang Lee/, American Chemical Society, 1985, 215-228.
18. Duvaut, G. and J.L.Lions: Les Inéquations en Mécanique et en Physique, Dunod, Paris 1972.
19. Elliot, C.M. and J.R.Ockendon: Weak and Variational Methods for Moving Boundary Problems, Pitman, London 1982.
20. Fasano, A. and M.Primiceno /Eds/: Free Boundary Problems: Theory and Applications, Vol.I and II, Pitman, London 1983.
21. Friedman, A.: Variational Principles and Free-Boundary Problems, John Wiley and Sons, New York 1982.
22. Fung, Y.C.: Biomechanics: Mechanical Properties of Living Tissue, Springer-Verlag, New York-Heidelberg-Berlin 1981.
23. Galin, L.A.: Development of the Theory of Contact Problems in USSR, Izd.Nauka, Moskva 1976./in Russian/.
24. Galin, L.A.: Contact Problems of the Theory of Elasticity and Viscoelasticity, Izd.Nauka, Moskva 1980./in Russian/
25. Gladwell, G.M.L.: Contact Problems in the Classical Theory of Elasticity, Sijthoff and Noordhoff, Alphen aan den Rijn 1980.
26. Hayes, W.C., L.M.Keer, G.Herrmann and L.F.Mockros: A mathematical analysis for indentation tests of articular cartilage, J.Biomechanics, 5/1972/, 541-551.
27. Hori, R.Y. and L.F.Mockros: Indentation tests of human articular cartilage, J.Biomechanics, 9/1976/, 259-268.
28. Huiskes, R.: Some fundamental aspects of human joint replacement, Acta Orthop.Scand., Supplementum, 1980, 1-208.
29. Huiskes, R.: Design Fixation and Stress Analysis of Permanent Orthopedic Implants: the Hip Joint, in:

Functional Behavior of Orthopedic Biomaterials, Vol.II: Applications /Eds P.Ducheyne and G.W.Hastings/, CRC Press Inc., Boca Ration, Florida 1984, 121-162.
30. Huiskes, R.: Biomechanics of Bone-Implant Interactions, in: Frontiers in Biomechanics /Eds G.W.Schmid-Schönbein, S.L.-Y.Woo and B.W.Zweifach/, Springer-Verlag, New York-Berlin-Heidelberg-Tokyo 1986, 245-262.
31. Huiskes, R. and E.Y.S.Chayo: A survey of finite element analysis in orthopedic biomechanics: the first decade, J.Biomechanics, 16/1983/, 385-409.
32. Johnson, K.L.: Contact Mechanics, Cambridge University Press, Cambridge 1985.
33. Kalker, J.J.: A survey of the mechanics of contact between solid bodies, ZAMM 57/1977/, T3-T17.
34. Kikuchi, N. and J.T.Oden: Contact Problems in Elasticity, SIAM Studies in Applied Mathematics, Philadelphia 1986.
35. Kinderlehrer, D. and G.Stampacchia: An Introduction to Variational Inequalities and Their Applications, Academic Press, New York 1980.
36. Mow, V.C. and W.M.Lai: Mechanics of animal joints, Ann. Rev.Fluid Mech.,11/1979/, 247-288.
37. Mow, V.C. and W.M.Lai: Recent developments in synovial joint biomechanics, SIAM Review 23/1980/, 275-317.
38. Mow, V.C., M.H.Holmes and W.M.Lai: Fluid transport and mechanical properties of articular cartilage: a review, J.Biomechanics, 17/1984/, 377-394.
39. Murase, K., R.D.Crowninshield, D.R.Pedersen and T.S. Chang: An analysis of tibial component design in total knee arthoplasty, J.Biomechanics, 16/1983/, 13-22.
40. Myers, E.R. and V.C.Mow: Biomechanics of Cartilage and its Response to Biomechanical Stimuli, in: Cartilage, vol.1 /Ed. B.K.Hall/, Academic Press 1983.
41. Naumann, J.: Einführung in die Theorie parabolischer Variationsungleichungen, Teubner-Texte zur Mathematik, Bd. 64, Leipzig 1984.
42. Oh, K.P.: The formulation of the mixed lubrication problem as a generalized nonlinear complementarity problem, Trans.ASME, J.Tribology, 108/1986/, 598-604.
43. Nečas, J.N. and I.Hlaváček: Mathematical Theory of Elastic and Elastico-Plastic Bodies: An Introduction, Elsevier, Amsterdam 1981.
44. Panagiotopoulos, P.D.: Inequality Problems in Mechanics and Applications. Convex and Nonconvex Energy Functions. Birkhäuser Verlag Boston-Basel-Stuttgart 1985.
45. Pinkus, O: The Reynolds centennial: a brief history of the theory of hydrodynamic lubrication, J.Tribology, 109/1987/, 2-20.
46. Rodrigues, J.-F.: Obstacle Problems in Mathematical Physics, North-Holland Mathematics Studies, 134, Am-

sterdam 1987.
47. Rvachev, V.L. and V.S.Protzenko: Contact Problems of the Theory of Elasticity for Non-Classical Domains, Naukova Dumka, Kiev 1977. /in Russian/
48. Saho, S. and S.Pal: Improvement of mechanical properties of acrylic bone cement by fibre reinforcement, J. Biomechanics, 17/1984/, 467-478.
49. Santare, M.H., L.M.Keer and J.L.Lewis: Cracks emanating from a fluid filled void loaded in compression: application to the bone-implant interface, J.Biomech.Eng. 109/1987/, 55-59.
50. Selvadurai, A.P.S. and G.Z.Voyiadjis /Eds/: Mechanics of Material Interfaces, Elsevier, Amsterdam 1986.
51. Seymov, V.M.: Dynamic Contact Problems, Naukova Dumka, Kiev 1976. /in Russian/
52. Stormont, T.J., K.N.An, B.F.Morrey and E.Y.Chao:Elbow joint contact study: comparison of techniques, J.Biomechanics,18/1985/, 329-336.
53. Strozzi, A.: An assessment of the numerical solution of the elastohydrodynamic problem for soft contacts, Wear 115/1987/, 53-61.
54. Süsskind, C.: Heinrich Hertz: Man and Scientist, in: Contact Mechanics and Wear of Rail / Wheel Systems, Proc.of the Int.Symp. held at University of British Columbia, July 6-9,1982,/Eds J.Kolousek, J.Dukkipat and G.M.L.Gladwell/, University of Waterloo Press 1983, 149-157.
55. Telega, J.J.: Variational methods in contact problems of mechanics, Uspekhi Mekhaniki Adv. in Mechanics, in print /in Russian/.
56. Tribology - Friction, Lubrication and Wear. Fifty Years On, Vol.I and II, International Conference, 1-3 July 1987, London, Published for the Institution of Mechanical Engineers by Mechanical Engineering Publications Ltd, London 1987.
57. Yang, R.J., K.K.Choi, R.D.Crowninshield and R.A.Brand: Design sensitivity analysis: a new method for inplant design and a comparison with parametric finite element analysis, J.Biomechanics, 17/1984/, 849-854.

CHAPER 2

58. Akagaki, T. and K.Kato: Plastic flow process of surface layers in flow wear under boundary lubricated conditions, Wear, 117/1987/, 179-196.
59. Aleksandrovich, A.I., B.S.Vekshin and I.N.Potapov: Tensor of friction coefficients of anisotropic surfaces, Trenye i Iznos, 6/1985/, 996-1004 /in Russian/.
60. Antoniou, S.S., A.Cameron and C.R.Gentle: The friction--speed relation from stick-slip data, Wear, 36/1976/, 235-254.

61. Armstrong, C.G. and V.C.Mow: Friction, Lubrication and Wear of Synovial Joints, in: Scientific Foundations of Orthopaedics and Traumatology /Eds R.Owen, J.W.Goodfellow and P.G.Bullough/, William Heineman, London 1980, 223-232.
62. Ascione, L. and D.Bruno: An analysis of the unilateral contact problem with friction of beams and plates on an elastic half-space, University of Calabria, Department of Structures, Report 77, 1985.
63. Ascione, L., D.Bruno and A.G.Grimaldi: Some Static and Dynamical Contact Problems Between a Mindlin Plate and an Elastic Foundation, in: Euromech 219, Refined Dynamical Theories of Beams, Plates and Shells and Their Applications, September 23-26, 1986, Kassel; Springer-Verlag, in print.
64. Atack, D. and D.Tabor: The friction of wood, Proc.Roy. Soc.London A246/1958/, 539-555.
65. Aybinder, S.B. and E.L.Tyunina: An Introduction to Theory of Friction of Polymers, Riga 1978 /in Russian/.
66. Ball, J.M., J.C.Currie and P.J.Olver: Null Lagrangians, weak continuity and variational problems of arbitrary order, J.Funct.Analysis, 41/1981/, 135-174.
67. Banerjee, A.K.: Influence of kinetic friction on the critical velocity of stick-slip motion, Wear, 12/1986/, 107-116.
68. Barquins, M.: Adherence, Friction and Wear of Rubber - Like Materials, in: Tribology - Friction, Lubrication and Wear. Fifty Years On, Vol.I Inst.Conf.,1-3 July 1987, London, Published for the Institution of Mechanical Engineers by Mechanical Engineering Publications Ltd, London 1987, 227-238.
69. Bartenev, G.M. and Yu.V.Zelenev: Physics and Mechanics of Polymers, Vyschaya Schola, Moskva 1983. /in Russian/
70. Bay, N. and T.Wanheim: Real area of contact and friction stress of high pressure sliding contact, Wear 38/1976/, 201-209.
71. Bereznyakov, A.I., S.V.Ventzel, V.N.Mamayev and S.V. Merkulova: Temperature dependence of friction force, Problemy Trenya i Iznaschyvanya, 27/1985/, 11-13 /in Russian/.
72. Barthe, D. and Ph.Vergne: An elastic approach to rough contact with asperity interactions, Wear, 117/1987/, 211-222.
73. Bielski, W.R. and J.J.Telega: A note on duality for von Kármán plates in the case of the obstacle problem, Arch.Mech. 37/1985/, 135-141.
74. Bielski, W.R. and J.J.Telega: A contribution to contact problems for a class of solids and structures,Arch.Mech. 37/1985/, 303-320.
75. Bikerman, J.J.: Adhesion in friction, Wear, 39/1976/,

1-13.
76. Blok, H.: Thermo-Tribology-Fifty Years On: in:Tribology-Friction, Lubrication and Wear, Fifty Years On, Vol. I, Int.Conference, 1-3 July 1987, Proc.of the Institution of Mechanical Engineers, Mechanical Engineering Publications Ltd, London 1987, 1-8.
77. Bowden, E.P. and D.Tabor: Friction and Lubrication of Solids, Vol.I Clarendon, Oxford 1950, Vol.II. 1964.
78. Buckley, D.H.: Surface Effects in Adhesion, Friction, Wear and Lubrication, Elsevier Scientific Publ.Co., Amsterdam 1981.
79. Burton, R.A.: Thermal deformation in frictionally heated contact, Wear, 59/1980/, 1-20.
80. Buttazzo, G.: Semicontinuity, relaxation and integral representation problems in the calculus of variations, Textos e Notas 34, CMAF Universidades de Lisboa 1986.
81. Campos, L.T., J.T.Oden and N.Kikuchi: A numerical analysis of a class of contact problems with friction in elastostatics, Comp.Meth.Appl.Mech.Eng. 34/1982/, 821-845.
82. Capuzzo Dolcetta, I. and M.Matzeu: Duality for implicit variational problems and numerical applications, Università degli Studi, Roma, Istituto Matematico "G.Castelnuovo" 6 marzo 1980, 13-49.
83. Challen, J.M. and P.L.B.Oxley: An explanation of the different regimes of friction and wear using asperity deformation models, Wear, 53/1979/, 229-243.
84. Ciarlet, P.G. and J.Nečas: Unilateral problems in nonlinear three-dimensional elasticity, ARMA 87/1985/, 319-338.
85. Clarke, F.H.: Optimization and Nonsmooth Analysis, John Wiley and Sons, New York 1983.
86. Cocu, M.: Existence of solutions of Signorini problems with friction, Int.J.Eng.Sci., 22/1984/, 567-575.
87. Cocu, M. and A.R.Radoslovescu: Regularity properties for the solutions of a class of variational inequalities, Nonlinear Analysis, Theory, Methods and Appl. 11/1987/, 221-230.
88. Curnier, A.: A theory of friction, Int.J.Solids Struct., 20/1984/, 637-647.
89. Curnier, A.R. and R.L.Taylor: A thermomechanical formulation and solution of lubricated contacts between deformable solids, Trans. ASME, J.Lubric. Technology 104/1982/, 109-117.
90. De Celis, B.: Theoretical analysis of dry friction in brittle and ductile materials, Wear, 116/1987/,287-298.
91. Demkowicz, L.: On some results concerning the reciprocal formulation for the Signorini's problem, Comp.and Math. with Appl., 8/1982/, 57-74.
92. Demkowicz, L. and J.T.Oden: On some existence and

uniqueness results in contact problems with nonlocal friction, Nonlinear Anal.Theory, Methods Applic.,6/1982/ 1075-1093.

93. Dowson, D.: The History of Tribology, Longmans, London 1978.

94. Dowson, D., A.Unsworth, A.F.Cooke and D.Gvozdanovic: Lubrication of Joints, in: An Introduction to the Biomechanics of Joints and Joint Replacement, /Eds D.Dowson and V.Wright/, Mechanical Engineering, London 1981, 120-145.

95. Duvaut, G.: Équilibre d'un solide élastique avec contact unilateral et frottement de Coulomb, C.R.Acad.Sci. Paris, A290/1980/, 263-265.

96. Duvaut, G.: Elasticité avec frottement visqueux, C.R. Acad.Sci., Paris, A291/1980/, 511-514.

97. Ekeland, I. and R.Temam: Convex Analysis and Variational Problems, North Holland, Amsterdam 1976.

98. Endo, T., J.T.Oden, E.B.Becker and T.Miller: A numerical analysis of contact and limit-point behavior in a class of problems of finite elastic deformation, Comp. and Structures, 18/1984/, 899-910.

99. Felder, E.F.: Formulation thermodynamique des interactions superficielle entre deux corps, J.Méc.Théor. Appl., 4/1985/, 283-303.

100. Funabashi, K. and T.Nakamura: Microscopic normal displacement of contacting bodies with tangential loads, Wear, 114/1987/, 339-354.

101. Fusciardi, A., U.Mosco, F.Scarpini and A.Schiaffino: A dual method for the numerical solution of some variational inequalities, J.Math.Anal.Applic. 40/1972/, 471-493.

102. Galanov, B.A.: On nonlinear boundary equations of contact mechanics of rough surfaces, Prikl.Mat.Mekhanika, 50/1986/, 470-474.

103. Gaylord, E.W. and H.Shu: Coefficients of static friction under statically and dynamically applied loads, Wear, 4/1961/, 401-412.

104. Giltrow, J.P. and J.K.Lancaster: Friction and wear properties of carbon fibre-reinforced metals, Wear, 12/1968/, 91-105.

105. Goldstein, R.V., A.F.Zazovskij, A.A.Spektor and R.P. Fedorenko: Solutions of three-dimensional rolling problems with slip and adhesion by variational methods, Uspekhi Mekhaniki /Advances in Mechanics/, 5/1982/, 61-102. /in Russian/

106. Greenwood, J.A., K.L.Johnson and E.Matsubara: A surface roughness parameter in Hertz contact, Wear, 100/1984/, 47-57.

107. Halaunbrenner, M.: Directional effects in friction, Wear, 3/1960/, 421-425.

108. Halling, J.: A contribution to the theory of friction, Wear, 37/1976/, 169-184.
109. Haslinger, J. and I.Hlaváček: Approximation of the Signorini problem with friction by a mixed finite element method, J.Math.Anal.Applic. 86/1982/, 99-122.
110. Haslinger, J. and P.D.Panagiotopoulos: The reciprocal variational approach to the Signorini problem with friction. Approximation results, Proc.Roy.Soc.Edinb. 98A/1984/, 365-383.
111. Haslinger, J. and M.Tvrdý: Approximation and numerical solution of contact problems with friction, Apl.Matematiky, 28/1983/, 55-71.
112. Higginson, G.R.: Elastohydrodynamic lubrication in human joints, Proc.Inst.Mech.Eng., 191/1977/, 217-223.
113. Hayer, M.W., E.C.Klang and D.E.Cooper: The effects of pin elasticity clearance and friction on the stresses in a pin loaded orthotropic plate, J.Comp.Materials, 21/1987/, 190-206.
114. Ivanov, A.P.: On well-posedness of the fundamental problem in systems with friction, Prikl.Mat.Mekh., 50/1986/, 712-716. /in Russian/
115. Jarušek, J.: Contact problems with bounded friction, Coercive case, Czech.Math.J., 33/1983/, 237-261; Semicoercive case, ibid, 34/1984/, 619-629.
116. Kalker, J.J.: Mathematical models of friction for contact problems in elasticity, Wear, 113/1986/, 61-77.
117. Kalker, J.J.: The principle of virtual work and its dual for contact problems, Ing.-Archiv., 56/1986/, 453-467.
118. Kanatani, K.-I.: A theory of contact force distribution in grannular materials, Powder Technology, 28/1981/, 167-172.
119. Kayaba, T., K.Kato and K.Hokkirigawa: Theoretical analysis of the plastic yielding of a hard asperity sliding on a soft flat surface, Wear, 87/1983/, 151-161.
120. Kikuchi, N.: A Class of Signorini's Problems by Reciprocal Variational Inequalities, in: Computational Techniques for Interface Problems, AMD, Vol.30 /Eds K.C.Park and D.K.Gartling/, Am.Soc.Mech.Eng., New York 1978, 135-153.
121. Kikuchi,N.: Beam bending problems on a Pasternak foundation using reciprocal variational inequalities, Quart.Appl.Math., 37/1980/, 91-108.
122. Kikuchi, N.: Friction Contact Problems by Using Penalty Regularization Methods, in: Contact Mechanics and Wear of Rail/Wheel Systems, /Eds J.Kolousek, R.V.Dukkipati and G.M.L.Gladwell/, University of Waterloo Press, 1983, 37-59.
123. Kikuchi, N. and J.T.Oden: Contact Problems in Elasticity, SIAM, Philadelphia 1986.

124. Kishida, H. and M.Uesugi: Tests of the interface between sand and steel in the simple shear apparatus, Géotechnique, 37/1987/, 45-52.
125. Klarbring, A.: Contact problems in linear elasticity. Friction laws and mathematical programming applications, Linköping Studies in Science and Technology, Dissertations No 133, 1985, Linköping University, Sweden.
126. Klarbring, A.: The Influence of Slip Hardening and Interface Compliance on Contact Stress Distributions. A Mathematical Programming Approach, in: Mechanics of Material Interfaces /Eds A.P.S.Selvadurai and G.Z. Voyiadjis/, Elsevier Science Publishers, Amsterdam 1986, 43-59.
127. Klarbring, A.: General contact boundary conditions and the analysis of frictional systems, Int.J.Solids Structures, 22/1986/, 1377-1398.
128. Klarbring, A.: A mathematical programming approach to three dimensional contact problems with friction,Comp. Meth.Appl.Mech.Eng., 58/1986/, 175-200.
129. Kleiber, M. and B.Raniecki: Elastic-Plastic Materials at Finite Strains, in: Plasticity Today. Modelling, Methods and Applications /Eds A.Sawczuk and G.Bianchi/ Elsevier Applied Science Publishers, London-New York 1985, 3-46.
130. Kravchuk, A.S.: A contribution to the theory of contact problems with taking into account of friction on the surface of contact, Prikl.Mat.Mekh., 44/1980/,122-129 /in Russian/.
131. Laurent, P.J.: Approximation et Optimisation, Herrmann, Paris 1972.
132. Leszek, W. and W.Zwierzycki: Modern interpretation of friction and wear notions - Part I: General description of solid friction, Trybologia, 18/1987/, 4-7 /in Polish/.
133. Licht, C.: Un problème d'élasticité avec frottement visqueux non linéaire, J.Méc.Théor.Appl. 4/1985/,15-26.
134. Ling, F.F. and S.L.Pu: Probable interface temperatures of solids in sliding contact, Wear, 7/1964/, 23-34.
135. Ludema, K.C. and D.Tabor: The friction and visco-elastic properties of polymeric solids, Wear, 9/1966/, 329-348.
136. Mackie, A.G.: Complementary variational inequalities, IMA J.Appl.Math., 36/1986/, 293-305.
137. Madakson, P.B.: The frictional behaviour of materials, Wear, 87/1983/, 191-206.
138. Martins, J.A.C. and J.T.Oden: A numerical analysis of a class of problems in elastodynamics with friction, Comp.Meth.Appl.Mech.Eng., 40/1983/, 327-360.
139. Michałowski, R. and Z.Mróz: Associated and non-associated sliding rules in contact friction problems, Arch. Mech.Stos., 39/1981/, 259-276.

140. Moreau, J.J.: La Convexité en Statique, in: Analyse Convexe et ses Applications /Ed. J.Aubin/, Lecture Notes in Economics and Mathematical Systems, Vol.102, Springer-Verlag, Berlin 1974, 141-167.
141. Moreau, J.J.: Un formulation du contact à frottement sec; application au calcul numérique, C.R.Acad.Sci., Paris, Série II, 302/1986/, 799-801.
142. Mosco, U.: Dual variational inequalities, J.Math.Anal. Appl., 40/1972/, 202-206.
143. Mosco, U.: Implicit Variational Problems and Quasi-Variational Inequalities, in: Nonlinear Operators and the Calculus of Variations, Lecture Notes in Math., Vol.543, Springer-Verlag, Berlin 1976, 83-156.
144. Mow, V.C. and A.F.Mak: Lubrication of Diarthroidal Joints, in: The Bioengineering Handbook /Eds R.Skalak and S.Chein/, McGraw Hill Publishers, New York 1984.
145. Nečas, J., J.Jarušek and J.Haslinger: On the solution of the variational inequality to the Signorini problem with small friction, Boll.Unione Mat.Ital.,17-B/1980/, 796-811.
146. Nowacki, W.: Theory of Elasticity, Państwowe Wydawnictwo Naukowe, Warszawa 1970 /in Polish/.
147. Oden, J.T. and J.A.C.Martins: Model of computational methods for dynamic friction phenomena, Comp.Meth. Appl.Mech.Eng., 52/1985/, 527-634.
148. Oden, J.T. and E.B.Pires: Nonlocal and nonlinear friction laws and variational principles for contact problems in elasticity, Trans.ASME, J.Appl.Mech., 50/1983/, 67-76.
149. Oden, J.T. and E.B.Pires: Algorithms and numerical results for finite element approximations of contact problems with non-classical friction laws, Comp.Struct., 19/1984/, 137-147.
150. Oden, J.T., E.B.Becker, T.L.Lin and L.Demkowicz: Formulation and Finite Element Analysis of a General Class of Rolling Contact Problems with Finite Elastic Deformations, in: Mathematics of Finite Elements with Applications /Ed. J.R.Whiteman/, Academic Press 1985.
151. Pires, E.B. and J.T.Oden: Analysis of contact problems with friction under oscillating loads, Comp.Meth.Appl. Mech.Eng., 39/1983/, 337-362.
152. Pozharytskii, G.K.: Extension of Gauss principle to systems with dry friction, Prikl.Mat.Mekh., 25/1961/, 391-406. /in Russian/
153. Rabier, P., J.A.C.Martins, J.T.Oden and L.Campos: Existence and local uniquenes of solutions to contact problems in elasticity with nonlinear friction laws, Int. J.Eng.Sci., 24/1986/, 1755-1768.
154. Raniecki, B.: Thermodynamic Aspects of Cyclic and Monotone Plasticity, in: The Constitutive Law in Thermoplasti-

city /Ed. Th.Lehmann/, Springer-Verlag, Wien 1984, 251-321.
155. Richardson, R.S., Nolle H.: Surface friction under time-dependent loads, Wear, 37/1976/, 87-101.
156. Rigney, D.A. and J.P.Hirth: Plastic deformation and sliding friction of metals, Wear, 53/1979/, 345-370.
157. Rockafellar, R.T.: Convex Analysis, Princeton University Press, Princeton 1970.
158. Rockafellar, R.T.: Integral Functionals, Normal Integrands and Measurable Selections, in: Nonlinear Operators and Calculus of Variations, Lecture Notes in Mathematics, Vol.543, Springer-Varlag, Berlin 1976, 157-207.
159. Rothert, H., H.Idelberger, W.Jacobi and L.Niemann: On geometrically nonlinear contact problems with friction, Comp.Meth.Appl.Mech.Eng.,51/1985/,139-155.
160. Simonov,I.V.: On behaviour of solutions of dynamical problems in the vicinity of boundary of contact zone of elastic bodies, Prikl.Mat.Mekh., 51/1987/, 85-94. /in Russian/
161. Simons, J.W. and P.G.Bergan: A finite element formulation of three dimensional contact problems with slip and friction, Comp.Mech., 1/1986/, 153-164.
162. Spector, A.A.: Variational methods in three dimensional problems of transient interaction of elastic bodies with friction, Prikl.Math.Mekh. 51/1987/,76-83 /in Russian/.
163. Suh, N.P. and H.-C.Sin: The genesis of friction,Wear, 69/1981/, 91-114.
164. Tabor, D.: Friction - the present state of our understanding, Trans.ASME, J.Lubr.Techn.,103/1981/,169-179.
165. Tabor, D.: Friction and Wear - Development over the Last Fifty Years, in: Tribology - Friction,Lubrication and Wear, Fifty Years On, Vol.I, Int.Conf.1-3 July 1987, Proc.of the Institution of Mechanical Engineers, Mechanical Engineering Publications Ltd, London 1987, 157-172.
166. Tabor, D. and D.E.Wynne Williams: The effect of orientation of the friction of polyetrafluoroethylene, Wear, 4/1961/, 391-400.
167. Telega, J.J.: Limit analysis theorems in the case of Signorini's boundary conditions and friction , Arch. Mech., 37/1985/, 549-562.
168. Telega, J.J.: Variational Inequalities in Contact Problems, in: Mechanics of Contact of Solid Bodies, Ossolineum, in press /in Polish/.
169. Trip,J.H., L.G.Houpert, E.Ioannides and A.A.Lubrecht: Dry and Lubricated Contact of Rough Surfaces, in: Tribology-Friction, Lubrication and Wear. Fifty Years On, Vol.I, Int.Conf.,1-3 July, Proc.of the Institution of

170. Uesugi, M. and H.Kishida: Frictional resistance at yield between dry sand and mild steel, Soil and Foundations, 26/1986/, 139-149.
171. Vovkushevskii, A.V.: On variational formulation of the Signorini's problem with friction, Mekh.Tv.Tela, 6/1984/, 73-78.
172. Woo, K.L.and T.R.Thomas: Contact of rough surfaces: a review of experimental work, Wear, 58/1980/,331-340.
173. Zmitrowicz,A.: A theoretical model of anisotropic dry friction, Wear, 73/1981/,9-39.
174. Zmitrowicz, A: A thermodynamical model of contact friction and wear: I. Governing equations, II. Constitutive equations for materials and linearized theories, III. Constitutive equations for friction, wear and frictional heat, Wear 114/1987/,135-168, 169-197, 199-221.

CHAPTER 3

175. Anzellotti, G. and S.Luckhaus: Dynamical evolution of elasto-perfectly plastic bodies, Appl.Math.and Optim., 15/1987/, 121-140.
176. Attouch, H. and R.J.-B. Wets: A convergence theory for saddle functionals, Trans.Amer.Math.Soc.,1, 280 /1983/.
177. Arutyunyan, N.H. and V.B.Kolmanovskii: Theory of Creep of Nonhomogeneous Bodies, Nauka, Moskva 1983 /in Russian/.
178. Arutyunyan, N.H. and B.A.Shoykhet: Asymptotic behaviour of solutions of boundary value problem of the theory of creep of nonhomogeneous ageing bodies with unilateral constraints, Mekh.Tv.Tela, 3/1981/, 31-48 /in Russian/.
179. Bogomolny, A.: Variational formulation of the roller contact problem, Math.Meth.in the Appl.Sci., 6/1984/, 84-96.
180. Bogomolnii,A., G.Eskin and Zuchowizkii, Numerical solution of the stamp problem, Comp.Meth.Appl.Mech.Eng., 15/1978/, 149-159.
181. Bouc, R., G.Geymonat, M.Jean and B.Nayroles: Hilbertian Unilateral Problems in Viscoelasticity, in:Applications of Methods of Functional Analysis to Problems in Mechanics, /Eds P.Germain and B.Nayroles/, Lect.Notes in Mathematics, vol.503,Springer-Verlag, Berlin 1975, 219-234.
182. Bouc, R., G.Geymonat, M.Jean and B.Nayroles: Cauchy and periodic unilateral problems for ag i.ng linear viscoelastic materials, J.Math.Anal.Appl., 61/1977/, 7-39.
183. Boucher, M.: Signorini's Problem in Viscoelasticity, in: The Mechanics of the Contact Between Deformable

Bodies /Eds A.D.de Pater and J.J.Kalker/, Delft University Press 1975.
184. Christiansen, E.: Limit analysis in plasticity as a mathematical programming problem, Calcolo, 17/1980/, 41-65.
185. Christiansen, E.: Limit analysis for plastic plates, SIAM J.Math.Anal., 11/1980/, 514-522.
186. Christiansen,E.: Examples of collapse solutions in limit analysis, Utilitas Mathematica, 22/1982/,77-102.
187. Christiansen, E.: On the collapse solution in limit analysis, ARMA, 91/1986/, 119-135.
188. Demengel, F.: Problèmes variationneles en plasticité parfaite des plaques, Numer.Funct.Anal.and Optim., 6/1983/, 73-119.
189. Do, C.: On the Dynamic Deformation of a Bar Against an Obstacle, in: Variational Methods in the Mechanics of Solids /Ed. S.Nemat-Nasser/, Pergamon Press 1980, 237-241.
190. Do, C., Raupp, A. and R.A.Feijóo: The dynamics of a bar in the presence of obstacles, Bol.Soc.Bras.Mat.,11/1980/ 55-78.
191. Duvaut, G.: Étude d'un problème dynamique en élastoviscoplasticité et plasticité parfaite avec conditions de frottement à la frontière, Sc.Techn.Armement, 47/1973/, 219-227.
192. Duvaut, G.: Problèmes de Contact Entre Corps Solides Deformables, in: Applications of Methods of Functional Analysis to Problemes in Mechanics, /Eds P.Germain and B.Nayroles/, Lect.Notes in Mathematics, Vol.503, Sprinber-Verlag, Berlin 1976, 317-327.
193. Drozd, M.S., M.M.Mamlich and Yu.I.Sidyakin: Engineering Calculations of Elastic-Plastic Contact Deformation, Mashinostroenye, Moskva 1986 /in Russian/.
194. Endahl, N.: Elastoplastic identation problems, Linköping Studies in Science and Technology Dissertations, No 128, Linköping University 1985.
195. Eskin, G.: Boundary Value Problems for Elliptic Pseudodifferential Equations, Transl.of Math.Monographs,AMS, vol.52, 1981.
196. Fredriksson, B.: Finite element solution of surface nonlinearities in structural mechanics with special emphasis to contact and fracture mechanics problems, Comp.Structures, 6/1976/, 281-290.
197. Fredriksson, B.B.: Elastic Contact Problems in Fracture Mechanics, in: Fracture 1977, vol.3, ICF 4, Waterloo, Canada, June 19-24,1977, 427-435.
198. Frémond, M.: Conditions unilatérales et non linéarité en calcul à la rupture, Mat.Aplic.Comp., 2/1983/, 237-256.
199. Gawęcki, A.: Elastic-plastic beams and frames with uni-

lateral boundary conditions, J.Struct.Mech., 14/1986/, 53-76.

200. Haslinger, J. and Hlaváček I.: Contact between elastic perfectly plastic bodies, Apl.Matematiky, 27/1982/, 27-45.

201. Hill, R.: Constitutive dual potentials in classical plasticity, J.Mech.Phys.Solids, 35/1987/, 23-33.

202. Hludnev, A.M.: On solution of boundary value problems for generalized equations of creep with a unilateral condition at the boundary, Prikl.Mat.Mekh.,48/1984/, 44-49.

203. Hunter, S.C.: The rolling contact of a rigid cylinder with a viscoelastic half space, Trans.ASME, J.Appl. Mech., 28/1961/, 611-617.

204. Jean, M. and B.Nayroles: Signorini Problem in Linear Viscoelasticity, in: Duality and Complementarity in Mechanics of Solids /Ed.A.Borkowski/, Ossolineum, Wrocław, 1979, 419-481.

205. Johnson, G.: An elasto-plastic contact problem, RAIRO Anal.Numer., 12/1978/, 59-74.

206. Johnson, K.L.: Aspects of Contact Mechanics, in: Tribology - Friction, Lubrication and Wear, Fifty Years On, Vol.II, Int.Conference, 1-3 July, London, Published for the Institution of Mechanical Engineers by Mechanical Engineering Publications Ltd, London, 919-932.

207. Joo, J.W. and B.M.Kwak: Analysis and applications of elasto-plastic contact problems considering large deformation, Comp.Struct., 24/1984/, 953-961.

208. Kikuchi, N. and K.Skalski: An elasto-plastic rigid punch problem using variational inequalities, Arch. Mech., 33/1981/, 865-877.

209. Kohn, R. and R.Temam: Dual spaces of stresses and strains with applications to Hencky plasticity, Appl. Math.Optim., 10/1983/, 1-35.

210. König, J.A.: Shakedown of Elastic-Plastic Structures, Państwowe Wydawnictwo Naukowe, Warszawa, Elsevier, Amsterdam 1987.

211. Kravtchuk, A.S.: On Hertz problem for linear and nonlinear elastic bodies, Prikl.Mat.Mekh., 41/1977/, 329-337 /in Russian/.

212. Kravchuk, A.S.: Formulation of contact problem for several deformable bodies as a nonlinear programming problem, Prikl.Mat.Mekh., 42/1978/, 466-474 /in Russian/.

213. Kravchuk, A.S.and V.A.Vasil'ev: Numerical methods of solving the contact problem for linear and nonlinear bounded elastic bodies, Prikl.Mekh., 16/1980/, 9-15, /in Russian/.

214. Kuz'menko, V.I.: On a variational approach in the the-

ory of contact problems for nonlinear layered elastic bodies, Prikl.Mat.Mekh., 43/1979/,893-901 /in Russian/

215. Kuz'menko, V.I.: On contact problems of the theory of plasticity under complex loading, Prikl.Mat.Mekh., 48/1984/, 473-481 /in Russian/.

216. Kuz'menko, V.I.: On unloading process in the case of contact interaction, Prikl.Mat.Mekh., 49/1985/, 445-452 /in Russian/.

217. Kuz'menko, V.I.: Contact problems for an elastic-plastic strip under complex loading, Mekh.Tv.Tela, 6/1985/ 128-135 /in Russian/.

218. Kuz'menko, V.I.: On inverse contact problems of the theory of plasticity, Prikl.Mat.Mekh., 50/1986/, 475-482 /in Russian/.

219. Kuz'menko, V.I.: Contact problems of plasticity with taking account of contact friction, Trenye i Iznos, 8/1987/, 45-52 /in Russian/.

220. Léné, F. and G.Loppin: Sur quelques problèmes de contact entre corps solides deformables, Colloque Franco-Brésilien sur les Méthodes Numériques de l'Ingénieur, Août 1976.

221. Liolios, A.A.: Upper and lower solution estimates in unilateral viscoelastodynamics, Acta Mechanica, 66/1987/, 275-278.

222. L'vov, G.I.: Variational formulation of the contact problem for linear elastic and physically nonlinear shallow shells, Prikl.Mat.Mekh., 46/1982/, 841-846, /in Russian/.

223. L'vov, G.I.: Contact problems of creep of shallow shells, Mekh.Tv.Tela, 5/1984, 116-124, /in Russian/.

224. Maier,G. and F.Andreuzzi: Elastic and elasto-plastic analysis of submarine pipelines as unilateral contact problems, Comp.Struct., 8/1978/, 421-431.

225. Maier, G. and J.Munno: Mathematical programming applications to engineering plastic analysis, Appl.Mech. Revs., 35/1982/, 1631-1643.

226. Maier, G., F.Andreuzzi, F.Giannessi, L.Jurina and F.Taddei: Unilateral contact, elastoplasticity and complementarity with reference to offshore pipeline design, Comp.Meth.Appl.Mech.Eng., 17/18/1979/, 465-495.

227. McLinden, L.: A minimax theorem, preprint, Department of Mathematics University of Illinois at Urbana - Champaign, 1979.

228. Moreau, J.J.: On Unilateral Constraints, Friction and Plasticity, in: New Variational Techniques in Mathematical Physics /Eds G.Capriz and G.Stampacchia/, Edizioni Cremonese, Roma 1974, 173-322.

229. Nagaraj, H.S.: Elastoplastic contact of bodies with friction under normal and tangential loading, Trans.

ASME, J.Tribology, 106/1984/, 519-526.
230. Naumann, J.: Periodic solutions to certain evolution inequalities, Czech.Math.J., 27/1977/, 424-433.
231. Nečas, J. and I.Hlaváček: Solution of Signorini's contact problem in the deformation theory of plasticity by secant modules method, Apl.Matematiky, 28/1983/, 199-214.
232. Nguyen, Q.S.: Uniqueness, Stability and Bifurcation of Standard Systems, in: Plasticity Today, Modelling, Methods and Applications /Eds A.Sawczuk and G.Bianchi/, Elsevier Applied Science Publishers, London and New York 1985, 399-412.
233. Nowacki, W.: Theory of Creep, Arkady, Warszawa 1963 /in Polish/.
234. Oden, J.T. and Lin T.L.: On the general rolling contact problem for finite deformations of a viscoelastic cylinder, Comp.Meth.Appl.Mech.Eng., 57/1986/,297-367.
235. Oden, J.T., E.B.Becker, T.L.Lin and K.T.Hsieh: Numerical Analysis of Some Problems Related to the Mechanics of Pneumatic Tires: Finite Deformation/Rolling Contact of a Viscoelastic Cylinder and Finite Deformation of Cord-Reinforced Rubber Composites, in: Research in Structures and Dynamics - 1984, NASA Conference Publication 2335, Symposium held of Washington, D.C., October 22-25, 1984, 297-307.
236. Panek, C. and J.J. Kalker: Three-dimensional contact of rigid roller traversing a viscoelastic half space, J.Inst.Maths Applics, 26/1980/, 299-313.
237. Paymushin, B.N. and V.A.Firsov: Equation of nonlinear theory of contact interaction of thin shells with deformable fundations of variable thickness, Mekh.Tv. Tela, 3/1985, 119-128 /in Russian/.
238. Ponter, A.R.S., A.D.Hearle and K.L.Johnson: Application of the kinematical shakedown theorem to rolling and sliding point contacts, J.Mech.Phys.Solids, 33/1985/, 339-362.
239. Raous, M.: On Two Variational Inequalities Arising from a Periodic Viscoelastic Unilateral Problem, in: Variational Inequalities and Complementarity Problems. Theory and Applications /Eds R.W.Cottle, F.Giannesi and J.L.Lions/, John Wiley and Sons, Chichester - New York 1980, 285-302.
240. Raous, M.: Comportement d'un solide fissuré sous charges alternatives en viscoélasticié non linéaire avec écrouissage, J.Méc.Théor.Appl., Numéro spécial, 1982, 125-146.
241. Raous, M.: Contacts Unilatéraux avec Frottement en Viscoélasticite, in: Unilateral Problems in Structural Analysis /Eds G.Del Piero and F.Maceri/, Springer-Verlag, Wien-New York 1985, 269-297.

242. Rońda, J., R.Bogacz and M.Brzozowski: Infinitesimal and large strain in rolling contact problems, Ing.-Archiv., 56/1986/, 241-253.
243. Rońda, J., O.Mahrenholtz, R.Bogacz and M.Brzozowski: The rolling contact problem for an elastic-plastic strip and a rigid roller, Mech.Res.Comm., 13/1986/, 119-132.
244. Rydholm, G.: On inequalities and shakedown in contact problems, Linköping Studies in Science and Technology, Dissertations No 61, Linköping University, Institute of Technology, Linköping 1981.
245. Rydholm, G. and B.Fredriksson: Shakedown Analysis in Rolling Contact Problems, in: Conf.on Mechansms of Deformation and Fracture, Sept.20-22, 1878, Luleå, Sweden.
246. Seregin, G.A.: Variational problems and evolution variational inequalities in non-reflexive spaces with applications to problems of geometry and plasticity, Izv.Akad.Nauk SSSR, Ser.Mat., 48/1984/, 420-445 /in Russian/.
247. Simo, J.C., P.Wriggers, K.H.Schweizerhof and R.L.Taylor: Finite deformation post-buckling analysis involving inelasticity and contact constraints, Int.J.Numer.Meth.Eng., 23/1986/, 779-800.
248. Taylor, L.M. and E.B.Becker: Some computational aspects of large deformation, rate-dependent plasticity problems, Comp.Meth.Appl.Mech.Eng., 41/1983/,251-277.
249. Telega, J.J.: Variational Methods and Extremum Principles for Some Non-classical Problems of Plasticity: Unilateral Boundary Conditions, Friction,Discontinuities, in: Variational Method in Engineering /Ed. C.A. Brebbia/, Springer-Verlag, Berlin 1985,/8-27/-/8-36/.
250. Temam, R.: Mathematical Problems in Plasticity, Gauthier-Villars, Bordas, Paris 1985.
251. Temam, R. and G.Strang: Duality and relaxation in the variational problems of plasticity, J.Méc., 19/1980/, 493-527.

CHAPTER 4

252. Attouch, H.: Variational Convergence for Functions and Operators, Pitman, Boston-London-Melbourne 1984.
253. Attouch, H. and F.Murat: Homogenization of fissured elastic materials, Publications AVAMAC, No 85-03, Université de Perpignan, 1985.
254. Bakhvalov, N.S. and G.P.Panasenko: Averaging of Processes in Periodic Media, Nauka, Moskva 1984, /in Russian/.
255. Bensoussan, A., J.-L.Lions and G.Papanicolaou: Asymptotic Analysis for Periodic Structures, North-Holland, Amsterdam 1978.
256. Bergman, D., J.L.Lions, G.Papanicolaou, F.Murat, L.Tar-

tar and E.Sanchez-Palencia: Les Méthodes de l'Homogénéisation: Théorie et Applications en Physique, Editions Eyrolles, Paris 1985.
257. Brézis, H.: Operateur Maximaux Monotones, North-Holland Publishing Company, Amsterdam 1973.
258. Brézis, H.: Analyse Fonctionnelle, Masson, Paris 1983.
259. Bruch, J.C.:Coupled variational inequalities for flow from a non-symmetric ditch, Contemporary Mathematics, 11/1982/, 49-70.
260. Burrige, R., S.Childress and G.Papanicolaou /Eds/: Macroscopic Properties of Disordered Media, Lecture Notes in Physics, Vol.154, Springer-Verlag, Berlin 1982.
261. Ciarlet, P.G. and E.Sanchez-Palencia /Eds/:Applications of Multiple Scaling in Mechanics, Masson, Paris 1987.
262. Eriksen, J.L., D.Kinderlehrer, R.Kohn and J.L.Lions /Eds/: Homogenization and Effective Moduli of Materials and Media, Springer-Verlag, New York-Berlin-Heidelberg-Tokyo 1986.
263. Grisvard, P.: Elliptic Problems in Nonsmooth Domains, Pitman, Boston-London-Melbourne 1985.
264. Karlsson, T.: Wiener's criterion and obstacles problems for vector valued functions, Arkiv f.Mat., 23/1986/, 315-325.
265. Leguillon, D. and E.Sanchez-Palencia: On the behaviour of a cracked elastic body with /or without/ friction, J.Méc.Théor.Appliquée, 1/1982/, 195-209.
266. Léné, F.: Contribution à l'étude des matériaux composites et de leur endommagement, Thèse de Doctorat d'Etat, Université Pierre et Marie Curie, 1984.
267. Lewiński, T. and J.J.Telega: On homogenization of fissured elastic plates, Mech.Res.Comm., 12/1985/, 271-281.
268. Lewiński, T. and J.J.Telega: Asymptotic method of homogenization of fissured elastic plates, J.Elasticity, 1987.
269. Lewiński, T. and J.J.Telega: Homogenization of fissured Reissner-like plates, Part 1. Method of two-scale asymptotic expansion; Part III. Some particular cases and illustrative example, Arch.Mech., 40/1988/, in print. Part 3. see Ref. 277 .
270. Lions, J.L.: Problèmes aux Limites dans les Equations aux Dérivées Partielles, Les Presses d l'Université de Montréal, 1962.
271. Lurie, K. and A.W.Cherkaiev: Effective characteristics of composites and optimum structural design, Uspekhi Mekhaniki /Advances in Mechanics/, 9/1986/, 3-81 /in Russian/.
272. Sanchez-Palencia, E.: Non-Homogeneous Media and Vibration Theory, Lecture Notes in Physics, 127, Springer-

-Verlag 1980.
273. Sanchez-Palencia, E.: Homogenization Method for the Study of Composite Media, in: Asymptotic Analysis, Lecture Notes in Mathematics, Vol.985, Springer-Verlag, Berlin 1983, 192-214.
274. Sanchez-Palencia, E. and A.Zaoni /Eds/: Homogenization Techniques for Composite Media, Lecture Notes in Physics, 272, Springer-Verlag 1987.
275. Suquet, P.M.: Approach by Homogenization of Some Linear and Nonlinear Problems in Solid Mechanics, in: Plastic Behaviour of Anisotropic Solids /Ed.J.-P. Boehler/, Editions du CNRS, Paris 1985, 77-117.
276. Telega, J.J.: Homogenization of fissured solids in the presence of unilateral conditions and friction, in preparation.
277. Telega, J.J. and T.Lewiński: Homogenization of fissured Reissner-like plates. Part 2. Convergence, Arch. Mech. 40/1988/, in print. Part 1 and 3 see Ref. 269 .

SUPPLEMENT

278. Barros-Neto, J.: Inhomogeneous boundary value problems in a half space, Annali della Scuola Normale Superiore di Pisa, Serie III, 19/1965/,331-365.
279. Boucher, M.: Problème de Signorini dans l'hypothèse d'un support déformable, Thèse 3-ème cycle, Université de Paris VI, 1973.
280. Bouchitté, G: Convergence et relaxation de fonctionnelles du calcul des variations à croissance linéaire. Application a l'homogénéisation en plasticité,Publications AVAMAC, Université de Perpignan 85-10.
281. Bouchitté, G.: Representation integrale de fonctionnelles convexes sur un espace de mesures - I, Publications AVAMAC, Université de Perpignan, 86-11.
282. Castaing, C. and Valadier M.: Convex Analysis and Measurable Multifunctions, Springer-Verlag, Berlin-Heidelberg-New York 1977.
283. Haraux, A.: Nonlinear Evolution Equations - Global Behavior of Solutions, Springer-Verlag, Berlin-Heidelberg-New York 1981.
284. Bouchitté, G.: Homogénéisation sur BV($\Omega$) de fonctionnelles intégrales à croissance linéaire. Application à un problème d'analyse limite en plasticité, C.R. Acad.Sc.Paris, Série I, 301/1985/, 785-788.
285. Ogden, R.W.: Non-Linear Elastic Deformations, Ellis Horwood Limited Publishers,Chichester 1984.
286. Dal Maso, G. and L.Modica: A general theory of variational functionals, in:Topics in Functional Analysis, Scuola Normale Superiore di Pisa, Pisa 1982, 149-221.
287. Klarbring,A.,Mikelić A.and Shillor M.:On friction problems with normal compliance,Nonlinear Anal.Theory Meth.Appl., submitted.

Printed in the United States
By Bookmasters